STUDENT WORKBOOK FOR

Essentials of Anatomy and Physiology

EIGHTH EDITION

STUDENT WORKBOOK FOR

Essentials of Anatomy and Physiology

EIGHTH EDITION

Valerie C. Scanlon, PhD
College of Mount Saint Vincent
Riverdale, New York

Tina Sanders
Medical Illustrator
Castle Creek, New York
Formerly Head Graphic Artist
Tompkins Cortland Community College
Dryden, New York

Philadelphia

F. A. Davis Company
1915 Arch Street
Philadelphia, PA 19103
www.fadavis.com

Copyright © 2019 by F. A. Davis Company

Copyright © 2015, 2011, 2007, 2003, 1999, 1995, and 1991 by F. A. Davis Company. All rights reserved. This book is protected by copyright. No part of it may be reproduced, stored in a retrieval system, or transmitted in any form or by any means, electronic, mechanical, photocopying, recording, or otherwise, without written permission from the publisher.

Printed in the United States of America

Last digit indicates print number: 10 9 8 7 6 5 4 3 2 1

Publisher, Nursing: Lisa B. Houck
Manager of Project and eProject Management: Catherine H. Carroll
Senior Content Project Manager: Shana Murph
Design and Illustration Manager: Carolyn O'Brien

As new scientific information becomes available through basic and clinical research, recommended treatments and drug therapies undergo changes. The author(s) and publisher have done everything possible to make this book accurate, up to date, and in accord with accepted standards at the time of publication. The author(s), editors, and publisher are not responsible for errors or omissions or for consequences from application of the book, and make no warranty, expressed or implied, in regard to the contents of the book. Any practice described in this book should be applied by the reader in accordance with professional standards of care used in regard to the unique circumstances that may apply in each situation. The reader is advised always to check product information (package inserts) for changes and new information regarding dose and contraindications before administering any drug. Caution is especially urged when using new or infrequently ordered drugs.

ISBN: 978-0-8036-6938-3

Authorization to photocopy items for internal or personal use, or the internal or personal use of specific clients, is granted by F. A. Davis Company for users registered with the Copyright Clearance Center (CCC) Transactional Reporting Service, provided that the fee of $.25 per copy is paid directly to CCC, 222 Rosewood Drive, Danvers, MA 01923. For those organizations that have been granted a photocopy license by CCC, a separate system of payment has been arranged. The fee code for users of the Transactional Reporting Service is: 978-0-8036-6938-3/18 0 + $.25.

To the Student

The purpose of this student workbook is to help you learn the material in the basic anatomy and physiology course you are about to take. If your goal is a career in the health professions, anatomy and physiology will be foundation sciences for your other courses. How much you learn and how well you learn it is up to you, so it is important to remember that real and lasting learning depends on many factors.

1. **Keep in mind that two of the best ways to learn something are to take a test on it and to teach it to someone.** This workbook has several different kinds of exercises that you can consider tests. Yes, the answers are in the back of the book, but do the exercises first, without looking, especially for the end-of-chapter tests. Then pay attention to anything you may have gotten wrong and try to figure out why (if you can figure out why you made a particular mistake, you will be very unlikely to make that same mistake again). And if a classmate asks you for help, don't pass up the chance to describe or explain something you know (and listen to yourself as you do).

2. **Time.** Try to give yourself enough study time to learn effectively. This may not always be easy, but it is worth striving for. Cramming the night before an exam may work once in a while, but setting aside some time for study every day is usually a much more successful strategy. Keep in mind that our brains consolidate memories while we sleep, so getting a good night's sleep before an exam can be very helpful.

3. **Self-discipline.** Try not to be discouraged by the amount of material you have to learn. Keep at it, and you will find that the more you learn, the easier it becomes to learn. At first you will be assembling a small foundation, and then you will build upon it. The more extensive your foundation, the more you can add to it. Wherever possible, make connections between new material and previous material. This will broaden your foundation. The more you practice making connections, the easier it becomes to see and learn new ones.

4. **Know when to memorize and when to understand.** Some material simply must be memorized, but other material requires true comprehension. For example, each bone in the body has a name, and you will have to memorize the name of each bone and its location in the body. Will this memorization serve a purpose? Indeed it will, for when you learn the actions of muscles, you will have to know the bones to which each muscle is attached. Memorization is often the foundation for comprehension. And if you have made a mental picture of the skeleton, so much the better, adding a visual dimension to learning.

5. **Ask questions—of your teacher, your fellow students, and yourself.** When you don't understand something, ask your teacher or a classmate, and you'll probably get an answer. But what kind of questions can you ask yourself? Very simple yet very important ones. If you are studying a part of the body, ask yourself: What is its name? Where is it? What is it made of? What does it do? For such things as cells, tissues, or organs, we can summarize these questions as name, location, structure, and function. For a process that takes place within the body, the questions will be a little different, but still simple. Ask yourself: What is happening? What is the purpose for this? If you can answer these questions, you truly comprehend and understand the process.

Continued

6. **Keep your brain intrigued.** When you study, try to avoid doing the same thing again and again. For example, do not simply read your class notes over and over. The human brain loves novelty but is easily bored with repetition and will tune it out (it may begin thinking about what it might like for dinner). Human brains tend to be curious. Keep your brain engaged; make your brain do something different with the information. This workbook is a start, but do not underestimate your own creativity. If you like music, use a favorite song to name the bones of the skull or the steps in the process of protein synthesis or muscle contraction. If you like to dance, then dance to help you learn the muscles of the body.

Drawing your own simple illustrations is an effective way to help you form mental pictures. They do not have to be great art, merely be meaningful to you. You cannot take textbook illustrations into a test, but you can bring with you as many mental pictures as you have created.

This book was written for you, to help you to learn, to help you continue your education, and to help you become a successful member of the profession you have chosen. If you have comments or suggestions, I would be pleased to hear from you. Please send your comments to me in care of: F. A. Davis Company, 1915 Arch Street, Philadelphia, PA 19103.

Valerie C. Scanlon

How To Use This Workbook

Each chapter in this workbook corresponds to the same chapter in the textbook. The workbook chapters are organized into sections that correspond to the major sections of the textbook's chapters so that you may proceed from topic to topic before completing the entire chapter in the textbook.

TYPES OF EXERCISES

The types of exercises within each section are completion (fill-in) questions, matching columns, and diagrams with labeling and coloring exercises. In addition, each chapter includes a clinical applications section and a crossword puzzle.

Completion Questions. For these questions, you are asked to supply the missing word or words in a statement. For example:

All the bones of the body are collectively called the ____________________.

On the line provided, you would write: skeleton.

When you have filled in all the blanks (some sentences will have more than one), read the sentence aloud. Yes, out loud (unless you're in the library, perhaps), so that you can actually hear the information. If you hear yourself saying something, and not just think it, that something becomes easier to recall.

Matching Columns. In these exercises, you are asked to match the terms in one column with the descriptive phrases in another column. For example:

Match each holiday with its proper custom. Use each letter once.

1) Valentine's Day ____________________
2) Halloween ____________________
3) July 4th ____________________
4) Thanksgiving ____________________

A. Fireworks
B. Costumes and pumpkins
C. Turkey dinner
D. Flowers and candy

In this matching column, there is only one letter per answer line: 1) D 2) B 3) A 4) C

For some matching columns, there will be more than one letter on some answer lines. When the directions say "Use each letter once," they mean that every letter belongs on one or another of the answer lines. There will never be any extra answers. For example:

Match each part of a week with the proper days. Use each letter once. One answer line will have two correct letters, and the other will have three.

1) Weekend ____________________
2) Weekdays ____________________

A. Sunday
B. Tuesday
C. Thursday
D. Saturday
E. Wednesday

To answer this question completely, you should write: 1) A, D 2) B, C, E

Occasionally, a letter or letters may be used more than once in a matching column. When this happens, the directions will clearly state this. Remember to read the directions carefully, and remember also that there will never be any unused or leftover answers.

Will reading your completed matching column out loud be helpful? Do you recall the suggestion about singing? It may be worth a try.

Diagram Exercises. Many diagrams have been included in this workbook to give you opportunities for the visual application of your knowledge and help in forming your own mental pictures. In labeling exercises, you will provide names for the parts indicated by the numbered leader lines on the diagram.

Some of the diagrams will be even more useful if you color certain parts. Colors help add emphasis to a diagram and may help you retain a mental picture of that part of the body. Some of the diagrams include a color key below the illustration, on which you may indicate the colors you have used for each part. In most cases, the choice of colors is up to you, but there are some traditional colors. Arteries, for example, are always red in anatomic illustrations. Veins are blue, lymphatic structures are green, and nerves are yellow. These colors may be used for other structures, of course, but they should always be used for the structures mentioned above.

Even if no color key is included, you may find it helpful to color certain diagrams. Some of us are visual learners, and a few colors help us keep the picture clearly in mind and link the picture to the appropriate words.

Crossword Puzzles. The crossword puzzle in each chapter consists of words taken from the New Terminology lists at the beginning of that chapter in the textbook. Try the puzzle only after you have read the entire chapter in the text so that all the new terms will be familiar to you. You may wish to refer to the lists in the textbook as you do the puzzles.

Clinical Applications. The situations and questions in this section of each chapter have been taken from the text material on diseases and disorders and, especially, from the discussions that appear in the boxes in each chapter. These are simple applications and are meant to make the text material more relevant and meaningful for you.

CHECKING YOUR ANSWERS

When you have completed a chapter section in the workbook (or, in the case of the crossword puzzles, when you have completed the whole chapter), check your answers against the Answer Key included at the end of this workbook. If you have answered a question incorrectly, reread the appropriate section in the textbook as a review. Try to figure out why you made the mistake. Was it simply not knowing the answer? Not reading the question carefully? Confusing two terms that sound alike? Whatever the reason, if you can determine why you made the mistake, you probably won't make it again (when it counts on a test). Remember also that once you have completed a section and made any necessary corrections, that section becomes a concise summary of the essential information on that topic and may be used for a quick review or more intensive studying.

END-OF-CHAPTER TESTS

When you complete each chapter, you will find three comprehensive chapter exams. All of these exams are in a multiple-choice format, though each is a different variation of that format. Answers to them can be found in the Answer Key at the back of the workbook.

Multiple Choice Test #1. This test consists of traditional multiple-choice questions for which you are asked to select the correct answer from four choices. For example:

Which of these is a plant?

a) earthworm b) fern c) frog d) dog

Circle or underline the correct answer, which is b.

This type of question tests your knowledge and your ability to recognize the correct answer. When you take this test, answer all the questions. If you are not sure, take a guess and make a note for yourself about why you have made this particular choice. Then, when you check your answers, you can pay special attention to these to see if your thinking was correct. As always, if you do make a mistake, try to determine why so that you will not make the same mistake again.

Multiple Choice Test #2. The questions on this test evaluate your level of knowledge, reading ability, and thinking ability. Each question asks you to select the statement that is NOT true about a particular topic, and then to correct the false statement to make it true. For example:

Which statement is NOT true of dogs?

a) Dogs have four legs and two ears.

b) Many dogs can be taught to do simple tricks.

c) Many dogs are friendly and loyal.

d) Very young dogs are called kittens.

Reword your choice to make it a correct statement.

In this example, choice d is not true and is therefore the correct answer. You may want to underline the part of the answer choice that is not true (in this case, kittens) to help you focus your thinking. Then, rewrite the statement (or just the incorrect part) to make it true: Very young dogs are called puppies.

Will you have to read carefully and think clearly when you take a test like this? Yes, but if you do well on a self-test such as this, you can be confident that you know and understand the material.

Multiple Choice Test #3. The questions in this test are in a multiple response format (also called multiple selection or group multiple choice). One question consists of several statements that are all concerned with a particular topic. The directions will tell you to select all of the statements that are correct (true). Any number and combination of statements (including none or all) may be true. As always, you should read carefully. For example:

Which of the following statements are true of the human skeleton? (Select all correct answers.)

a) The spine is made of individual bones called vertebrae.

b) The skull protects the eye and ear receptors from mechanical injury.

c) The rib cage has 14 pairs of ribs.

d) Another name for the kneecap is the calcaneus.

e) The shoulder joint is formed by the scapula and radius.

f) The femur is the shinbone.

As you see, this type of question is simply a true–false test on one topic. In this case, only a and b are correct.

If you work with friends or in a study group, make up questions like this and test one another. The more you practice (and keep your brain interested and engaged with new work), the more familiar the material will become, and the more comfortable you will be at test time, regardless of the format used for questions.

Contents

Chapter 1

Organization and General Plan of the Body

This chapter describes how the body is organized, from simple to more complex levels. The terminology used to describe the body is also presented in this chapter.

LEVELS OF ORGANIZATION

Complete each statement and give appropriate examples.

1. a) The simplest level of organization is the ________________ level.
   b) Give three examples of organic chemicals. ________________, ________________, and ________________
   c) Give three examples of inorganic chemicals. ________________, ________________, and ________________
2. a) The most complex level of organization is the ________________ level.
   b) Give three examples of this level. ________________, ________________, and ________________
3. a) The simplest living level of organization is the ________________ level.
   b) When cells with similar structure and function are grouped and work together, they form a ________________.
   c) A group of tissues that is arranged in a particular way to accomplish specific functions is called an ________________.
4. Match each group of tissues with its function (a letter) and its example in the body (a number).

   Each letter and number are used once.

Epithelial tissue ________________________

Connective tissue ________________________

Muscle tissue ________________________

Nerve tissue ________________________

Functions

A. Supports, transports, or stores materials
B. Contracts and brings about movement
C. Transmits impulses that regulate body functions
D. Covers or lines surfaces

Examples

1. The lining of the stomach and the epidermis of the skin
2. The heart and skeletal muscles
3. Bone, blood
4. Spinal cord, brain

5. a) The structure of a tissue or cell or organ is called its ________________. The function of the tissue, cell, or organ is called its ________________.

   b) Describe one way the physiology of a bone (an organ) is related to its anatomy.

   ____

6. a) The collective name for all of the beneficial bacteria (and other microorganisms) that live on or in human beings is ________________ or ________________.

   b) The different environmental sites the human body provides, and the bacteria that inhabit them, are called ________________. Give three examples: ________________, ________________, and ________________

METABOLISM AND HOMEOSTASIS

Complete each statement.

1. a) Metabolism is all of the ________________ reactions and ________________ changes that take place in the body.

   b) The amount of heat and energy produced by the body per unit of time is the ________________.

2. A person who is in a state of homeostasis may also be said to be in a state of good ________________.
3. Homeostasis means that despite constant changes, the body remains relatively ________________.
4. Changes that affect the body may take place where? ________________ or ________________
5. For each of the changes listed, describe in simple terms what will happen to maintain homeostasis. (Note: Some responses may include conscious decisions.)

   1) Eating lunch ________________________________
   2) Inhaling ________________________________
   3) Having a headache ________________________________
   4) Cutting your finger ________________________________
   5) Going outside on a cold day ________________________________

6. Number the following events as they would occur in a negative feedback mechanism.

   ____1____ Stimulus
   ________ Stimulus occurs again
   ________ Response by the body increases
   ________ Response by the body decreases
   ________ Stimulus is decreased

7. Briefly explain how a positive feedback mechanism differs from a negative feedback mechanism.

   ____

8. The following diagram depicts negative and positive feedback mechanisms.

Label the parts indicated.

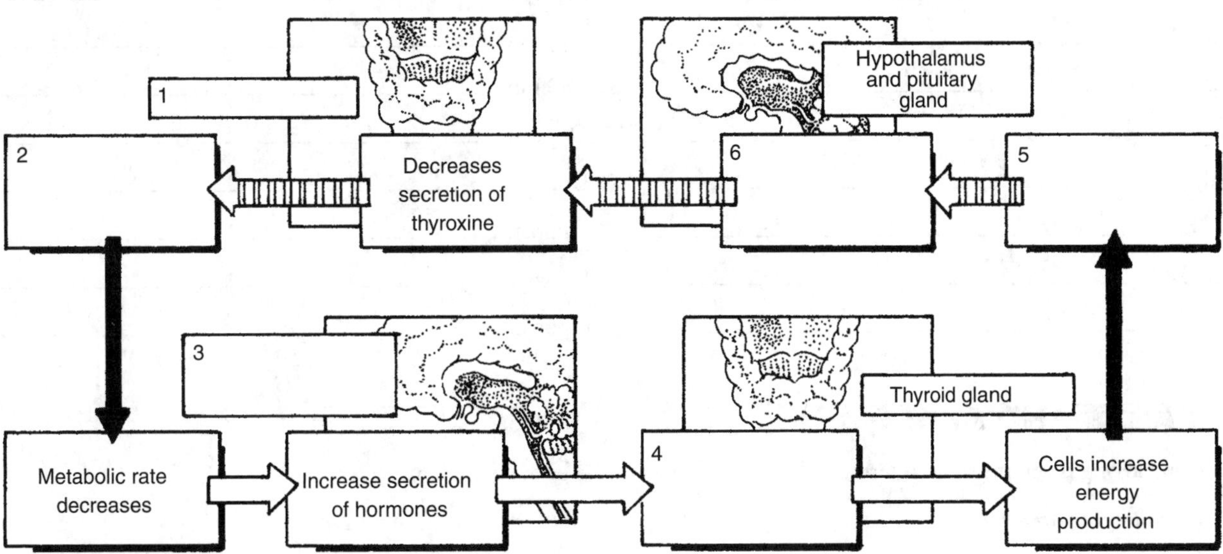

A. Negative feedback mechanism

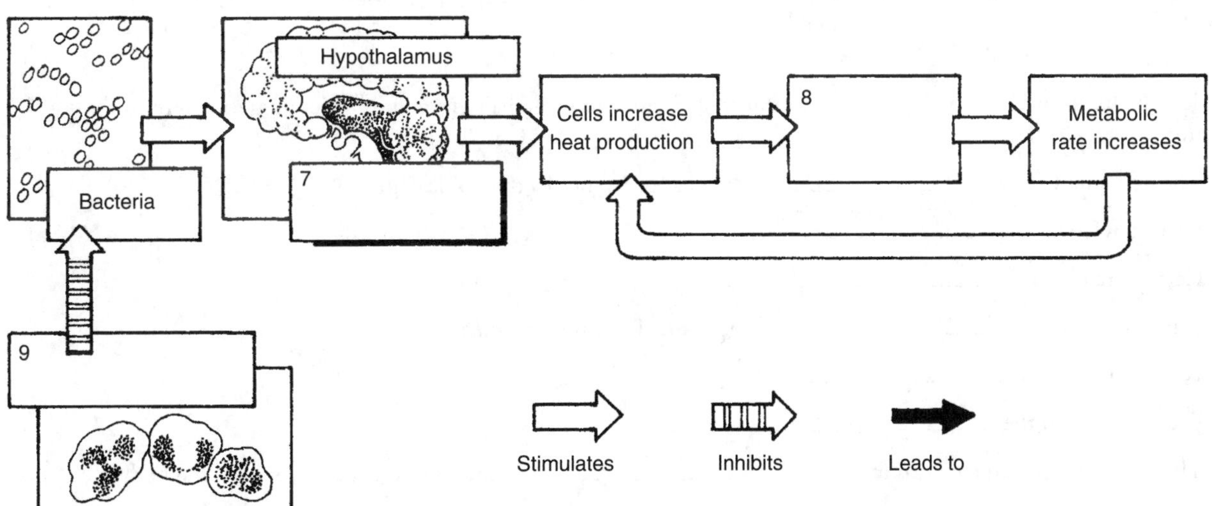

B. Positive feedback mechanism

TERMS OF LOCATION

Define each term given and provide the term with the opposite meaning.

1. Inferior: ______________________
   Opposite: ______________________
2. Proximal: ______________________
   Opposite: ______________________
3. Lateral: ______________________
   Opposite: ______________________
4. Peripheral: ______________________
   Opposite: ______________________
5. Superficial: ______________________
   Opposite: ______________________
6. Anterior: ______________________
   Opposite: ______________________
7. Dorsal: ______________________
   Opposite: ______________________
8. Internal: ______________________
   Opposite: ______________________

BODY PARTS AND AREAS

1. The accompanying diagrams show anterior and posterior views of the body in anatomic position.

   Label the following areas:

Frontal	Brachial	Inguinal	Popliteal	Femoral	Volar
Temporal	Buccal	Iliac	Cervical	Deltoid	Plantar
Nasal	Orbital	Gluteal	Mammary	Occipital	Patellar
Pectoral	Parietal	Lumbar	Cranial	Sacral	Perineal
Axillary	Umbilical				

2. Complete the following by choosing the correct directional term from the pair given and by defining each term for a body area.

   1) a) The deltoid area is ______________ (medial or lateral) to the cervical area.
      b) Deltoid refers to the ______________.
      c) Cervical refers to the ______________.
   2) a) The femoral area is ______________ (proximal or distal) to the patellar area.
      b) Femoral refers to the ______________.
      c) Patellar refers to the ______________.
   3) a) The occipital and lumbar areas are on the ______________ (anterior or posterior) side of the body.
      b) Occipital refers to the ______________.
      c) Lumbar refers to the ______________.
   4) a) The pectoral and umbilical areas are on the ______________ (dorsal or ventral) side of the body.
      b) Pectoral refers to the ______________.
      c) Umbilical refers to the ______________.

5) a) The volar area is ________________ (superior or inferior) to the plantar area.

b) Volar refers to the ________________.

c) Plantar refers to the ________________.

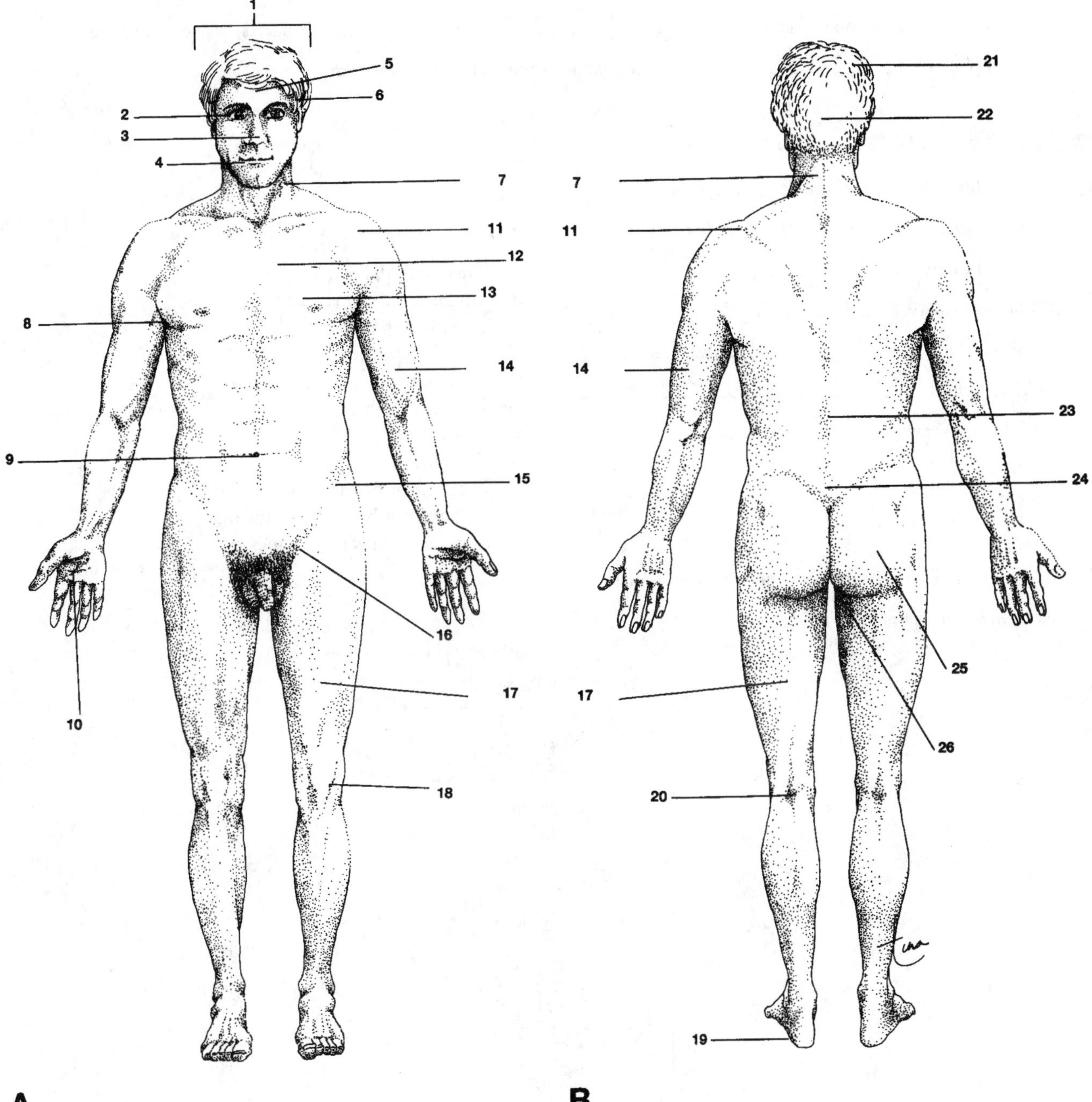

6) a) The cutaneous area is ______________ (superficial or deep) to the muscles.

b) Cutaneous refers to the ______________.

7) a) The ribs are ______________ (internal or external) to the pulmonary area.

b) Pulmonary refers to the ______________.

8) The nerves in the hand are part of the ______________ (central or peripheral) nervous system, and the brain is part of the ______________ (central or peripheral) nervous system.

BODY CAVITIES

1. Match each statement with the body cavity(ies) it pertains to.

Some answer lines will have more than one correct letter. Use each letter once; one letter is used twice.

1) Cranial cavity ______________
2) Thoracic cavity ______________
3) Spinal cavity ______________
4) Pelvic cavity ______________
5) Abdominal cavity ______________

A. Lined with the parietal pleura
B. Contains the spinal cord
C. Lined with meninges
D. Contains the heart and lungs
E. Lined with the peritoneum
F. Contains the internal reproductive organs and urinary bladder
G. Contains the brain
H. The inferior boundary is the diaphragm
I. The superior boundary is the diaphragm
J. Contains the liver and pancreas

2. The following diagram depicts the major body cavities.

Label each numbered cavity.

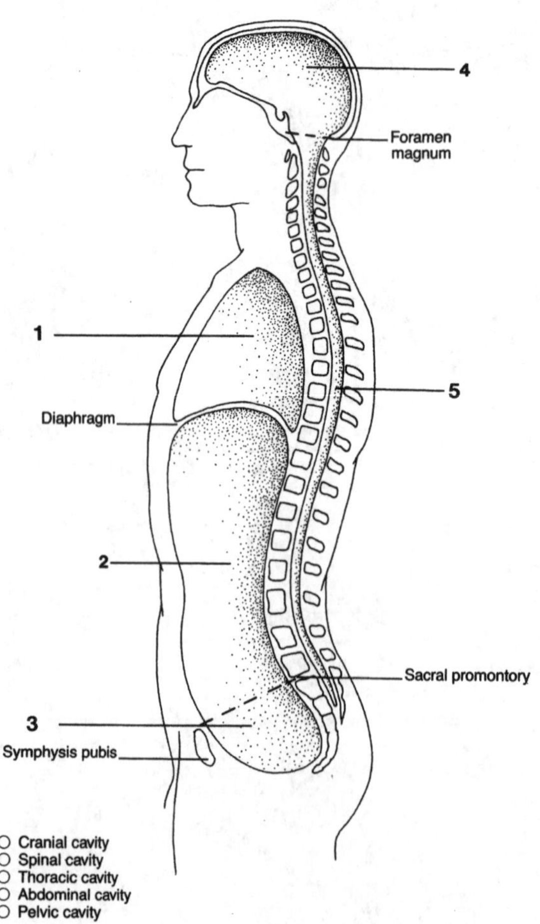

BODY SECTIONS

1. Match the following sections with the correct descriptions.

   Use each letter once.

   1) Sagittal section ____________
   2) Cross section ____________
   3) Frontal section ____________
   4) Transverse section ____________
   5) Longitudinal section ____________

   A. A plane along the long axis of an organ
   B. A plane that divides the body into right and left portions
   C. A plane that divides the body into upper and lower portions
   D. A plane perpendicular to the long axis of an organ
   E. A plane that divides the body into front and back portions

2. On the following diagrams, label each plane or section.

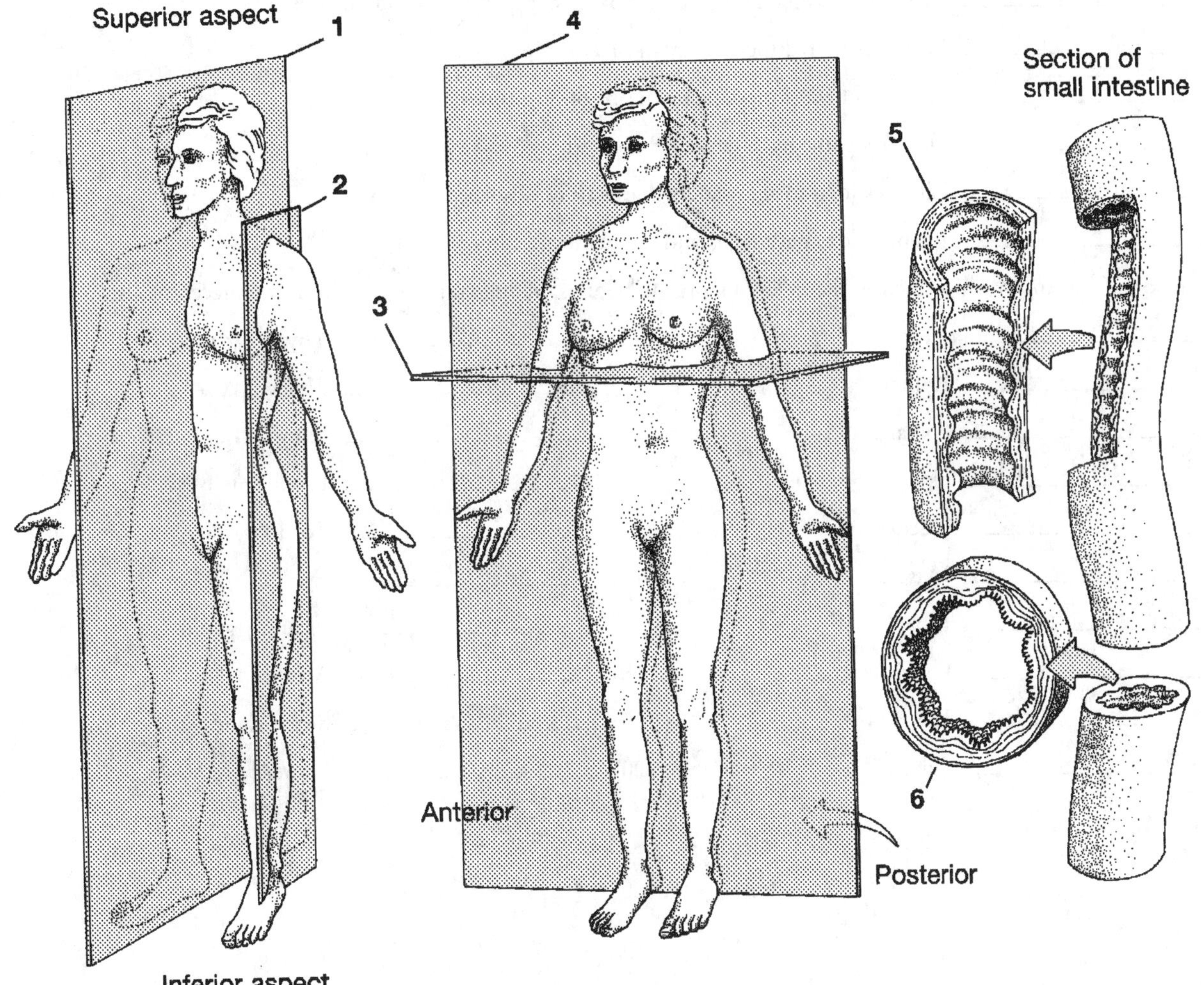

ORGAN SYSTEMS

Integumentary	Nervous	Circulatory	Respiratory	Urinary
Skeletal	Endocrine	Lymphatic	Digestive	Reproductive
Muscular				

1. Take your answers from the preceding list and name the organ system described by each statement.

 1) ______________ Moves the skeleton and produces heat
 2) ______________ Transports oxygen and nutrients to tissues
 3) ______________ Produces egg or sperm to continue the human species
 4) ______________ Supports the body and protects internal organs
 5) ______________ Removes waste products from the blood
 6) ______________ Exchanges oxygen and carbon dioxide between the blood and air
 7) ______________ Regulates body functions by means of impulses
 8) ______________ Regulates body functions by means of hormones
 9) ______________ Destroys pathogens that enter the body
 10) ______________ Changes food to simpler chemicals to be absorbed
 11) ______________ Is a barrier to pathogens and to chemicals

2. Name the organ system to which each of the following organs belongs, taking your answers from the preceding list.

 1) ______________ Heart
 2) ______________ Kidneys
 3) ______________ Spinal cord
 4) ______________ Skin
 5) ______________ Trachea
 6) ______________ Muscles
 7) ______________ Ribs
 8) ______________ Ovaries
 9) ______________ Pancreas
 10) ______________ Spleen
 11) ______________ Thyroid gland
 12) ______________ Testes
 13) ______________ Lungs
 14) ______________ Esophagus
 15) ______________ Tendons
 16) ______________ Eyes
 17) ______________ Pituitary gland
 18) ______________ Arteries
 19) ______________ Sweat glands
 20) ______________ Stomach

CROSSWORD PUZZLE

ACROSS

5. The serous membranes of the thoracic cavity (two words)
6. The smallest living unit of the body
7. Lines the abdominal cavity
11. Covers the abdominal organs
12. Membranes that cover the organs of the central nervous system
13. To cut or divide the body or a part
14. Stability of the internal environment of the body
15. Chemicals that contain carbon
16. A compartment within the body (two words)
18. Study of how the body functions
20. Chemicals that do not contain carbon

DOWN

1. A group of organs that work together to perform specific functions (two words)
2. Detailed x-ray images are produced using this method (two words)
3. Imaginary flat surface that separates the body
4. Study of disorders of functioning
8. A group of cells with similar structure and function
9. Membranes that surround the heart
10. Study of body structure
17. A group of tissues precisely arranged to accomplish specific functions
19. Magnetic resonance imaging (initials)

CLINICAL APPLICATIONS

1. Surgery involving the intestines requires an incision through the abdominal wall. Name the layers that must be cut to enter the abdominal cavity.

   1) ______________________ (part of the integumentary system)

   2) ______________________ (provides for movement of the trunk)

   3) ______________________ (the lining of the abdominal cavity)

2. Meningitis is a serious disease caused by certain bacteria. Name the membranes affected and the organs they cover.

   Membranes: ______________________ Organs: ______________________

3. Briefly explain each of the following and include the common name for each anatomic term.

   1) Renal failure: ______________________

   2) Cardiac arrest: ______________________

   3) Pulmonary vein: ______________________

   4) Hepatic coma: ______________________

   5) Gastric ulcer: ______________________

4. Look at the digestive organs depicted in the following diagram and name the quadrant(s) in which each of these organs is (are) found. Some may be in more than one.

   1) Stomach ______________________

   2) Liver ______________________

   3) Small intestine ______________________

   4) Large intestine ______________________

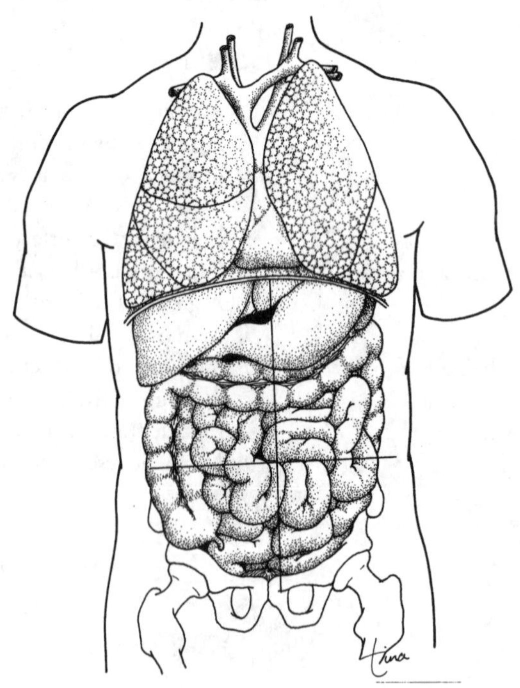

MULTIPLE CHOICE TEST #1

Choose the correct answer for each question.

1. A synonym is a word that has the same meaning as another word. A synonym for ventral is:
   a) anterior b) dorsal c) posterior d) lateral

2. An antonym is a word that has the opposite meaning of another word. An antonym for medial is:
   a) distal b) dorsal c) peripheral d) lateral

3. The most superior body cavity is the:
   a) abdominal cavity b) spinal cavity c) thoracic cavity d) cranial cavity

4. The diaphragm separates the:
   a) thoracic and abdominal cavities
   b) abdominal and pelvic cavities
   c) cranial and spinal cavities
   d) thoracic and pelvic cavities

5. The body is cut in a mid-transverse section and is now separated into two parts that may be called:
   a) anterior-posterior b) superior-inferior c) dorsal-ventral d) medial-lateral

6. A cross section of an artery would look like a:
   a) rectangle b) hollow circle c) cylinder d) hollow cube

7. The most inferior body cavity is the:
   a) thoracic cavity b) abdominal cavity c) pelvic cavity d) cranial cavity

8. The most superior body system is the:
   a) nervous system b) skeletal system c) muscular system d) integumentary system

9. A tissue found on the surface of an organ must be a(n):
   a) epithelial tissue b) connective tissue c) muscle tissue d) nerve tissue

10. A tissue that causes contractions of the stomach to mix food with gastric juice must be a type of:
    a) epithelial tissue b) connective tissue c) muscle tissue d) nerve tissue

11. The tissue that enables you to think and answer all of these questions is:
    a) epithelial tissue b) connective tissue c) muscle tissue d) nerve tissue

12. A tissue that supports body parts or transports materials within the body is a type of:
    a) epithelial tissue b) connective tissue c) muscle tissue d) nerve tissue

13. Homeostasis means that the internal environment of the body is:
    a) exactly the same at all times
    b) relatively stable in spite of constant changes
    c) constantly changing drastically
    d) not changing at all

14. The two organ systems that regulate body functions by means of impulses and hormones are the:
    a) skeletal and muscular
    b) nervous and endocrine
    c) digestive and respiratory
    d) urinary and circulatory

15. The two organ systems that are *most* responsible for protecting the body against pathogens are the:
    a) integumentary and lymphatic
    b) skeletal and circulatory
    c) digestive and urinary
    d) muscular and endocrine

16. Which of the following is NOT an aspect of the anatomy of an organ?
    a) size b) type of cells present c) shape d) function

17. To describe abdominal locations more precisely, the abdomen may be divided into (the division most often used clinically):
    a) two halves b) four quadrants c) eight sections d) 12 sections

18. A feedback mechanism that contains its own brake is called:
    a) independent b) positive c) negative d) dependent

19. A feedback mechanism that requires an external brake is called:
    a) dependent b) positive c) negative d) independent

20. The disadvantage of a positive feedback mechanism is that it may:
    a) stop before it completes its function
    b) slow down unless an external event keeps it going
    c) not always start, even with a strong stimulus
    d) become a self-perpetuating cycle that causes harm

21. Which area is not part of the upper limb?
    a) palmar b) brachial c) antecubital d) inguinal

22. Which area is not part of the lower limb?
    a) popliteal b) patellar c) lumbar d) femoral

23. Which area is not part of the trunk of the body?
    a) umbilical b) frontal c) lumbar d) pectoral

24. The metabolism of the body includes:
    a) breathing b) any chemical reaction c) any physical change d) all of these

25. With respect to the normal values of metabolism, the best way to describe them is with:
    a) a range of possible values, from low to high
    b) the highest possible value
    c) the lowest possible value
    d) the average of the possible values

MULTIPLE CHOICE TEST #2

Read each question and the four answer choices carefully. When you have made a choice, follow the instructions to complete your answer.

1. The chemical level of organization of the body includes all of the following *except*:
   a) water b) oxygen c) protein d) muscles

   For your choice, name the level of organization to which it does belong.

   ___

2. The organ level of organization includes all of the following *except* the:
   a) arteries b) liver c) blood d) kidneys

   For your choice, name the level of organization to which it does belong.

   ___

3. Which sequence lists the levels of organization in the proper order of increasing complexity?
   a) chemicals, cells, tissues, organs, organ systems
   b) chemicals, tissues, organs, cells, organ systems
   c) cells, chemicals, organs, tissues, organ systems
   d) cells, chemicals, tissues, organs, organ systems

4. Which type of tissue is NOT paired with its correct general function?
   a) muscle—specialized to contract
   b) connective—specialized to support, transport, or store materials
   c) nerve—specialized to protect body parts
   d) epithelial—specialized to cover or line body surfaces

   For your choice, state its correct general function.

   ___

5. Which statement is NOT true of homeostasis?
   a) It is a state of good health.
   b) External changes bring about specific responses by the body.
   c) Internal changes have no effect on homeostasis.
   d) The proper functioning of all the organ systems contributes to homeostasis.

   Reword your choice to make it a correct statement.

   ___

6. Which membrane is NOT paired with its proper location?
   a) peritoneum—lines abdominal organs
   b) meninges—lines the cranial cavity
   c) visceral pleura—covers the lungs
   d) parietal pleura—lines the thoracic cavity

   For your choice, state its proper location.

   ____

7. A midsagittal section of the head would cut through all of the following *except* these two:
   a) nose
   b) eyes
   c) mouth
   d) brain
   e) ears

   For your choices, explain why they would not be cut by this section.

   ____

8. A midfrontal section of the body would NOT separate the:
   a) inguinal area from the gluteal area
   b) frontal area from the occipital area
   c) pectoral area from the lumbar area
   d) cervical area from the femoral area

   For your choice, name the type of section that would separate these two areas.

   ____

9. Which statement is NOT true of cells?
   a) The human body contains many different types of cells.
   b) Cells are made of inorganic and organic chemicals.
   c) Cells are the smallest living subunits of structure and function.
   d) A group of cells with similar structure and function is called an organ.

   Reword your choice to make it a correct statement.

   ____

10. Which statement is NOT true of body cavities?
    a) The diaphragm separates the thoracic and pelvic cavities.
    b) The cranial and spinal cavities are enclosed by bone.
    c) The thoracic cavity contains the pericardial cavity for the heart.
    d) The pleural membranes are the serous membranes of the thoracic cavity.

    Reword your choice to make it a correct statement.

    ____

11. Which statement is NOT true of metabolism?
    a) It describes body functioning as a whole.
    b) External changes do not affect it.
    c) Energy and heat production are called metabolic rate.
    d) Physical changes are part of it.

    Reword your choice to make it a correct statement.

    ____

12. Which statement is NOT true of body parts and locations?
    a) The muscular system is external to the skeletal system.
    b) The orbital area is inferior to the oral area.
    c) The pectoral area is on the anterior side of the body.
    d) The cervical area is medial to the deltoid areas.

    Reword your choice to make it a correct statement.

    ____

MULTIPLE CHOICE TEST #3

Each question is a series of statements concerning a topic in this chapter. Read each statement carefully and select all of the correct statements.

1. Which of the following statements are true of homeostasis?
    a) All aspects of metabolism contribute to homeostasis.
    b) Positive feedback mechanisms are numerous in the body because they work rapidly to decrease body functions.
    c) A negative feedback mechanism contains its own brake, and each step always decreases a specific body function.
    d) Changes inside the body may bring about further changes.
    e) Changes outside the body may bring about changes inside the body.
    f) Many aspects of body function have a range of normal values.

2. Which of the following statements are true of body parts and areas (with the body in anatomic position)?
    a) The patellar area is proximal to the femoral area.
    b) The umbilical area is lateral to the pectoral area.
    c) The deltoid area is distal to the antecubital area.
    d) The brachial area is to the upper limb as the femoral area is to the lower limb.
    e) The scapular area is both dorsal and posterior.
    f) The pulmonary area is inferior to the hepatic area.
    g) The palmar area is to the lower limb as the plantar area is to the upper limb.
    h) The temporal areas are anterior to the occipital area.
    i) A midsagittal section of the trunk would separate the two iliac areas.
    j) A midfrontal section of the head would separate the two parietal areas.

3. Which of the following statements are true of the levels of organization of the body?
    a) If an organ is shaped like a tube, it probably transports something.
    b) The simplest living level of organization is the protein level.
    c) The nervous system is one of the major regulatory systems of the body.
    d) There are only four different kinds of human cells, but they can be arranged into many kinds of tissues.
    e) The chemical level of organization includes minerals and organic chemicals.
    f) The trachea is an organ, but an artery is not.

Chapter 2

Some Basic Chemistry

This chapter presents the simple chemistry necessary to an understanding of anatomy and physiology. The functions of the important inorganic and organic chemicals are described as they relate to the working of the human body.

ELEMENTS, ATOMS, AND BONDS

1. A substance made of only one type of atom is called an ________________.
2. Name the element represented by each chemical symbol.

Fe ____________	Na ____________	Cl ____________
Ca ____________	I ____________	Cu ____________
O ____________	P ____________	Zn ____________
C ____________	Mg ____________	Co ____________
H ____________	N ____________	Mn ____________
K ____________	S ____________	F ____________

3. The subunits of atoms are ______________, ______________, and ______________.
4. a) The atomic subunit that has a negative charge is the ______________, and the subunit that has a positive charge is the ______________.
   b) The subunit that has no charge is the ______________.
   c) Which of these subunits are found in the atomic nucleus? ______________ and ______________
   d) Which of these subunits are present in equal numbers in an atom? ______________ and ______________
   e) Which of these subunits gives an atom its bonding capabilities? ______________
5. An atom that has lost or gained electrons and now has a positive or negative charge is called an ______________.
6. The bonding of two or more atoms results in the formation of a ______________.
7. The bond formed when one atom loses electrons that are gained by other atoms is called an ______________ bond.
8. The bond formed when two or more atoms share electrons is called a ______________ bond.
9. a) Name a molecule that exists naturally as a gas. ______________
   b) Name a molecule that exists naturally as a liquid. ______________
   c) Name a molecule that exists naturally as a solid. ______________

10. The type of bond that may be weakened in an aqueous (water) solution is the ________________ bond.
11. The atoms of organic molecules such as carbohydrates and proteins are bonded by ________________ bonds.
12. The following diagram depicts the formation of a molecule of sodium chloride.

Label the following: the protons in each atom, the electron orbitals in each atom, and the transfer of an electron.

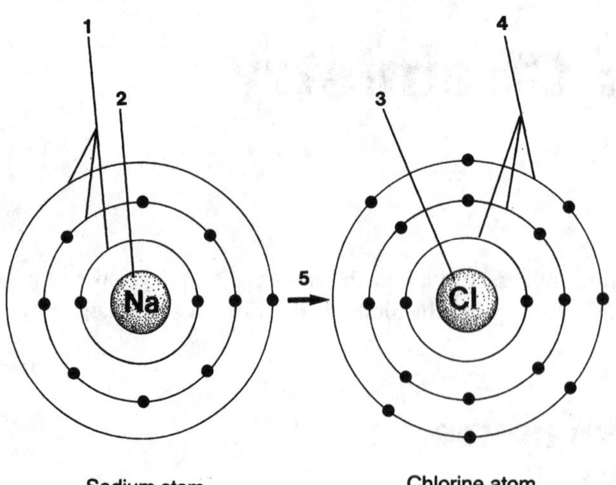

13. The special covalent bonds found only in proteins such as insulin that help maintain their three-dimensional shape are called ________________ bonds.
14. The weak bonds that maintain the three-dimensional shape of proteins and nucleic acids are called ________________ bonds.
15. a) A reaction in which smaller molecules bond to form a new, larger molecule is called a ________________ reaction.

    b) A reaction in which bonds are broken and a large molecule is changed to smaller ones is called a ________________ reaction.

INORGANIC COMPOUNDS

Complete the following:

1. The most abundant compound in the human body is ________________.
2. a) Water is a solvent, which means that many substances ________________ in water.

   b) State a specific way in which water as a solvent is important to the body.

   ____

3. a) Water is a lubricant, which means that it prevents ________________ between surfaces.

   b) State a specific way in which water as a lubricant is important to the body.

   ____

4. Water absorbs a great deal of heat as it evaporates. State why this is important to the body.

   ____

5. a) Water found in blood vessels is called (NOT blood) ______________________.

   b) Water found within cells is called ______________________.

   c) Water found in lymph vessels is called ______________________.

   d) Water found between cells is called ______________________.

   e) The following diagram depicts the water compartments.

   Name each of the specific forms of water within each compartment.

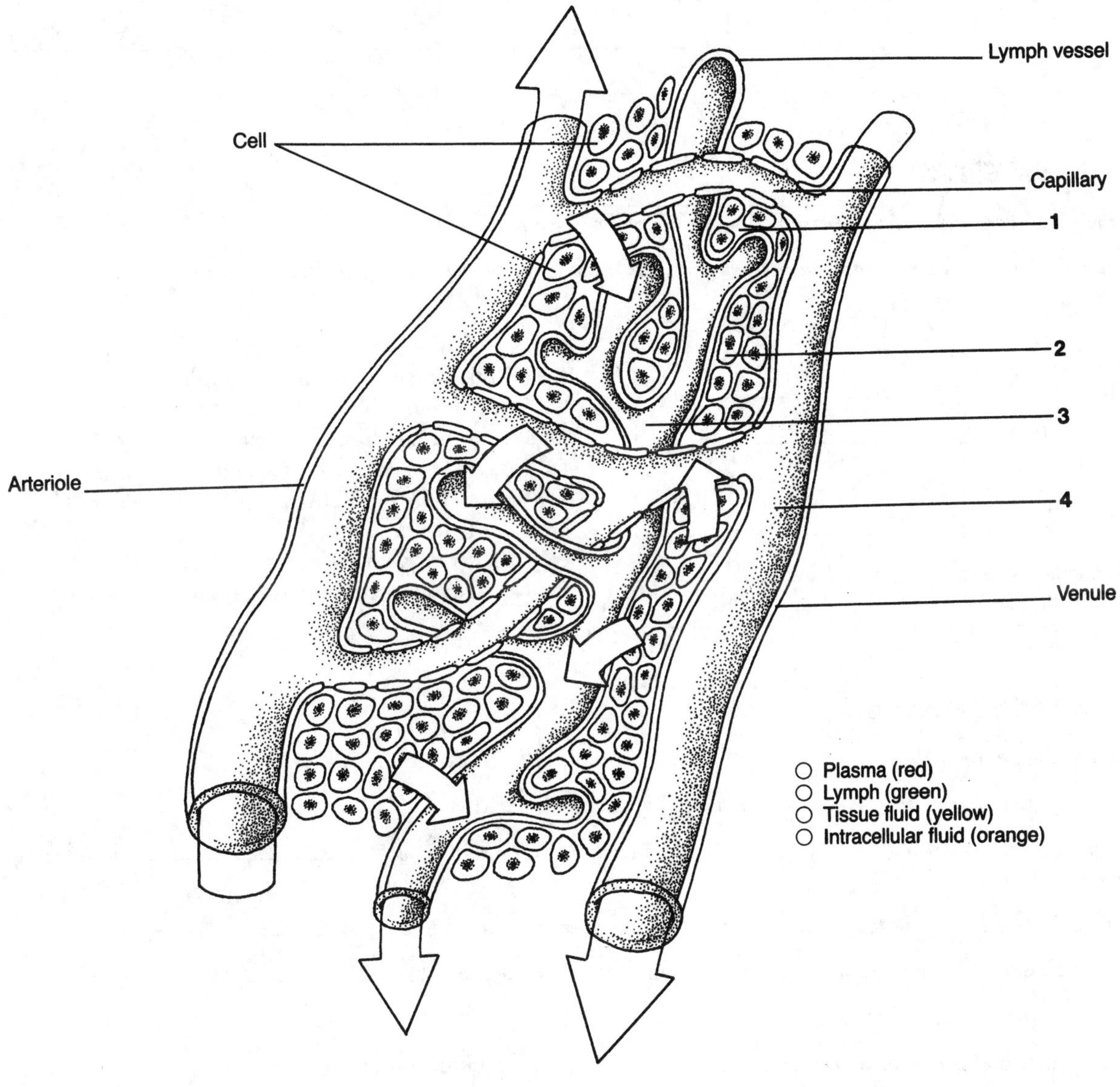

6. a) In what form is oxygen found in the atmosphere? ______________________

   b) State the chemical formula of oxygen ______________________.

   c) Within the body, oxygen is essential for the process called ______________________, which produces ______________________ for cellular processes that require energy.

7. a) Carbon dioxide is produced in the body as a waste product of the process of ______________________.

   b) State the chemical formula of carbon dioxide ______________________.

   c) If excess CO_2 is present in body fluids, these fluids will become too ______________________.

CELL RESPIRATION

1. Complete the summary reaction of cell respiration by naming the four products of the reaction:

   $C_6H_{12}O_6$ (glucose) + $O_2 \rightarrow$ ________________ + ________________ + ________________ + ________________

2. State what happens to each of the two molecular products.

   ____

   ____

3. State the purpose of each of the two energy forms produced.

   ____

   ____

TRACE ELEMENTS

1. Match these elements with their proper functions in the body.

   Each letter is used at least once, and two letters are used twice. Some answer lines will have more than one correct letter.

   1) Iron ________________
   2) Sulfur ________________
   3) Calcium ________________
   4) Iodine ________________
   5) Phosphorus ________________
   6) Sodium and potassium ________________
   7) Cobalt ________________

   A. Part of some proteins such as insulin
   B. Provides strength in bones and teeth
   C. Part of hemoglobin in red blood cells
   D. Part of the hormone thyroxine
   E. Necessary for muscle contraction
   F. Part of DNA and RNA
   G. Part of vitamin B_{12}
   H. Necessary for blood clotting
   I. Necessary for nerve impulse transmission

ACIDS, BASES, AND pH

1. a) State the range of the pH scale: ________________ to ________________.
   b) In this scale, the number that indicates a neutral pH is ________________.
   c) *Neutral* means that there are as many ________________ ions as there are ________________ ions in the solution.
   d) The portion of the pH scale from 0 to 6.99 represents solutions that are ________________.
   e) The portion of the pH scale from 7.01 to 14 represents solutions that are ________________.
2. a) State the normal pH range of blood. ________________
   b) This pH range means that blood is slightly ________________ (acidic or alkaline).
3. Chemicals in body fluids that help prevent drastic pH changes are called ________________.
4. The bicarbonate buffer system may buffer either strong acids or strong bases in body fluids.
   a) H_2CO_3 (name: ________________) may buffer a strong ________________ (acid or base).
   b) $NaHCO_3$ (name: ________________) may buffer a strong ________________ (acid or base).

5. When a strong acid such as HCl is buffered by the bicarbonate buffer system, the following reaction takes place:

   $HCl + NaHCO_3 \rightarrow NaCl + H_2CO_3$

   a) Name the products of this reaction. ________________ and ________________

   b) HCl is a strong acid, which means it would have a ________________ (great or slight) effect on the pH of body fluids.

   c) One of the products of this reaction is the salt ________________, which has what effect on pH?

      ________________ (great, slight, or no effect)

   d) The other product is the weak acid ________________, which has what effect on pH?

      ________________ (great, slight, or no effect)

ORGANIC COMPOUNDS

1. Structure—Match each structural description with its proper organic compound.

   Use each letter once.

   1) Glucose ________________
   2) Pentose sugar ________________
   3) Sucrose ________________
   4) Starch ________________
   5) Glycogen ________________
   6) Cellulose ________________
   7) True fat ________________
   8) Phospholipid ________________
   9) Cholesterol ________________
   10) Amino acids ________________
   11) Proteins ________________

   A. A polysaccharide made by plants for energy storage
   B. Made of one glycerol and one, two, or three fatty acids
   C. A 6-carbon monosaccharide or hexose sugar
   D. A steroid
   E. Made of many amino acids
   F. A polysaccharide for energy storage in animal cells
   G. The molecular subunits of proteins
   H. A 5-carbon monosaccharide
   I. A diglyceride that includes a phosphate group
   J. A polysaccharide that is part of plant cell walls
   K. A disaccharide

2. Functions—Match each description of function with its proper organic compound.

   Use each letter once. Three lines will have two correct letters each.

   1) Glucose ________________
   2) Pentose sugars ________________
   3) Sucrose ________________
   4) Starch ________________
   5) Glycogen ________________
   6) Cellulose ________________
   7) True fats ________________
   8) Phospholipid ________________
   9) Cholesterol ________________
   10) Amino acids ________________
   11) Proteins ________________

   A. Changed to vitamin D in the skin on exposure to sunlight
   B. A polysaccharide that is digested to glucose and used for energy production
   C. An energy storage molecule in subcutaneous tissue
   D. The primary energy source for cells
   E. Bonded by peptide bonds to form proteins
   F. A diglyceride that is part of cell membranes
   G. Part of DNA and RNA
   H. These include enzymes and antibodies
   I. A sugar that is digested to monosaccharides to produce energy
   J. The storage form for glucose in the liver and muscles
   K. A polysaccharide that promotes peristalsis in the colon
   L. Used to synthesize the steroid hormones
   M. These include hemoglobin and insulin
   N. Also called fiber

CHEMICAL STRUCTURE—REVIEW

1. The following diagram depicts an organic molecule. Name the elements in this molecule.

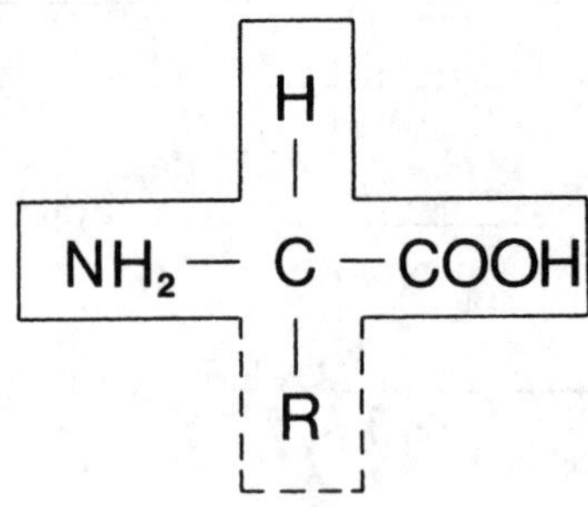

________________, ________________, ________________, and ________________

2. a) Name this molecule. ________________

   b) Name the type of bond that holds its elements together. ________________

3. a) Name the larger molecules that are made of many of this type of smaller molecule. ________________

   b) State three functions of these larger molecules in the body.

      1) ________________

      2) ________________

      3) ________________

4. The following diagram depicts another important molecule. Name the elements in this molecule.

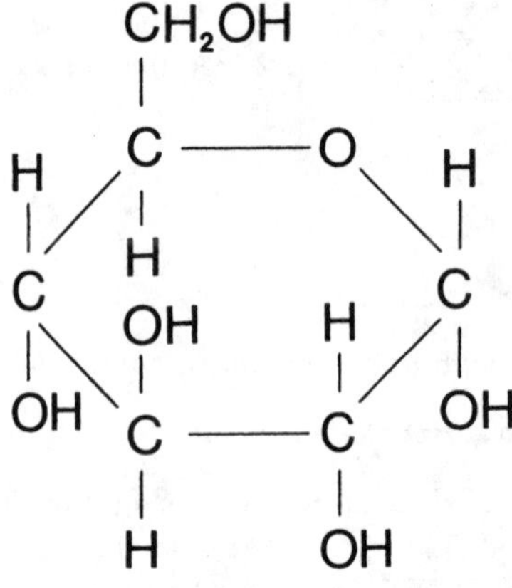

________________, ________________, and ________________

5. State the chemical formula of this molecule. ________________ (If you do not recognize it, simply count the number of each type of atom.)

6. a) Name this molecule. ________________

   b) State its function in the body. ________________

   c) Name three larger molecules that are made of many of this smaller molecule, and state a function of each.

      1) ________________

      2) ________________

      3) ________________

ENZYMES

1. Enzymes are proteins that increase the rate of reactions without the need for additional energy. Another way to say this is that enzymes are ____________________ in the body.
2. The following diagram depicts the active site theory of enzyme functioning in synthesis and decomposition reactions and in two examples of disrupted functioning.

Label the parts indicated.

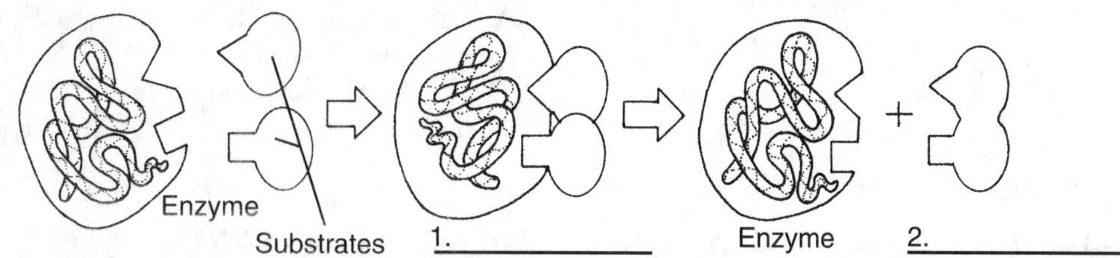

A. Synthesis

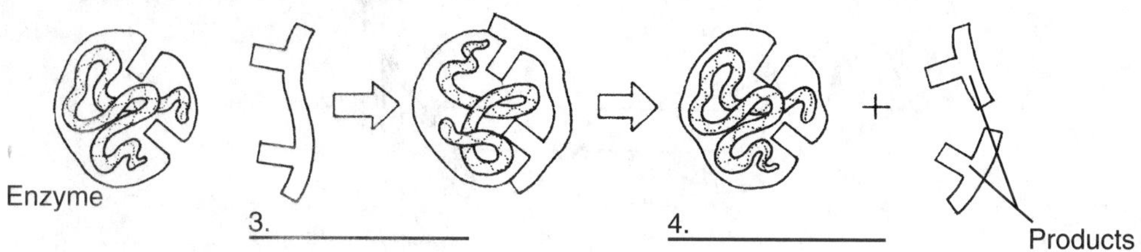

B. Decomposition

C. Effect of heat

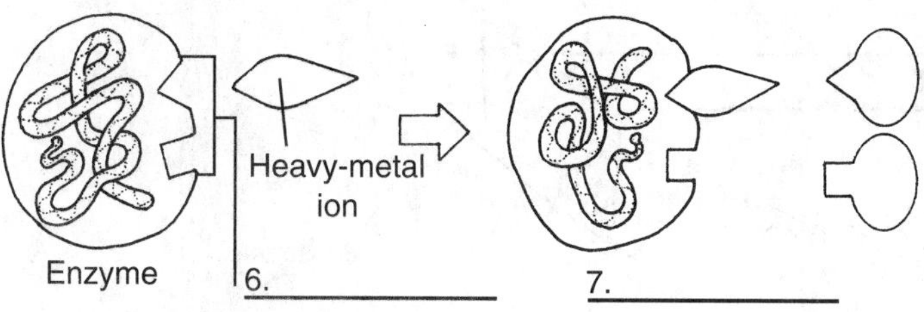

D. Effect of heavy-metal ions

3. An increase in body temperature, such as a fever, may inactivate enzymes because the higher temperature changes the ____________________ of the enzymes.
4. A state of acidosis may inactivate enzymes because excess ____________________ ions block the ____________________ of these enzymes.

DNA, RNA, AND ATP

1. Both DNA and RNA are made of subunits called nucleotides. A nucleotide consists of three smaller molecules:

   ____________________, ____________________, and ____________________.

2. Structure—Match each statement with the proper nucleic acid molecule.

   Use each letter once. Each answer line will have two correct letters.

   1) DNA ______________________________
   2) RNA ______________________________
   3) ATP ______________________________

   A. A single strand of nucleotides
   B. A double strand of nucleotides
   C. A single nucleotide with three phosphate groups
   D. Contains the nitrogenous bases A, T, C, and G
   E. Made from ADP and phosphate
   F. Contains the nitrogenous bases A, U, C, and G

3. The following diagram depicts the DNA double helix.

   Complete the DNA nucleotides according to the complementary base pairing found in DNA.

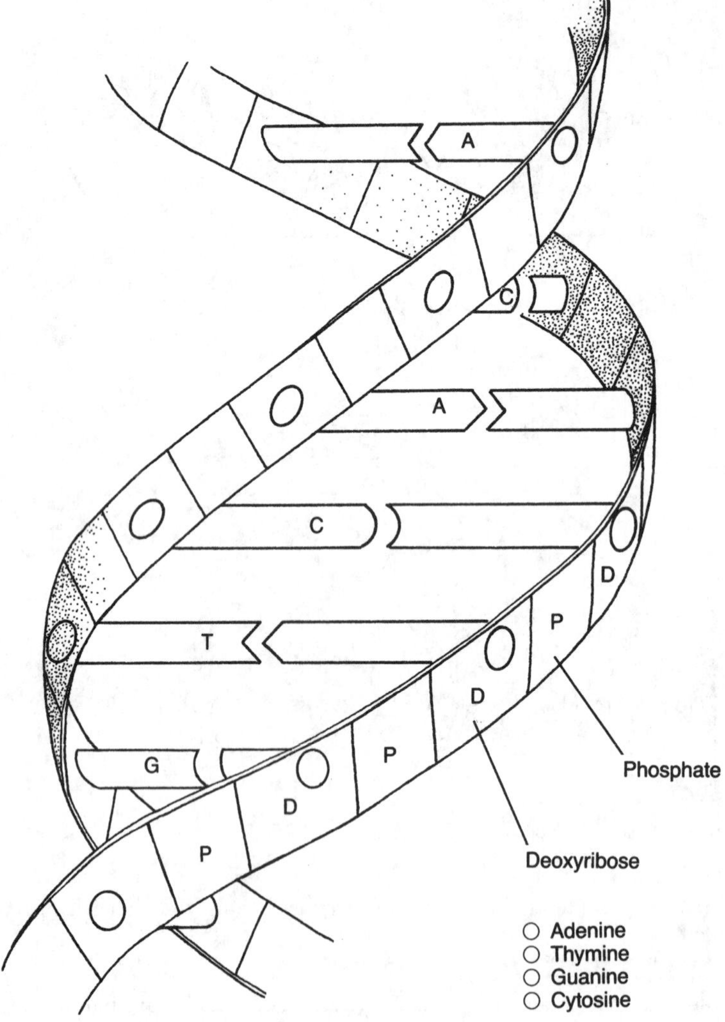

4. Functions—Match each statement with the proper nucleic acid molecule.

   Use each letter once. Two answer lines will have two correct letters.

   1) DNA ______________________________

   2) RNA ______________________________

   3) ATP ______________________________

   A. Directly involved in protein synthesis
   B. Formed when energy is released in cell respiration
   C. Makes up the chromosomes in the nucleus of a cell
   D. Provides energy for cellular reactions
   E. The genetic code for hereditary characteristics

5. The following diagram depicts the cycle of ATP and ADP formation.

   Label the parts indicated.

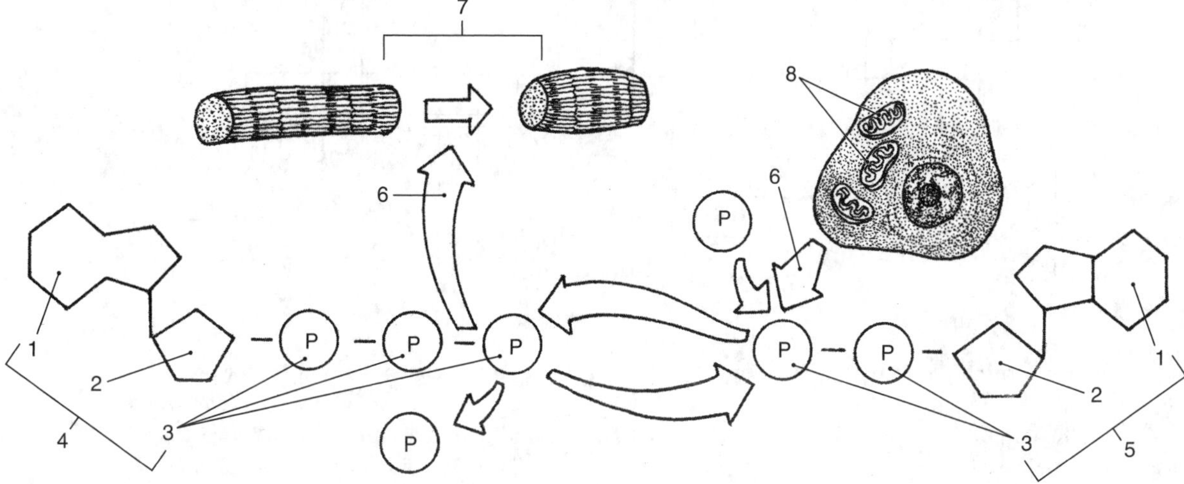

CROSSWORD PUZZLE

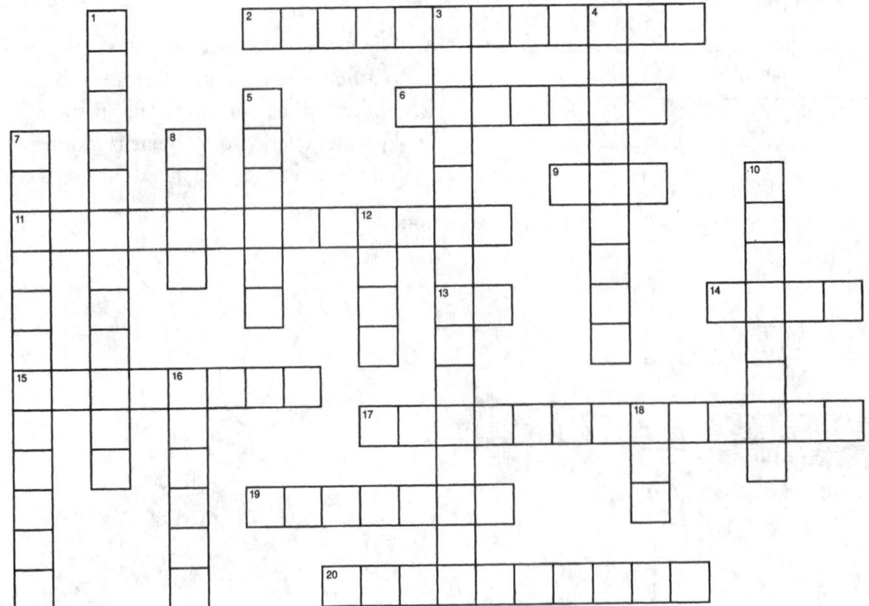

ACROSS

2. Allows ions to take part in other chemical reactions
6. Made of only one type of atom
9. Water within cells (initials)
11. Include sugars, starch, and glycogen
13. The value that indicates acidity or alkalinity
14. Decreases the concentration of H^+ ions
15. Speeds up the rate of a chemical reaction
17. Inorganic chemicals needed by the body in small amounts (two words)
19. Made of amino acids
20. The subunits of proteins (two words)

DOWN

1. Minimizes changes in pH (two words)
3. Energy production within cells (two words)
4. Bond that holds ions together (two words)
5. Will catalyze only one type of reaction
7. DNA and RNA are _______________ (two words)
8. Smallest part of an element
10. Sharing of electrons between atoms forms a _______________ bond
12. Increases the concentration of H^+ ions
16. Include true fats and steroids
18. 35% of the body's total water (initials)

CLINICAL APPLICATIONS

1. a) Mr. B has severe emphysema and cannot exhale efficiently. As CO_2 accumulates, the pH of his body fluids will _______________ (increase or decrease), and he will be in the state called _______________.

   b) In this state, Mr. B's blood pH might be:
   a) 7.32 b) 7.38 c) 7.40 d) 7.42 e) 7.50

2. Mrs. T is 72 years old and has osteoporosis, a condition in which the bones have become fragile and tend to fracture more easily. Which mineral is NOT present in sufficient amounts in Mrs. T's bones?
   a) iron b) sodium c) iodine d) calcium e) potassium

3. Following a car accident, Mr. M. is hospitalized and is receiving an intravenous solution of dextrose (glucose) in water. The direct purpose of this solution is to (choose one answer):
   a) replace the RBCs lost in hemorrhage
   b) provide an energy source for cell respiration
   c) provide the molecules needed for protein synthesis
   d) provide the molecules to repair tissues

4. Ms. C. is 16 years old, and her doctor has advised her to eat foods rich in iron. The most probable reason for this recommendation is to (choose one answer):
   a) provide for adequate hemoglobin synthesis to prevent anemia
   b) promote the growth of bones
   c) provide for the synthesis of thyroxine by the thyroid gland
   d) provide a primary energy source

MULTIPLE CHOICE TEST #1

Choose the correct answer for each question.

1. The direct source of energy for cells is usually:
   a) protein b) glucose c) DNA d) fats
2. An atom that has a charge after losing or gaining electrons is called:
   a) a molecule b) an ion c) a compound d) a substance
3. The smallest part of an element is:
   a) an atom b) a molecule c) an ion d) a compound
4. The normal pH range of blood is:
   a) 7.05–7.25 b) 7.40–7.50 c) 7.15–7.35 d) 7.35–7.45
5. Which formulas depict *ions* of sodium and chlorine?
   a) Na^+, Cl^- b) Na, Cl c) Na, Cl^+ d) Na^{+2}, Cl^{-2}
6. Which formula shows a molecule with covalent bonds?
   a) $CaCl_2$ b) H_2O c) NaCl d) Ca^{+2}
7. Which two compounds are important for energy storage in the body?
   a) proteins and phospholipids c) glycogen and true fats
   b) glucose and cholesterol d) DNA and cellulose
8. Intracellular fluid is the name for water found in:
   a) tissue spaces b) arteries and veins c) cells d) lymph vessels
9. Enzymes are molecules that catalyze reactions and are all:
   a) carbohydrates b) steroids c) fats d) proteins
10. The trace element most essential for oxygen transport in the blood is:
    a) calcium b) iron c) iodine d) sodium
11. Which one of the following sets of pH values is correct?
    a) acid 5, neutral 7, alkaline 9 c) acid 7, neutral 8, alkaline 9
    b) acid 1, neutral 3, alkaline 5 d) acid 6, neutral 8, alkaline 10
12. Which of the following organic molecules is NOT paired with its correct subunits?
    a) glycogen—glucose c) DNA and RNA—nucleotides
    b) proteins—monosaccharides d) true fats—fatty acids and glycerol
13. An enzyme may become inactive when:
    a) body temperature rises excessively d) all of the above
    b) the pH of body fluids becomes too acidic e) a and b only
    c) homeostasis is being maintained f) b and c only
14. Which of the following is the genetic code in our cells?
    a) glucose in cell respiration c) RNA in chromosomes
    b) proteins in enzymes d) DNA in chromosomes
15. The energy-transfer molecule formed in cell respiration is:
    a) DNA b) protein c) glucose d) ATP

16. The waste product of cell respiration is:
    a) water b) heat c) carbon dioxide d) ATP

17. The two elements that provide strength in bones and teeth are:
    a) Fe and Na b) I and Cl c) Na and K d) Ca and P

18. A carbohydrate is made of the elements:
    a) oxygen, calcium, nitrogen
    b) carbon, nitrogen, chlorine
    c) hydrogen, nitrogen, oxygen
    d) carbon, hydrogen, oxygen

19. The polysaccharide that is NOT an energy source is:
    a) glycogen b) cellulose c) starch d) sucrose

20. The molecule used to synthesize the hormones estrogen and testosterone is:
    a) cholesterol b) DNA c) proteins d) phospholipids

21. Hydrogen bonds help maintain the three-dimensional shape of:
    a) proteins and DNA
    b) DNA and glucose
    c) glucose and true fats
    d) true fats and proteins

22. The trace element that helps maintain the shape of some proteins by forming bonds is:
    a) cobalt b) iron c) sulfur d) sodium

23. All of the following are functions of proteins *except*:
    a) muscle contraction
    b) energy storage
    c) antibodies to pathogens
    d) structures such as tendons

24. An organic molecule that is an important part of cell membranes is:
    a) cellulose b) glycogen c) RNA d) phospholipid

25. Blood plasma is an example of:
    a) an extracellular fluid b) a transporting fluid c) a solvent d) all of these

MULTIPLE CHOICE TEST #2

Read each question and the four answer choices carefully. When you have made a choice, follow the instructions to complete your answer.

1. Which of the following is NOT an element?
   a) carbon b) water c) oxygen d) iron

   Explain why your choice is not an element.

   ____

2. Which statement is NOT true of chemical bonds?
   a) Covalent bonds involve the sharing of electrons between atoms.
   b) Covalent bonds are often weakened when in a water solution.
   c) Ionic bonds involve the loss of electron(s) by one atom and the gain of this (these) electron(s) by another atom.
   d) Inorganic molecules such as salts are formed by ionic bonds.

   Reword your choice to make it a correct statement.

   ____

3. Which statement is NOT true of water compartments?
   a) Tissue fluid is water found surrounding cells.
   b) Plasma is water found in blood vessels.
   c) Lymph is water found in lymph vessels.
   d) Intracellular fluid is water found between cells.

   Reword your choice to make it a correct statement.

   ____

4. Which statement is NOT true of cell respiration?
   a) Oxygen is required for the complete breakdown of a glucose molecule.
   b) The oxygen required comes from breathing.
   c) Glucose is a fat molecule obtained from food.
   d) The purpose of cell respiration is to produce ATP to provide energy for cellular processes.

   Reword your choice to make it a correct statement.

   ____

5. Which statement is NOT true of the products of cell respiration?
   a) The heat energy produced has no purpose in the body.
   b) The CO_2 formed is exhaled.
   c) The water formed becomes part of intracellular fluid.
   d) The ATP produced is used for cellular functions that require energy.

   Reword your choice to make it a correct statement.

   ____

6. Which trace element is paired with its proper function?
   a) Iron is part of bones and teeth.
   b) Sulfur is part of some proteins such as insulin.
   c) Calcium is part of the hormone thyroxine.
   d) Iodine is part of hemoglobin.

   Rearrange the three incorrect pairings to make them correct.

   ____

7. Which statement is NOT true of pH and buffer systems?
   a) A buffer system prevents drastic changes in the pH of body fluids.
   b) The normal pH range of blood is 7.0 to 8.0.
   c) The neutral point on the pH scale is 7.0.
   d) An acidic solution contains more H^+ ions than OH^- ions.

   Reword your choice to make it a correct statement.

   ____

8. Which statement is NOT true of the functions of water in the body?
   a) The sense of taste depends upon the solvent ability of water.
   b) In sweating, excess body heat evaporates water on the skin surface.
   c) Water is a lubricant and prevents friction as food moves through the digestive tract.
   d) The excretion of waste products in urine depends upon the lubricant action of water.

   Reword your choice to make it a correct statement.

   ____

9. Which organic compound is paired with its proper function?
   a) glycogen—a form of energy storage in adipose tissue
   b) true fats—part of DNA and RNA
   c) cholesterol—used to synthesize the steroid hormones
   d) pentose sugars—a form of energy storage in the liver

   Rearrange the three incorrect pairings to make them correct.

   ____

10. Which of the following is NOT a function of proteins?
    a) hemoglobin transport for nutrients in RBCs
    b) muscle structure and contraction
    c) part of the structure of skin and tendons
    d) enzymes to catalyze reactions

    Reword your choice to make it a correct function.

    ____

11. Which statement is NOT true of enzymes and their functioning?
    a) All enzymes are carbohydrates.
    b) The theory of enzyme functioning is called the active site theory.
    c) The shape of an enzyme is related to the type of reaction it will catalyze.
    d) Changes in body temperature or pH may affect the functioning of enzymes.

    Reword your choice to make it a correct statement.

    ___

12. Which statement is NOT true of the nucleic acids?
    a) DNA makes up the chromosomes of cells.
    b) RNA is a single strand of nucleotides.
    c) RNA is the genetic code for our hereditary characteristics.
    d) RNA functions in the process of protein synthesis.

    Reword your choice to make it a correct statement.

    ___

13. Which statement is NOT true of ATP?
    a) It is a specialized nucleotide.
    b) It is a product of cell respiration.
    c) It is needed for energy-requiring cellular reactions.
    d) It contains one phosphate group.

    Reword your choice to make it a correct statement.

    ___

14. Which statement is NOT true of chemical reactions?
    a) Synthesis reactions involve the breaking of bonds.
    b) Synthesis reactions usually require energy.
    c) Decomposition reactions involve the breaking of bonds.
    d) Decomposition reactions change large molecules to smaller ones.

    Reword your choice to make it a correct statement.

    ___

15. Which statement is NOT true of chemical bonds and molecules?
    a) Disulfide bonds maintain the shape of some proteins.
    b) Glucose is an organic molecule because it contains carbon.
    c) Water molecules are cohesive because of the presence of oxygen bonds.
    d) Hydrogen bonds help maintain the shape of DNA.

    Reword your choice to make it a correct statement.

    ___

16. Which statement is NOT true of lipids?
    a) Saturated fatty acids have the maximum number of carbon atoms.
    b) Cholesterol is part of cell membranes.
    c) Trans fats are believed to contribute to atherosclerosis.
    d) Unsaturated fatty acids are often found in vegetable oils.

    Reword your choice to make it a correct statement.

    ___

MULTIPLE CHOICE TEST #3

Each question is a series of statements concerning a topic in this chapter. Read each statement carefully and select all of the correct statements.

1. Which of the following statements are true of inorganic compounds?
   a) Water is considered the solvent within cells, tissues, and blood vessels.
   b) The most useful product of cell respiration is oxygen.
   c) Iodine is a trace element and is part of the hormone insulin.
   d) The mineral necessary for blood clotting is calcium.
   e) Cell respiration is the series of reactions that breaks down molecules of carbon dioxide.
   f) The mineral iron is part of vitamin B_{12}.
   g) Ionic bonds are found in molecules of salts.
   h) Oxygen is one of the reactant molecules in cell respiration.

2. Which of the following statements are true of organic compounds?
   a) All organic compounds contain the elements C, H, O, and P.
   b) Some lipid and carbohydrate molecules are storage forms for energy.
   c) The genetic code is contained in DNA, a type of nucleic acid.
   d) Cell membranes contain phospholipids, steroids, and proteins.
   e) Disaccharide molecules are important because they are energy sources.
   f) Proteins may be structural molecules in cells.
   g) The synthesis of proteins requires RNA.
   h) Cholesterol is used to make the steroid hormone insulin.
   i) All amino acids contain the element nitrogen as well as C, H, O, and P.
   j) The covalent bonds of a glucose molecule involve the loss and gain of electrons.
   k) Disulfide bonds are important to maintain the shape of steroids.
   l) Oligosaccharides are markers of "self" on cell membranes.

3. Which of the following statements are true of reactions and enzymes?
   a) A successful enzymatic reaction depends upon the shapes of the enzyme and the substrate(s).
   b) All enzymes are proteins.
   c) The active site of an enzyme is the part that is adaptable and varies to fit different substrates.
   d) At the end of a decomposition reaction, both the substrate and the enzyme have been broken down.
   e) A synthesis reaction forms larger molecules from small ones.
   f) Enzymes may be inactivated by excessive heat or a very low pH, both of which may alter the shape of the active site of the enzyme.

Chapter 3

Cells

Cells are the smallest living units of the structure and function of the body. This chapter describes cellular structure and some of the general activities that cells carry out within the body.

CELL STRUCTURE

1. Match the following major parts of a cell with their proper descriptions.

   Use each letter once. Each answer line will have two or more correct letters.

   1) Cell membrane ______________________________
   2) Cytoplasm ______________________________
   3) Nucleus ______________________________

   A. Contains the chromosomes of a cell
   B. Made of phospholipids, protein, and cholesterol
   C. Is the control center of a cell because of the genes it contains
   D. A watery solution of minerals, organic molecules, and gases
   E. Forms the outermost boundary of a cell
   F. Contains the nucleolus
   G. Permits certain substances to enter or leave the cell
   H. Found between the cell membrane and the nucleus
   I. Is selectively permeable
   J. Mature red blood cells lack this structure

2. a) Name the three organic molecules that make up cell membranes. ____________________,

   ____________________, and ____________________

   b) Use your answers above to complete the following:

   1) Provides stability for the cell membrane. ____________________
   2) Form pores and transporters in the cell membrane. ____________________
   3) Are antigens, in combination with oligosaccharides. ____________________
   4) Permit the diffusion of lipid-soluble substances into or out of the cell. ____________________
   5) Are receptor sites for hormones. ____________________

3. The following diagram depicts a typical cell. Label all of the structures indicated.

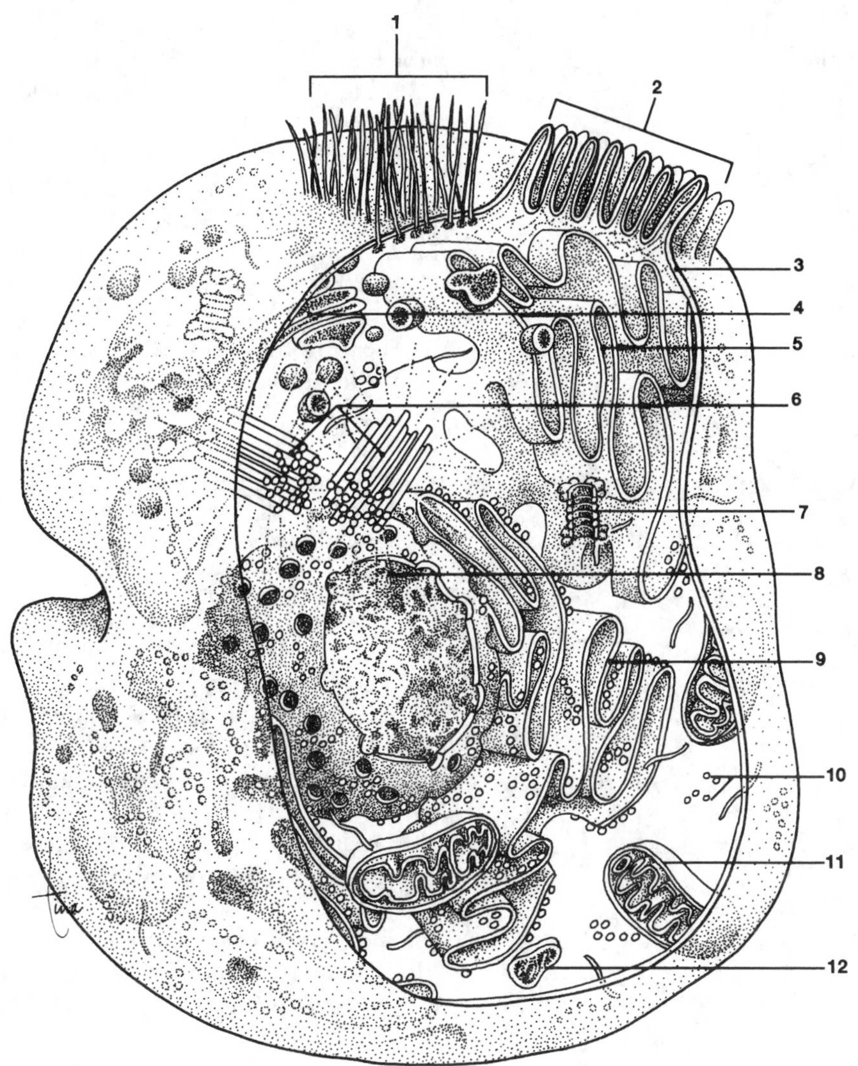

4. Match each cell organelle with its proper structure (a letter statement) and its proper function (a number statement).

Use each letter or number only once.

1) Endoplasmic reticulum ____________________

2) Ribosomes ____________________

3) Golgi apparatus ____________________

4) Mitochondria ____________________

5) Lysosomes ____________________

6) Centrioles ____________________

7) Motile cilia ____________________

8) Primary cilium ____________________

9) Flagellum ____________________

10) Microvilli ____________________

11) Proteasome ____________________

Structure

A. Made of protein and ribosomal RNA
B. Double-membrane structures; the inner membrane has folds called cristae
C. Two rod-shaped structures perpendicular to one another
D. A single, long, threadlike projection from the cell
E. An extensive series of membranous tubules that extend from the nuclear membrane to the cell membrane; may be rough or smooth
F. Short, threadlike projections through the cell membrane
G. Single-membrane structures that contain tissue-digesting enzymes
H. A single, short, threadlike projection from the cell
I. Folds of the cell membrane on the free surface
J. A series of flat, membranous sacs
K. Barrel-shaped enzymatic structure

Function

1. The site of protein synthesis
2. The site of destruction of damaged proteins
3. The site of ATP production
4. Digest worn-out cell parts or ingested bacteria
5. Provides motility for a sperm cell
6. Sweep materials across the cell surface
7. Passageway for transport of materials within the cell
8. Organize the spindle fibers during cell division
9. Synthesize carbohydrates and secrete materials from the cell
10. Increase surface area for absorption by the cell
11. Detect chemical or mechanical changes important for cellular communication

CELLULAR TRANSPORT MECHANISMS

1. Match each cellular transport process with its proper definition (a letter statement) and the proper example of the process in the body (a number statement).

Use each letter or number only once.

1) ____________________ Diffusion

2) ____________________ Osmosis

3) ____________________ Facilitated diffusion

4) ____________________ Active transport

5) ____________________ Filtration

6) ____________________ Phagocytosis

7) ____________________ Pinocytosis

Definition

A. Diffusion of molecules requiring carrier enzymes or transporters

B. The engulfing of something by a stationary cell

C. The movement of molecules from an area of greater concentration to an area of lesser concentration

D. The engulfing of something by a moving cell

E. The use of energy to move molecules from an area of lesser concentration to an area of greater concentration

F. The diffusion of water through a semipermeable membrane

G. Water and dissolved materials move through a membrane from an area of higher pressure to an area of lower pressure

Example

1. The absorption of water by the small intestine or kidneys
2. The movement of oxygen from the blood to the cells of the body
3. Blood pressure in capillaries forces plasma out to become tissue fluid
4. The engulfing of bacteria by white blood cells
5. The intake of glucose by cells of the body
6. The reabsorption of small proteins by cells of the kidney tubules
7. The absorption of amino acids or glucose by the cells of the small intestine

2. The following diagram represents one alveolus (air sac) in the lung; the alveolus is surrounded by a pulmonary capillary. Indicate the relative concentrations of oxygen and carbon dioxide on the basis of your knowledge of diffusion.

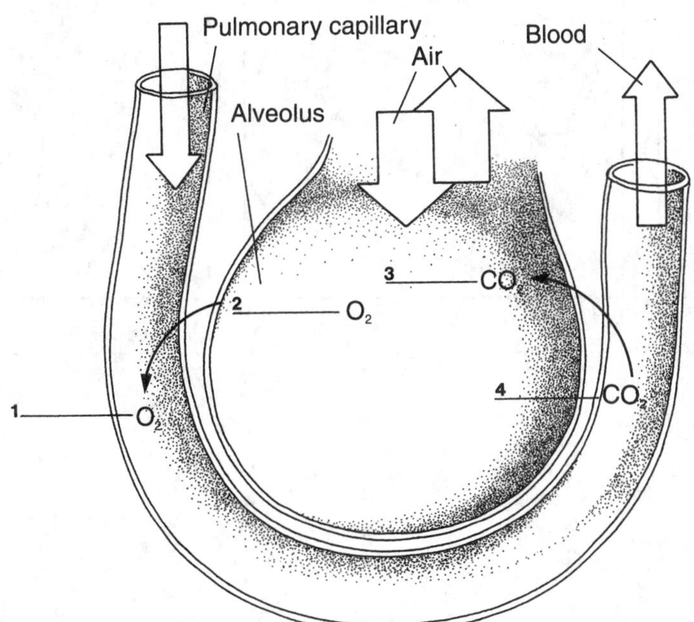

3. The following diagram represents a small section of the lining of the small intestine, with certain food molecules within the cavity of the intestine. For each nutrient, name the mechanism by which it is absorbed.

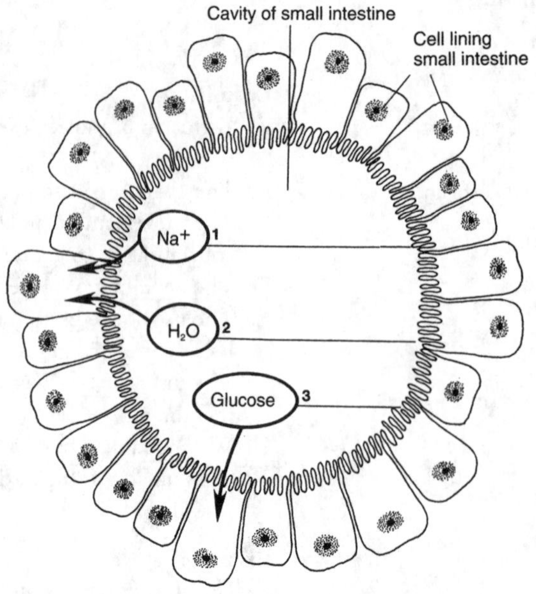

4. Using human cells as a reference point, complete each statement by using the phrase "a higher," "a lower," or "the same."

   1) An isotonic solution has ______________ concentration of dissolved material as does a cell.

   2) A hypertonic solution has ______________ concentration of dissolved material than does a cell.

   3) A hypotonic solution has ______________ concentration of dissolved materials than does a cell.

5. The following diagrams depict human red blood cells in three different solutions. Notice what, if anything, has happened to the cells.

   On the basis of your knowledge of osmosis, indicate which solution is hypertonic, which is hypotonic, and which is isotonic to the cells.

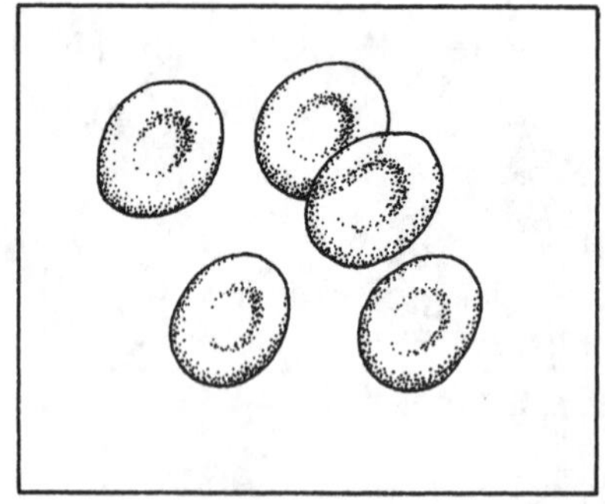

1) ______________

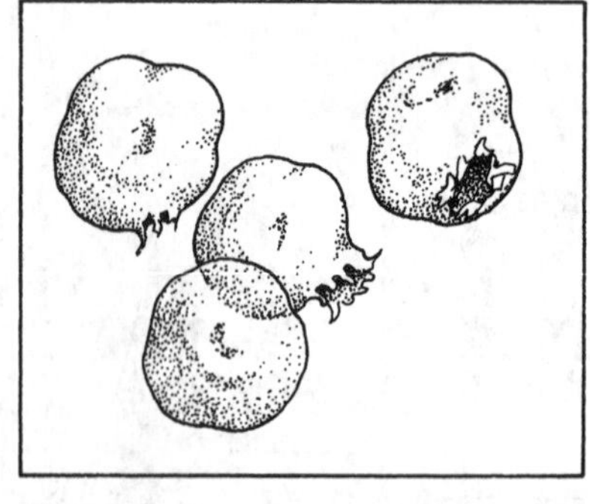

2) ______________

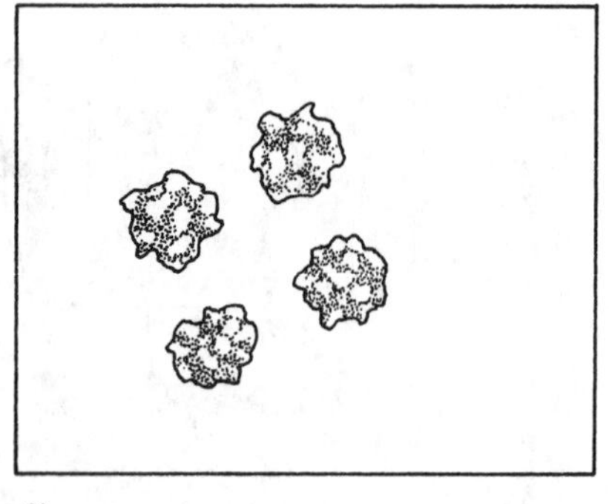

3) ______________

DNA AND THE GENETIC CODE

1. a) DNA makes up the ____________________ of a cell, which are found in the nucleus.

   b) Human cells contain how many of these DNA structures? ____________________

2. A DNA molecule is made of ____________________ (one, two, three, or four) strands of nucleotides twisted into a spiral called a ____________________.

3. The complementary base pairing of DNA means that adenine is always paired with ____________________, and cytosine is always paired with ____________________.

4. A gene is a segment of DNA that is the genetic code for one ____________________.

5. A protein is made of the smaller molecules called ____________________.

6. a) The DNA code for one amino acid consists of how many bases? ____________________

   b) Therefore, the name for this code may also be the ____________________ code.

   c) The other name for a triplet is a ____________________.

RNA AND PROTEIN SYNTHESIS

1. An RNA molecule consists of how many strands of nucleotides? ____________________

2. One type of RNA copies the genetic code of a DNA gene; this type is called ____________________ RNA, which may be abbreviated ____________________.

3. The bases found in mRNA are adenine, guanine, cytosine, and ____________________.

4. An mRNA molecule is synthesized from the DNA in the ____________________ (part) of a cell but then moves to the cytoplasm and becomes attached to which type of cell organelle? ____________________

5. A second type of RNA picks up amino acids in the cytoplasm of a cell; this type is called ____________________ RNA, which may be abbreviated ____________________.

6. A tRNA has a triplet of bases called an anticodon, which matches the ____________________ on the mRNA and ensures that the amino acid is positioned in its proper place in the protein.

7. Ribozyme (rRNA) to form ____________________ bonds between amino acids is contained in the ____________________ to which the mRNA is attached.

8. a) The expression of the genetic code that gives us our hereditary characteristics is summarized in the sequence below. Fill in the missing parts of the sequence.

DNA→1. ____________________ →Proteins→2. ____________________ →Hereditary characteristics

3. ______________ →Catalyze reactions ↑

   b) In this sequence, which part is transcription? ____________________ Which part is translation?

   ____

9. The following diagram depicts the process of protein synthesis taking place in a cell.

Label each of the following: nucleus, DNA, mRNA, tRNA, ribosome, amino acids, and peptide bond.

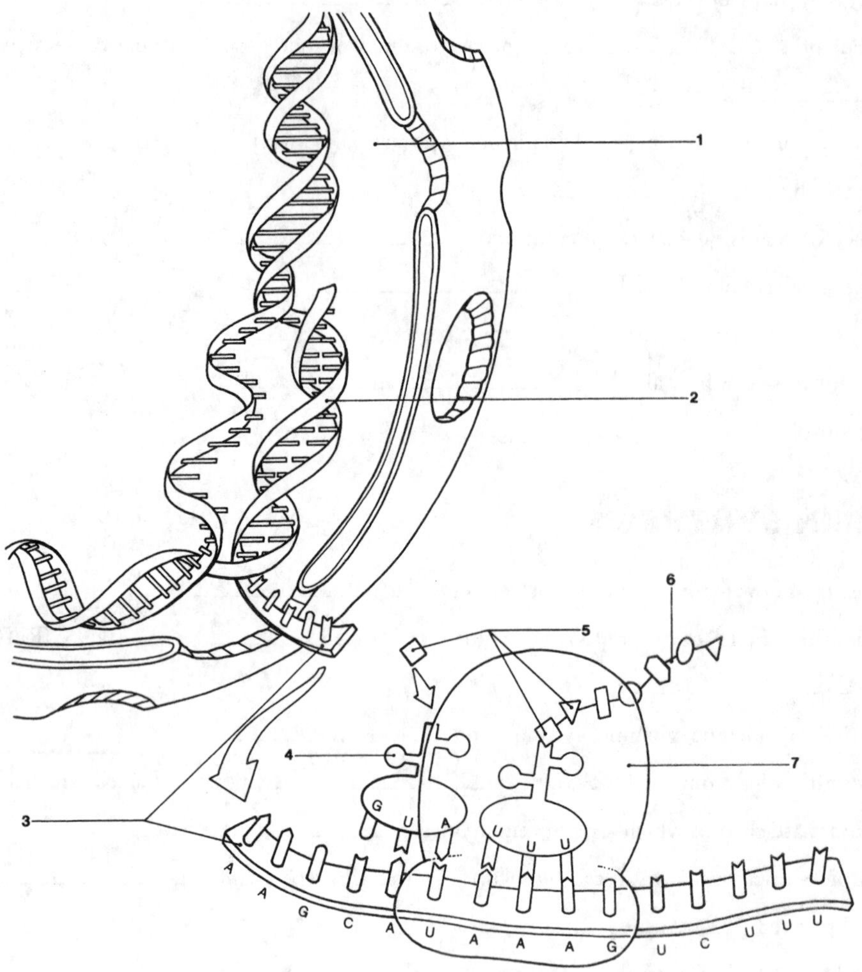

MITOSIS AND MEIOSIS

1. With respect to the processes of cell division, match the following statements.

Use each letter once. Each answer line will have four correct letters.

1) Mitosis ______________________

2) Meiosis ______________________

A. Two identical cells are produced.
B. One cell with the diploid number of chromosomes divides once.
C. Four cells are produced.
D. One cell with the diploid number of chromosomes divides twice.
E. Each cell produced has the haploid number of chromosomes.
F. The cells produced are egg or sperm cells.
G. The cells produced are needed for the growth and repair of tissues.
H. Each cell produced has the diploid number of chromosomes.

2. Before mitosis or meiosis takes place, a cell is said to be in a stage called *interphase.* During this time, DNA replication takes place. Explain what happens in DNA replication.

____

3. Match the stages of mitosis with the events that take place during each.

Use each letter once. Each answer line will have more than one correct letter.

1) Prophase ______________________

2) Metaphase ______________________

3) Anaphase ______________________

4) Telophase ______________________

A. The pairs of chromatids line up on the equator of the cell.
B. The centrioles organize the formation of the spindle fibers.
C. The spindle fibers pull each set of chromosomes toward opposite poles of the cell.
D. The chromosomes coil up and become visible as short rods; each is a pair of chromatids connected at the centromere.
E. Each chromatid is now considered a separate chromosome.
F. The centromeres divide.
G. The nuclear membrane disappears.
H. Each set of chromosomes uncoils, and nuclear membranes are re-formed.
I. Cytokinesis, the division of the cytoplasm, follows.

4. Name two sites in the human body where mitosis takes place constantly and explain why this rapid mitosis is necessary.

a) Site ______________ Reason ______________

b) Site ______________ Reason ______________

5. Name one type of adult human cell that usually does not reproduce itself by mitosis and explain the significance of this.

a) Type of cell ______________

b) Significance ______________

6. The following diagrams depict the stages of mitosis in a cell with a diploid number of four.

Label each cell as one of the following: interphase, prophase, metaphase, anaphase, telophase, cytokinesis. Then arrange the letters in the proper sequence.

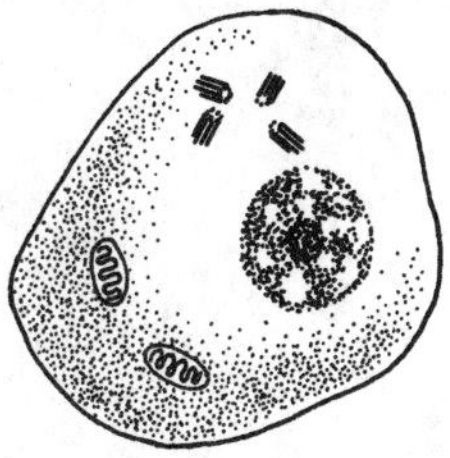

a ______________

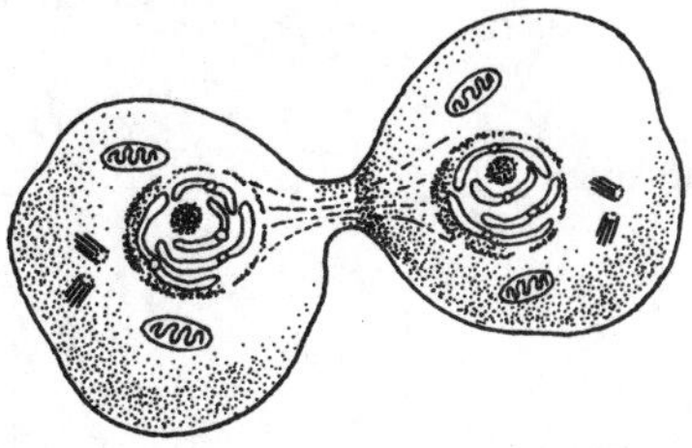

b ______________

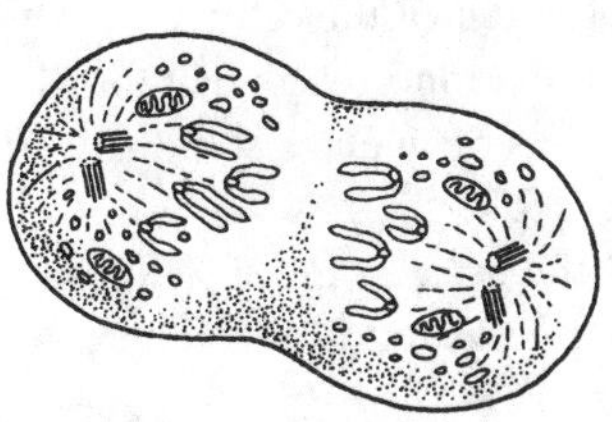

c ______________

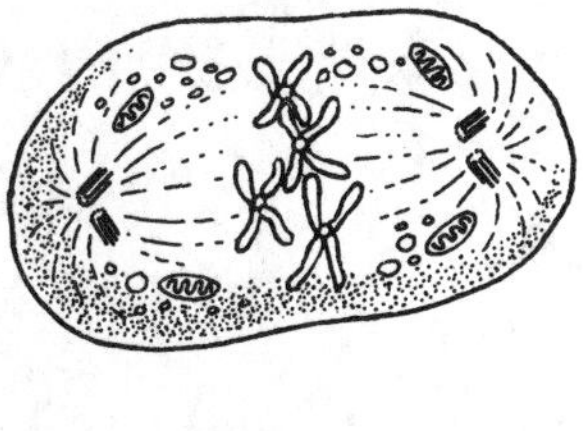

d ______________

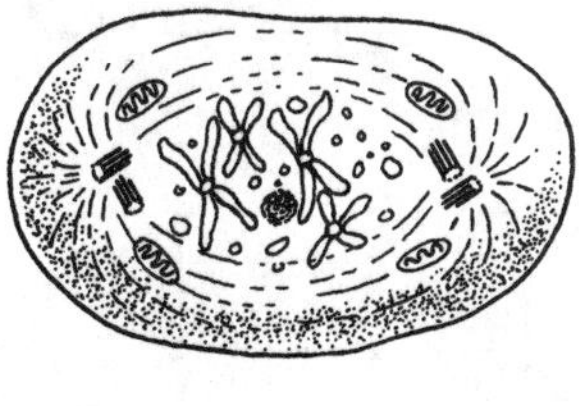

e ______________

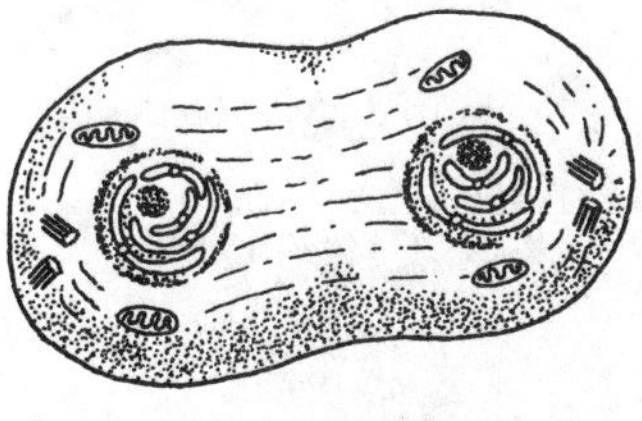

f ______________

Sequence: ________, ________, ________, ________, ________, ________.

CROSSWORD PUZZLE

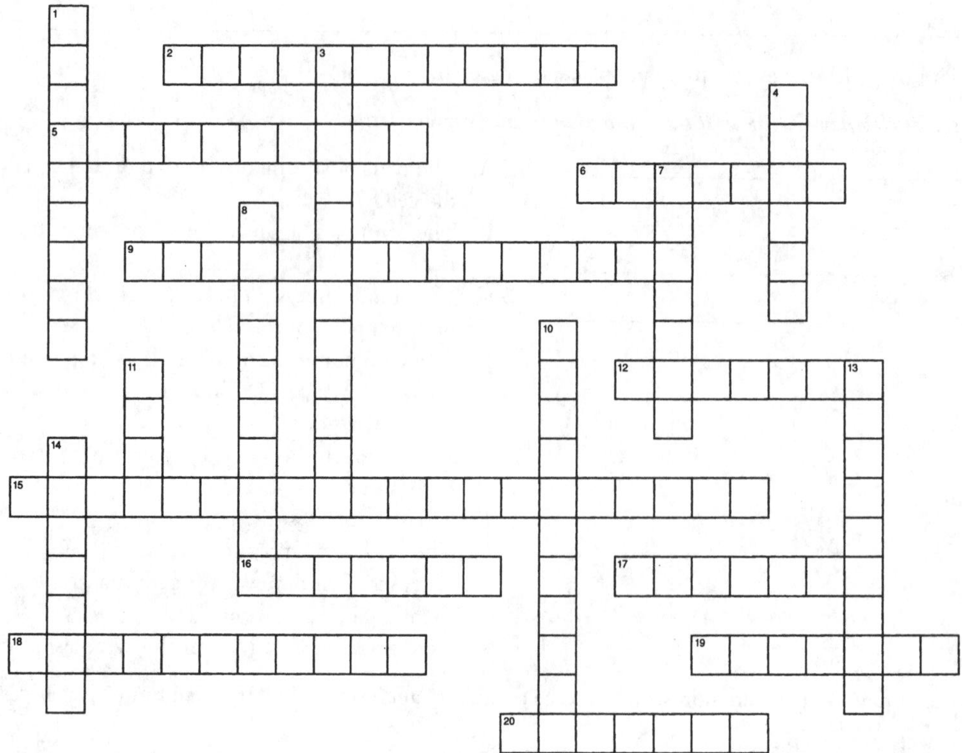

ACROSS

2. The organelles in which cell respiration takes place
5. Intracellular structures with specific roles in cellular functioning
6. Egg and sperm cells
9. Movement of molecules against a concentration gradient (two words)
12. The usual number of chromosomes within a cell
15. A membrane that permits only certain substances to pass through (two words)
16. Control center of the cell
17. The cell division process that forms gametes
18. The process by which a stationary cell takes in small particles
19. Half of the usual number of chromosomes
20. Diffusion of water through a membrane

DOWN

1. Found between the cell membrane and the nucleus
3. Made of phospholipids, cholesterol, and proteins (two words)
4. A statement that best explains the available evidence
7. Type of cell division essential for repair of tissues
8. The process in which blood pressure creates tissue fluid
10. A human cell has 46 of these
11. Genetic code for one protein
13. Movement of molecules with or along a concentration gradient
14. Oxygen-requiring

CLINICAL APPLICATIONS

1. Mr. D is receiving chemotherapy for cancer. This medication inhibits the process of mitosis, which slows the production of malignant cells. Also affected, however, are other cells that undergo rapid mitosis.

   Mr. D may develop anemia if the ____________________ is affected in this way.

2. a) Intravenous solutions are often isotonic, which means that their concentrations of water and dissolved materials are

   ____________________ as the concentrations found in the blood plasma.

   b) If distilled water were mistakenly administered intravenously, the red blood cells at the site would

   ____________________ (gain or lose) water by the process of ____________________ (filtration or osmosis)

   and would eventually ____________________ (swell or shrivel).

3. a) Certain antibiotics that are used to treat bacterial infections inhibit the synthesis of the nucleic acids DNA and RNA. Although this may slow the growth of bacteria, the same processes in human cells may also be inhibited.

   Without DNA synthesis, what cellular process could not take place? ____________________

   b) Without RNA synthesis, what cellular process would not take place? ____________________

MULTIPLE CHOICE TEST #1

Choose the correct answer for each question.

1. The cell organelles most directly associated with cell division are the:
   a) Golgi apparatus b) centrioles c) ribosomes d) lysosomes

2. The hereditary material of cells is ____________________, which is found in the ____________________ of the cell.
   a) protein/ribosomes b) DNA/chromosomes c) glucose/nucleus d) RNA/ribosomes

3. The major structural parts of a cell are the:
   a) cell membrane, nucleus, cytoplasm
   b) nucleus, nucleolus, mitochondria
   c) cytoplasm, nucleolus, endoplasmic reticulum
   d) ribosomes, cell membrane, Golgi apparatus

4. The cell organelle most directly associated with the production of cellular proteins is:
   a) mitochondria b) lysosomes c) Golgi apparatus d) ribosomes

5. The cellular transport mechanism that depends upon transporters or carrier enzymes in the cell membrane is:
   a) filtration b) osmosis c) diffusion d) facilitated diffusion

6. The cellular transport mechanism that depends upon blood pressure is:
   a) diffusion b) filtration c) phagocytosis d) active transport

7. A water-salt solution with the same salt concentration as in cells is called:
   a) isotonic b) hypertonic c) hypotonic d) lemon-lime tonic

8. Mitosis produces cells that have:
   a) only one chromosome each
   b) the diploid number of chromosomes, 46 for people
   c) the haploid number of chromosomes, 23 for people
   d) twice the usual number of 46 chromosomes for people

9. Meiosis produces cells that have:
   a) the diploid number of chromosomes, 23 for people
   b) the haploid number of chromosomes, 46 for people
   c) the haploid number of chromosomes, 23 for people
   d) double the number of chromosomes, 92 for people

10. The organic molecules in the cell membrane that form pores and receptor sites for hormones are:
a) cholesterol b) phospholipids c) proteins d) carbohydrates

11. Meiosis is necessary to produce:
a) skin cells b) bone cells c) muscle cells d) egg or sperm cells

12. Mitosis is necessary for:
a) constant production of new nerve cells
b) continuation of the human species by reproduction
c) growth and repair of tissues
d) production of egg and sperm cells

13. The function of motile cilia is to:
a) sweep materials across a surface
b) provide locomotion (movement) for human cells
c) move the chromosomes during mitosis
d) support cells and keep them stable

14. The nucleus of a cell:
a) forms the outer boundary of a cell and protects the cell
b) contains most of the cell organelles
c) regulates the activities of a cell by means of the genetic material it contains
d) regulates what enters or leaves the cell to protect the cell from pathogens

15. The function of the Golgi apparatus of a cell is to synthesize:
a) proteins b) lipids c) new genes d) carbohydrates

16. Diffusion is defined as the movement of molecules:
a) from an area of lesser concentration to an area of greater concentration
b) from an area of greater pressure to an area of lower pressure
c) from an area of lower energy to an area of higher energy
d) from an area of greater concentration to an area of lesser concentration

17. A gene is the genetic code for one:
a) cell b) protein c) organ d) glucose

18. The complementary base pairing of DNA is:
a) A-T and G-C b) A-C and G-T c) A-G and C-T d) A-U and C-G

19. The complementary base pairing of DNA with mRNA is:
a) A-C and G-U b) A-U and G-C c) A-G and C-U d) A-T and U-C

20. Human cells that usually do not undergo mitosis in an adult are:
a) stomach lining cells and muscle cells
b) nerve cells and muscle cells
c) bone marrow cells and skin cells
d) skin cells and nerve cells

21. The cell organelle most directly associated with the destruction of damaged cellular proteins is the:
a) ribosome b) proteasome c) mitochondrion d) centrosome

22. The function of the microvilli of a cell is to:
a) sweep materials across the cell surface
b) provide motility for the cell
c) increase the area of the chromosomes
d) increase the surface area of the cell membrane

23. In the body, the process of diffusion is responsible for the movement of:
a) food molecules b) proteins c) oxygen and CO_2 d) fats

24. A cell that has the potential to develop into several different kinds of cells is a:
a) blast cell b) stem cell c) multipurpose cell d) specialized cell

25. The cells lining the small intestine absorb glucose and amino acids by the process of:
a) active transport b) osmosis c) filtration d) diffusion

MULTIPLE CHOICE TEST #2

Read each question and the four answer choices carefully. When you have made a choice, follow the instructions to complete your answer.

1. Which cellular transport mechanism is NOT paired with its correct definition?
   a) osmosis—the diffusion of water through a membrane
   b) filtration—the movement of water and dissolved materials through a membrane from an area of higher pressure to an area of lower pressure
   c) active transport—the movement of molecules from an area of greater concentration to an area of lesser concentration
   d) phagocytosis—the engulfing of something by a moving cell

   For your choice, state the correct definition.

   ____

2. Which cell organelle is NOT paired with its proper function?
   a) mitochondria—the site of cell respiration and ATP production
   b) lysosomes—contain enzymes to digest worn-out cell parts
   c) endoplasmic reticulum—membranous tubules that are passageways within the cell
   d) ribosomes—the site of carbohydrate synthesis

   For your choice, state its correct function.

   ____

3. Which statement is NOT true of DNA?
   a) DNA makes up the chromosomes of cells.
   b) DNA exists as a single strand of nucleotides called a double helix.
   c) DNA is the genetic code for hereditary characteristics.
   d) A gene is the genetic code for one protein.

   Reword your choice to make it a correct statement.

   ____

4. Which statement is NOT true of the process of protein synthesis?
   a) The site of protein synthesis is the ribosomes in the cytoplasm.
   b) Transfer RNA molecules pick up amino acids and bring them to the proper mRNA triplet.
   c) Amino acids are bonded to one another by ionic bonds.
   d) Messenger RNA is formed in the nucleus as a copy of a DNA gene.

   Reword your choice to make it a correct statement.

   ____

5. Which stage of mitosis is NOT paired with a correct event of that stage?
   a) telophase—a nuclear membrane re-forms around each new set of chromosomes
   b) metaphase—the pairs of chromatids line up on the equator of the cell
   c) prophase—the chromosomes become visible and spindle fibers form
   d) anaphase—each pair of chromatids becomes attached to a spindle fiber

   For your choice, state what does happen during this stage.

   ____

6. Which statement is NOT true of the cell membrane?
   a) The phospholipids permit diffusion of lipid-soluble substances.
   b) It forms the outer boundary of the cell.
   c) It is impermeable, meaning that only certain substances may pass through.
   d) Some of the proteins form pores to permit the entry or exit of molecules.

   Reword your choice to make it a correct statement.

   ____

7. Which statement is NOT true of mitosis in an adult?
   a) Most nerve cells reproduce only when there is damage to the brain.
   b) Skin cells reproduce to replace those worn off the skin surface.
   c) Red blood cells are continuously replaced by the red bone marrow.
   d) Heart muscle cells seem to be unable to reproduce themselves.

   Reword your choice to make it a correct statement.

   ____

8. Which cellular transport mechanism is NOT paired with its proper function in the body?
   a) pinocytosis—the reabsorption of small proteins by the kidney tubules
   b) diffusion—the exchange of gases between the air in the lungs and the blood
   c) osmosis—the absorption of salts by the small intestine
   d) filtration—the formation of tissue fluid as plasma is forced out of capillaries

   For your choice, state a correct function of the mechanism.

   ____

9. Which statement is NOT true of solutions?
   a) Human cells in a hypertonic solution would remain undamaged.
   b) A hypotonic solution has a lower concentration of dissolved materials than do human cells.
   c) Blood plasma is isotonic to red blood cells.
   d) Human cells in a hypotonic solution would swell.

   Reword your choice to make it a correct statement.

   ____

10. Which statement is NOT true of cells?
    a) The cell membrane has receptor sites for hormones.
    b) The only cells that undergo meiosis are those that will produce egg or sperm cells.
    c) Some human cells are capable of movement.
    d) Proteasomes have enzymes to destroy misfolded sugars.

    Reword your choice to make it a correct statement.

    ____

MULTIPLE CHOICE TEST #3

Each question is a series of statements concerning a topic in this chapter. Read each statement carefully and select all of the correct statements.

1. Which of the following statements are true of cellular structures?
   a) Microvilli enable a cell to absorb more material.
   b) The sites of protein synthesis are the proteasomes.
   c) Lysosomes have enzymes to destroy damaged cell parts.
   d) The cell membrane has many different receptor sites for hormones and other signalling molecules.
   e) Chromosomes are made of DNA and protein, and they are found in the nucleus of a cell.
   f) Ribosomes are the site for the synthesis of ATP.
   g) White blood cells move by means of their cilia.
   h) The endoplasmic reticulum is a series of transport tunnels in the cytoplasm.
   i) The primary cilium of a cell is a sensory structure.
   j) The cell membrane is made of phospholipids, cholesterol, and proteins.

2. Which of the following statements are true of cellular transport processes?
   a) Both diffusion and osmosis depend on concentration gradients.
   b) Phagocytosis is an essential process for red blood cells.
   c) Active transport is the pumping out of excess water by cells.
   d) Filtration is the removal of air pollution by cells in the lungs.
   e) Most cells have transporters for the facilitated diffusion of glucose.
   f) Osmosis is simply the diffusion of water.
   g) A cell capable of pinocytosis can absorb a small protein molecule.
   h) Gas exchange in the lungs depends on diffusion.

3. Which of the following statements are true of mitosis and meiosis?
   a) Both processes produce two cells from one cell.
   b) Cells produced by meiosis have the diploid number of chromosomes.
   c) Cells produced by mitosis have the haploid number of chromosomes.
   d) Meiosis is necessary for the growth and repair of tissues.
   e) The production of gametes requires two successive mitotic divisions.
   f) Stem cells are highly specialized cells that carry out meiosis frequently.

Chapter 4

Tissues and Membranes

This chapter describes the tissues, which are groups of cells with similar structure and functions. The four major groups of tissues are epithelial tissue, connective tissue, muscle tissue, and nerve tissue. Each of these groups has very specific characteristics and purposes.

EPITHELIAL TISSUE

1. The following diagrams depict some of the types of epithelial tissue.

   Label each type with its complete name.

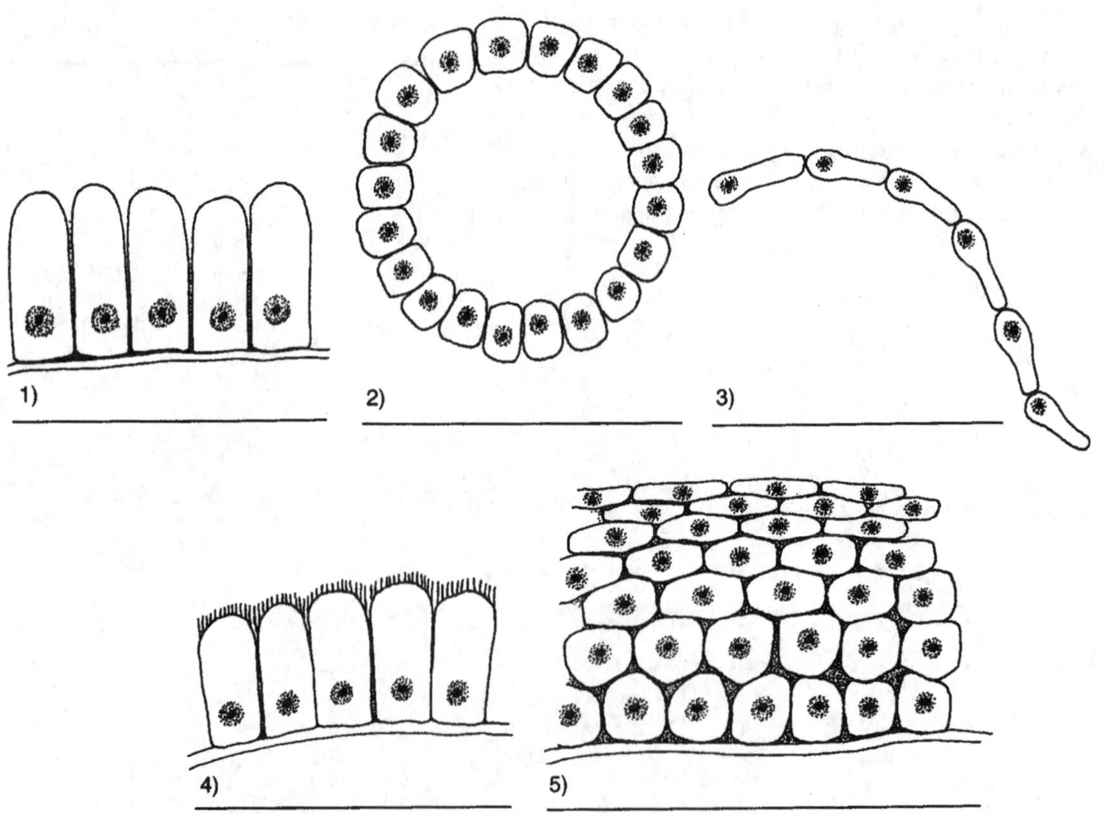

2. Match each epithelial tissue with its proper structure (a letter statement) and functions in the body (one or more number statements).

Use each letter and number once. Each answer line will have one correct letter and may have more than one correct number.

1) Simple squamous epithelium ____________________

2) Stratified squamous epithelium ____________________

3) Transitional epithelium ____________________

4) Cuboidal epithelium ____________________

5) Columnar epithelium ____________________

6) Ciliated epithelium ____________________

Structure

A. Many layers of cells; surface cells are flat
B. Columnar cells with cilia on their free surfaces
C. One layer of cells that are taller than they are wide
D. One layer of cube-shaped cells
E. One layer of flat cells
F. Many layers of cells; surface cells are alternately rounded or flat

Function

1. Forms the alveoli of the lungs and permits diffusion of gases
2. Secretes the hormones of the thyroid gland
3. Forms the epidermis of the skin
4. Forms the stomach lining and secretes gastric juice
5. Forms capillaries to permit exchanges of materials
6. Permits stretching of the urinary bladder as it fills
7. Lines arteries and veins and is smooth to prevent abnormal blood clotting
8. Forms the lining of the mouth and the esophagus
9. Lines the trachea and sweeps mucus and bacteria toward the pharynx
10. Forms the lining of the small intestine and absorbs nutrients
11. Lines the fallopian tubes to sweep an ovum toward the uterus
12. May have microvilli to increase the surface area for absorption

3. Glands are made of epithelial tissue, and there are several different categories. Match each type of gland with its proper structure (letter statements) and an example in the body (a number statement).

Use each letter and number once. One answer line will have two correct letters.

1) Unicellular glands ____________________

2) Exocrine glands ____________________

3) Endocrine glands ____________________

Structure

A. Consist of only one cell
B. Have no ducts; their secretions enter capillaries
C. Have ducts to take their secretions to their site of action
D. Their secretions are called hormones

Example

1. The thyroid gland and pituitary gland
2. Goblet cells that secrete mucus
3. The salivary glands and sweat glands

CONNECTIVE TISSUE

1. Match each connective tissue with its proper structure (a letter statement) and functions in the body (one or more number statements).

Use each letter and number once. Each answer line will have one correct letter and may have more than one correct number.

1) Blood ______________________

2) Areolar connective tissue ______________________

3) Adipose tissue ______________________

4) Fibrous connective tissue ______________________

5) Elastic connective tissue ______________________

6) Bone ______________________

7) Cartilage ______________________

Structure

A. Made primarily of elastin fibers
B. Made of cells in the fluid matrix called plasma
C. Made of osteocytes in a matrix of calcium salts and collagen
D. Made of cells specialized to store fat
E. Made of fibroblasts in a matrix of tissue fluid, collagen, and elastin fibers
F. Made of chondrocytes in a matrix that is smooth and flexible
G. Made primarily of collagen fibers

Function

1. Beneath the skin and the epithelium of mucous membranes; has white blood cells to destroy pathogens
2. Forms tendons that connect muscles to bones
3. Surrounds the alveoli of the lungs and contributes to normal exhalation
4. Transports oxygen and nutrients and contains cells to destroy pathogens
5. Supports the body
6. Provides a smooth surface to prevent friction on joint surfaces
7. Stores excess energy in subcutaneous tissue
8. The cells are produced in red bone marrow
9. Forms ligaments that connect bone to bone
10. Forms rings to keep the trachea open
11. In the walls of the large arteries where it helps maintain blood pressure
12. Protects some internal organs from mechanical injury

2. The following diagrams depict some of the types of connective tissue.

Label each type with its complete name.

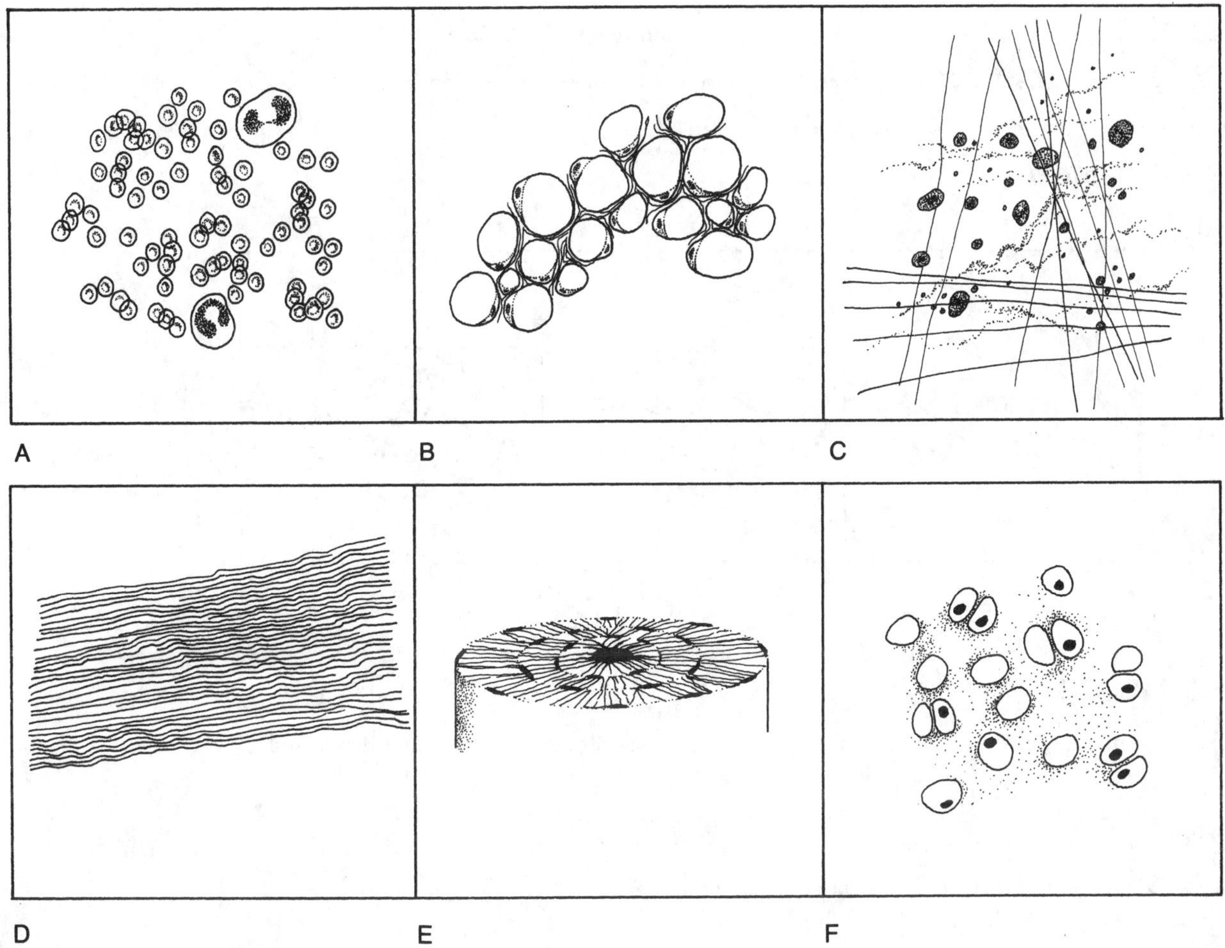

MUSCLE TISSUE

1. All three types of muscle tissue are specialized to contract and bring about movement of some kind.

Each of the following statements describes a structural or functional aspect of one of the three types of muscle tissue. Indicate the type to which the statement applies by writing SK for skeletal muscle, SM for smooth muscle, or C for cardiac muscle on the line before the statement.

1) ____________________ The cells are tapered and have one nucleus each.

2) ____________________ Attached to bones, moves the skeleton.

3) ____________________ Enables arteries to constrict or dilate to maintain blood pressure.

4) ____________________ The cells are branched and each has one nucleus.

5) ____________________ Also called striated muscle tissue because the cylindrical cells appear to have striations.

6) ____________________ Produces a significant amount of body heat.

7) ____________________ Forms the walls of the chambers of the heart; its function is to pump blood.

8) ____________________ Also called visceral muscle because it is found in many internal organs.

9) ____________________ Produces involuntary waves of contraction, called peristalsis, in the intestines.

10) ____________________ Each cell has several nuclei.

11) ____________________ The cells contract by themselves; nerve impulses regulate only the rate of contraction.

12) ____________________ Also called voluntary muscle because nerve impulses are required for contraction.

13) ____________________ In the iris of the eye, it will constrict or dilate the pupil.

14) ____________________ Has intercalated discs for rapid impulse transmission from cell to cell.

2. The following diagrams depict the three types of muscle tissue.

   Label each type with its complete name.

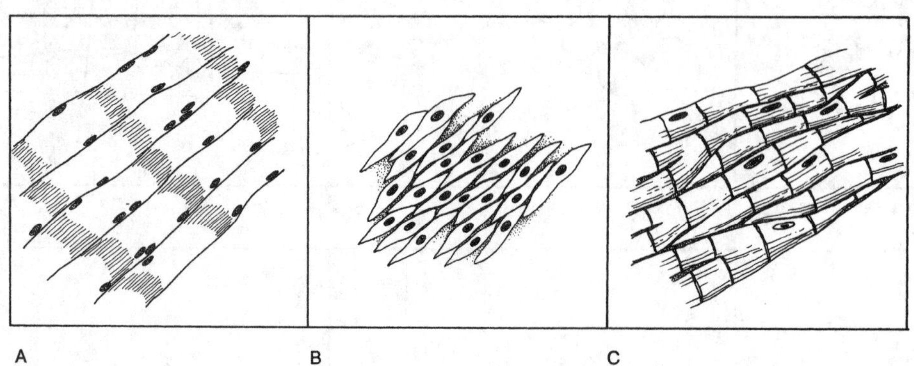

NERVE TISSUE

1. The name for nerve cells is ____________________, and these cells are specialized to generate and transmit ____________________.
2. The following diagram depicts a neuron.

   Label the following structures: cell body, nucleus, axon, dendrites.

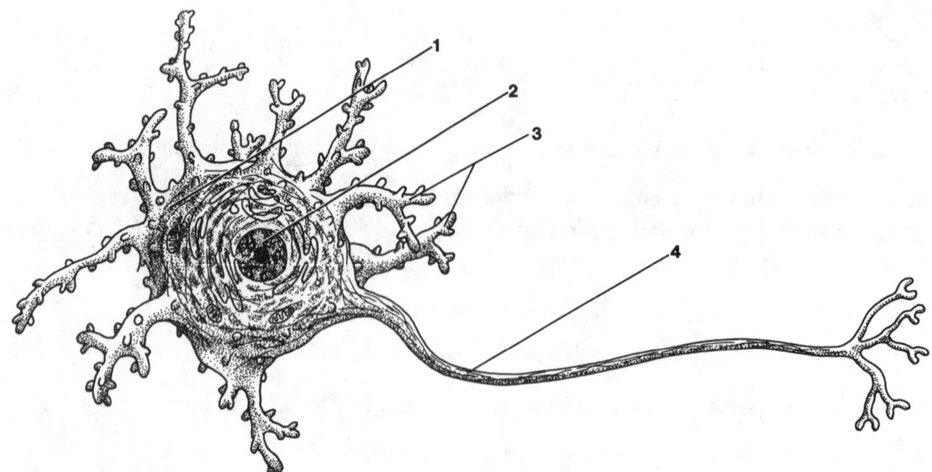

3. a) The axon of a neuron carries impulses ____________________ (toward or away from) the cell body.

   b) The dendrites of a neuron carry impulses ____________________ (toward or away from) the cell body.

4. a) In the peripheral nervous system, the specialized cells that form the myelin sheath are called ____________________.

   b) In the central nervous system, the specialized cells are called ____________________.

5. a) The space between the axon of one neuron and the dendrites or cell body of the next neuron is called the ____________________.

   b) Here, the transmission of nerve impulses depends upon chemicals called ____________________.

6. Name two organs made of nerve tissue. ____________________ and ____________________

7. State two general functions of nerve tissue in these organs or the nervous system as a whole. ____________________ and ____________________

MEMBRANES

1. Match each epithelial membrane with its proper locations and functions.

   Use each letter once. One answer line will have seven correct letters, and the other will have five correct letters.

   1) Serous membranes ____________________________

   2) Mucous membranes ____________________________

   A. Line the respiratory and digestive tracts
   B. Line closed body cavities
   C. Made of simple squamous epithelium
   D. Cover organs in closed body cavities
   E. Line the urinary and reproductive tracts
   F. Secrete serous fluid to prevent friction
   G. Secrete mucus to keep the living surface cells wet
   H. Include the pleural membranes
   I. Line body tracts that open to the environment
   J. Include the peritoneum and mesentery
   K. May contain goblet cells
   L. Include the pericardial membranes

2. a) The serous membranes that are found in closed body cavities are shown in the following diagrams.

Label both membranes in each pair, and then complete the statements using proper terminology.

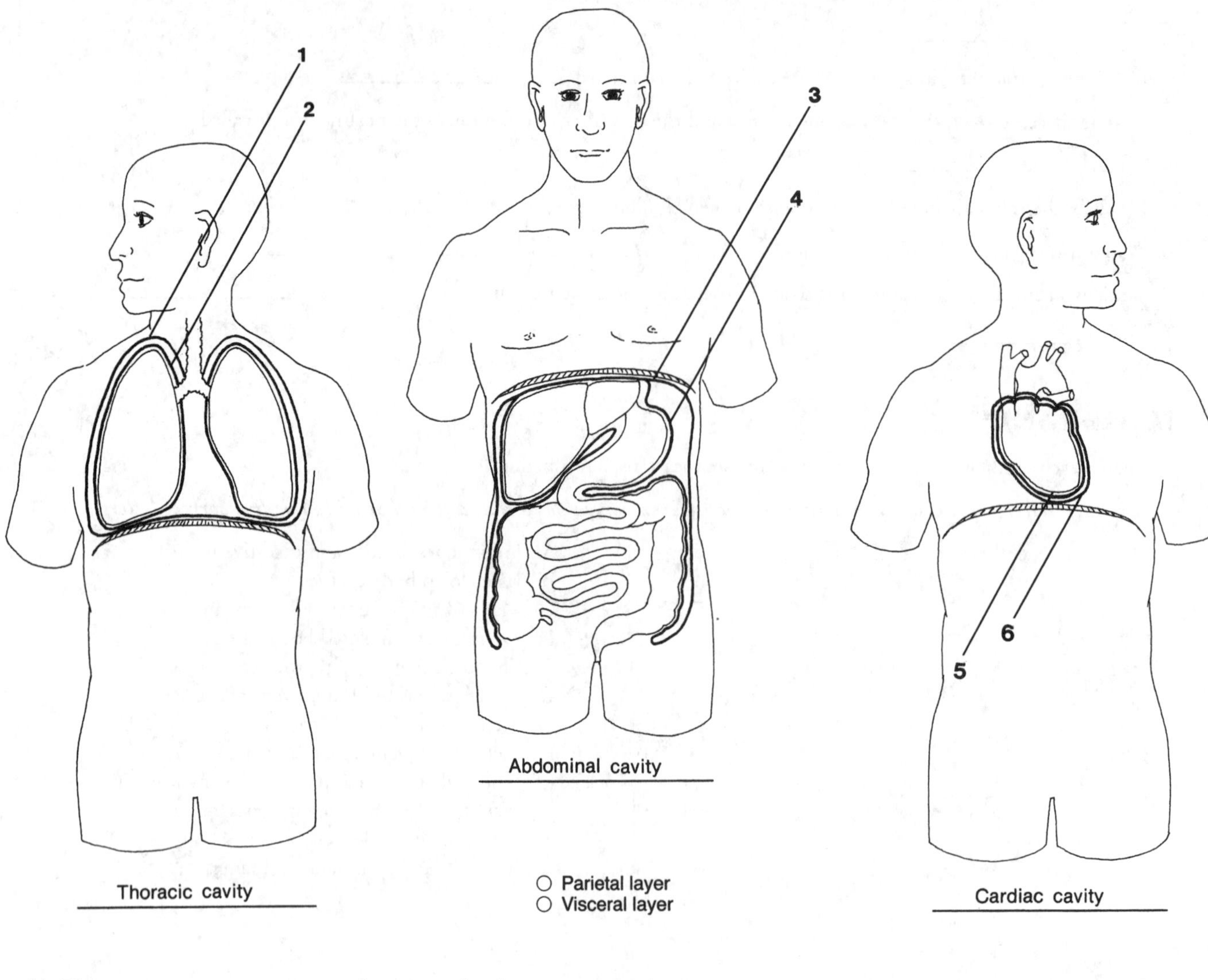

b) The ____________________ pleura lines the chest cavity, and the ____________________ pleura covers the lungs.

c) The ____________________ lines the abdominal cavity, and the ____________________ covers the abdominal organs.

d) The ____________________ pericardium lines the fibrous pericardium, and the ____________________ pericardium covers the heart muscle.

3. Match each connective tissue membrane with the statement that describes its location and function.

Use each letter once.

1) Superficial fascia ________________

2) Deep fascia ________________

3) Synovial membrane ________________

4) Fibrous pericardium ________________

5) Perichondrium ________________

6) Periosteum ________________

7) Meninges ________________

A. Lines joint cavities and secretes fluid to prevent friction when joints move
B. Forms a sac around the heart
C. Covers cartilage and contains capillaries
D. Covers the brain and spinal cord and contains cerebrospinal fluid
E. Covers bone and contains blood vessels that enter the bone
F. Between the skin and the muscles; contains adipose tissue
G. Covers each skeletal muscle and anchors tendons

CROSSWORD PUZZLE

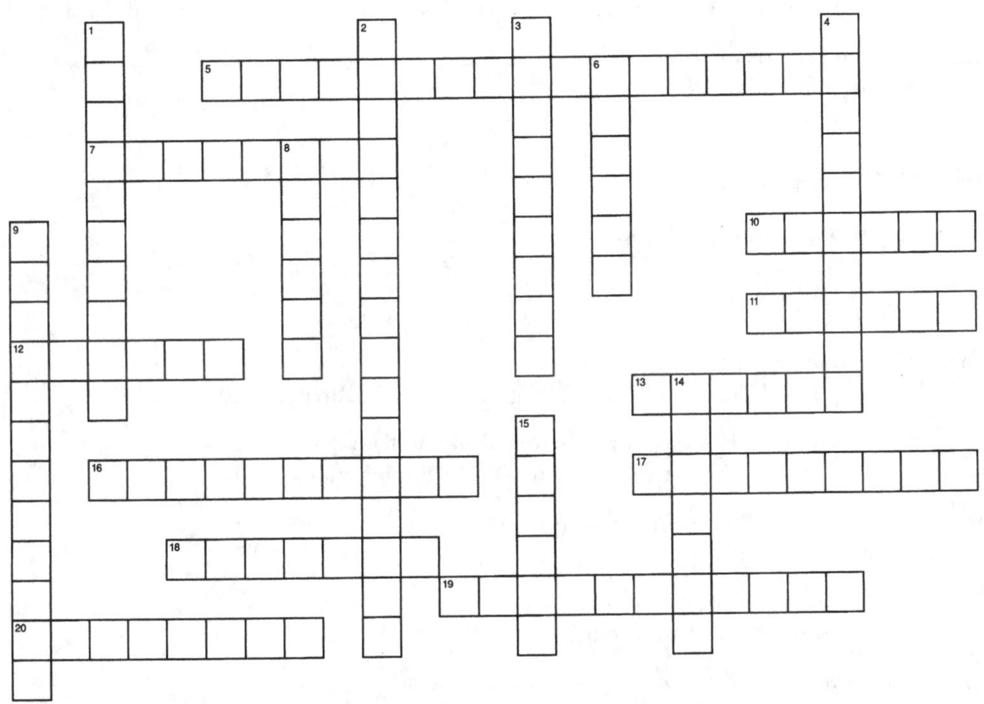

ACROSS

5. Chemicals that transmit impulses at synapses
7. Protein fibers that are very strong
10. Membranes that line body tracts open to the environment
11. Structural network of nonliving intercellular material
12. Nerve cell
13. Membranes that line closed body cavities
16. Tissue found on body surfaces
17. Tissue found on the joint surface of bones
18. Small space between two neurons
19. Blood-forming tissue
20. Glands that have ducts

DOWN

1. Cardiac muscle
2. A tissue that contains matrix and cells (two words)
3. Ductless glands
4. Bone cells
6. The tissue capable of contraction
8. Organs that produce secretions
9. Cartilage cells
14. Protein fibers that are elastic
15. Matrix of blood

CLINICAL APPLICATIONS

1. A 9-year-old boy has a simple fracture of the humerus, the bone of the upper arm. This fracture will heal relatively rapidly because bone has a good ________________ to transport nutrients and oxygen to the site of repair.
2. A 26-year-old football player has torn cartilage in his knee joint. Such damage will be repaired slowly or not at all because cartilage itself has no ________________.
3. a) A victim of a diving accident has had his spinal cord severed in the lower cervical region, and no nerve impulses pass below this level. As a result, the ________________ muscles below the neck are paralyzed because they no longer receive nerve impulses to initiate contraction.
   b) However, the ________________ continues to contract because cardiac muscle cells are able to contract without the stimulus of nerve impulses.
4. A child with a ruptured appendix is receiving antibiotics to treat peritonitis. This serious infection involves the ________________, the membrane that lines the ________________ cavity.
5. An elderly man has arthritis of the knees. The joint pain that accompanies arthritis may be due to inflammation of the ________________ membrane that lines the joint cavities or due to damage to the ________________ (type of tissue) that covers the joint surfaces of bones.

MULTIPLE CHOICE TEST #1

Choose the correct answer for each question.

1. An endocrine gland has:
   a) a duct b) no duct c) no secretion d) no blood supply
2. An example of an exocrine gland is the:
   a) salivary gland b) thyroid gland c) pituitary gland d) adrenal gland
3. The type of epithelium in which the surface cells alternate from round to flat is:
   a) cuboidal b) columnar c) stratified squamous d) transitional
4. The type of connective tissue with a liquid matrix called plasma is:
   a) cartilage b) bone c) adipose d) blood
5. Axon, dendrite, and cell body are the three parts of:
   a) the brain b) nerve tissue c) a neuron d) the central nervous system
6. The type of connective tissue with a solid matrix made of calcium salts is:
   a) areolar b) bone c) cartilage d) fibrous
7. The type of muscle tissue also known as voluntary muscle is:
   a) smooth b) cardiac c) visceral d) skeletal
8. The membrane that lines the digestive tract is a:
   a) serous membrane c) mucous membrane
   b) synovial membrane d) fascia
9. The serous membrane that lines the thoracic cavity is the:
   a) visceral pleura b) peritoneum c) parietal pleura d) mesentery
10. In the fallopian tube, an egg cell is moved toward the uterus by:
    a) ciliated epithelium b) striated muscle c) nerve tissue d) cuboidal epithelium
11. To increase their surface area for absorption, columnar cells in the small intestine have:
    a) microvilli b) cilia c) goblet cells d) ducts
12. The strong tissue that forms tendons and ligaments is:
    a) skeletal muscle b) fibrous connective tissue c) bone d) elastic connective tissue
13. The type of epithelium that makes up the outer layer of skin is:
    a) simple squamous b) stratified columnar c) stratified squamous d) simple columnar

14. The tissue that is thin enough to form capillaries and permit exchanges of materials is:
 a) smooth muscle c) areolar connective tissue
 b) elastic connective tissue d) simple squamous epithelium

15. The type of muscle tissue that produces a significant amount of body heat is:
 a) skeletal b) smooth c) cardiac d) visceral

16. Cardiac muscle is found in:
 a) the heart and arteries b) arteries only c) the heart only d) arteries, veins, and the heart

17. The membranes that cover the brain and spinal cord are the:
 a) visceral cranial membranes b) periosteum c) synovial membranes d) meninges

18. The space between two neurons where a neurotransmitter carries the impulse is called a:
 a) cell body b) matrix c) Schwann cell d) synapse

19. The unicellular glands that secrete mucus in the respiratory tract are:
 a) goblet cells b) endocrine glands c) microvilli d) serous glands

20. The tissue that transports nutrients and oxygen throughout the body is:
 a) nerve tissue b) blood c) areolar connective tissue d) serous tissue

21. The type of muscle tissue that provides peristalsis in the intestines is:
 a) skeletal b) voluntary c) striated d) smooth

22. The type of connective tissue that stores excess energy in the form of fat is:
 a) fibrous b) cartilage c) elastic d) adipose

23. The membrane that lines a joint cavity and produces fluid is the ___________ membrane.
 a) mucous b) synovial c) serous d) pleural

24. The tissue in the wall of the trachea that keeps it open is:
 a) bone b) fibrous tissue c) cartilage d) areolar tissue

25. The type of connective tissue beneath mucous membranes that contains many white blood cells is:
 a) areolar b) fibrous c) elastic d) cartilage

MULTIPLE CHOICE TEST #2

Read each question and the four answer choices carefully. When you have made a choice, follow the instructions to complete your answer.

1. Which tissue does NOT contribute to the functioning of the trachea?
 a) Ciliated epithelium sweeps mucus and pathogens to the pharynx.
 b) Cartilage rings keep the trachea open.
 c) Goblet cells produce mucus.
 d) Columnar epithelium absorbs nutrients.

 For your choice, state the correct location of the tissue with this function.

 ____

2. Which tissue does NOT contribute to the functioning of an artery?
 a) Simple squamous epithelium forms the lining and prevents abnormal clotting.
 b) Cardiac muscle pumps blood.
 c) Elastic connective tissue helps maintain normal blood pressure.
 d) Smooth muscle tissue helps maintain normal blood pressure.

 For your choice, state the correct location of the tissue with this function.

 ____

3. Which epithelial membrane is NOT paired with its proper location?
   a) peritoneum—lines the thoracic cavity
   b) mucous membrane—lines the urinary tract
   c) mesentery—covers the abdominal organs
   d) visceral pleura—covers the lungs

   For your choice, state its correct location.

   ___

4. Which of the following does NOT contribute to the structure and function of bones?
   a) The periosteum is a membrane that covers the bone.
   b) Calcium salts in the bone matrix provide strength.
   c) Cartilage on joint surfaces is smooth to prevent friction.
   d) Bones are moved by smooth muscle.

   Reword your choice to make it a correct statement.

   ___

5. Which statement is NOT true of glands?
   a) Exocrine glands have ducts to transport their secretions to other sites.
   b) The secretions of endocrine glands are called hormones.
   c) Endocrine glands have no ducts, and their secretions enter capillaries.
   d) An example of an exocrine gland is the thyroid gland.

   Reword your choice to make it a correct statement.

   ___

6. Which statement is NOT true of muscle tissue?
   a) Skeletal muscle in the iris of the eye changes the size of the pupil.
   b) Cardiac muscle forms the heart and pumps blood.
   c) Smooth muscle provides peristalsis in the intestines.
   d) Skeletal muscle moves the skeleton.

   For your choice, name the muscle tissue that does have this function.

   ___

7. Which statement is NOT true of nerve tissue?
   a) Transmission of impulses at synapses depends upon chemicals called neurotransmitters.
   b) Nerve tissue makes up the peripheral nerves, spinal cord, and brain.
   c) Schwann cells produce the myelin sheath for peripheral neurons.
   d) The axon of a neuron carries impulses toward the cell body.

   Reword your choice to make it a correct statement.

   ___

8. Which statement is NOT true of blood?
   a) White blood cells destroy pathogens and provide immunity.
   b) Nutrients and waste products are transported by red blood cells.
   c) Red blood cells contain hemoglobin to carry oxygen.
   d) Platelets are important for clotting to prevent blood loss.

   Reword your choice to make it a correct statement.

   ___

9. Which statement is NOT true of the connective tissues?
   a) Adipose tissue stores protein as a potential energy source.
   b) Fibrous connective tissue makes up tendons that connect muscle to bone.
   c) Areolar connective tissue is found between the skin and the muscles.
   d) Elastic connective tissue around the alveoli contributes to normal exhalation.

   Reword your choice to make it a correct statement.

   ___

10. Which statement is NOT true of the epithelial tissues?
    a) Transitional epithelium permits expansion of the urinary bladder.
    b) Simple cuboidal epithelium in the salivary glands secretes saliva.
    c) Stratified squamous epithelium of the outer layer of skin has living cells on the surface.
    d) Simple squamous epithelium in the alveoli permits exchange of gases.

    Reword your choice to make it a correct statement.

    ____

MULTIPLE CHOICE TEST #3

Each question is a series of statements concerning a topic in this chapter. Read each statement carefully and select all of the correct statements.

1. Which of the following statements are true of epithelial tissues?
    a) Cuboidal epithelium is found on the skin surface as dead cells.
    b) Simple squamous epithelium lines the heart; its smoothness prevents abnormal blood clotting.
    c) Transitional epithelium permits the lining of the urinary bladder to stretch.
    d) Stratified squamous epithelium forms the lining of the stomach.
    e) Columnar epithelium lines the small intestine and forms capillaries.
    f) Epithelial tissues have capillaries only if they are on an inner body surface.
    g) Both unicellular and multicellular glands are made of epithelial cells and tissues.
    h) The function of ciliated epithelium is to sweep materials across a surface.
2. Which of the following statements are true of connective tissues?
    a) Fibrous connective tissue forms ligaments that connect bone to bone.
    b) Adipose tissue is an important storage site for glycogen.
    c) Excess calcium is stored in bone tissue.
    d) The blood cells that contribute to clotting are the platelets.
    e) Areolar connective tissue is found subcutaneously and contains white blood cells.
    f) Elastic connective tissue enables the diaphragm to expand and contract.
    g) Blood plasma transports most nutrients and oxygen.
    h) Cartilage forms smooth surfaces on many joints.
    i) Brown fat is a heat-producing tissue.
    j) The major supporting tissue of the body is fibrous connective tissue.
3. Which of the following statements are true of muscle tissues?
    a) Arteries contain smooth muscle that contributes to maintaining blood pressure.
    b) Only skeletal muscle can be called voluntary muscle.
    c) Cardiac muscle must receive nerve impulses in order to contract.
    d) The iris of the eye has smooth muscle fibers that focus light rays on the retina.
    e) Cardiac muscle forms the walls of the chambers of the heart.
    f) A significant amount of body heat is produced by cardiac muscle.
4. Which of the following statements are true of nerve tissue?
    a) The electrical nerve impulse is carried by the neuron's cell membrane.
    b) The myelin sheath assists impulse transmission across synapses.
    c) Neurotransmitters are produced by dendrites.
    d) The cell body of a neuron contains the nucleus.
    e) A synapse is the space between two axons.
    f) Schwann cells are found only in the peripheral nervous system.
5. Which of the following statements are true of membranes?
    a) The meninges cover the brain and spinal cord.
    b) The visceral pleura covers the lungs.
    c) The periosteum is fibrous connective tissue that covers a bone.
    d) The heart has both serous and fibrous pericardial layers.
    e) The membrane that lines joint cavities is the synovial membrane.
    f) The peritoneum lines the abdominal cavity.

Chapter 5

The Integumentary System

This chapter describes the integumentary system, which consists of the skin and the subcutaneous tissue. Each part is made of specific tissues that have very specific functions.

GENERAL STRUCTURE AND FUNCTIONS

1. Match each major part of the integumentary system with its proper structure (a letter statement) and function (a number statement).

   Use each letter and number once.

   1) Epidermis ____________________

   2) Dermis ____________________

   3) Subcutaneous tissue ____________________

   Structure

   A. Made of areolar connective and adipose tissue
   B. The outer layer of the skin, made of stratified squamous keratinizing epithelium
   C. The inner layer of the skin, made of modified fibrous connective tissue

   Function

   1. Contains the accessory structures of the skin, such as receptors and sweat glands
   2. Connects the dermis to the muscles and stores fat
   3. Mitosis constantly renews this layer, which forms a barrier between the body and the external environment

EPIDERMIS

1. The epidermis is the outer layer of the skin and may be further subdivided into layers of its own.
   a) Name the epidermal layer in which mitosis takes place. ____________________
   b) Name the epidermal layer that is composed of dead cells. ____________________
2. a) In the stratum corneum, all that is left of the cells is the protein ____________________.
   b) The stratum corneum is an effective barrier that prevents the loss of ____________________ from the body and also prevents the entry of ____________________ and ____________________.
3. If the skin is subjected to constant pressure, a thicker area of epidermis will be formed. More cells are produced by the process of ____________________ in the ____________________ (layer).
4. If the skin is subjected to friction, layers of the epidermis may be separated, and ____________________ will collect in the area and form a ____________________.

5. a) Langerhans cells are found in the epidermis but have come from the ____________________.

   b) When Langerhans cells phagocytize pathogens, they carry them to ____________________ (type of WBC) found in ____________________.

   c) In response to the pathogen, the lymphocytes initiate an immune reaction such as the production of ____________________.

   d) Antibacterial chemicals produced in the epidermis are called ____________________.

6. a) Melanocytes produce the protein ____________________, which is a pigment.

   b) What is the stimulus for increased melanin production? ____________________

   c) Explain the function of melanin (NOT to give the skin color).

   ____

7. Merkel cells are found in the ____________________ and are receptors for the sense of ____________________.

8. The following diagram is a section through the epidermis.

   Label the parts indicated.

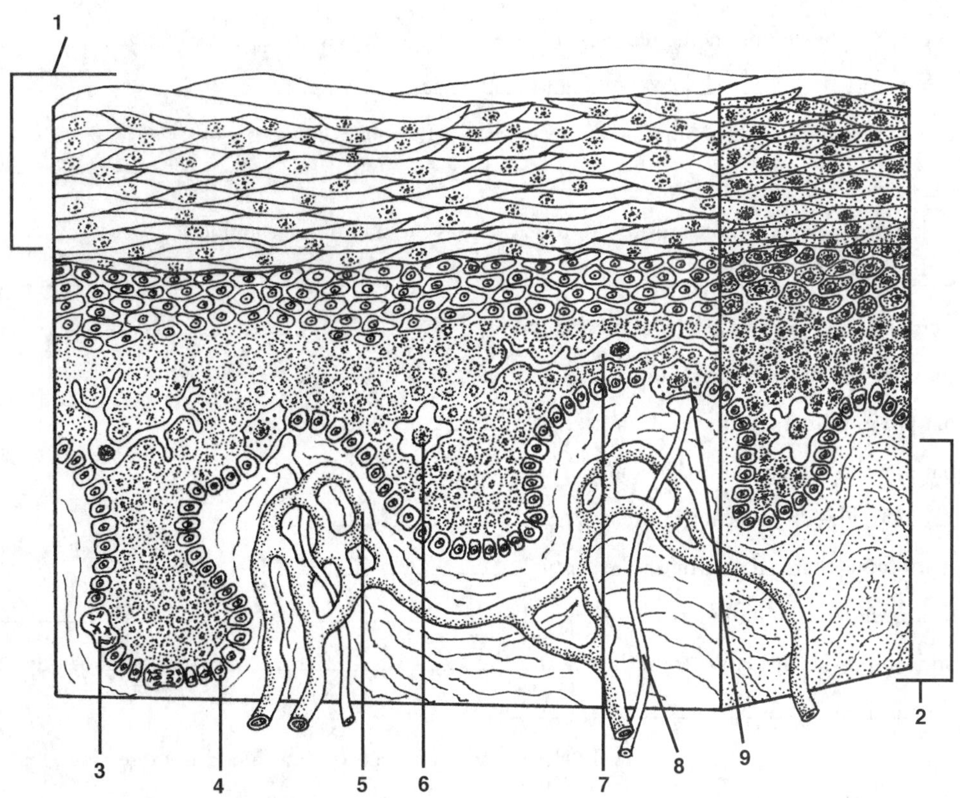

DERMIS

1. Match the following parts of the dermis with their proper descriptions.

Use each letter once.

1) Fibroblasts ____________

2) Collagen fibers ____________

3) Elastin fibers ____________

4) Papillary layer ____________

5) Hair follicles ____________

6) Nail follicles ____________

7) Sebaceous glands ____________

8) Eccrine sweat glands ____________

9) Apocrine sweat glands ____________

10) Ceruminous glands ____________

11) Receptors ____________

A. Contains capillaries to nourish the stratum germinativum of the epidermis
B. Provide information about changes in the external environment
C. The protein that gives the dermis its strength
D. Produce a lipid substance called sebum
E. The cells that produce collagen and elastin
F. Mitosis at the root produces the hair shaft
G. Produce their secretion in times of stress or strong emotions
H. Produce their secretion during exercise or in a warm environment
I. Mitosis at the root produces the nail
J. The protein that gives the dermis elasticity
K. Produce cerumen, or earwax

2. a) The free nerve endings in the dermis are the receptors for the cutaneous senses of ____________, ____________, ____________, and ____________.

b) The encapsulated nerve endings in the dermis are the receptors for the cutaneous senses of ____________ and ____________.

c) Explain why the skin of the palm is more sensitive to touch than is the skin of the shoulder.

____

3. a) The ends of fingers and toes are protected from mechanical injury by ____________.

b) Name the protein these structures are made of. ____________

4. a) The secretion that prevents drying of the eardrum is ____________.

b) The secretion that prevents drying of the skin and hair is ____________.

c) The secretion that helps lower body temperature is ____________.

d) The secretion that inhibits bacterial growth on the skin is ____________.

5. a) One function of human hair is to keep dust out of the ____________ or ____________.

b) Another function is to provide insulation from the cold for the ____________ (part of the body).

c) Name the protein hair is made of. ____________

6. In stress situations, the ____________ in the dermis will constrict to shunt blood to more vital organs.

7. a) The vitamin formed in the skin is vitamin ____________, which is made from ____________ when the skin is exposed to ____________.

b) The function of this vitamin is to promote the absorption of ____________ and ____________ in the small intestine.

8. The following diagram is a section through the skin and subcutaneous tissue.

   Label the parts indicated.

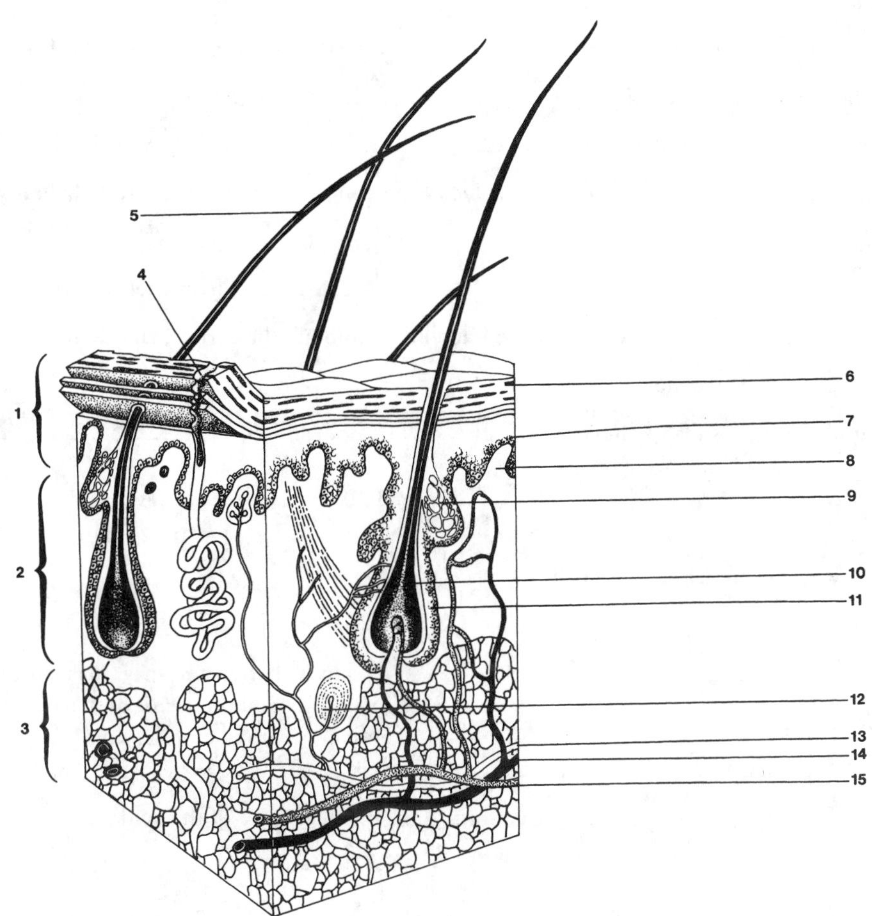

SUBCUTANEOUS TISSUE

1. The other name for subcutaneous tissue is the ____________________.

2. a) The subcutaneous tissue is located between the __________ and the __________.

   b) Name the two types of connective tissue in this layer. __________ and __________

3. a) The areolar connective tissue contains many white blood cells that destroy __________ that have entered __________.

   b) Areolar connective tissue also contains mast cells that produce __________ when tissue damage occurs; this substance contributes to the process of __________.

4. a) The adipose tissue contains cells that are specialized to store __________ as a source of potential __________.

   b) State two other functions of subcutaneous fat. __________ and __________

MAINTENANCE OF BODY TEMPERATURE

1. The role of the skin in the maintenance of body temperature depends upon the ________________ glands and the small arteries called ________________.
2. a) In a ________________ environment, the eccrine glands secrete more sweat onto the skin surface.
   b) Excess body heat is then lost in the process of ________________ of the sweat.
3. a) In a cold environment, the arterioles in the dermis will ________________ (constrict or dilate).
   b) This will ________________ (increase or decrease) blood flow through the dermis, and body heat will be ________________ (lost or retained).
4. a) In a warm environment, the arterioles in the dermis will ________________ (constrict or dilate).
   b) This will ________________ (increase or decrease) blood flow through the dermis, and body heat will be ________________ (lost or retained).
5. The tissue in the walls of the arterioles that permits vasoconstriction or vasodilation is ________________.

BURNS

1. Match each type of burn with the proper description.

   Use each letter once.

   1) First-degree burn ________________________
   2) Second-degree burn ________________________
   3) Third-degree burn ________________________

   A. The skin is charred and may not be painful at first.
   B. The skin is painful but not blistered.
   C. The skin is painful and blistered.

2. a) Extensive third-degree burns may be very serious because of the loss of which layer of the epidermis?

   ____

   b) State the two potentially serious problems for patients with extensive third-degree burns. ________________ and ________________

CROSSWORD PUZZLE

ACROSS

2. ____________ sweat gland, maintains normal body temperature
5. Produced in follicles in the scalp
7. Covers the surface of the body
8. Tissue that connects the skin to the muscles
10. Protects living skin from exposure to UV rays
12. Protects the end of a finger
13. Small arteries
16. Detect changes in the environment
17. Waterproof protein of the epidermis
18. Sunshine vitamin
19. Produces melanin

DOWN

1. Stratum ____________, produces new epidermal cells
3. Gland that secretes cerumen
4. Decreases blood flow through arterioles
6. Increases blood flow through arterioles
7. Gland that secretes sebum
9. Uneven junction of the dermis with the epidermis (two words)
11. Outer layer of the skin
14. Inner layer of the skin
15. Stratum ____________, prevents entry of pathogens to the body

CLINICAL APPLICATIONS

1. a) Mr. S is 55 years old and has just had a small squamous cell carcinoma removed from his forehead. A carcinoma is a form of ____________________, and this type is most often caused by overexposure to ____________________.

   b) Mr. S's hobby is gardening, and he can continue to enjoy working outdoors by taking one very sensible precaution (besides wearing a hat). What is this precaution? ____________________

2. a) Mrs. B burned her hand in a cooking accident. She will require a skin graft for the damaged area, which indicates that this is a ____________________ (first-, second-, or third-degree) burn.

   b) The burned area is a possible site for ____________________ because of the loss of the stratum corneum.

   c) Because the burn covers only a small area of the body surface, it will probably not lead to the other serious complication of extensive burns, which is ____________________.

3. Mrs. M is 72 years old and does not have an air conditioner in her apartment. Her doctor has advised her to be sure to use a fan in hot weather. This is important because the ____________________ glands of elderly people are not as active, and the body temperature might ____________________ sharply during hot weather.

4. Three-year-old Donald occasionally develops a rash that is very itchy. His mother has noticed that the rash often follows times when Donald has eaten eggs. This rash is probably ____________________, which is a form of ____________________ reaction.

MULTIPLE CHOICE TEST #1

Choose the correct answer for each question.

1. The outer layer of the skin is the:
   a) papillary layer b) dermis c) subcutaneous tissue d) epidermis

2. The mechanism of heat loss that depends upon evaporation is:
   a) fat storage b) sweating c) vasodilation in the dermis d) vasoconstriction in the dermis

3. The protein in epidermal cells that makes the skin relatively waterproof is:
   a) keratin b) collagen c) melanin d) elastin

4. The tissue that stores fat in subcutaneous tissue is:
   a) adipose tissue c) areolar connective tissue
   b) fibrous connective tissue d) stratified squamous epithelium

5. The glands of the skin that are most concerned with the maintenance of body temperature are:
   a) apocrine b) eccrine c) sebaceous d) ceruminous

6. In the dermis, the receptors for pain are:
   a) encapsulated nerve endings b) axons c) free nerve endings d) Langerhans cells

7. Vitamin D is formed in the skin when the skin is exposed to:
   a) pressure b) friction c) stress d) ultraviolet rays

8. The layer of the dermis that contains capillaries to nourish the stratum germinativum of the epidermis is the:
   a) follicle layer b) papillary layer c) collagen layer d) subcutaneous layer

9. The part of a hair follicle that undergoes mitosis to form the hair is the:
   a) hair root b) hair shaft c) generative layer d) keratin portion

10. The part of the epidermis that undergoes mitosis is the:
    a) stratum corneum b) papillary layer c) stratum germinativum d) stratum melanin

11. The protein that protects inner living skin from the damaging effects of ultraviolet rays is:
    a) keratin b) collagen c) elastin d) melanin

12. The layer of skin that, if unbroken, prevents the entry of most pathogens is the:
a) stratum germinativum b) papillary layer c) stratum corneum d) collagen layer

13. The type of burn that is characterized by painful blisters is the:
a) first-degree b) second-degree c) third-degree d) fourth-degree

14. Cells that increase their secretion when stimulated by ultraviolet rays are:
a) fibroblasts b) Langerhans cells c) keratinocytes d) melanocytes

15. The secretion that prevents drying of the eardrum is:
a) cerumen b) sebum c) sweat d) tissue fluid

16. At the ends of fingers and toes, nails are produced in structures called:
a) fibroblasts b) nail glands c) follicles d) mitosis

17. Many white blood cells, which destroy pathogens that enter breaks in the skin, are found in the:
a) adipose cells b) areolar connective tissue c) stratum corneum d) keratinized layer

18. The dermis is strong because of the presence of:
a) keratin b) elastin fibers c) collagen fibers d) areolar connective tissue

19. For a person with extensive third-degree burns, serious potential problems are infection and:
a) dehydration b) loss of tissue fluid c) both of these d) neither of these

20. Some human hair functions to keep dust out of the:
a) nose and mouth b) eyes and mouth c) ears and mouth d) eyes and nose

21. All of the following are part of subcutaneous tissue except:
a) areolar connective tissue b) keratin c) adipose tissue d) tissue fluid

22. The defensins produced in the epidermis provide protection against:
a) pressure b) ultraviolet rays c) bacteria d) changes in temperature

23. In cold weather, the arterioles in the dermis will:
a) constrict to conserve heat c) constrict to release heat
b) dilate to conserve heat d) dilate to release heat

24. The cells that are able to pick up pathogens and transport them to lymph nodes are:
a) keratinocytes b) fibroblasts c) melanocytes d) Langerhans cells

25. In the subcutaneous tissue, histamine is produced by ________ and contributes to the process of ________.
a) melanocytes/tanning c) fat cells/energy storage
b) fibroblasts/collagen synthesis d) mast cells/inflammation

MULTIPLE CHOICE TEST #2

Read each question and the four answer choices carefully. When you have made a choice, follow the instructions to complete your answer.

1. Which statement is NOT true of the subcutaneous tissue?
   a) White blood cells in the areolar connective tissue destroy pathogens that enter through breaks in the skin.
   b) Adipose tissue stores carbohydrates as a form of potential energy.
   c) It is found between the dermis and the muscles.
   d) Adipose tissue cushions some bony prominences.

   Reword your choice to make it a correct statement.

   ____

2. Which statement is NOT true of the tissues of the skin?
   a) The epidermis is made of simple squamous epithelium.
   b) The dermis is made of a modified fibrous connective tissue.
   c) Fibroblasts in the dermis produce collagen fibers and elastin fibers.
   d) In the epidermis, the surface layers of cells are dead.

   Reword your choice to make it a correct statement.

   ____

3. Which statement is NOT true of the glands in the dermis?
   a) Cerumen is produced by ceruminous glands in the ear canals.
   b) Apocrine sweat glands increase their secretions in response to emotions.
   c) Sebum from sebaceous glands prevents drying and cracking of the skin.
   d) Secretion of sweat by eccrine sweat glands is important to excrete excess body water.

   Reword your choice to make it a correct statement.

   ___

4. Which statement is NOT true of the epidermis?
   a) The stratum corneum consists of keratinized dead cells that prevent the loss or entry of water.
   b) Mitosis takes place in the stratum germinativum, and new cells are pushed toward the skin surface.
   c) Melanocytes produce melanin, which protects living skin layers from bacteria.
   d) An unbroken stratum corneum prevents the entry of pathogens and many chemicals.

   Reword your choice to make it a correct statement.

   ___

5. Which statement is NOT true of the role of the skin in the maintenance of body temperature?
   a) In a cold environment, vasoconstriction in the dermis conserves body heat.
   b) In a warm environment, eccrine glands secrete more sweat onto the skin surface.
   c) In a warm environment, vasodilation in the dermis promotes heat loss.
   d) In sweating, excess body heat is lost in the process of osmosis of sweat.

   Reword your choice to make it a correct statement.

   ___

6. Which statement is NOT true of the accessory skin structures?
   a) Hair helps insulate the head from cold.
   b) Nails protect the ends of the fingers and toes from mechanical injury.
   c) The receptors in the epidermis are for the cutaneous senses of pain, heat, cold, pressure, touch, and itch.
   d) Eyelashes and eyebrows keep dust and perspiration out of the eyes.

   Reword your choice to make it a correct statement.

   ___

7. Which statement is NOT true of the dermis?
   a) Collagen fibers provide strength.
   b) The papillary layer has capillaries to nourish the stratum germinativum.
   c) Elastin fibers permit the skin to be stretched and then returned to its original shape.
   d) It is located between the stratum corneum and the subcutaneous tissue.

   Reword your choice to make it a correct statement.

   ___

8. Which of the following does NOT happen when the sun's ultraviolet rays strike the skin?
   a) Melanocytes increase their production of melanin.
   b) Cholesterol in the skin is converted to vitamin C.
   c) The epidermal cells take in the melanin produced and become darker in color.
   d) Living cells in the stratum germinativum may be damaged.

   Reword your choice to make it a correct statement.

   ___

9. Which statement is NOT true of the skin?
   a) The skin covers the body and is considered an organ.
   b) The epidermis is the inner layer, and the dermis is the outer layer.
   c) Functions include maintenance of body temperature and prevention of the entry of pathogens.
   d) Functions include detection of changes in the external environment and production of vitamin D.

   Reword your choice to make it a correct statement.

   ___

10. Which of the following is NOT made of the protein keratin?
    a) The fingernails and toenails
    b) The hair shafts
    c) The pigment of hair and skin
    d) The stratum corneum

    For your choice, state what it is made of.

    ____

MULTIPLE CHOICE TEST #3

Each question is a series of statements concerning a topic in this chapter. Read each statement carefully and select all of the correct statements.

1. Which of the following statements are true of the epidermis?
    a) It is made of stratified squamous epithelium.
    b) In the stratum germinativum, new cells are produced by the process of meiosis.
    c) The stratum corneum consists of many layers of dead cells.
    d) The protein of the stratum corneum is keratin.
    e) The stratum germinativum is an excellent barrier to pathogens.
    f) Langerhans cells phagocytize pathogens and produce antibodies to them.
    g) Melanocytes produce melanin when stimulated by fluorescent light.
    h) Vitamin D is produced here and then is part of the process of blood clotting.

2. Which of the following statements are true of the dermis?
    a) The arterioles constrict in cold weather in order to keep heat close to the body surface.
    b) The protein collagen provides strength.
    c) The papillary layer has capillaries to nourish the stratum corneum.
    d) Within a hair follicle, mitosis at the hair root produces the hair shaft.
    e) With aging, both collagen and elastin fibers break down and are not replaced.
    f) Many cutaneous receptors are located in the dermis.

3. Which of the following statements are true of the subcutaneous tissue?
    a) This layer is made of adipose tissue and fibrous connective tissue.
    b) This layer is between the dermis and the visceral muscles.
    c) Triglycerides are the energy storage form in adipocytes.
    d) White blood cells are present to intercept pathogens that enter breaks in the skin.
    e) Inflammation in this layer is a response to damage.
    f) This layer can vary considerably in thickness.

4. Which of the following statements are true of the accessory structures of the skin?
    a) Eccrine glands contribute to heat gain and heat loss.
    b) Drying and cracking of the skin are prevented by sebum.
    c) The eardrum is kept pliable by cerumen.
    d) The sensory receptors for pain and pressure are free nerve endings.
    e) Fingernails are useful for scratching and for picking up small objects.
    f) Both scalp and body hair provide thermal insulation.

Chapter 6

The Skeletal System

The skeletal system consists of the bones that make up the skeleton, the ligaments that connect bone to bone, and the cartilage that is a structural part of most joints. This chapter describes the anatomy of the skeleton and relates anatomy to the functioning of the skeleton within the body as a whole.

FUNCTIONS OF THE SKELETON

1. Some bones contain and protect the __________________, the principal hemopoietic tissue that produces the blood cells.

2. Bones are a storage site for excess __________________ (mineral), which is essential for blood clotting and bone structure.

3. a) The skeleton is a framework that __________________ the body.

   b) Attached to the skeleton are the __________________ that move the bones.

4. a) Some bones protect internal organs from ____________________________________.

   b) State a specific example of this function. ____________________________________

BONE TISSUE

1. Match the following types and parts of bone tissue with the proper descriptive statements.

   Use each letter once. Two answer lines will have two correct letters.

   1) Compact bone ________________________

   2) Spongy bone ________________________

   3) Osteocytes ________________________

   4) Bone matrix ________________________

   A. Regulate the amount of calcium in the bone matrix
   B. Made of calcium carbonate and calcium phosphate
   C. Made of haversian systems, which are cylindrical arrangements of osteocytes within matrix
   D. Contains osteocytes and bone matrix, but these are not arranged in haversian systems
   E. Bone cells
   F. Often contains red bone marrow

CLASSIFICATION OF BONES

1. Match each type of bone with the proper example (a letter statement) and the proper aspects of its structure (a number statement).

Use each letter once and use each number as many times as it is correct. Each answer line will have one correct letter and two correct numbers.

1) Long bones ______________________

2) Short bones ______________________

3) Flat bones ______________________

4) Irregular bones ______________________

Example

A. Pelvic bone, cranial bones, ribs
B. Vertebrae, facial bones
C. Wrist and ankle bones
D. Bones of the arms, legs, hands, and feet

Structure

1. Made of spongy bone covered with a thin layer of compact bone.
2. Each consists of a diaphysis made of compact bone and epiphyses made of spongy bone.
3. The spongy bone contains red bone marrow.
4. The marrow canal contains yellow bone marrow.

2. The tissue that covers the joint surfaces of bones is ______________, which provides a smooth surface when joints are moved.

3. The membrane that covers the rest of a bone is called the ______________ and is made of ______________ connective tissue.

4. The periosteum contains ______________ that enter the bone itself.

5. a) The periosteum anchors the ______________ that connect muscle to bone, and the ______________ that connect bone to bone.

   b) Both of these connecting structures are made of what type of tissue? ______________

6. The following diagram depicts a long bone.

Label the parts indicated.

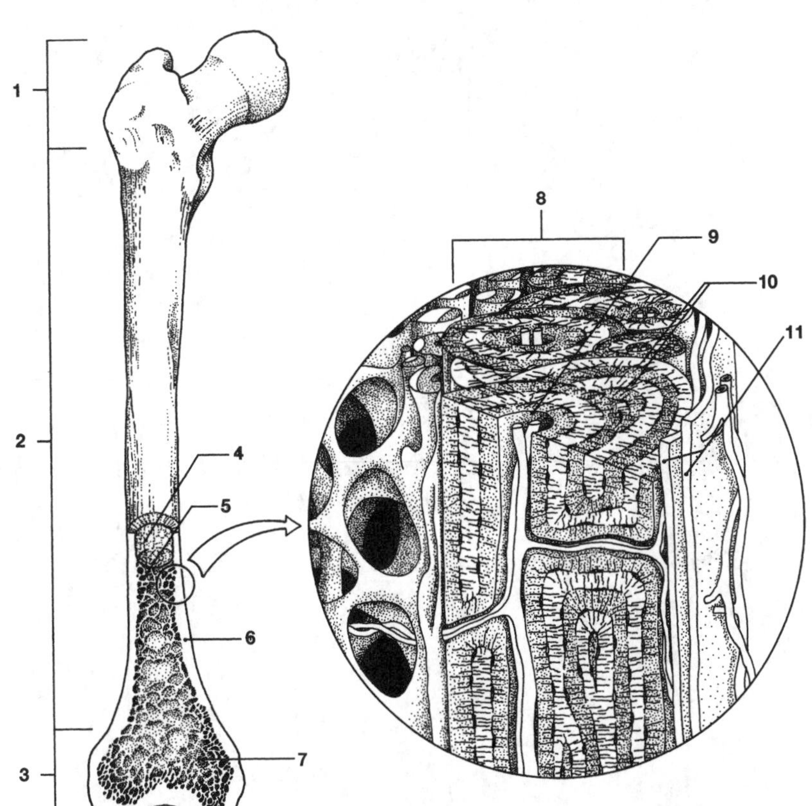

EMBRYONIC GROWTH OF BONE

1. a) The skeleton of the embryo is first formed of other tissues that are gradually replaced by bone. The process of bone replacement of another tissue is called ____________________.

   b) The production of bone matrix is accomplished by cells called ____________________.

2. a) In the embryo, the cranial and facial bones are first made of which tissue? ____________________

   b) The process of ossification begins in the ____________________ month of gestation, when osteoblasts differentiate from ____________________ in the centers of ossification in these bones.

3. a) At birth, ossification of the bones of the skull is not complete, and areas of fibrous connective tissue called ____________________ remain between the bones.

   b) The following diagram is a lateral view of an infant skull.

   Label the bones and fontanels indicated.

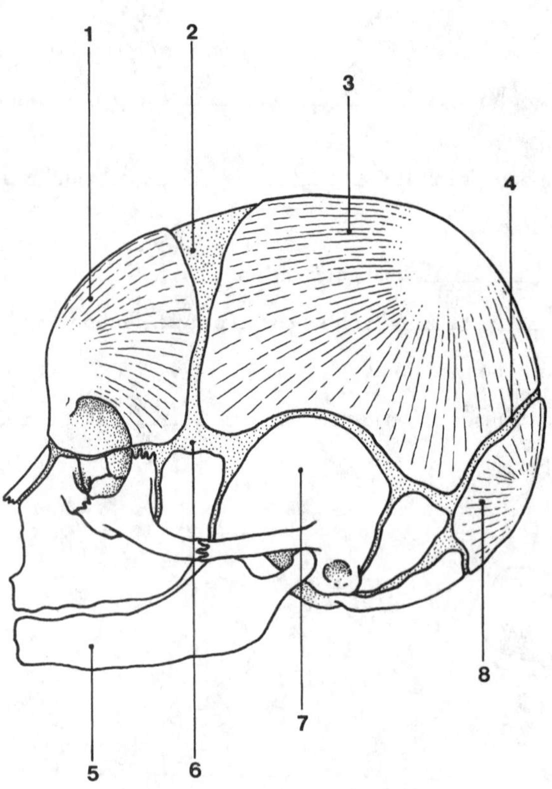

4. Explain the purpose of fontanels.

   ____

5. In the embryo, bones of the trunk and extremities are first made of which tissue? ____________________

6. In a long bone, several centers of ossification develop: one in the ____________________ and one or more in each ____________________ of the bone.

7. At birth, ossification of these bones is not complete. In long bones, growth occurs at the sites of the ____________________, which are made of cartilage.

8. The following diagram shows a growing long bone. At each site indicated, state whether cartilage is being produced or whether bone is replacing cartilage.

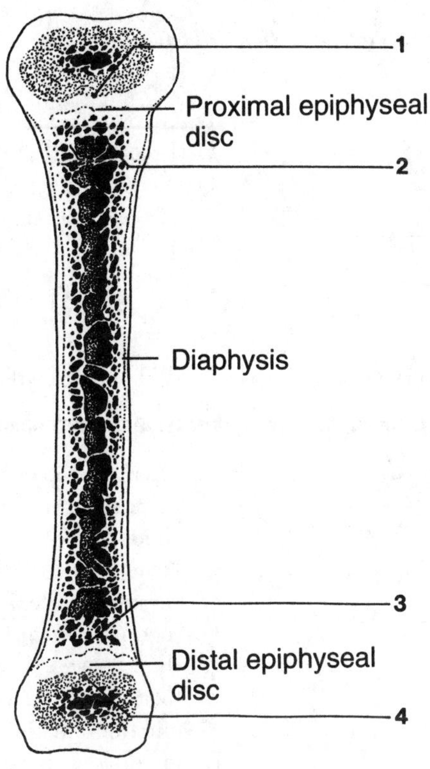

9. Closure of the epiphyseal discs means that all of the ____________________ of the discs has been replaced by ____________________, and growth in length stops.

10. a) In long bones, the marrow canal is formed by cells called ____________________ that reabsorb bone matrix.

    b) After birth, the marrow canal contains ____________________ bone marrow, which is mostly ____________________ tissue.

FACTORS THAT AFFECT BONE GROWTH AND MAINTENANCE

1. Nutrients—Match each nutrient with its function related to bone growth.

   Two letters are used once only, and two letters are used twice. Each answer line will have only one correct letter.

   1) Vitamin D ____________
   2) Vitamin C ____________
   3) Vitamin A ____________
   4) Calcium ____________
   5) Phosphorus ____________
   6) Protein ____________

   A. Necessary for the *process* of bone formation
   B. Becomes part of the collagen in bone matrix
   C. Becomes part of the salts of bone matrix
   D. Necessary for the absorption of calcium and phosphorus in the small intestine

2. Hormones—Match each hormone with its specific function as related to bone growth.

   Use each letter once. In the space indicated, name the gland that secretes each hormone.

   1) Growth hormone ____________
      Gland ____________
   2) Insulin ____________
      Gland ____________
   3) Calcitonin ____________
      Gland ____________
   4) Thyroxine ____________
      Gland ____________
   5) Parathyroid hormone ____________
      Gland ____________
   6) Estrogen or testosterone ____________
      Gland ____________
      or ____________

   A. Decreases the reabsorption of calcium from bones
   B. Increases protein synthesis and energy production from all food types
   C. Helps maintain a stable bone matrix and promotes closure of the epiphyses of long bones
   D. Increases protein synthesis and the rate of mitosis of chondrocytes in cartilage
   E. Increases the reabsorption of calcium by the small intestine and kidneys
   F. Increases energy production from glucose

   Name the hormone (from the preceding list) with each of these functions:

   7) Lowers the blood calcium level ____________
   8) Raises the blood calcium level ____________
   9) Causes long bones to stop growing ____________

3. a) Heredity—A person's height is a genetic characteristic that is regulated by genes inherited from ____________ and ____________.

   b) You already know that genes are the genetic codes for proteins. The genes for height are probably for the ____________ that are needed for the production of ____________ and ____________ (tissues).

4. a) Exercise or stress—For bones, exercise or stress means ____________.

   b) Without this normal stress, bones will lose ____________ faster than it is replaced.

   c) Describe what may happen to affected bones.

   ____

THE SKELETON

1. The human skeleton has two divisions: a) the ____________ skeleton, which consists of the ____________, vertebrae, and ____________; and b) the ____________ skeleton, which consists of the bones of the ____________ and ____________ and the shoulder and pelvic girdles.

2. a) Name the part of the skeleton that protects the heart, lungs, spleen, and liver from mechanical injury. ____________

   b) Name the part of the skeleton that protects the brain from mechanical injury. ____________

   c) Name the bones that attach the legs to the axial skeleton. ____________

   d) Name the bones that attach the arms to the axial skeleton. ____________ and ____________

3. The following diagram is of the full skeleton.

   Label the bones indicated.

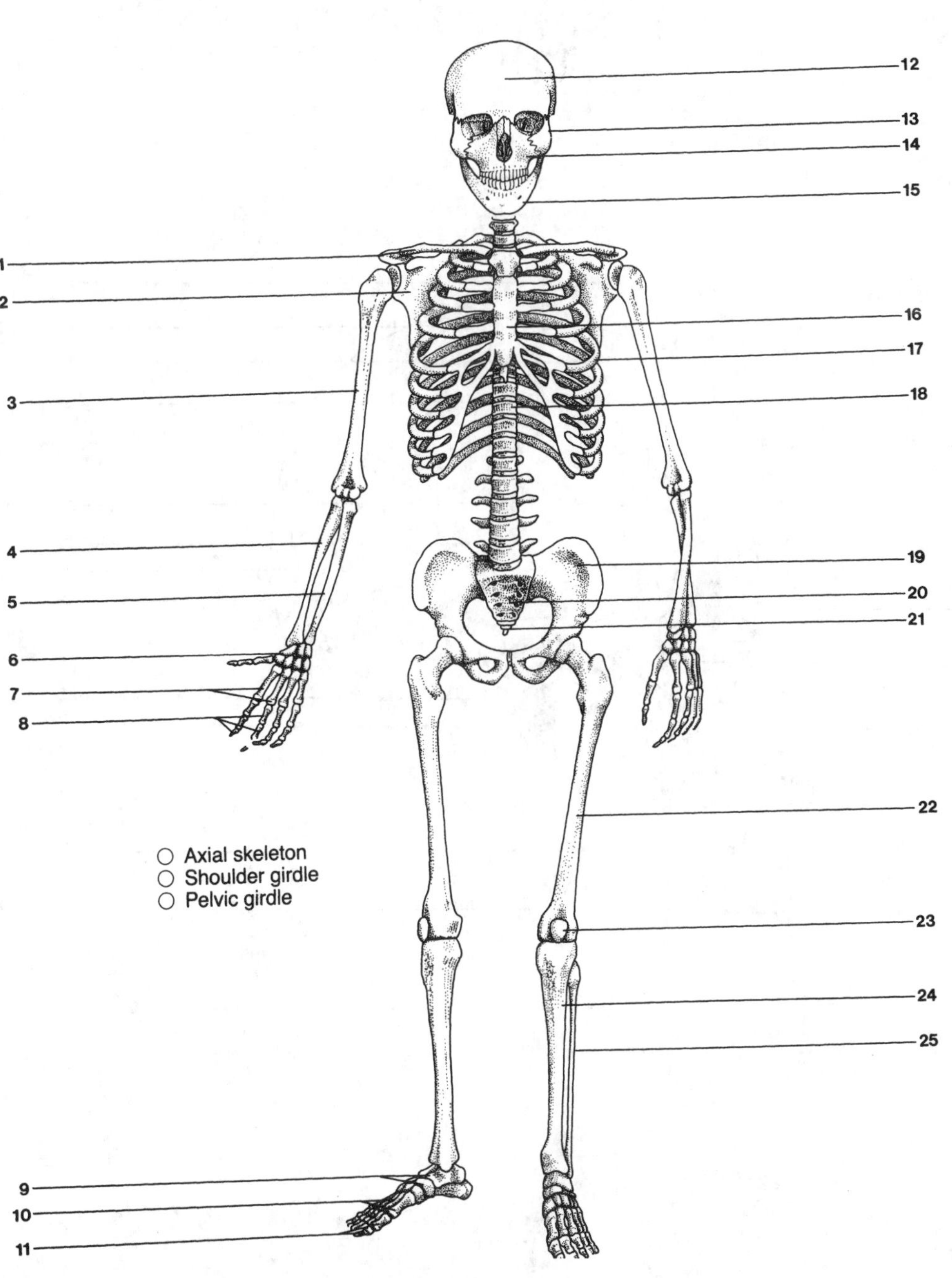

THE SKULL

1. The following diagrams depict an anterior and a lateral view of the bones of the skull.

   Label the bones indicated.

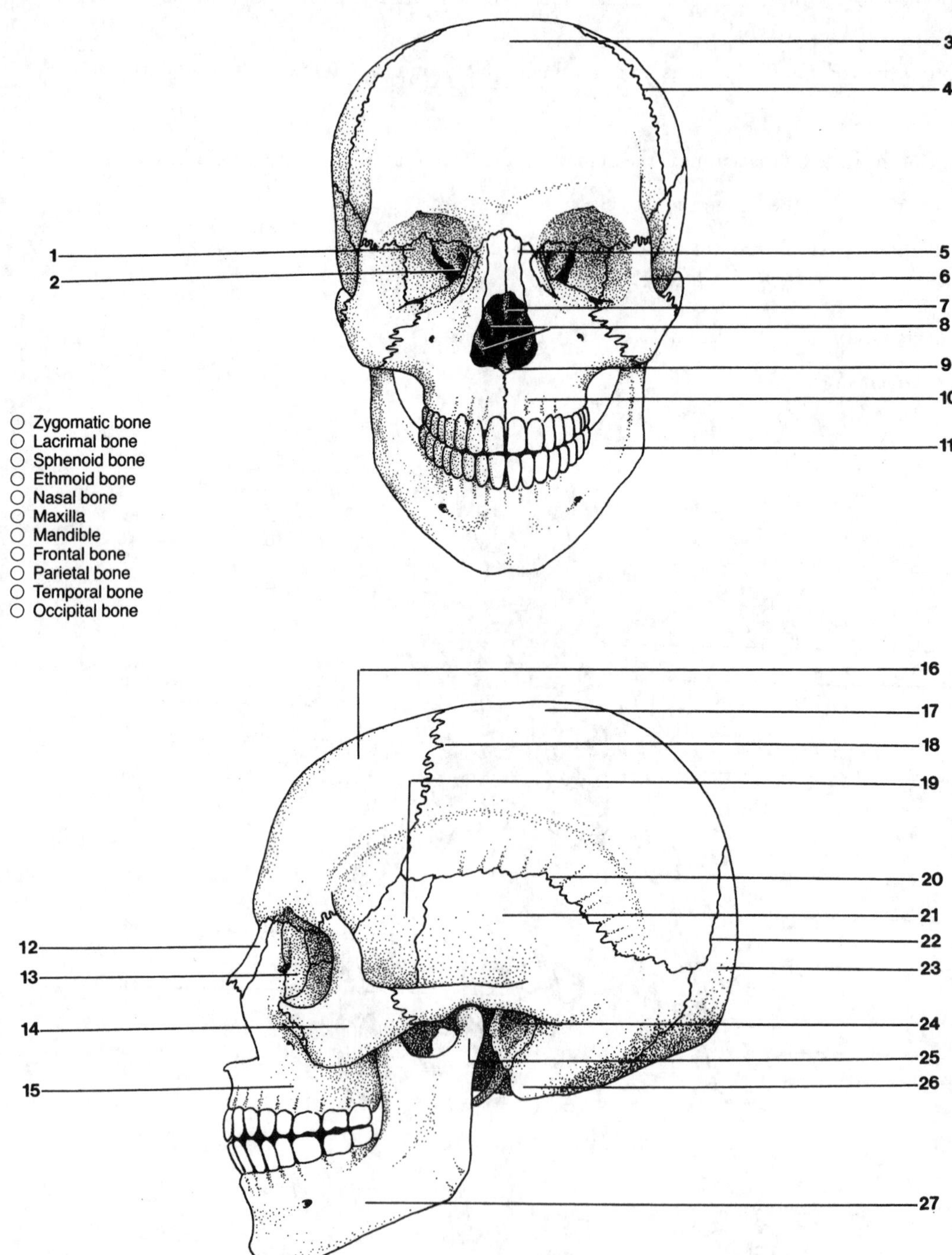

2. Name the bone or bones of the skull with each of these functions:

    1) Contain sockets for teeth. ____________________ and ____________________

    2) Contains the external auditory meatus. ____________________

    3) Forms the point of the cheek. ____________________

    4) Form the hard palate (roof of the mouth). ____________________ and ____________________

    5) Contains openings for the olfactory nerves. ____________________

    6) Form the nasal septum. ____________________ and ____________________

    7) Contains the nasolacrimal duct. ____________________

    8) Contains the foramen magnum ____________________, which in turn contains the ____________________.

    9) Protects the pituitary gland. ____________________

    10) Form the bridge of the nose. ____________________

    11) a) Contain paranasal sinuses. ____________________, ____________________, ____________________,

        and ____________________

        b) State the two functions of paranasal sinuses.

        ____

3. The squamosal suture is between the ____________________ and ____________________ bones.

4. The lambdoidal suture is between the ____________________ and ____________________ bones.

5. The coronal suture is between the ____________________ and ____________________ bones.

6. The sagittal suture is between the two ____________________ bones.

7. a) The three auditory bones in each middle ear cavity are the ____________________, ____________________,

    and ____________________.

    b) The vibrations of these bones are concerned with which sensation? ____________________

THE VERTEBRAL COLUMN

1. The functions of the vertebral column are to ____________________ the trunk and head and to protect the ____________________ from mechanical injury.

2. The vertebral column is shown in the diagram at the right.
   a) Label the parts indicated.
   b) State the number of each type of vertebrae.
      1) Cervical vertebrae ____________________
      2) Thoracic vertebrae ____________________
      3) Lumbar vertebrae ____________________
      4) Sacral vertebrae ____________________
         fused into one bone called the ____________________
      5) Coccygeal vertebrae ____________________
         fused into one bone called the ____________________

1
2
8
3
9
4
5
10
11
6
12
7

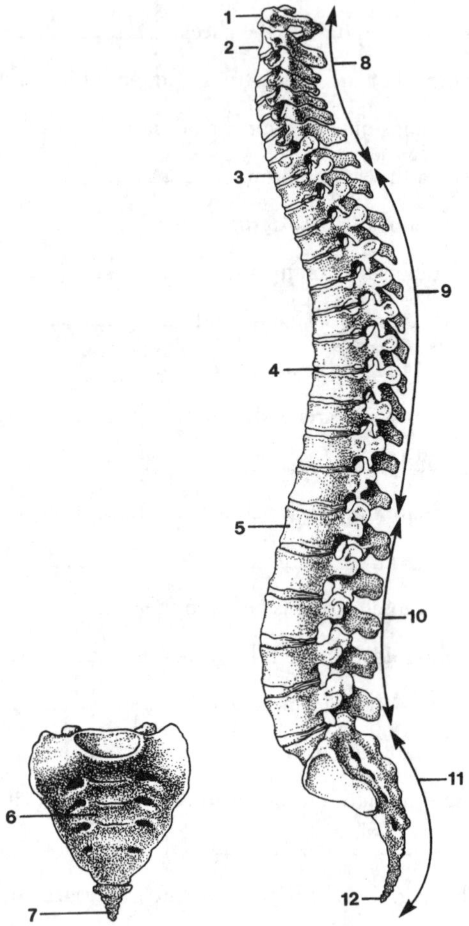

3. a) The first cervical vertebra is called the ____________________, and the second cervical vertebra is called the ____________________.
   b) These two vertebrae form a ____________________ joint that permits ____________________ movement of the head.

4. The ____________________ ends of the ribs articulate with the ____________________ vertebrae.

5. The two hip bones articulate with the ____________________ of the vertebral column.

6. a) The supportive part of each vertebra is called the ____________________.
   b) The bodies of adjacent vertebrae are separated by discs of ____________________.
   c) State the two functions of these discs. ____________________ and ____________________
   d) Name the type of joint between two vertebrae. ____________________

THE RIB CAGE

1. The following diagram depicts the rib cage.

   Label the parts indicated.

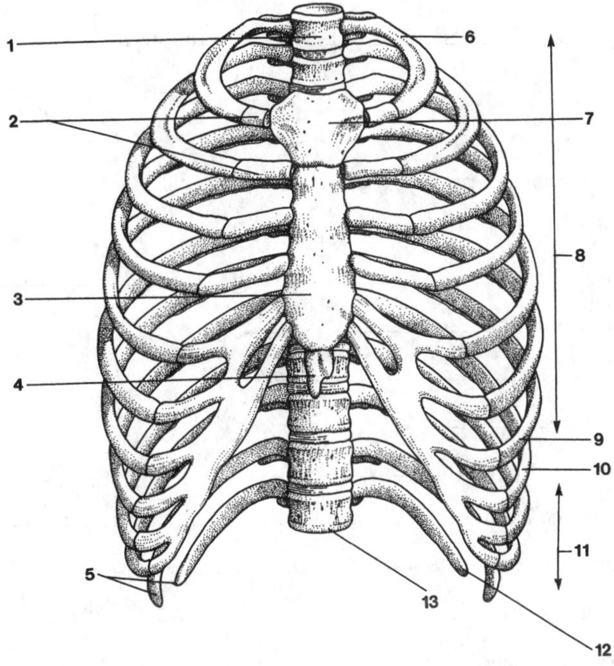

2. There are ____________ pairs of true ribs, ____________ pairs of false ribs, and ____________ pairs of floating ribs.
3. Name two organs in the thoracic cavity that are protected from mechanical injury by the rib cage.

   ____________ and ____________
4. Name two organs in the upper abdominal cavity that are protected from mechanical injury by the rib cage.

   ____________ and ____________
5. During inhalation, the ribs are pulled ____________ and ____________ by the external intercostal muscles to ____________ the chest cavity and bring about inflation of the ____________

THE SHOULDER AND ARM

1. The shoulder girdle that attaches the arm to the axial skeleton consists of two bones, the ____________ and the ____________.
2. Which of these two bones forms the socket for the humerus? ____________
3. a) The clavicle articulates medially with the ____________ of the sternum and laterally with the ____________.

   b) Explain the function of the clavicle. ____________
4. The ____________ of the humerus fits into the ____________ of the scapula to form a ____________ joint at the shoulder.
5. a) The two bones of the forearm are the ____________ and ____________.

   b) Which of these is part of the elbow joint? ____________
6. a) The elbow is the articulation of the ____________ and ____________.

   b) What type of joint is this? ____________
7. a) At their proximal ends, the radius and ulna form a ____________ joint.

   b) This joint permits what motion of the hand? ____________
8. a) At their distal ends, the radius and ulna articulate with the ____________, the bones of the wrist.

   b) How many carpals are found in each wrist? ____________

   c) What kind of joint is found between carpals? ____________
9. In each hand there are ____________ (number) metacarpals that articulate proximally with the ____________ and distally with the ____________.
10. a) The carpometacarpal joint of the thumb is what type of joint? ____________

    b) The joint permits what type of motion of the thumb? ____________
11. a) How many phalanges are present in the thumb? ____________

    b) How many phalanges are present in each other finger? ____________

    c) How many phalanges are present in one hand? ____________

    d) What type of joint is found between phalanges? ____________

12. The following diagram depicts the bones of the shoulder and arm.

Label the parts indicated.

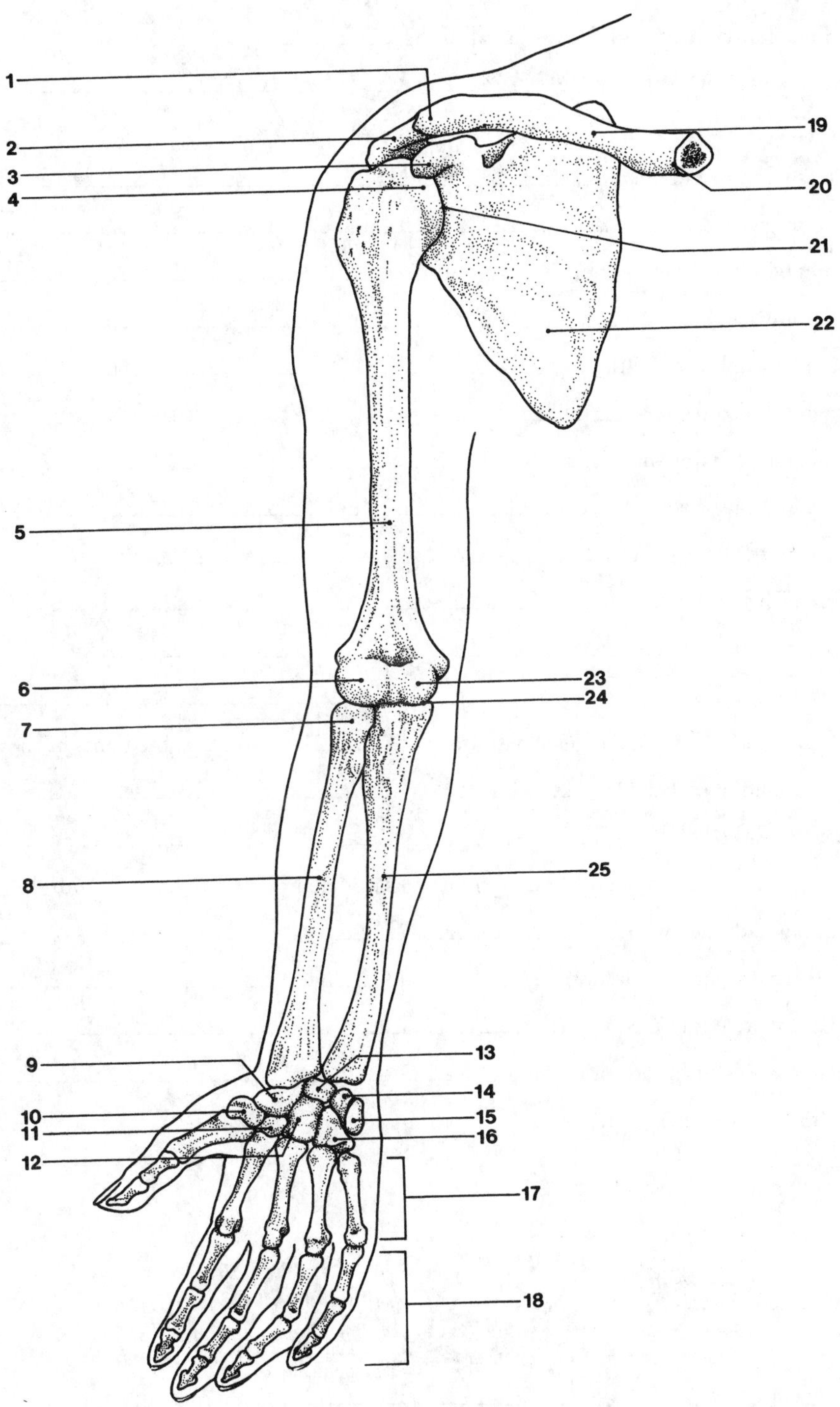

THE HIP AND LEG

1. The pelvic girdle that attaches the legs to the axial skeleton consists of one ____________ on each side, which together may be called the pelvic bone.
2. The socket in the hip bone for the head of the femur is called the ____________.
3. a) The three major parts of each hip bone are the ____________, ____________, and ____________.
   b) Which part articulates with the sacrum? ____________
   c) What name is this joint given? ____________
   d) Which part articulates with the other hip bone? ____________
   e) What name is this joint given? ____________
   f) Which part supports the trunk when sitting? ____________
4. a) The bone of the thigh is called the ____________.
   b) This bone forms what type of joint with the hip bone? ____________
5. a) The two bones of the lower leg are the ____________ and ____________.
   b) Which of these is part of the knee joint? ____________
   c) What is the function of the fibula? ____________
6. a) The knee is the articulation of the ____________ and ____________.
   b) What type of joint is this? ____________
7. a) At their distal ends, the tibia and fibula articulate with the ____________, the bones of the ankle.
   b) How many tarsals are found in each ankle? ____________
   c) Name the tarsal that forms the heel ____________
8. In each foot there are ____________ (number) metatarsals that articulate proximally with the ____________ and distally with the ____________.
9. a) How many phalanges are present in each foot? ____________
   b) How many phalanges are present in the big toe? ____________
   c) How many phalanges are present in each other toe? ____________
10. The following diagram depicts the pelvic bone.

    Label the parts indicated.

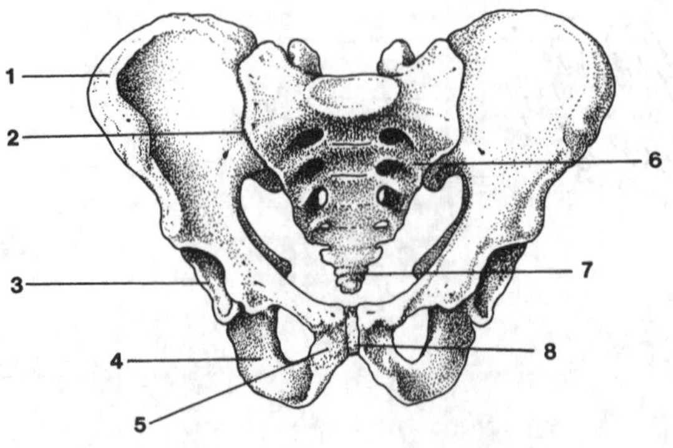

11. The following diagram depicts the bones of the hip and leg.

Label the parts indicated.

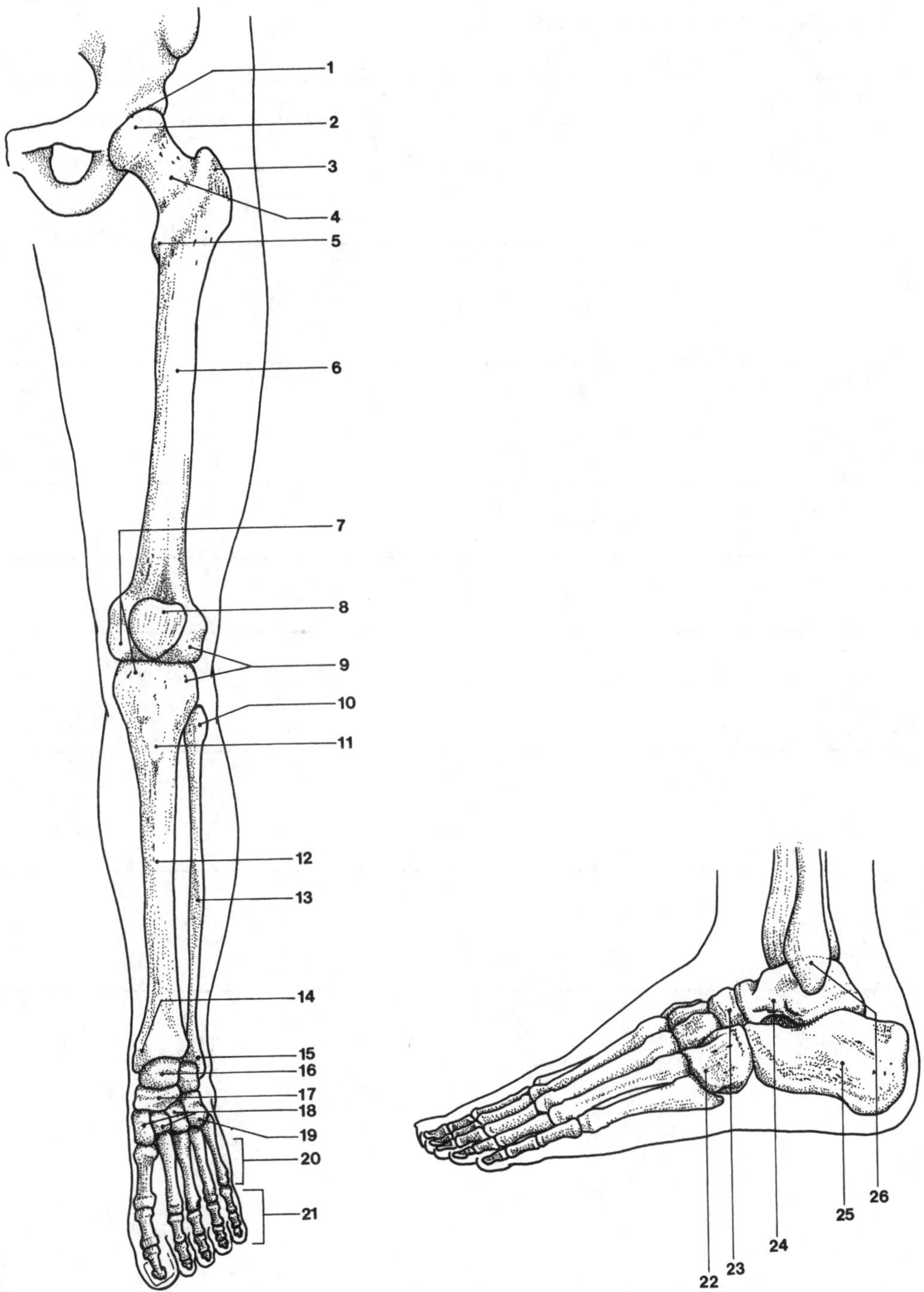

JOINTS (ARTICULATIONS)

1. The types of joints are listed in the following chart.

   For each, describe the movement possible at the joint, and give one or two examples by naming specific bones. Some answers have been filled in to get you started.

Name of Joint	Movement Possible	Examples (name two bones)
1) Suture		a. frontal and parietal bones b.
2) Symphysis	Slight movement	a. b.
3) Hinge		a. b.
4) Gliding	Sliding movement	a.
5) Saddle		a.
6) Ball and socket		a. scapula and humerus b.
7) Pivot		a. b.
8) Condyloid	Movement in one plane with some lateral movement	a.

2. The following diagram depicts several types of joints.

Label the parts indicated and, at each bracket, name the type of joint.

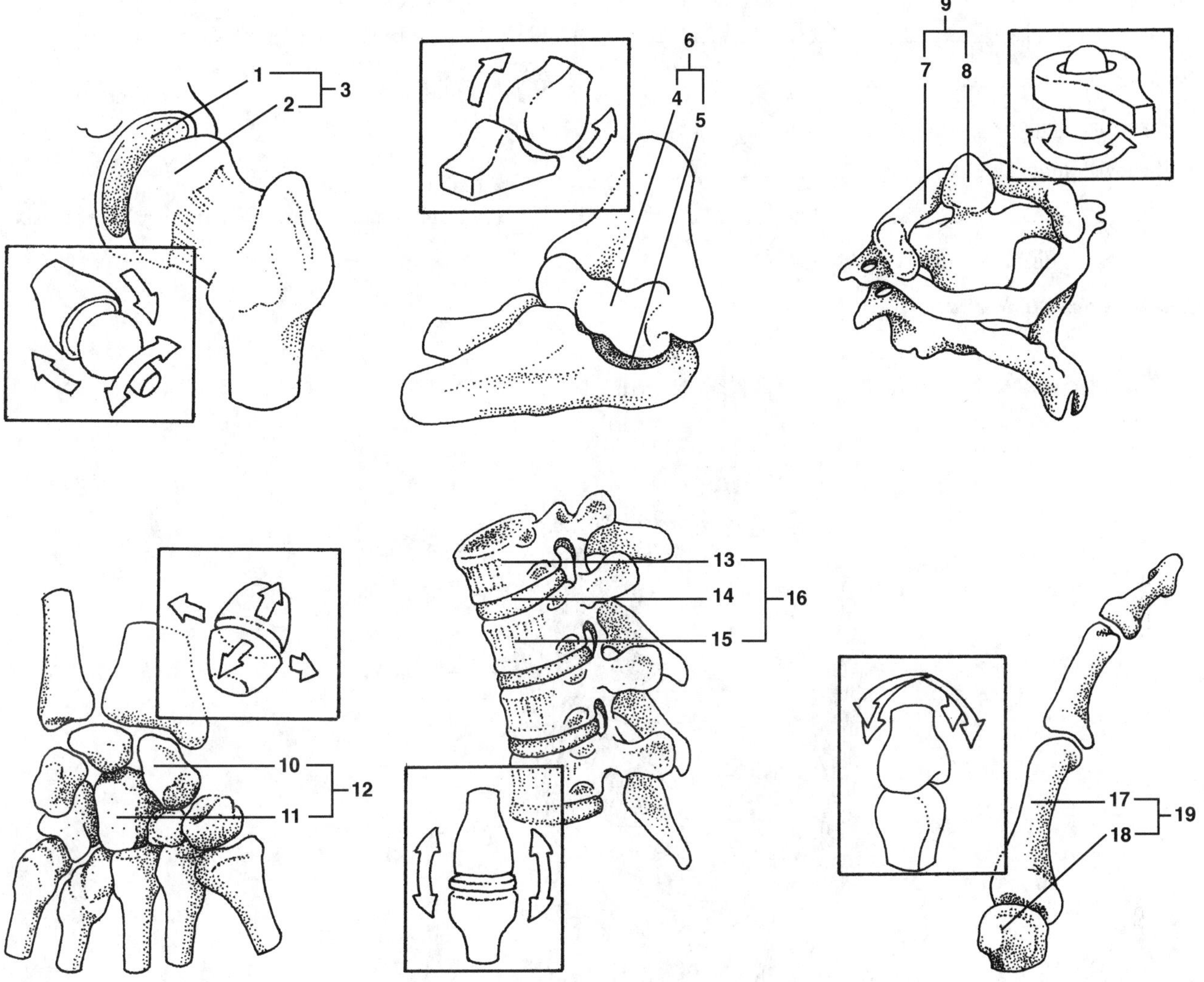

SYNOVIAL JOINT STRUCTURE

1. Match each part of a synovial joint with the proper descriptive statement.

Use each letter once.

1) Articular cartilage ______________________

2) Joint capsule ______________________

3) Synovial membrane ______________________

4) Synovial fluid ______________________

5) Bursae ______________________

A. Lines the joint capsule and secretes synovial fluid
B. Prevents friction within the joint cavity
C. Made of fibrous connective tissue; encloses the joint like a sleeve
D. Sacs of synovial fluid that permit tendons to slide easily across a joint
E. Provides a smooth surface on the joint surfaces of bones

2. The following diagram is of a typical synovial joint.

Label the parts indicated.

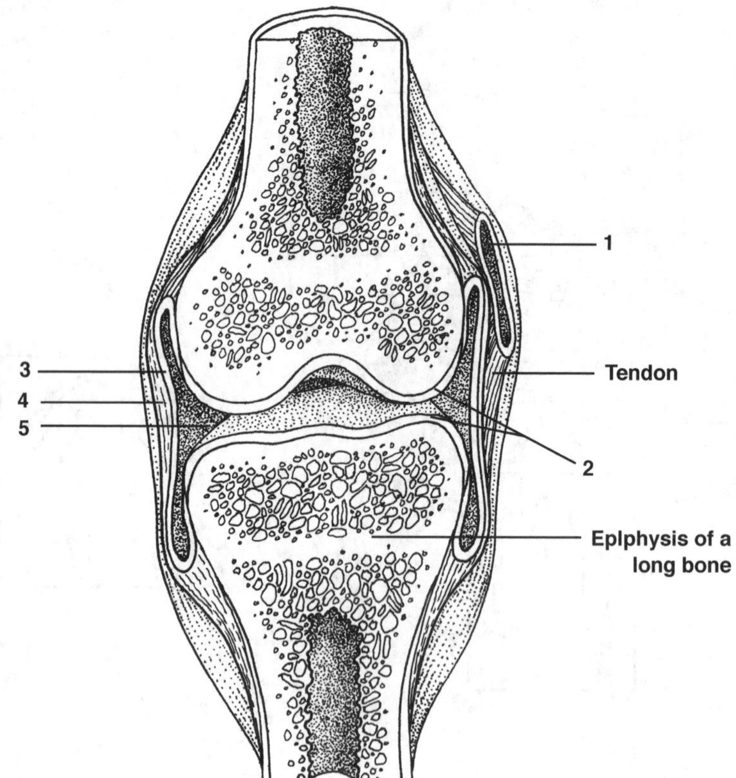

CROSSWORD PUZZLE

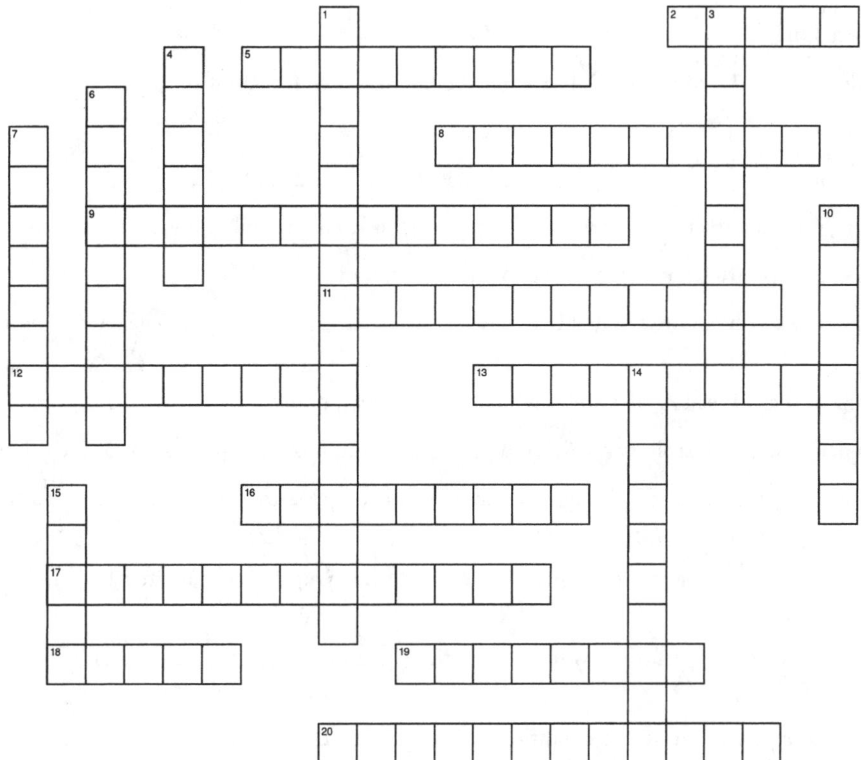

ACROSS

2. Where two bones meet
5. Shaft of a long bone
8. The disc at which the growth of a long bone takes place
9. Air cavity that opens into nasal cavity (two words)
11. Division of the skeleton with the arms and legs
12. End of a long bone
13. Anchors tendons and ligaments
16. Cartilage that is smooth; on joint surfaces
17. Produces red blood cells (three words)
18. Division of the skeleton that includes the vertebrae
19. Fluid that prevents friction as the bones move
20. Production of bone matrix

DOWN

1. Compact bone is made of ________________ (two words)
3. Produces bone matrix
4. Immovable joint between cranial bones
6. Type of joint with a disc of fibrous cartilage between two bones
7. "Soft spot" in an infant's skull
10. Connects one bone to another bone
14. Reabsorbs bone matrix
15. Sac of synovial fluid between a joint and tendons

CLINICAL APPLICATIONS

1. a) A 6-year-old boy has a compound fracture of the distal end of the humerus. A compound fracture means that the broken bone has pierced the ____________________.

   b) The boy's doctor orders antibiotics to prevent infection at the fracture site, because if the ____________________ were to be damaged by infection, the bone would cease to grow.

   c) To repair the fracture, new bone will be produced by cells called ____________________.

2. a) A 78-year-old woman suffered a fractured hip when getting out of bed one morning. Such a fracture is called a ____________________ fracture because there was no apparent trauma.

   b) In this case, the bones had become more fragile because of ____________________, a degenerative bone disorder that often affects elderly women.

   c) The term "broken hip" actually means a fracture of which part of which bone? ____________________

3. a) An 8-year-old girl is diagnosed with scoliosis, which is a ____________________ curvature of the spine.

   b) If not corrected, scoliosis may displace the rib cage on one side and compress the ____________________ (organ) on that side.

4. Mr. P sees his doctor because of severe back and leg pain. He says that his symptoms began after he helped his sister's family move furniture.

   a) The doctor diagnoses a herniated lumbar disc, which means that an ____________________ disc has been compressed and has ruptured.

   b) Explain why pain in the leg may occur in this condition. ____________________

5. On patients' charts, fractures are indicated using the anatomically correct name of the broken bone, which may not be a familiar term for some people. How would you explain each of the fractures below to patients or their families?

   1) A fracture of the zygomatic bone ____________________

   2) A fracture of the occipital bone ____________________

   3) A fracture of the distal tibia ____________________

   4) A fracture of the patella ____________________

   5) A fracture of the clavicle and manubrium ____________________

   6) A fracture of the mandible ____________________

   7) A fracture of the phalanges (hand) ____________________

MULTIPLE CHOICE TEST #1

Choose the correct answer for each question.

1. The largest bones of the arm and the leg are the:
   a) ulna/fibula b) radius/femur c) radius/tibia d) humerus/femur

2. The nutrients that become part of the bone matrix are:
   a) calcium and vitamin C c) phosphorus and vitamin A
   b) calcium and phosphorus d) protein and vitamin D

3. The cells that produce bone matrix and those that reabsorb bone matrix are called:
   a) osteoclasts/fibroblasts c) fibroblasts/osteoblasts
   b) osteoblasts/osteoclasts d) osteoblasts/fibroblasts

4. Red bone marrow is found in the:
   a) compact bone in long bones c) compact bone that covers flat bones
   b) marrow cavity of long bones d) spongy bone in flat and irregular bones

5. Red bone marrow produces these cells:
   a) red blood cells
   b) white blood cells
   c) platelets
   d) a, b, and c
   e) a and b only
   f) a and c only

6. All of the following are irregular bones except:
   a) vertebrae b) zygomatic bones c) metacarpals d) maxillae

7. The malleus, incus, and stapes are within the ________ bone and are concerned with ________.
   a) occipital bone/vision b) temporal bone/hearing c) maxillae/chewing d) mandible/taste

8. Sutures are the ________ joints of the ________.
   a) immovable/skull b) symphysis/vertebrae c) immovable/vertebrae d) gliding/skull

9. The fontanels of a baby's skull are made of:
   a) fibrous connective tissue b) bone c) cartilage d) elastic connective tissue

10. The vitamin necessary for efficient absorption of calcium and phosphorus by the small intestine is:
    a) C b) D c) A d) B_{12}

11. An example of a ball-and-socket joint is the one between the:
    a) femur and tibia b) scapula and ulna c) femur and hip bone d) tibia and fibula

12. For bones, the terms "exercise" or "stress" mean:
    a) becoming fatigued
    b) bearing weight
    c) running 5 miles every day
    d) not having enough protein

13. These bones are all part of the axial skeleton except the:
    a) sternum b) skull c) hip bone d) sacrum

14. In the embryo, the bones of the arms and legs are first made of:
    a) fibrous connective tissue b) collagen fibers c) bone d) cartilage

15. Which bone is not part of the braincase that protects the brain?
    a) mandible b) ethmoid c) sphenoid d) frontal bone

16. Which hormone increases the rate of mitosis in growing bones?
    a) thyroxine b) insulin c) growth hormone d) parathyroid hormone

17. Which of these joints is not a freely movable joint (diarthrosis)?
    a) hinge b) symphysis c) saddle d) pivot

18. In the long bones of a child, the part of the bone that is actually growing is the:
    a) articular cartilage b) epiphyseal disc c) diaphysis d) synovial membrane

19. The hormones that regulate blood calcium level by regulating calcium intake or output from bones are:
    a) growth hormone and thyroxine
    b) calcitonin and parathyroid hormone
    c) insulin and parathyroid hormone
    d) calcitonin and growth hormone

20. The structure not directly involved in the functioning of a synovial joint is the:
    a) synovial membrane b) marrow canal c) joint capsule d) articular cartilage

21. The largest and strongest vertebrae are the:
    a) cervical b) thoracic c) lumbar d) coccygeal

22. Which of the following bones do not protect an internal organ?
    a) vertebrae b) ribs c) sternum d) phalanges

23. The ligaments that connect bones are anchored to the:
    a) synovial membrane b) periosteum c) osteons d) matrix

24. The pituitary gland is directly protected by the:
    a) sphenoid bone b) occipital bone c) atlas d) zygomatic bone

25. The intervertebral discs are made of:
    a) cartilage b) adipose tissue c) spongy bone d) elastic tissue

MULTIPLE CHOICE TEST #2

Read each question and the four answer choices carefully. When you have made a choice, follow the instructions to complete your answer.

1. Which statement is NOT true of the tissues of the skeletal system?
   a) Bone matrix provides strength and stores excess calcium.
   b) The periosteum is made of fibrous connective tissue and anchors tendons and ligaments.
   c) Blood cells are produced by yellow bone marrow found in spongy bone.
   d) Ligaments are made of fibrous connective tissue and connect bone to bone.

   Reword your choice to make it a correct statement.

   ____

2. Which statement is NOT true of hormones and bone growth?
   a) Growth hormone increases mitosis and protein synthesis in growing bones.
   b) Thyroxine increases protein synthesis and energy production from all foods.
   c) Parathyroid hormone increases absorption of calcium by the small intestine.
   d) Calcitonin increases the reabsorption of calcium from bones.

   For the hormone of your choice, state its correct function.

   ____

3. Which statement is NOT true of growing bones?
   a) Estrogen or testosterone secretion stops the growth of long bones.
   b) New bone matrix is produced by cells called osteoclasts.
   c) A long bone grows as more cartilage is produced in the epiphyseal discs.
   d) Vitamins A and C are necessary for the production of bone matrix.

   Reword your choice to make it a correct statement.

   ____

4. Which statement is NOT true of the nutrients needed for bone growth?
   a) Vitamin C is needed for the ossification process.
   b) Protein is needed for the synthesis of collagen in bone matrix.
   c) Calcium and phosphorus are needed to become part of the bone matrix.
   d) Vitamin D is necessary for the absorption of calcium by the kidneys.

   For the nutrient of your choice, state the correct function.

   ____

5. Which statement is NOT true of the protective functions of bones?
   a) The vertebral column protects the spinal cord.
   b) The rib cage protects the small and large intestines.
   c) The skull protects the brain, eyes, and ears.
   d) Spongy bone in flat and irregular bones protects the red bone marrow.

   Reword your choice to make it a correct statement.

   ____

6. Which statement is NOT true of synovial joints?
   a) Joint surfaces are smooth because of the presence of articular cartilage.
   b) Synovial fluid increases friction within the joint cavity.
   c) Bursas permit tendons to slide easily as a joint is moved.
   d) The fibrous connective tissue joint capsule encloses the joint.

   Reword your choice to make it a correct statement.

   ____

7. Which statement is NOT true of types of joints?
   a) The hinge joint between femur and fibula permits movement in one plane at the knee.
   b) The pivot joint between atlas and axis permits the head to be turned from side to side.
   c) The ball-and-socket joint between scapula and humerus permits movement in all planes at the shoulder.
   d) The saddle joint of the carpometacarpal of the thumb permits the hand to grip.

   Reword your choice to make it a correct statement.

   ____

8. Which statement is NOT true of the skull?
   a) Sutures are the immovable joints between cranial bones and between facial bones.
   b) The only movable joint is the condyloid joint between the temporal bone and the maxillae.
   c) The spinal cord passes through the foramen magnum in the occipital bone.
   d) Sockets for the teeth are found in the mandible and the maxillae.

   Reword your choice to make it a correct statement.

   ____

9. Which statement is NOT true of the vertebral column?
   a) Discs of cartilage between vertebrae absorb shock and form symphysis joints.
   b) The vertebral column supports the trunk and head.
   c) Thoracic vertebrae articulate with the posterior ends of the ribs.
   d) The hip bones articulate with the lumbar vertebrae.

   Reword your choice to make it a correct statement.

   ____

10. Which statement is NOT true of the bones of the leg and hip?
    a) The femur and tibia form a hinge joint at the knee.
    b) The only weight-bearing bone in the lower leg is the fibula.
    c) The calcaneus bears weight because it is the heel bone.
    d) The hip, knee, and ankle joints are all weight-bearing joints.

    Reword your choice to make it a correct statement.

    ____

11. Which statement is NOT true of the rib cage?
    a) The seven pairs of true ribs articulate with the thoracic vertebrae and sternum.
    b) During inhalation, the ribs are pulled down and in to expand the lungs.
    c) The rib cage protects the heart and liver from mechanical injury.
    d) The two pairs of floating ribs articulate only with the thoracic vertebrae

    Reword your choice to make it a correct statement.

    ____

12. Which statement is NOT true of the bones of the arm and shoulder?
    a) The joints between phalanges are hinge joints.
    b) The hinge joint between humerus and ulna permits movement in one plane at the elbow.
    c) The clavicle braces the scapula and keeps the shoulder joint stable.
    d) There are 6 carpals in each wrist and 16 phalanges in each hand.

    Reword your choice to make it a correct statement.

    ____

MULTIPLE CHOICE TEST #3

Each question is a series of statements concerning a topic in this chapter. Read each statement carefully and select all of the correct statements.

1. Which of the following statements are true of the axial skeleton?
   a) The rib cage is pulled down and out to bring about inhalation.
   b) The largest vertebrae are the cervical vertebrae that support the head.
   c) The pituitary gland is located in a depression in the frontal bone.
   d) The paranasal sinuses make the skull lighter in weight.
   e) The vertebral canal contains the meninges and spinal cord.
   f) The ear canal is a tunnel into the parietal bone.
   g) The rib cage protects the liver and the lungs from mechanical injury.
   h) The parts of the sternum are the manubrium, body, and xylophone process.
   i) The eyes are protected by the bony sockets called orbits.
   j) The spinal cord enters the skull through the foramen magnum in the sphenoid bone.
   k) The rib cage has eight pairs of true ribs.
   l) The nasal bones form the nasal septum.

2. Which of the following statements are true of the appendicular skeleton?
   a) The radius articulates with the ulna and the metacarpals.
   b) One hand contains 14 phalanges and one foot contains 13 phalanges.
   c) The scapula and clavicle attach the arm to the axial skeleton.
   d) The two hip bones are fused at the pubic midline.
   e) The tibia and fibula are to the leg what the radius and ulna are to the arm.
   f) The ilium, ischium, and pubis all contribute to the acetabulum to attach the tibia.
   g) Both the humerus and femur have projections for muscle attachment.
   h) The calcaneus supports the base of the thumb.
   i) The proximal tibia is part of the knee joint.
   j) The foot contains eight carpals and the wrist contains seven tarsals.

3. Which of the following statements are true of joints?
   a) The pivot joint between the radius and humerus permits the hand to be turned palm up to palm down.
   b) The sacroiliac joint is a symphysis.
   c) Gliding joints between carpals permit movement at the wrist.
   d) The only movable joint in the skull is between the mandible and the two maxillae.
   e) The most movable type of joint is the ball-and-socket joint.
   f) Both the elbow and the knee are hinge joints.
   g) The pivot joint between the first and second thoracic vertebrae permits the head to be turned side to side.
   h) Sutures are immovable joints because the bones have been fused together.
   i) The synovial membrane lines the joint capsule and produces cartilage.
   j) The joint capsule is made of elastic connective tissue.
   k) The joints between vertebrae are symphysis joints.
   l) Saddle joints are part of the two hip bones.

Chapter 7

The Muscular System

This chapter presents the structure and functioning of the muscular system. Gross anatomy and physiology are discussed first, followed by microscopic anatomy and the biochemistry of muscle contraction. Also included are the locations and functions of the major skeletal muscles of the body.

MUSCLES AND MOVEMENT

1. Most of the more than 600 human skeletal muscles are attached to ________________ by fibrous connective tissue structures called ________________.
2. a) The most obvious function of the muscular system is to ________________.
   b) State the other function of the skeletal muscles. ________________
3. Name the other organ systems that contribute to movement with these functions.
   a) Exchanges oxygen and carbon dioxide between the air and blood ________________
   b) Forms a framework with movable joints that are moved by muscles ________________
   c) Transmits impulses to skeletal muscles to bring about contractions ________________
   d) Transports oxygen to muscles and takes carbon dioxide away ________________

MUSCLE STRUCTURE

1. One muscle is made of thousands of muscle fibers. A muscle fiber is simply another name for a ________________.
2. a) When muscle fibers contract, how does their length change? ________________
   b) In a muscle, the more muscle fibers that contract, the ________________ (more or less) work the muscle is able to do.
3. Tendons are made of fibrous connective tissue and anchor muscles to bones. The fibers of a tendon merge with the ________________, the fibrous connective tissue membrane around a muscle, and at the other end merge with the ________________, the fibrous connective tissue membrane that covers a bone.
4. a) A muscle usually has two points of attachment to two different bones, and the muscle itself crosses the ________________ formed by the two bones.
   b) The more stationary attachment (bone) is called the ________________ of the muscle.
   c) The more movable attachment (bone) is called the ________________ of the muscle.
   d) When the muscle contracts, it pulls on the ________________ and moves that bone at the joint.

MUSCLE ARRANGEMENTS

1. a) Antagonistic muscles have ______________ functions.

   b) Synergistic muscles have ______________ functions.

2. a) Antagonistic muscle arrangements are necessary because a contracting muscle can only ______________ (pull or push) a bone in one direction.

   b) When the muscle relaxes and lengthens, it does not exert force and cannot ______________ (pull or push) the bone the other way.

   c) Another muscle is needed to ______________ (pull or push) the bone back to its original position.

3. For each muscle action listed below, define the term and state the term that has the opposite meaning.

   1) Flexion: ______________

   Opposite: ______________

   2) Abduction: ______________

   Opposite: ______________

   3) Pronation: ______________

   Opposite: ______________

   4) Plantar flexion: ______________

   Opposite: ______________

4. a) The parts of the brain that generate the nerve impulses that initiate voluntary movement are the ______________ lobes of the ______________.

   b) The part of the brain that coordinates voluntary movement is the ______________.

5. The following diagrams depict the biceps brachii and triceps brachii.

Label the parts and actions indicated.

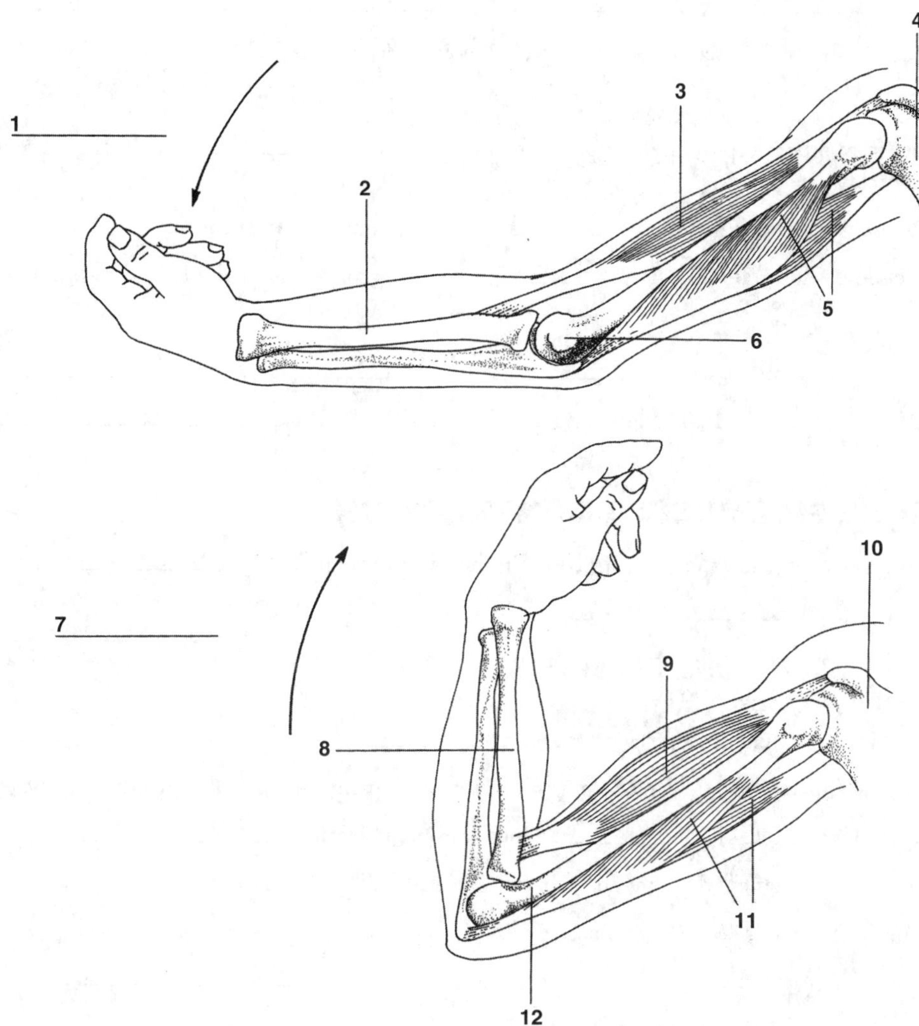

MUSCLE TONE AND EXERCISE

1. Muscle tone is the state of ____________________ that is usually present in healthy muscles.
2. The maintenance of muscle tone requires ATP. When ATP is produced in cell respiration, ____________________ energy is also released, which contributes to normal ____________________.
3. a) Isotonic exercise involves contraction of muscles and ____________________ of part of the body.
   b) During isometric exercise, muscles contract, but there is ____________________ of body parts.
   c) Which of these types of exercise (or both) will improve muscle tone? ____________________
   d) Which of these types of exercise (or both) may be considered aerobic? ____________________
   e) Besides strengthening skeletal muscles of the body, aerobic exercise also strengthens the ____________________ and the ____________________ muscles that enlarge the chest cavity.
4. State two reasons why good muscle tone is important.
   a) ____
   b) ____

MUSCLE SENSE

1. State a brief definition of muscle sense. ____

____

2. The stretching of muscles is detected by sensory receptors that may be called ______________ or ______________.

3. When sensory impulses from proprioceptors are interpreted by the ______________, a mental picture is created of where the muscle is.

4. a) Conscious muscle sense is integrated by the ______________ lobes of the cerebrum.

   b) Unconscious muscle sense is integrated by the ______________ of the brain and is used to promote good ______________.

5. Briefly explain the importance of muscle sense. ____

____

ENERGY SOURCES FOR MUSCLE CONTRACTION

1. a) Name the three sources of energy for muscle contraction, in the order they are used. ______________, ______________, and ______________

   b) The second and third energy sources are actually used to produce more ______________, which is the direct energy source for muscle contraction.

   c) The most abundant of the three energy sources is ______________.

2. When creatine phosphate is broken down to creatine plus phosphate plus energy, some of the creatine is converted to the nitrogenous waste product ______________, which is excreted by the ______________.

3. Glycogen is made of many molecules of which monosaccharide? ______________

4. a) Complete the summary reaction of cell respiration:

   Glucose + ______________ → ______________ + H_2O + ______________ + ______________

   b) The energy product used for muscle contraction is ______________.

5. a) The oxygen needed for cell respiration comes from breathing, and muscles have two sources of oxygen, both of which are proteins. The protein in muscle cells that stores oxygen is ______________.

   b) The protein in red blood cells that transports oxygen to muscles is ______________.

   c) The mineral that both of these proteins contain is ______________.

6. a) Muscles that lack sufficient oxygen are in a state called ______________.

   b) In this state, the intermediate molecule ______________ is produced and contributes to ______________ of the muscles.

7. The reversal of oxygen debt requires oxygen to enable the ______________ (organ) to convert lactic acid to a simple carbohydrate called ______________.

8. With respect to the other products of cell respiration (besides ATP), state what happens to, or the purpose of:

   a) Water ______________________________

   b) Heat ______________________________

   c) Carbon dioxide ______________________________

MUSCLE FIBER—MICROSCOPIC STRUCTURE

1. The neuromuscular junction is the junction of a ____________________ and a ____________________ (types of cells).

2. Match each part of the neuromuscular junction with the proper descriptions.

   Use each letter once. Each answer line will have two correct letters.

   1) Sarcolemma ____________________
   2) Axon terminal ____________________
   3) Synapse ____________________

   A. The space between the motor neuron and the muscle fiber
   B. The cell membrane of the muscle fiber
   C. Contains the inactivator cholinesterase
   D. The end of the motor neuron
   E. Contains the neurotransmitter acetylcholine
   F. An impulse is transmitted by the diffusion of acetylcholine

3. Match each internal structure of a muscle fiber with its proper function.

   Use each letter once.

   1) Sarcomeres ____________________
   2) Myosin and actin ____________________
   3) Troponin and tropomyosin ____________________
   4) Sarcoplasmic reticulum ____________________

   A. A reservoir for calcium ions
   B. The proteins that inhibit contraction when a muscle fiber is relaxed
   C. The proteins that contract when a muscle fiber receives a nerve impulse
   D. The structural units of contraction

4. The following diagram depicts the neuromuscular junction.

Label the parts indicated.

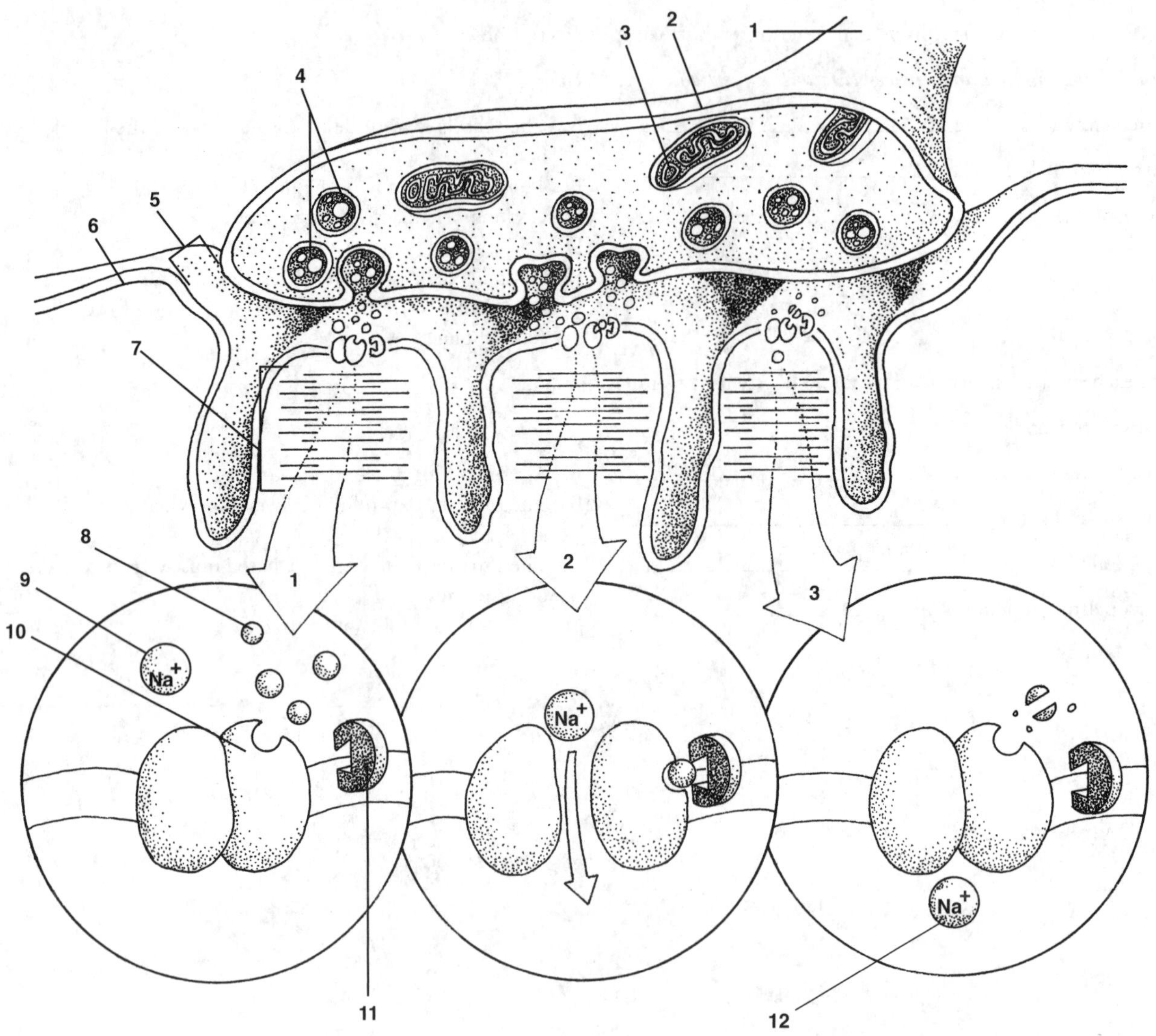

SARCOLEMMA—ELECTRICAL EVENTS

1. a) When a muscle fiber is relaxed, the sarcolemma has a ____________________ charge outside and a ____________________ charge inside.

   b) In this state of polarization, sodium ions are more abundant ____________________ the cell, and potassium ions are more abundant ____________________ the cell.

   c) The concentrations of these ions are maintained by the ____________________.

2. a) When a nerve impulse stimulates a muscle fiber, depolarization occurs, and ____________________ ions rush into the cell.

   b) The sarcolemma now has a ____________________ charge outside and a ____________________ charge inside.

   c) Depolarization is followed by ____________________, which is the exit of ____________________ ions from the cell.

   d) This completed impulse, depolarization followed by repolarization, is called an ____________________.

3. The following diagram shows the electrical events at the sarcolemma, with the important ions depicted as circles and ovals.

   Label the parts indicated.

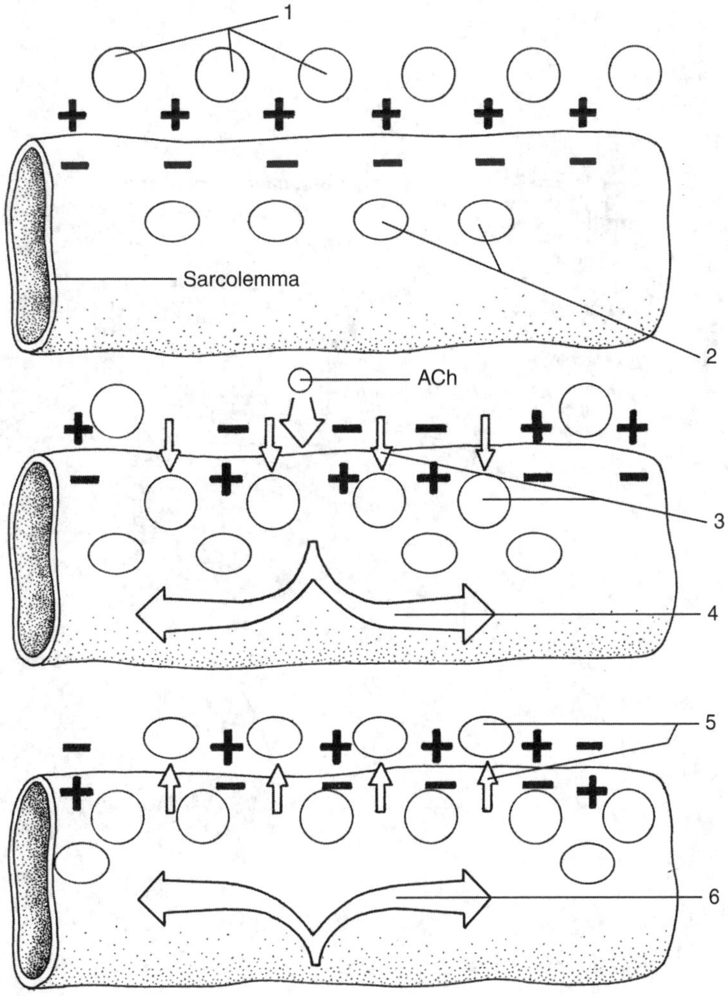

CONTRACTION—SLIDING FILAMENT MECHANISM

1. Match each part of the muscle contraction process with its proper function.

   Use each letter once.

   1) Acetylcholine ______
   2) Cholinesterase ______
   3) Sarcolemma ______
   4) Sarcoplasmic reticulum ______
   5) Myosin ______
   6) Actin ______
   7) Troponin and tropomyosin ______
   8) Sodium ions ______
   9) Potassium ions ______
   10) Calcium ions ______

   A. Releases energy from ATP, then pulls the actin filaments toward the center of the sarcomere
   B. More abundant inside the cell during polarization
   C. Inhibit the sliding of myosin and actin when a muscle fiber is relaxed
   D. The cell membrane of the muscle fiber; carries electrical charges
   E. Bond to troponin to permit contraction to take place
   F. Released by the axon terminal; makes the sarcolemma very permeable to sodium ions
   G. Pulled by myosin toward the center of the sarcomere
   H. Rush into the cell during depolarization
   I. Found at the sarcolemma to inactivate acetylcholine
   J. Releases calcium ions when stimulated by depolarization

2. The following diagrams depict sarcomeres, relaxed and contracted.

   Label the parts indicated.

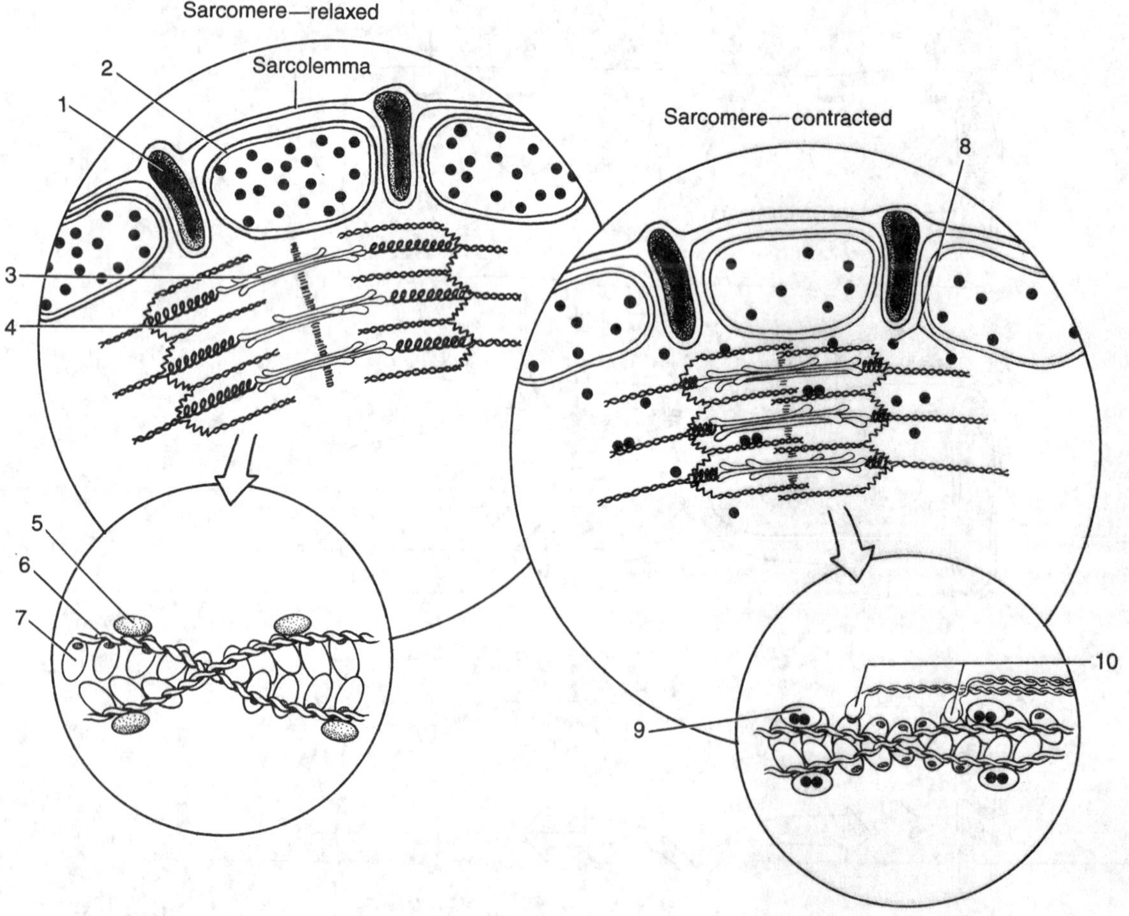

3. A sustained muscle contraction is called a ______________.

4. Explain the relationship between nerve impulses and tetanus.

____

5. Briefly explain the difference between the tetanus required for effective movements and the disease called tetanus.

Tetanus (movement) ______________________________

Tetanus (disease) ______________________________

EXERCISE—HOMEOSTASIS

1. Match each response to exercise with the purpose(s) of each.

Use each letter once. Two answer lines will have two correct letters.

1) Increased heart rate ______________

2) Increased sweating ______________

3) Increased respiration ______________

4) Increased cell respiration ______________

5) Vasodilation in muscle ______________

A. Produces more ATP for the muscle contraction process
B. Permits the loss of excess body heat
C. Circulates oxygen more rapidly to the muscles
D. Permits more oxygen to enter the blood
E. Circulates excess carbon dioxide more rapidly to the lungs to be exhaled
F. Permits excess carbon dioxide to be exhaled rapidly
G. Increases the blood flow within muscles

MAJOR MUSCLES OF THE BODY

1. The accompanying diagrams show the body's muscles.

   Label the muscles indicated.

   a) Anterior view

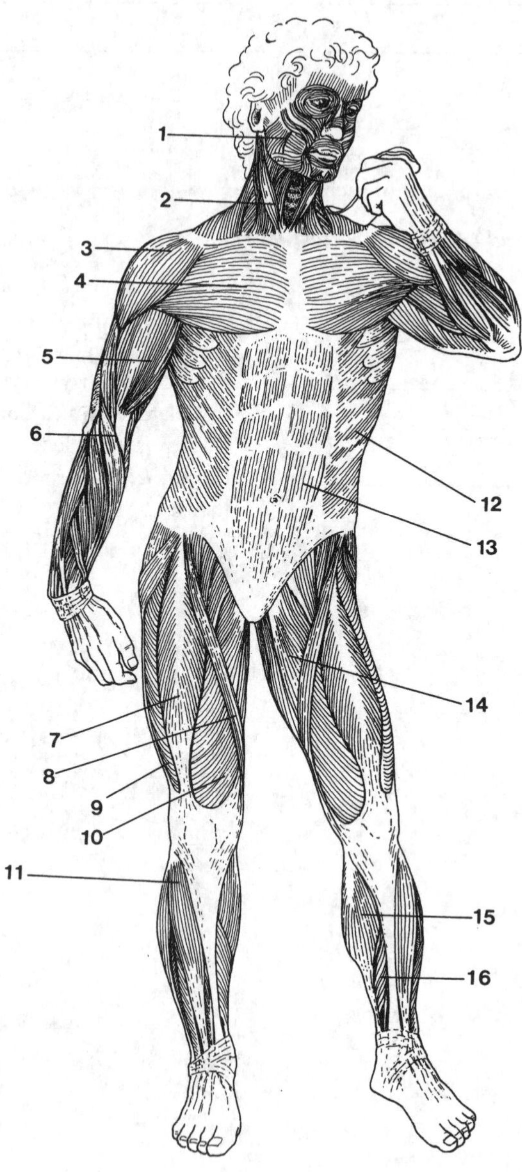

b) Posterior view

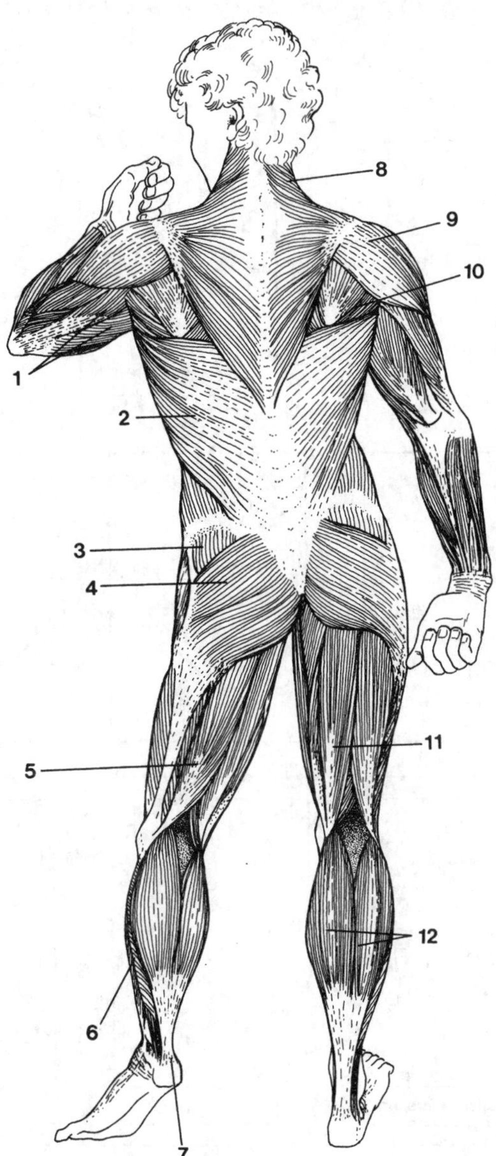

2. List three antagonistic pairs of muscles among those you have labeled on the preceding diagrams.

   a) ______________________________

   b) ______________________________

   c) ______________________________

3. Name three muscles that are common sites for intramuscular injections. ____________________,

   ____________________, and ____________________

4. Muscle actions and terminology

Name each muscle action shown in the following diagrams. Several of these have been completed to get you started.

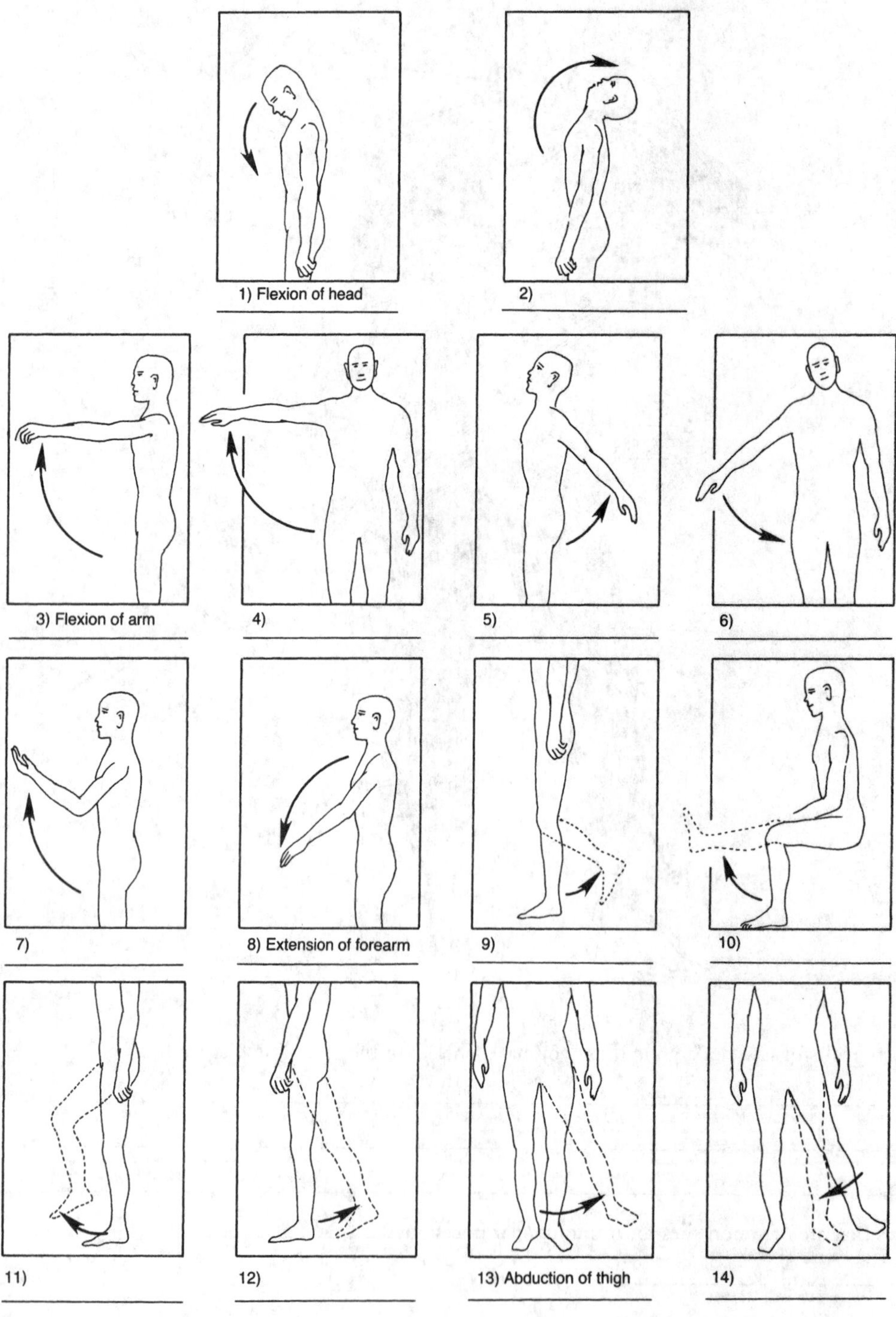

CROSSWORD PUZZLE

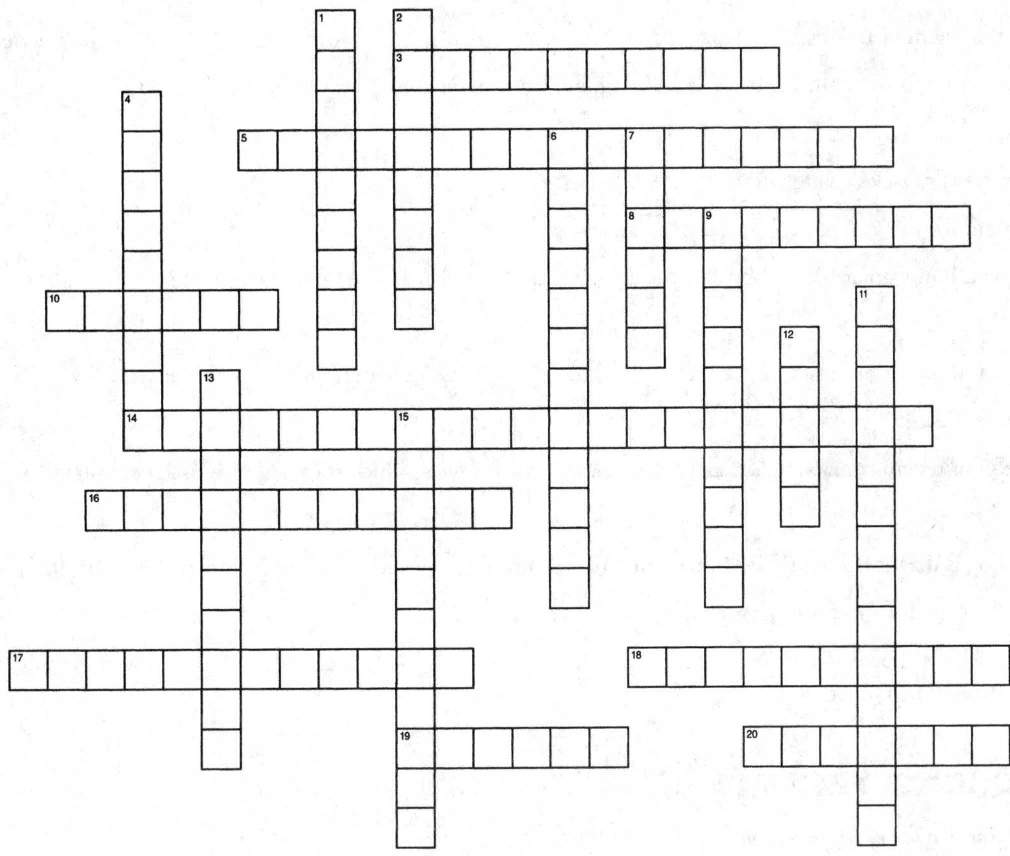

ACROSS

3. Contracting units in a muscle fiber
5. Secondary energy sources for muscle contraction are ____________ and glycogen
8. ____________ exercise, contraction without movement
10. The contracting protein in a sarcomere that releases energy from ATP
14. Where the motor neuron terminates on the muscle fiber (two words)
16. Knowing where muscles are without looking at them (two words)
17. Muscles that have opposite functions (adj.)
18. Without oxygen, glucose is converted into this (two words)
19. Attaches a muscle to a bone
20. Muscle ____________; caused by lactic acid

DOWN

1. Movable attachment of the muscle is called the ____________
2. ____________ exercise; contraction with movement
4. Protein in the muscle fibers that stores some oxygen
6. The state in which the sarcolemma has a (+) charge outside and a (–) charge inside
7. Stationary attachment of the muscle is its ____________
9. Muscle fibers run out of oxygen (two words)
11. The state in which the sarcolemma has a (+) charge inside and a (–) charge outside
12. A contracting protein that is pulled by myosin
13. A state of slight contraction (two words)
15. Muscles that work together (adj.)

CLINICAL APPLICATIONS

1. a) Following a severe compound fracture of his leg, Mr. M has difficulty extending his foot and moving his toes. It is possible that the extent of the fracture damaged the ________________ to the leg muscles for these functions.

   b) When a muscle is unable to contract because it does not receive nerve impulses, the muscle is said to be ________________.

   c) How will paralyzed muscles change in size? ________________

   d) What is the term for this? ________________

2. a) Both tetanus and botulism are caused by neurotoxins produced by what kind of microorganism? ________________

   b) The neurotoxin of botulism causes ________________ of muscles because it blocks the release of ________________ by motor neurons.

   c) The neurotoxin of tetanus causes ________________ of muscles, which means that muscles are unable to ________________.

3. a) A 2-year-old boy is diagnosed with Duchenne's muscular dystrophy, a genetic disease. From which parent did the child inherit the gene for this form of muscular dystrophy? ________________

   b) The child's muscles will be replaced by ________________ or by ________________, and the muscles will become unable to contract.

MULTIPLE CHOICE TEST #1

Choose the correct answer for each question.

1. The contracting proteins within a muscle fiber are:
   a) troponin and tropomyosin c) myosin and myoglobin
   b) hemoglobin and myoglobin d) myosin and actin

2. When a muscle contracts, it:
   a) shortens and pushes a bone c) shortens and pulls or pushes a bone
   b) lengthens and pulls a bone d) shortens and pulls a bone

3. Muscles are attached to bones by:
   a) ligaments b) fascia c) other muscles d) tendons

4. Some oxygen may be stored within muscle fibers bonded to the protein:
   a) myosin b) troponin c) myoglobin d) hemoglobin

5. An antagonist to a muscle that flexes the arm would be a muscle that:
   a) adducts the arm b) abducts the arm c) extends the arm d) rotates the arm

6. The axon terminal of a motor neuron releases:
   a) sodium ions b) cholinesterase c) acetylcholine d) potassium ions

7. Acetylcholine makes the sarcolemma more permeable to:
   a) oxygen b) sodium ions c) potassium ions d) calcium ions

8. The most abundant source of energy in a muscle fiber is:
   a) glycogen b) ATP c) fat d) creatine phosphate

9. The part of the brain that coordinates voluntary movement is the:
   a) frontal lobe b) cerebellum c) parietal lobe d) cerebrum

10. The neuromuscular junction refers to the:
    a) axon terminal, sarcolemma, and sarcomeres c) synapse, sarcomeres, and motor neuron
    b) sarcolemma, synapse, and sarcomeres d) axon terminal, synapse, and sarcolemma

11. Unconscious muscle sense is integrated by the:
  a) cerebellum, and we are unaware of it
  b) parietal lobes, and we are aware of it
  c) parietal lobes, and we are unaware of it
  d) cerebellum, and we are aware of it

12. The more movable attachment of a muscle to a bone is called the:
  a) origin b) tendon c) insertion d) wobbly part

13. Good muscle tone is important to:
  a) be able to do strenuous exercise
  b) get rid of carbon dioxide
  c) know where our muscles are without looking at them
  d) maintain posture and produce body heat

14. Muscle fatigue is the result of the lack of ____________ and the production of ____________.
  a) oxygen/lactic acid b) lactic acid/oxygen c) glucose/oxygen d) oxygen/glucose

15. The specific part of the brain that initiates muscle contraction is the:
  a) frontal lobe b) cerebellum c) cerebrum d) parietal lobe

16. During exercise, the blood flow within a muscle is increased by:
  a) vasoconstriction
  b) more nerve impulses to the muscle
  c) increased respiration
  d) vasodilation

17. The product of cell respiration that is considered a waste product is:
  a) water b) ATP c) carbon dioxide d) heat

18. The organ system that transports oxygen to muscles and removes carbon dioxide is the:
  a) skeletal system b) circulatory system c) respiratory system d) nervous system

19. The energy of ATP is released within muscle fibers by:
  a) calcium ions b) troponin c) actin d) myosin

20. In response to a nerve impulse, the electrical charges on the sarcolemma are reversed. This is called:
  a) polarization b) the sodium pump c) the potassium pump d) depolarization

21. Synergistic muscles are those that have the same:
  a) size b) shape c) function d) origin

22. An isometric contraction is one:
  a) with movement b) without movement c) without myosin d) without actin

23. The inhibiting proteins within a muscle fiber are:
  a) troponin and tropomyosin
  b) hemoglobin and myoglobin
  c) myosin and myoglobin
  d) myosin and actin

24. A nitrogenous waste product of muscle contraction is ____________, which is excreted by the ____________
  a) urea/skin b) creatinine/skin c) urea/kidneys d) creatinine/kidneys

25. The mineral released within sarcomeres to trigger contraction is:
  a) calcium b) iron c) copper d) potassium

MULTIPLE CHOICE TEST #2

Read each question and the four answer choices carefully. When you have made a choice, follow the instructions to complete your answer.

1. Which statement is NOT true of the role of the brain in the functioning of muscles?
  a) The cerebellum coordinates voluntary movement.
  b) The frontal lobes initiate voluntary movement.
  c) Conscious muscle sense is integrated by the parietal lobes.
  d) Unconscious muscle sense is integrated by the cerebrum.

  Reword your choice to make it a correct statement.

  ____

2. Which statement is NOT true of muscles and bones?
   a) The more stationary attachment of a muscle to a bone is called the origin.
   b) When a muscle contracts, it pulls a bone.
   c) Muscles are attached to bones by tendons, which are made of elastic connective tissue.
   d) The tendon of a muscle merges with the periosteum that covers the bone.

   Reword your choice to make it a correct statement.

   ____

3. Which statement is NOT true of muscle locations and functions?
   a) The deltoid is the shoulder muscle that adducts the arm.
   b) The quadriceps femoris group is on the front of the thigh and extends the lower leg.
   c) The triceps brachii extends the forearm; its antagonist is the biceps brachii.
   d) The rectus abdominis is on the front of the abdomen and flexes the vertebral column.

   Reword your choice to make it a correct statement.

   ____

4. Which statement is NOT true of muscle disorders?
   a) Duchenne's muscular dystrophy is a sex-linked genetic trait.
   b) Myasthenia gravis involves extreme muscular weakness after mild exertion.
   c) Muscles that are paralyzed will atrophy, which means to become larger from disuse.
   d) Botulism is characterized by muscle paralysis, and tetanus is characterized by muscle spasms.

   Reword your choice to make it a correct statement.

   ____

5. Which statement is NOT true of muscle tone?
   a) Muscle tone is a state of slight contraction.
   b) Good muscle tone improves coordination and helps maintain posture.
   c) The cerebellum regulates muscle tone.
   d) Muscle tone does not depend upon nerve impulses to muscle fibers.

   Reword your choice to make it a correct statement.

   ____

6. Which statement is NOT true of the energy sources for muscle contraction?
   a) Creatinine is a nitrogenous waste product formed when creatine phosphate is used for energy.
   b) The direct energy source for contraction is glycogen.
   c) When glycogen is used for energy, it is first broken down to glucose.
   d) Oxygen is required for the cell respiration of glucose to produce ATP.

   Reword your choice to make it a correct statement.

   ____

7. Which statement is NOT true of the neuromuscular junction?
   a) The axon terminal is the end of the motor neuron.
   b) Acetylcholine makes the sarcolemma very permeable to potassium ions.
   c) The sarcolemma contains cholinesterase to inactivate acetylcholine.
   d) The synapse is the small space between the axon terminal and the sarcolemma.

   Reword your choice to make it a correct statement.

   ____

8. Which statement is NOT true of exercise?
   a) Increased heart rate transports oxygen to muscles more rapidly.
   b) Isotonic exercise is aerobic because it involves contraction without movement.
   c) Increased cell respiration produces more heat, ATP, and carbon dioxide.
   d) Increased respiration is necessary to exhale excess carbon dioxide.

   Reword your choice to make it a correct statement.

   ____

9. Which statement is NOT true of oxygen and muscle contraction?
   a) Oxygen is brought to muscle fibers by hemoglobin in red blood cells.
   b) Some oxygen is stored in muscles by the protein myosin.
   c) Oxygen is needed for the complete breakdown of glucose in cell respiration.
   d) In the absence of oxygen, lactic acid is formed and contributes to fatigue.

   Reword your choice to make it a correct statement.

   ___

10. Which statement is NOT true of the sliding filament mechanism of muscle contraction?
    a) The sarcoplasmic reticulum releases calcium ions following depolarization.
    b) Myosin filaments pull actin filaments toward the center of the sarcomere.
    c) Troponin and tropomyosin are inhibiting proteins.
    d) During depolarization, the inside of the sarcolemma becomes negative.

    Reword your choice to make it a correct statement.

    ___

MULTIPLE CHOICE TEST #3

Each question is a series of statements concerning a topic in this chapter. Read each statement carefully and select all of the correct statements.

1. Which of the following statements are true of muscle structure?
   a) Antagonistic muscles are on opposite sides of the same joint.
   b) Tendons attach muscle to muscle across a joint.
   c) Oxygen is stored in muscle fibers by the protein myoglobin.
   d) A neuromuscular junction consists of a motor nerve and a muscle.
   e) One muscle is made of thousands of muscle fibers.
   f) The sarcoplasmic reticulum is the cell membrane of a muscle fiber.
   g) Troponin and tropomyosin are found on the sarcolemma.
   h) The synaptic cleft is between the axon terminal and the T tubules.

2. Which of the following statements are true of muscle physiology?
   a) The unit of contraction of a muscle fiber is the sarcomere.
   b) Acetylcholine is the transmitter at skeletal muscle neuromuscular junctions.
   c) The contracting proteins in a sarcomere are myosin and actin.
   d) A tetanus is a sustained contraction of one muscle fiber.
   e) Muscle tone is a state of slight contraction of a muscle.
   f) Muscle sense is the brain's mental picture of where muscles are.
   g) Acetylcholine causes depolarization of the sarcolemma.
   h) The energy of ATP is released by the protein myosin.

3. Which of the following statements are true of muscles and energy production?
   a) The direct source of energy for muscle contraction is ATP.
   b) A waste product of energy production is carbon dioxide.
   c) Isometric contraction does not result in movement and does not require energy.
   d) ATP is stored in muscles while they are at rest.
   e) The energy from creatine phosphate is used to make more glycogen.
   f) The production of ATP during strenuous exercise requires oxygen for the process of cell respiration of glucose.

4. Which of the following statements are true of muscle contraction?
   a) The nerve impulses for contraction originate in the parietal lobes of the cerebrum.
   b) Muscle contraction moves the skeleton.
   c) Coordination of voluntary movement is regulated by the frontal lobes.
   d) Muscle contraction requires delivery of oxygen by the circulatory system.
   e) Muscle tone may increase in a cold environment.
   f) Synergistic muscles work together.

5. Which of the following statements are true of the location and function of muscles?
   a) The rectus abdominis extends the vertebral column.
   b) The triceps brachii flexes the forearm.
   c) The quadriceps femoris extends the thigh.
   d) The gastrocnemius plantar flexes the foot.
   e) The gluteus maximus extends the thigh.
   f) The trapezius raises the shoulder.
   g) The latissimus dorsi flexes the upper arm.
   h) The masseter closes the jaw.
   i) The sternocleidomastoid extends the head.
   j) The orbicularis oculi opens the eye.
   k) The gluteus medius abducts the thigh.
   l) The sartorius extends the lower leg.

Chapter 8

The Nervous System

This chapter describes the anatomy and physiology of the nervous system, one of the regulating systems of the body. The general functions of the nervous system are to detect changes and feel sensations, initiate appropriate responses to changes, and organize information for immediate or later use. These functions of the nervous system are directly related to the normal physiology of other organ systems and to homeostasis of the body as a whole.

NERVOUS SYSTEM DIVISIONS

1. The central nervous system (CNS) consists of the ____________________ and ____________________.
2. The peripheral nervous system (PNS) consists of the ____________________ nerves and ____________________ nerves.
3. The autonomic nervous system (ANS) is part of which division? ____________________
4. The enteric nervous system is part of the peripheral nervous system and is found in the ____________________.

NERVE TISSUE

1. The term that means nerve cell is ____________________.
2. Name the three major parts of a neuron.
   a) ______________________________
   b) ______________________________
   c) ______________________________

3. The following diagram depicts two neurons.

   a) Label the major parts and put each with its proper function.

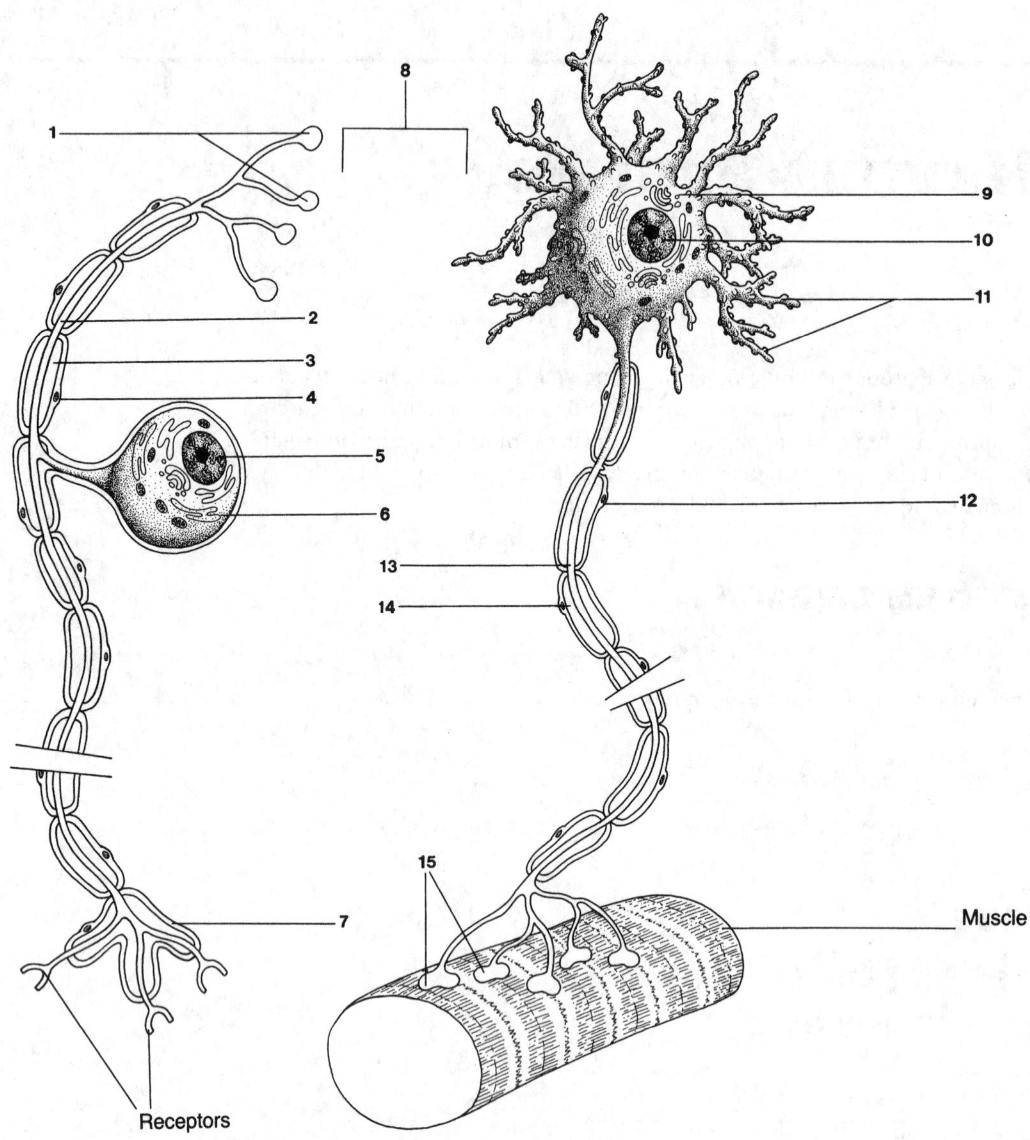

   b) ____________________ carries impulses away from the cell body.

   c) ____________________ contains the nucleus of the neuron.

   d) ____________________ carries impulses toward the cell body.

   e) The space between the axon of one neuron and the dendrites of the cell body of the next neuron is called the ____________________.

   f) Label the neuron that is the sensory neuron and the one that is the motor neuron.

4. a) In the PNS, the myelin sheath is made by cells called ____________________.

   b) In the CNS, the myelin sheath is made by cells called ____________________.

   c) State the function of the myelin sheath. ____________________

5. a) In the PNS, the neurolemma is formed by which parts of the Schwann cells?

   ____________________ and ____________________

   b) The neurolemma permits ____________________ of damaged axons or dendrites in the PNS.

6. a) At synapses, the nerve impulse is carried by a ____________________ that is released by the synaptic knobs of the ____________________ of a neuron.

   b) The postsynaptic neuron contains an ____________________ to prevent continuous impulses by inactivating the neurotransmitter.

   c) Name the inactivator for the neurotransmitter acetylcholine ____________________

7. The following diagram shows a synapse.

   Label the parts indicated.

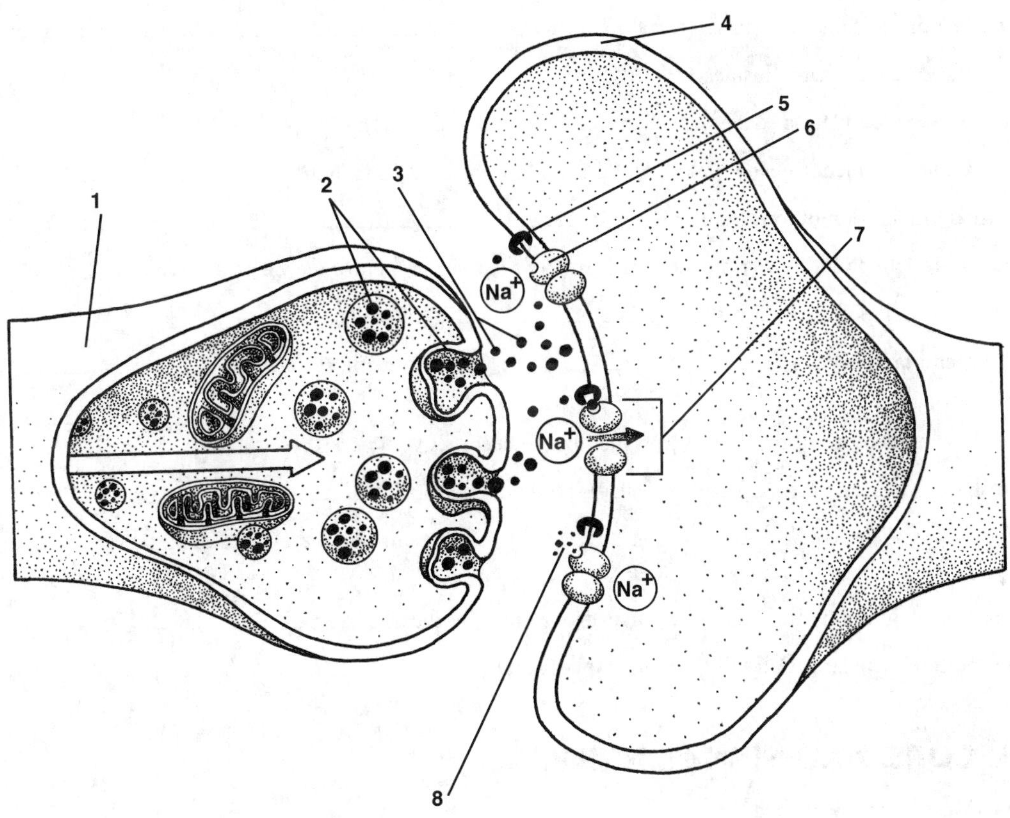

NEURONS, NERVES, AND NERVE TRACTS

1. Match each of the following structures with its proper description.

   Use each letter once. Two answer lines will have two correct letters.

   1) Sensory neurons ____________________
   2) Motor neurons ____________________
   3) Interneurons ____________________
   4) Mixed nerve ____________________
   5) Nerve tract ____________________

   A. Neurons found entirely within the CNS
   B. Carry impulses from receptors to the CNS
   C. Also called efferent neurons
   D. Also called afferent neurons
   E. A group of functionally related neurons within the CNS
   F. Carry impulses from the CNS to effectors
   G. Made of both sensory and motor neurons

THE NERVE IMPULSE

1. a) When a neuron is not carrying an impulse, its cell membrane has a ____________ charge outside and a ____________ charge inside.
   b) The ions that are more abundant outside the cell are ____________ ions.
   c) The ions that are more abundant inside the cell are ____________ and ____________ ions.
   d) The name for this distribution of ions (and charges) is ____________.
2. a) Depolarization is brought about by a stimulus, which makes the neuron membrane very permeable to ____________ ions, which rush ____________ the cell.
   b) As a result, the membrane now has a ____________ charge outside and a ____________ charge inside.
3. a) Immediately following depolarization, the neuron membrane becomes very permeable to ____________ ions, which rush ____________ the cell; this is called ____________.
   b) As a result, the membrane now has a ____________ charge outside and a ____________ charge inside.
   c) The sodium and potassium ions are returned to their proper sites by the ____________.
   d) The complete impulse, depolarization followed by repolarization, is an ____________.

THE SPINAL CORD AND SPINAL NERVES

1. State the two functions of the spinal cord.
   1) ______________________________
   2) ______________________________
2. a) The spinal cord is protected from mechanical injury by the ____________ (bones).
   b) In length, the spinal cord extends from the ____________ of the skull to the disc between the ____________ vertebrae.

3. The following diagram shows a cross section of the spinal cord and spinal nerve roots.

   a) Label the parts indicated and complete the statements.

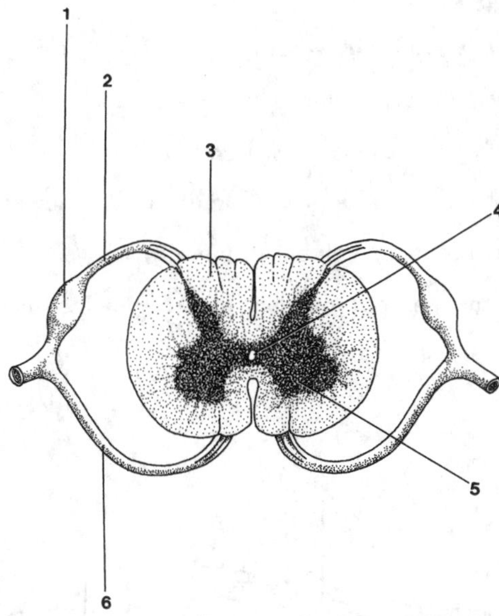

   b) The cell bodies of motor neurons and interneurons are located in the ________________.

   c) The cell bodies of sensory neurons are located in the ________________.

   d) The dorsal root may also be called the ________________ root because it is made of ________________ neurons.

   e) The ventral root may also be called the ________________ root because it is made of ________________ neurons.

   f) The ascending and descending tracts of the spinal cord are located in the ________________.

   g) Ascending tracts may be called ________________ tracts because they carry impulses ________________ the brain.

   h) Descending tracts may also be called ________________ tracts because they carry impulses ________________ the brain.

   i) Cerebrospinal fluid is found within the ________________.

4. There are 31 pairs of spinal nerves, which are named according to their locations. State the number of pairs of:

   1) Cervical spinal nerves ________________

   2) Thoracic spinal nerves ________________

   3) Lumbar spinal nerves ________________

   4) Sacral spinal nerves ________________

   5) Coccygeal spinal nerves ________________

5. Name the groups of spinal nerves that supply these parts of the body.

   1) Trunk of the body ________________

   2) Hips, legs, and pelvic cavity ________________

   3) Neck, shoulders, and arms ________________

   4) Diaphragm ________________

SPINAL CORD REFLEXES

1. Define reflex. ____

2. A reflex arc is the pathway nerve impulses travel during a reflex.

   Number the following parts of a reflex arc in proper sequence.

   _____1_____ Receptors that detect a change and generate impulses

   __________ An effector, which performs the reflex action

   __________ Motor neurons that transmit impulses from the CNS to an effector

   __________ The central nervous system, with one or more synapses

   __________ Sensory neurons that transmit impulses from receptors to the CNS

3. The following diagram depicts the patellar reflex arc.

   Label the parts indicated.

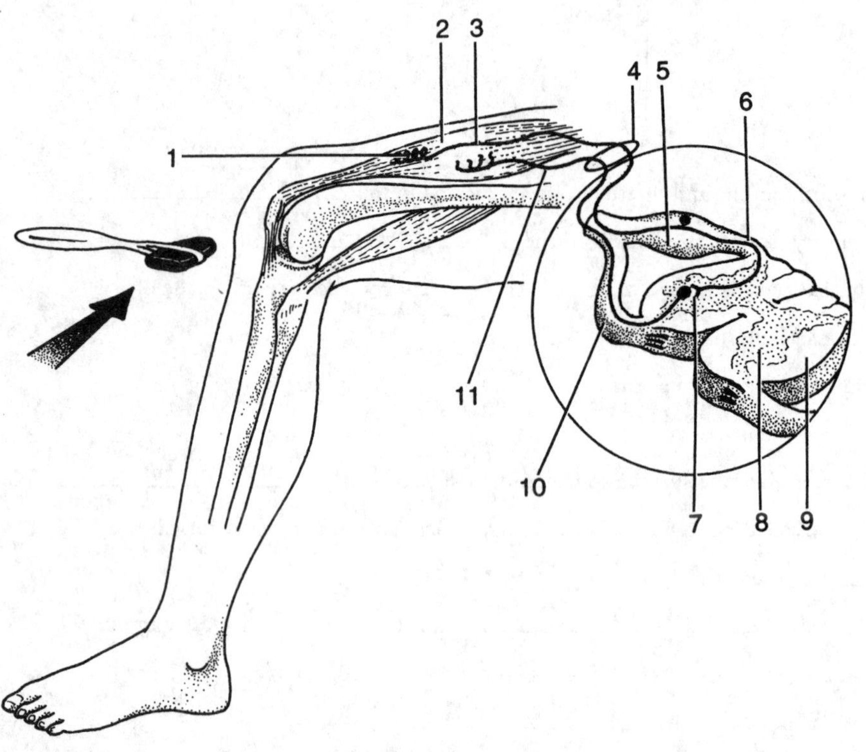

4. The patellar reflex is an example of a stretch reflex, which means that the stimulus is __________, and the response is __________.

5. a) Explain the everyday importance of stretch reflexes. __________

   ____

   b) Explain the everyday importance of flexor reflexes. __________

   ____

   c) Explain why it is important that these are spinal cord reflexes that do not depend directly on the brain.

   ____

THE BRAIN

1. Match each part of the brain with its proper function(s).

Use each letter once. Some answer lines will have two or more correct letters.

1) Ventricles ______
2) Medulla ______
3) Pons ______
4) Midbrain ______
5) Cerebellum ______
6) Hypothalamus ______
7) Thalamus ______
8) Basal ganglia ______
9) Corpus callosum ______
10) Cerebrum ______

A. Regulates visual and auditory reflexes
B. Regulates heart rate and respiration
C. Coordinates voluntary movement
D. Responsible for thinking and memory
E. Produces ADH and oxytocin
F. Connects the cerebral hemispheres
G. Regulates accessory movements
H. Four cavities within the brain
I. Regulates body temperature and eating
J. Regulates muscle tone and equilibrium
K. Is anterior to the medulla and helps regulate respiration
L. Regulates coughing and swallowing
M. Integrates sensations before relaying them to the cerebral cortex
N. Contain choroid plexuses that form cerebrospinal fluid
O. Regulates blood pressure
P. Encloses the cerebral aqueduct and helps maintain equilibrium
Q. Integrates the functioning of the autonomic nervous system
R. Suppresses unimportant sensations
S. Regulates the secretions of the anterior pituitary gland
T. Is the biological clock for the body's daily rhythms

2. Match the lobes of the cerebral cortex with the functional areas they contain.

Use each letter once. Three answer lines will have two correct letters.

1) Frontal lobes ______
2) Parietal lobes ______
3) Temporal lobes ______
4) Occipital lobes ______

A. General sensory areas for cutaneous sensations
B. Taste areas, which overlap the temporal lobes
C. Motor areas that initiate voluntary movement
D. Visual areas
E. Auditory areas
F. Motor speech area (left lobe only)
G. Olfactory areas

3. The following diagram shows a midsagittal section of the brain.

   Label the parts indicated.

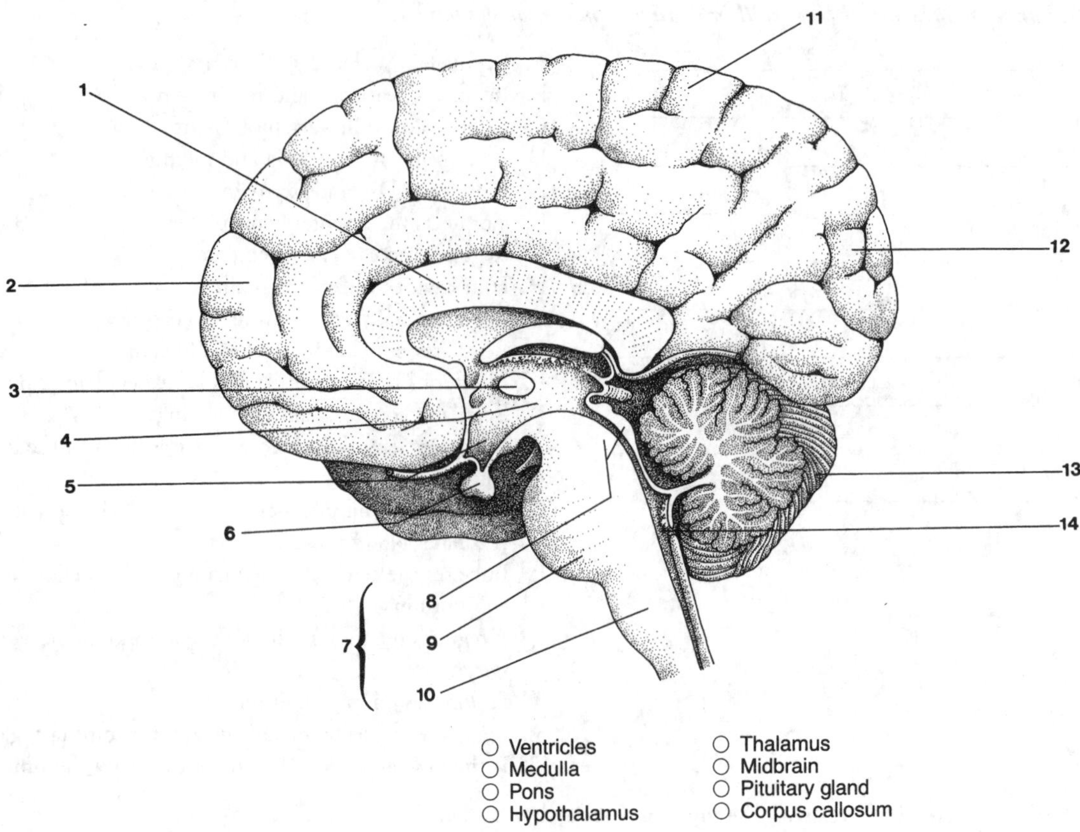

4. The following diagram shows the brain from the left side.

   Label the parts indicated.

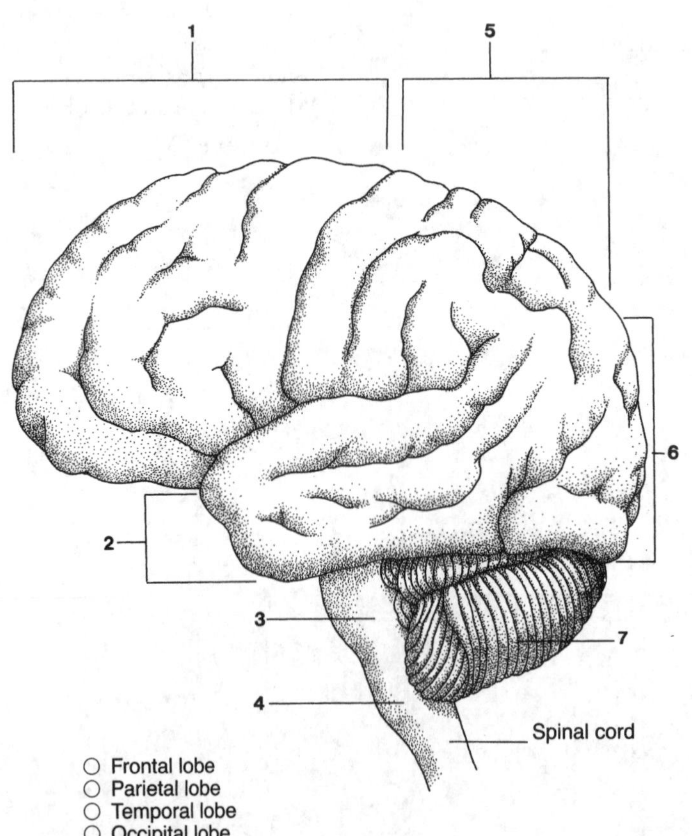

5. The gray matter on the surface of the cerebral hemispheres is called the ____________ and is made of the ____________ of neurons.

6. a) The two ventricles within the cerebral hemispheres are called ____________ ventricles.
   b) The ventricle between the cerebellum and the medulla-pons is called the ____________ ventricle.
   c) The ventricle within the hypothalamus and thalamus is called the ____________ ventricle.
   d) The tunnel through the midbrain called the ____________ connects the third to the fourth ventricle.

7. The following diagram shows the brain in a frontal (coronal) section.

   Label the parts indicated.

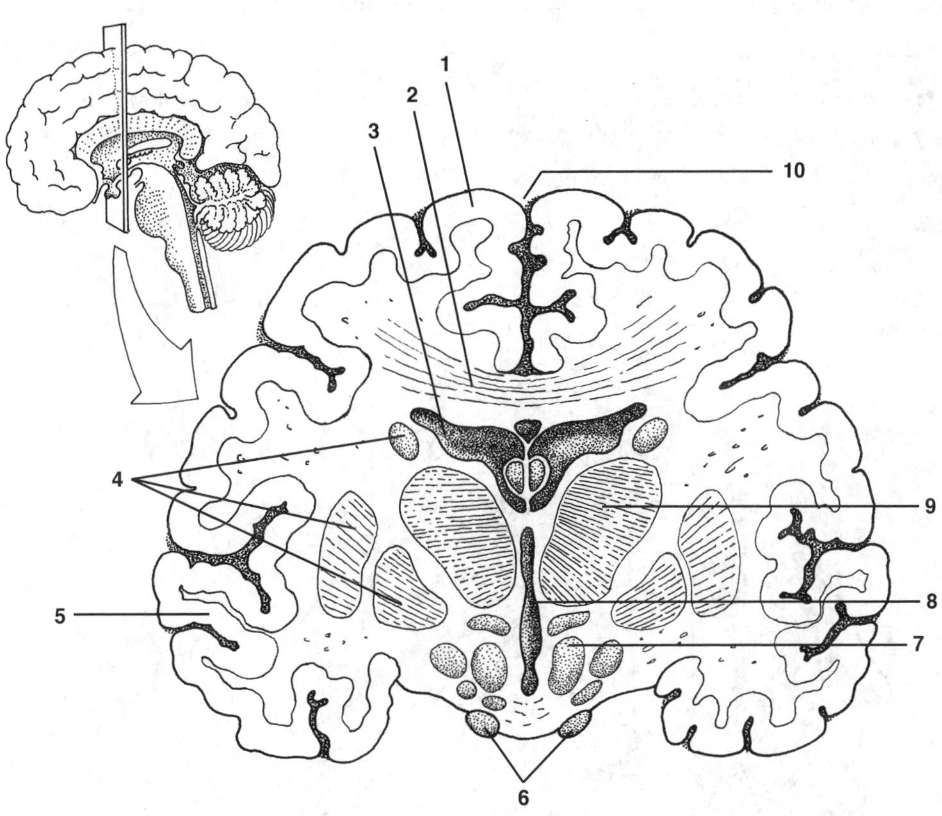

MENINGES AND CEREBROSPINAL FLUID

1. The meninges consist of how many layers of connective tissue? ____________

2. a) The outermost of the meninges is called the ____________.
   b) This layer lines the ____________ and ____________.
   c) The middle layer is called the ____________ membrane.
   d) The innermost layer is the ____________, which is on the surface of the ____________ and ____________.

3. The subarachnoid space is between the ____________ and the ____________ and contains ____________.

4. Cerebrospinal fluid is formed from blood plasma by capillary networks called ____________ that are found within the ____________ of the brain.

5. Cerebrospinal fluid is the circulating tissue fluid of the CNS. Its locations are the:

   1) ____________________ within the brain

   2) ____________________ around the brain

   3) ____________________ within the spinal cord

   4) ____________________ around the spinal cord

6. State the two functions of cerebrospinal fluid.

   1) ____

   2) ____

7. a) Cerebrospinal fluid is reabsorbed from the cranial subarachnoid space through the ____________________, into the blood in the ____________________.

   b) Normally, the rate of reabsorption is ____________________ (faster than, the same as, or slower than) the rate of production.

8. The following diagram shows the spinal cord and meninges.

   Label the parts indicated.

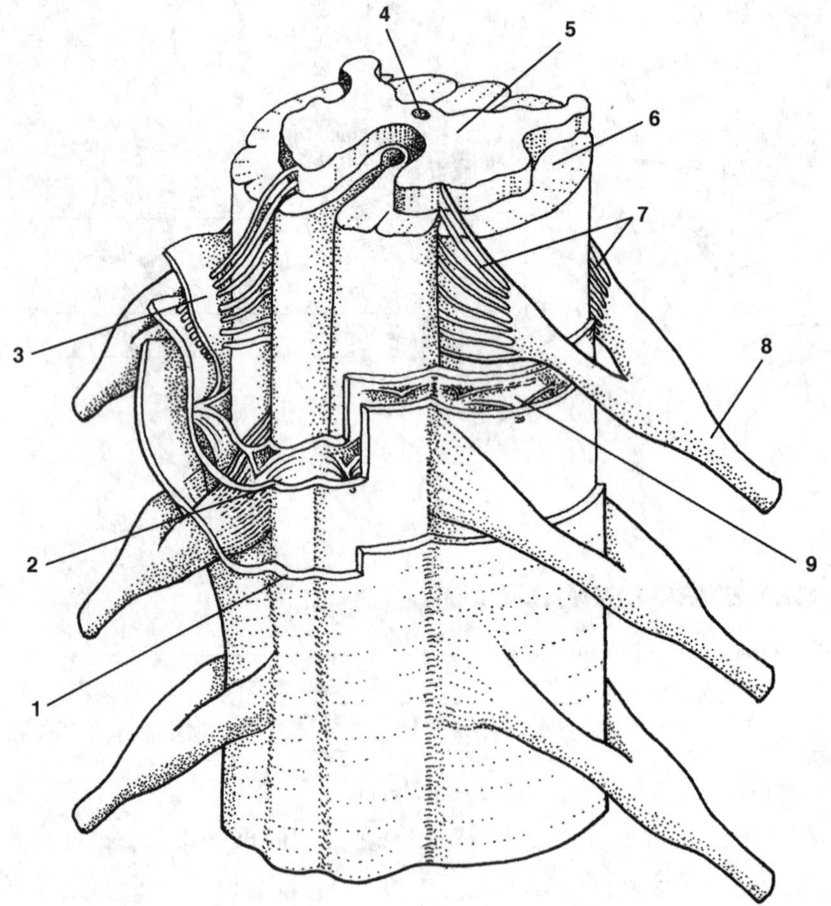

9. The following diagram shows the formation, circulation, and reabsorption of cerebrospinal fluid.

Label the parts indicated.

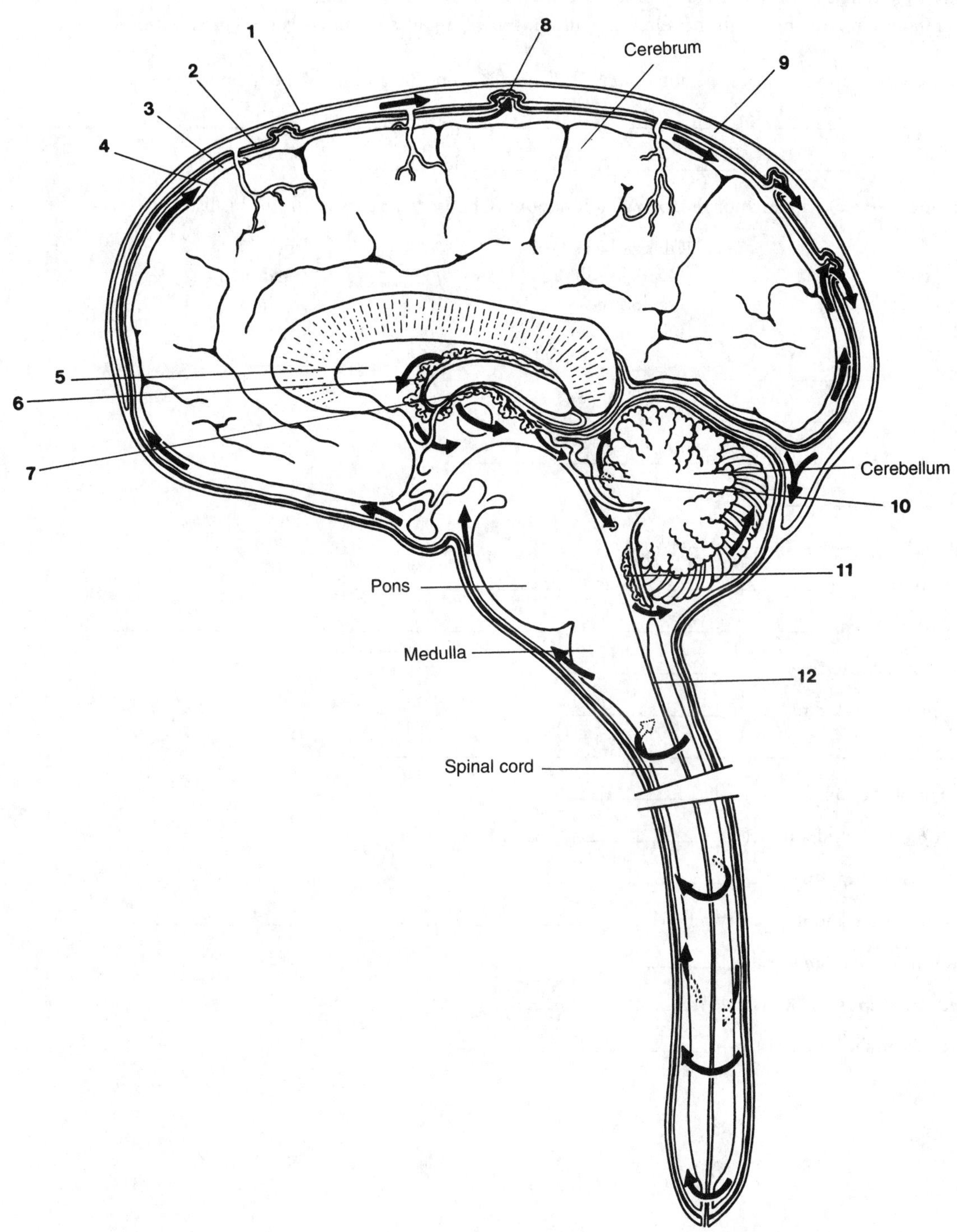

CRANIAL NERVES

The traditional mnemonic for remembering the names of the cranial nerves in order is: On Old Olympus Tiny Top A Finn And German Viewed A Hop. (Hops are vines with flowers that are used in making beer.)

A more contemporary mnemonic might be: On One Onion, Tiny Tomatoes Acquired For A Guest Vegetarian Are Heaped.

You may want to make up your own mnemonic; make it funny or even silly and it will be memorable.

O ________ O ________ O ________ T ________ T ________ A ________ F ________

A ________ G ________ V ________ A ________ H ________

1. Name the cranial nerve(s) with each of the following functions. Choose your answers from this list.

Abducens	Glossopharyngeal	Optic
Accessory	Hypoglossal	Trigeminal
Acoustic	Oculomotor	Trochlear
Facial	Olfactory	Vagus

1) Vision ________________
2) Hearing ________________
3) Taste ________________ and ________________
4) Smell ________________
5) Equilibrium ________________
6) Secretion of saliva ________________ and ________________
7) Movement of eyeball ________________, ________________, and ________________
8) Peristalsis of the intestine ________________
9) Decreasing the heart rate ________________
10) Speaking (larynx) ________________ and ________________
11) Movement of the tongue ________________
12) Contraction of shoulder muscles ________________
13) Constriction of pupil of eye ________________
14) Sensory in cardiac and respiratory reflexes ________________ and ________________
15) Sensation in the face and teeth ________________
16) Contraction of muscles of the face ________________
17) Contraction of chewing muscles ________________

2. The following diagram shows the cranial nerves by number and their destinations.

   Name each nerve.

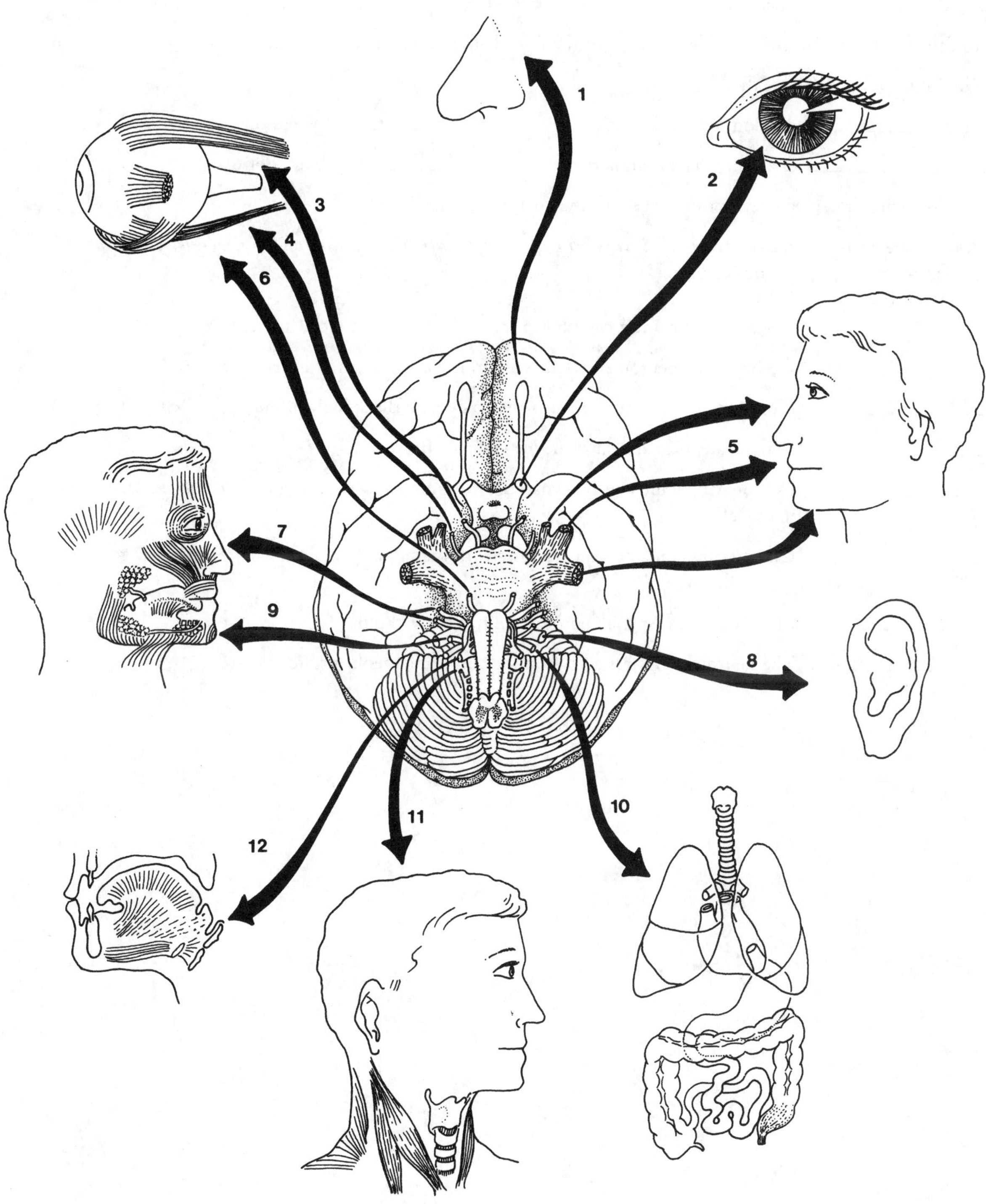

THE AUTONOMIC NERVOUS SYSTEM (ANS)

1. The autonomic nervous system (ANS) consists of motor neurons to visceral effectors, which are

   ____________________ muscle, ____________________ muscle, and ____________________.

2. The part of the brain that integrates the functioning of the ANS is the ____________________.

3. The two divisions of the ANS are the:

   a) ____________________ division, which dominates in ____________________ situations.

   b) ____________________ division, which dominates in ____________________ situations.

4. Each of the following statements refers to an aspect of the anatomy of the ANS.

   On the line before each statement, write "S" if it is true of the sympathetic division of the nervous system or "P" if it is true of the parasympathetic division.

   1) ____________________ Preganglionic neuron cell bodies are in the brain and sacral spinal cord

   2) ____________________ Preganglionic neuron cell bodies are in the thoracic and lumbar spinal cord

   3) ____________________ Most ganglia are located in two chains outside the spinal column

   4) ____________________ Ganglia are located near or in the visceral effector

   5) ____________________ One preganglionic neuron synapses with only a few postganglionic neurons, which all go to one effector

   6) ____________________ One preganglionic neuron synapses with many postganglionic neurons, which go to many effectors

   7) ____________________ All neurons release the neurotransmitter acetylcholine

   8) ____________________ The neurotransmitters acetylcholine and norepinephrine are released

5. The following diagram illustrates the pathways of the ANS.

Label the parts indicated.

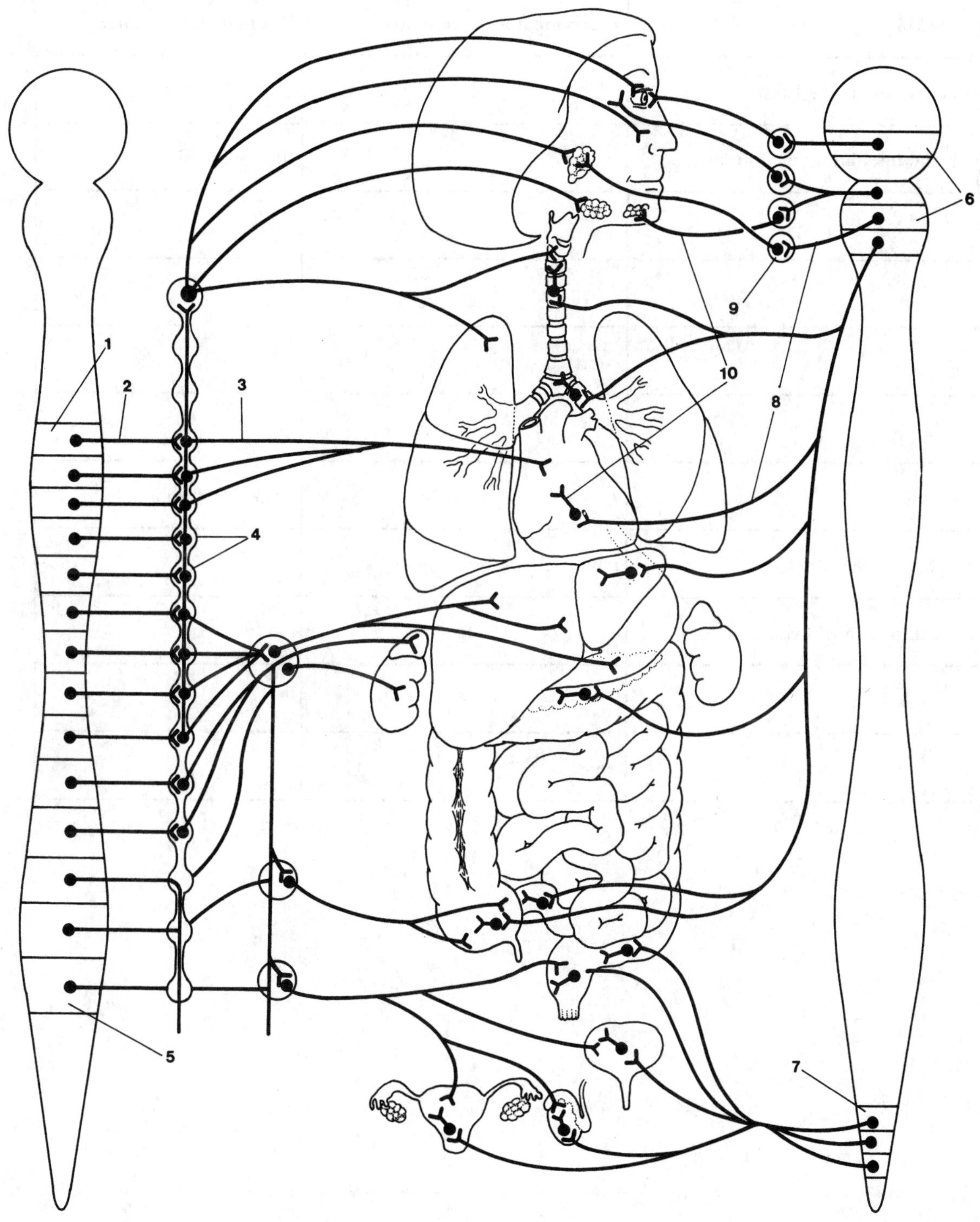

6. Complete the following chart by supplying the proper sympathetic and parasympathetic responses for the visceral effectors listed. Several have been included to get you started.

Visceral Effector	Parasympathetic Response	Sympathetic Response
1. Stomach and intestines (glands)		
2. Stomach and intestines (smooth muscle)		slow peristalsis
3. Heart		
4. Iris		
5. Urinary bladder	contracts	
6. Bronchioles		
7. Salivary glands		decrease secretion
8. Blood vessels in skeletal muscle	none	
9. Blood vessels in skin and viscera		
10. Sweat glands		
11. Liver	none	

CROSSWORD PUZZLE

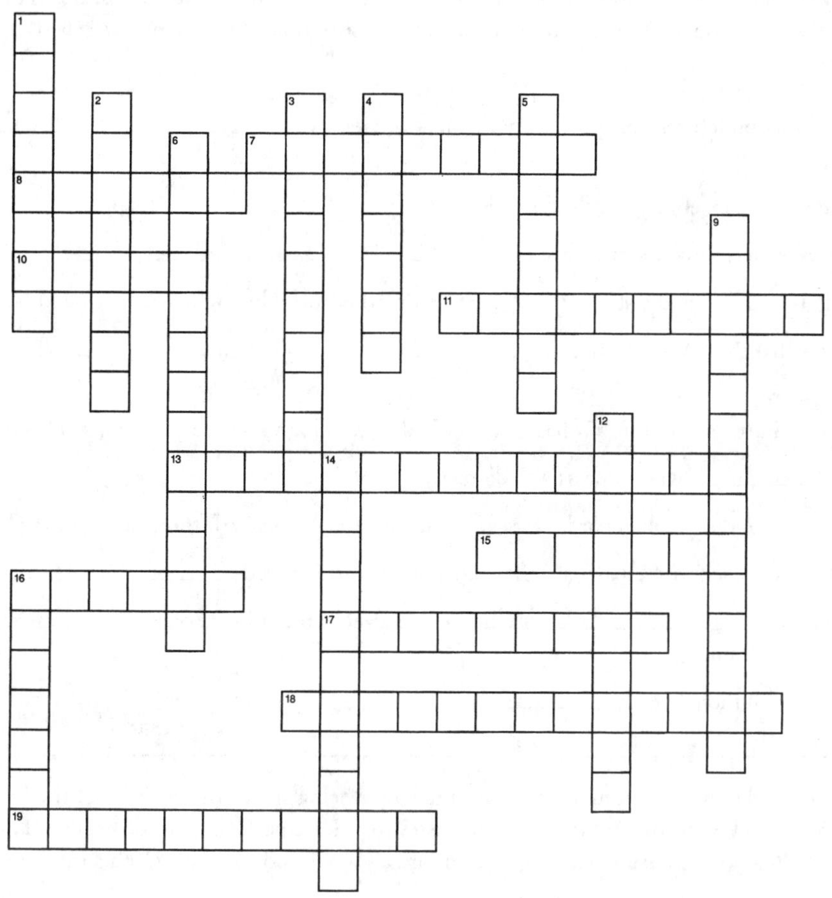

ACROSS

7. Specialized cells found only in the brain and spinal cord
8. Involuntary response to a stimulus
10. Group of axons and/or dendrites of many neurons
11. ________________ nervous system; consists of cranial nerves and spinal nerves
13. Division of the ANS that dominates in relaxed situations
15. ________________ nervous system, consists of the brain and spinal cord
16. ________________ nerves, 31 pairs
17. ________________ nervous system; consists of motor neurons to visceral effectors
18. Capillary network in each ventricle (two words)
19. "Horse's tail" spinal nerves (two words)

DOWN

1. Motor or ________________ neurons (syn.)
2. Sensory or ________________ neurons (syn.)
3. Nuclei and cytoplasm of the Schwann cells form this
4. ________________ nerves, 12 pairs; emerge from the brain
5. Sensory neurons from receptors in internal organs
6. Tissue of fluid of the CNS
9. Connects the cerebral hemispheres (two words)
12. Four cavities within the brain
14. Division of the ANS that dominates in stress situations
16. Sensory neurons from receptors in skin, skeletal muscles, and joints

CLINICAL APPLICATIONS

1. a) Mrs. C brings her 5-year-old son to the hospital because he has a high fever and a severe headache. The doctor suspects that these symptoms are being caused by a bacterial infection of the meninges called

   ____________________.

   b) The doctor orders a lumbar puncture. This procedure involves the removal of ____________________ from the meningeal sac.

   c) If meningitis were present, the cerebrospinal fluid would be ____________________ in appearance and on examination and culture would show the presence of ____________________ and ____________________.

2. a) Mr. H is 70 years old and has suffered a stroke (CVA). He can neither move his right arm nor bend his head to the right. This indicates that the CVA is in the ____________________ lobe on the ____________________ side of the brain.

   b) Mr. H understands everything said to him but has great difficulty speaking to answer questions. This indicates that the ____________________ was also partially damaged.

3. a) Ms. J was in a car accident, and her spinal cord was severed at the level of the 12th thoracic vertebra. The spinal cord reflexes below this level, such as the patellar reflex, are absent; this is called ____________________.

   b) Ms. J will also have loss of ____________________ below the level of the injury because impulses cannot get from the lower body to the brain.

   c) What other consequence will there be? ____________________

   d) Explain why. ____

4. a) Mr. W is a construction worker who is brought to the hospital after being struck on the back of the head by falling debris. He is awake and oriented, and the following assessments are performed to determine if any brain damage has occurred: Pulse, blood pressure, and respirations are all within normal limits. This indicates that the

   ____________________ is functioning properly.

   b) Vision is normal. This indicates that the ____________________ lobes have not been damaged.

   c) Coordination is good, and Mr. W can touch his nose or ear when asked to do so. This indicates that the

   ____________________ is functioning normally.

MULTIPLE CHOICE TEST #1

Choose the correct answer for each question.

1. The part of the brain that initiates voluntary movement is the:
   a) cerebellum b) parietal lobes c) frontal lobes d) hypothalamus

2. Neurons that carry impulses from receptors to the CNS are called:
   a) sensory b) mixed c) motor d) efferent

3. The cranial nerve that decreases heart rate is the:
   a) vagus b) abducens c) accessory d) hypoglossal

4. The part of the brain that regulates heart rate and blood pressure is the:
   a) cerebrum b) hypothalamus c) cerebellum d) medulla

5. Cerebrospinal fluid is formed by:
   a) dura mater b) choroid plexuses c) subarachnoid spaces d) pia mater

6. The gray matter on the surface of the cerebral hemispheres is called the:
   a) cerebral cortex b) interneurons c) nerve tract d) gray stuff

7. Neurons are electrically insulated by the presence of the:
   a) white matter b) myelin sheath c) astrocytes d) interneurons

8. The part of the brain that regulates body temperature and the autonomic nervous system is the:
   a) thalamus b) cerebrum c) medulla d) hypothalamus

9. During depolarization of a neuron, sodium ions:
   a) stay outside the cell b) rush out of the cell c) stay inside the cell d) rush into the cell

10. The spinal nerve root that is made of sensory neurons is the:
   a) internal root b) external root c) ventral root d) dorsal root

11. The part of a neuron that carries impulses away from the cell body is the:
   a) myelin sheath b) dendrite c) axon d) Schwann cell

12. At a synapse, a nerve impulse is carried by:
   a) depolarization b) the myelin sheath c) a neurotransmitter d) repolarization

13. Which of these is NOT a sympathetic response?
   a) decreased peristalsis c) decreased heart rate
   b) dilation of the bronchioles d) dilation of the pupils

14. The lobes of the cerebral cortex that contain the areas for cutaneous sensation are the:
   a) frontal lobes b) parietal lobes c) temporal lobes d) occipital lobes

15. Within the spinal cord, the cell bodies of motor neurons are contained within the:
   a) dorsal root ganglion b) white matter c) ventral root d) gray matter

16. A reflex is:
   a) an involuntary response to a stimulus c) a sensation to be interpreted
   b) a conscious decision to maintain homeostasis d) a stimulus to be felt

17. Keeping the body upright is the purpose of:
   a) brain reflexes b) stretch reflexes c) flexor reflexes d) cerebrum reflexes

18. The two cerebral hemispheres are connected by the:
   a) frontal lobes b) corpus callosum c) hypothalamus d) medulla

19. The correct description of the location of a layer of the meninges is:
   a) middle—dura mater c) outer—arachnoid membrane
   b) inner—pia mater d) middle—pia mater

20. Which part of the brain regulates muscle tone and coordination?
   a) frontal lobes b) medulla c) cerebellum d) hypothalamus

21. The cranial nerve for hearing and equilibrium is the:
   a) abducens b) trochlear c) acoustic d) vagus

22. Which of the following is NOT a parasympathetic response?
   a) decreased peristalsis c) contraction of the urinary bladder
   b) constriction of the pupils d) increased secretion of saliva

23. Which spinal nerve group is matched with its correct number of pairs?
   a) cervical—7 b) thoracic—5 c) lumbar—5 d) sacral—12

24. The descending tracts of the spinal cord are:
   a) made of white matter and carry impulses away from the brain
   b) made of white matter and carry impulses toward the brain
   c) made of gray matter and carry impulses away from the brain
   d) made of gray matter and carry impulses toward the brain

25. The cerebral cortex is made of the same parts of neurons as is the:
   a) white matter of the spinal cord c) corpus callosum
   b) gray matter of the spinal cord d) ventral roots of the spinal nerves

26. Preventing prolonged contact with harmful stimuli is the function of:
   a) the cerebrum b) stretch reflexes c) the cerebellum d) flexor reflexes

27. The blood–brain barrier is formed by anatomic features of the:
   a) brain capillaries and astrocytes c) microglia and cranial venous sinuses
   b) meninges and cerebrospinal fluid d) subarachnoid space and choroid plexus

28. The myelin sheath of CNS neurons is made by:
    a) astrocytes b) Schwann cells c) oligodendrocytes d) microglia

29. The visual areas of the brain are in the:
    a) frontal lobes b) parietal lobes c) temporal lobes d) occipital lobes

30. The parts of the brain that regulate accessory movements (such as gestures when speaking) are the:
    a) autonomic ganglia b) basal ganglia c) frontal lobes d) temporal lobes

MULTIPLE CHOICE TEST #2

Read each question and the four answer choices carefully. When you have made a choice, follow the instructions to complete your answer.

1. Which statement is NOT true of cerebrospinal fluid (CSF)?
   a) CSF is formed by choroid plexuses in the ventricles of the brain.
   b) CSF exchanges nutrients and wastes between the CNS and the blood.
   c) CSF is reabsorbed into the blood in the cranial arteries.
   d) CSF is a shock absorber around the brain and spinal cord.

   Reword your choice to make it a correct statement.

   ___

2. Which part of the brain is NOT paired with its proper function?
   a) pons—helps regulate respiration
   b) thalamus—suppresses unimportant sensations
   c) basal ganglia—regulate accessory movements
   d) cerebellum—initiates voluntary movement

   For your choice, state a correct function.

   ___

3. Which statement is NOT true of peripheral nerves and neurons?
   a) Motor neurons carry impulses from the CNS to effectors.
   b) A mixed nerve contains both sensory and motor neurons.
   c) Peripheral nerves are in the arms, legs, and trunk of the body.
   d) Sensory neurons may also be called efferent neurons.

   Reword your choice to make it a correct statement.

   ___

4. Which lobe of the cerebral cortex is NOT paired with its correct function?
   a) frontal lobe—initiates voluntary movement
   b) parietal lobe—cutaneous sensory area
   c) occipital lobe—hearing area
   d) temporal lobe—olfactory area

   For your choice, state a correct function.

   ___

5. Which statement is NOT true of a reflex arc?
   a) The effector responds and carries out the reflex act.
   b) The stimulus is detected by motor neurons.
   c) The CNS contains one or more synapses.
   d) Sensory neurons carry impulses from receptors to the CNS.

   Reword your choice to make it a correct statement.

   ___

6. Which part of the brain is NOT paired with its correct function?
   a) midbrain—taste and smell reflexes
   b) hypothalamus—regulates secretions of the anterior pituitary gland
   c) medulla—regulates respiration and heart rate
   d) cerebellum—regulates coordination and muscle tone

   For your choice, state a correct function.

   ___

7. Which statement is NOT true of the spinal cord?
   a) The gray matter is shaped like the letter H and contains cell bodies of motor neurons and interneurons.
   b) The white matter consists of myelinated neurons that form ascending and descending tracts.
   c) The spinal cord extends from the foramen magnum to the disc between the first and second sacral vertebrae.
   d) The spinal cord is protected from mechanical injury by the backbone and by cerebrospinal fluid.

   Reword your choice to make it a correct statement.

   ___

8. Which statement is NOT true of the spinal nerves?
   a) There are 12 thoracic pairs and 8 cervical pairs.
   b) Each spinal nerve has two roots: the dorsal root and the ventral, or sensory, root.
   c) The cauda equina refers to the lumbar and sacral spinal nerves that extend below the end of the spinal cord.
   d) The neck, shoulder, and arm are supplied by cervical spinal nerves.

   Reword your choice to make it a correct statement.

   ___

9. Which statement is NOT true of neuron structure?
   a) The myelin sheath provides electrical insulation for neurons.
   b) The axon carries impulses away from the cell body.
   c) The nucleus is located within the end of the axon.
   d) The neurolemma of the PNS is formed by Schwann cells and is necessary for regeneration of damaged axons or dendrites.

   Reword your choice to make it a correct statement.

   ___

10. Which statement is NOT true of the meninges?
    a) The cranial dura mater has two layers and contains cranial venous sinuses.
    b) The pia mater is the outermost layer and is on the surface of the brain and spinal cord.
    c) The subarachnoid space is between the arachnoid membrane and the pia mater and contains cerebrospinal fluid.
    d) Arachnoid villi are the pathway for reabsorption of CSF back into the blood.

    Reword your choice to make it a correct statement.

    ___

11. Which cranial nerve pair is NOT paired with its correct function?
    a) Vagus nerves—increasing peristalsis and decreasing heart rate
    b) Facial and glossopharyngeal nerves—secretion of saliva
    c) Hypoglossal nerves—movement of the tongue
    d) Olfactory nerves—sense of taste

    For your choice, state the correct function.

    ___

12. Which statement is NOT true of the autonomic nervous system?
    a) The sympathetic division dominates in stress situations.
    b) The parasympathetic division causes constriction of the bronchioles and decreased peristalsis.
    c) The sympathetic division causes the liver to change glycogen to glucose.
    d) The parasympathetic division includes four pairs of cranial nerves.

    Reword your choice to make it a correct statement.

    ____

13. Which statement is NOT true of the autonomic nervous system?
    a) The sympathetic division increases heart rate.
    b) The parasympathetic division dominates in relaxed situations.
    c) The sympathetic division contains the vagus nerves.
    d) The parasympathetic ganglia are near or in the visceral effectors.

    Reword your choice to make it a correct statement.

    ____

14. Which statement is NOT true of the nerve impulse?
    a) When not carrying an impulse, a neuron membrane has a positive charge outside and a negative charge inside.
    b) During depolarization, potassium ions rush into the cell.
    c) During depolarization, the neuron membrane has a negative charge outside and a positive charge inside.
    d) During repolarization, potassium ions rush out of the cell.

    Reword your choice to make it a correct statement.

    ____

MULTIPLE CHOICE TEST #3

Each question is a series of statements concerning a topic in this chapter. Read each statement carefully and select all of the correct statements.

1. Which of the following statements are true of nerves and nerve impulses?
    a) Depolarization is the result of the exit of K^+ ions from a neuron.
    b) A peripheral nerve is made of many neurons and Schwann cells.
    c) A nerve impulse is electrical when it travels along a neuron membrane.
    d) Nerve impulse transmission becomes chemical at synapses.
    e) A mixed nerve contains afferent neurons and interneurons.
    f) An action potential is the potential for Na^+ ions to move.

2. Which of the following statements are true of the spinal cord and spinal cord reflexes?
    a) Spinal nerves are mixed nerves, with both sensory and motor neurons.
    b) Motor neuron cell bodies are in the gray matter of the spinal cord.
    c) The superior end of the spinal cord is at the foramen magnum.
    d) The brain ensures that a flexor reflex is rapid.
    e) The 10 pairs of cervical spinal nerves supply the head and neck.
    f) The motor root of a spinal nerve is the dorsal root.
    g) A reflex arc requires both sensory and motor neurons.
    h) Stretch reflexes protect us from sharp objects.

3. Which of the following statements are true of the brain?
   a) The respiratory centers are in the hypothalamus.
   b) The thalamus regulates body temperature.
   c) The visual areas are in the frontal lobes, right behind the eyes.
   d) Coordination of voluntary movement is regulated by the midbrain.
   e) The corpus callosum connects the hemispheres of the cerebellum.
   f) Appetite and water balance are regulated by the medulla.
   g) The basal ganglia initiate voluntary movement.
   h) The auditory areas are in the occipital lobes.
   i) The frontal lobes contain the areas for the cutaneous senses.
   j) The temporal lobes contain areas for smell and taste.
   k) The cerebral cortex is the internal gray matter of the cerebrum.
   l) The pons assists the medulla with regulation of heart rate.

4. Which of the following statements are true of the cranial nerves?
   a) The sense of taste is a function of the hypoglossal nerves.
   b) The vagus nerves help increase intestinal peristalsis.
   c) Movement of the eyeball is a function of the abducens and optic nerves.
   d) Sensation in teeth is a function of the trigeminal nerves.
   e) Shoulder muscles are innervated by the trochlear nerves.
   f) Hearing and equilibrium are both functions of the acoustic nerves.
   g) The sense of smell requires the olfactory nerves in the upper oral cavity.
   h) Movement of the tongue is a function of the glossopharyngeal nerves.

5. Which of the following statements are true of the meninges and cerebrospinal fluid?
   a) The subarachnoid space is between the arachnoid and the pia mater.
   b) All three meninges are made of fibrous connective tissue.
   c) Cerebrospinal fluid (CSF) is produced in the ventricles of the brain.
   d) Cranial CSF circulates, but spinal CSF remains stationary.
   e) CSF is actually the tissue fluid of the central nervous system.
   f) The meninges are a shock absorber for the CNS because they are so thick.
   g) The reabsorption of CSF should equal its rate of production.
   h) The site of a lumbar puncture is the meningeal sac of the thoracic vertebrae.

6. Which of the following statements are true of the autonomic nervous system?
   a) The parasympathetic division increases heart rate.
   b) The sympathetic division has two neurotransmitters, norepinephrine and epinephrine.
   c) Ganglia near the visceral effectors are sympathetic.
   d) The parasympathetic division is the craniosacral division.
   e) The sympathetic division slows digestion.
   f) The vagus nerves are part of the sympathetic division.
   g) Dilation of the pupils is a parasympathetic response.
   h) Bronchial dilation is a sympathetic response.

Chapter 9

The Senses

The senses provide us with information about the constant changes that take place in the external and internal environments. This chapter describes the organs that contain sensory receptors and includes the role of the nervous system in sensation.

SENSORY PATHWAY

1. Name the four parts of a sensory pathway.

   1) ____________________ 3) ____________________

   2) ____________________ 4) ____________________

2. a) The purpose (or function) of receptors is to detect ____________ and then to generate ____________.

   b) Another way to say this is that receptors change the energy of a ____________ into the energy of ____________.

3. Sensory neurons transmit impulses from ____________ to the ____________.

4. Sensory tracts consist of white matter in the ____________ that transmits sensory impulses to a specific part of the ____________.

5. Most sensory areas are located in the ____________ of the brain. These areas feel, project, and interpret sensations.

CHARACTERISTICS OF SENSATIONS

1. Match each characteristic of sensations with the proper description.

   Use each letter once.

   1) Projection ____________
   2) Intensity ____________
   3) Contrast ____________
   4) Adaptation ____________
   5) After-image ____________

   A. Some sensations are felt more strongly than are others
   B. The effect of a current sensation may be exaggerated when the brain compares it with a previous sensation
   C. A stimulus is still present, but we become unaware of it
   D. The sensation seems to come from the area where the receptors were stimulated
   E. The stimulus stops, but we remain aware of the sensation

CUTANEOUS SENSES

1. a) State the location of the receptors for the cutaneous senses (be specific).

____

   b) These receptors give us information about ____________________.

2. a) For the senses of ____________________, ____________________, ____________________, and ____________________, the receptors are free nerve endings.

   b) For the senses of ____________________ and ____________________, the receptors are encapsulated nerve endings.

3. The sensory areas for the cutaneous senses are in the ____________________ lobes of the cerebrum.

4. Referred pain means that pain that originates in an ____________________ may be felt as ____________________ pain.

5. Give a specific example of referred pain by naming the organ where the pain originates: ____________________, and the area where the pain is felt: ____________________.

MUSCLE SENSE

1. The receptors for muscle sense are called ____________________, and they detect the ____________________ of muscles.

2. a) Name the lobe of the cerebral cortex that contains the sensory area for conscious muscle sense.

   ____

   b) Name the part of the brain that uses unconscious muscle sense to coordinate voluntary movement.

   ____

3. Briefly explain the importance of muscle sense.

____

____

HUNGER AND THIRST

1. Match each of these visceral sensations with the proper descriptive statements.

   Use each letter once. Each answer line will have three correct letters.

   1) Hunger ____________________

   2) Thirst ____________________

   A. Receptors in the hypothalamus detect changes in the water content of the body
   B. Receptors in the hypothalamus detect changes in blood nutrient levels and GI hormones
   C. Projection is to the stomach
   D. Projection is to the mouth and pharynx
   E. Adaptation does occur
   F. Adaptation does not occur

TASTE AND SMELL

1. Match the senses of taste and smell with the proper descriptive statements.

   Use each letter once. One answer line will have six correct letters, and the other will have four correct letters.

   1) Taste ______________________

   2) Smell ______________________

   A. Chemoreceptors are in the upper nasal cavities
   B. Chemoreceptors are in taste buds on the tongue
   C. Impulses are interpreted in a sensory area in the temporal lobes only
   D. Impulses are interpreted in a sensory area that overlaps the temporal-parietal lobes
   E. A cranial nerve that is sensory is the facial nerve
   F. A cranial nerve that is sensory is the olfactory nerve
   G. A cranial nerve that is sensory is the glossopharyngeal nerve
   H. The receptors detect chemicals in solution in saliva
   I. The receptors detect vaporized chemicals
   J. Much of what we think of as this sense is really the other sense

2. The following diagram depicts a midsagittal section of the head.

   Label the parts indicated.

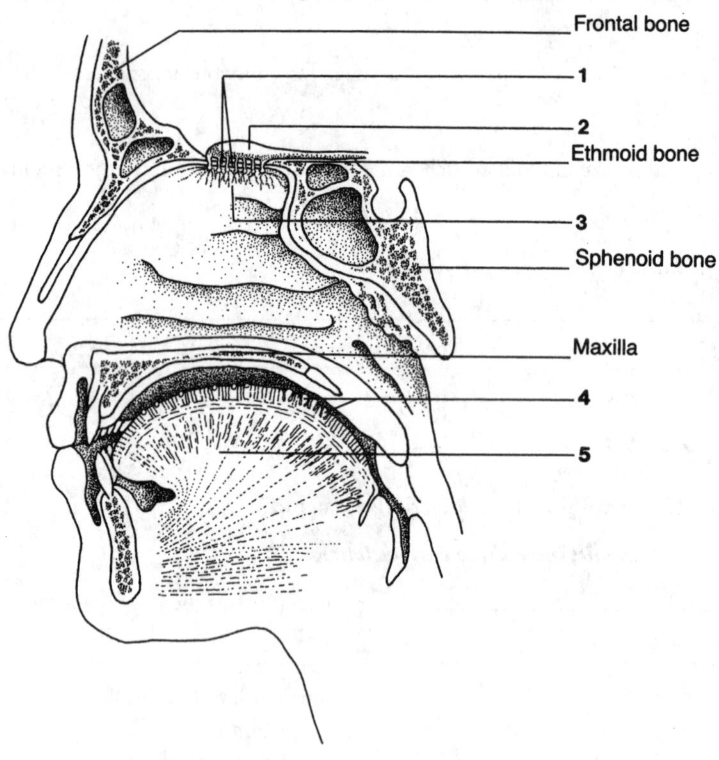

THE EYE—STRUCTURES OUTSIDE THE EYEBALL

1. Match each structure or substance with the proper descriptive statement.

Use each letter once.

1) Lacrimal glands ____________________
2) Eyelids ____________________
3) Conjunctiva ____________________
4) Eyelashes ____________________
5) Nasolacrimal duct ____________________
6) Lysozyme ____________________
7) Tears ____________________

A. Takes tears to the nasal cavity
B. Mostly water, to wash the front of the eyeball
C. Small hairs that help keep dust out of the eyes
D. Within the orbits; secrete tears
E. An enzyme in tears that inhibits the growth of bacteria on the front of the eyeball
F. A thin membrane that lines the eyelids and covers the whites of the eyes
G. Contain skeletal muscle; spread tears across the front of the eyeball

2. The following diagram is an anterior view of the orbit of the eye.

Label the parts indicated.

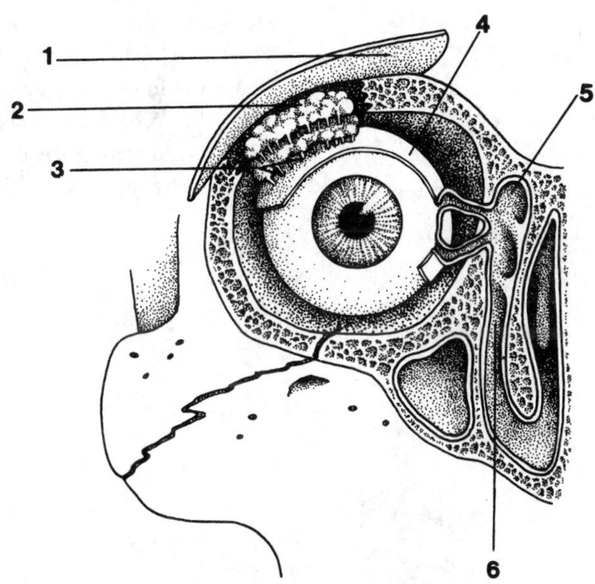

3. a) Also outside the eyeball are the extrinsic muscles of the eye. These are attached to the surface of the eyeball and to the ____________________.

   b) How many of these muscles are there for each eye? ____________________

   c) Describe the general function of these muscles. ____________________

4. The cranial nerves involved in movement of the eyeball are the ____________________, the ____________________, and the ____________________.

THE EYE—EYEBALL

1. a) The eyeball has three layers. The outer layer is the ____________________, which is made of ____________________ tissue.

   b) The middle layer is the ____________________.

   c) The inner layer is the ____________________, which is made of nerve tissue.

2. Match each structure or substance of the eyeball with the proper descriptive statement.

Use each letter once.

1) Optic nerve ________
2) Rods ________
3) Cones ________
4) Optic disc ________
5) Choroid layer ________
6) Ciliary body ________
7) Cornea ________
8) Extrinsic muscles ________
9) Iris ________
10) Lens ________
11) Fovea ________
12) Posterior cavity ________
13) Anterior cavity ________
14) Aqueous humor ________
15) Vitreous humor ________
16) Sclera ________
17) Retina ________
18) Canal of Schlemm ________

A. Contains a dark blue pigment to absorb light and prevent glare within the eyeball
B. The only adjustable part of the light-refracting system
C. The receptors that detect color
D. The part of the retina that contains only cones
E. Regulates the size of the pupil
F. Formed by ganglion neurons, transmits impulses from the retina to the brain
G. Between the cornea and the lens, contains aqueous humor
H. The transparent part of the sclera that refracts light
I. The layer of the eyeball that contains the visual receptors
J. Small veins that reabsorb aqueous humor
K. Between the lens and the retina; contains vitreous humor
L. The area of the retina where the optic nerve passes through; no rods or cones are present
M. Helps hold the retina in place
N. The white of the eye
O. The tissue fluid of the eye; nourishes the lens and cornea
P. Move the eyeball side to side or up and down
Q. Contracts to change the shape of the lens
R. Contain rhodopsin and detect the presence of light

3. The following diagram is a section through the eyeball.

Label the parts indicated.

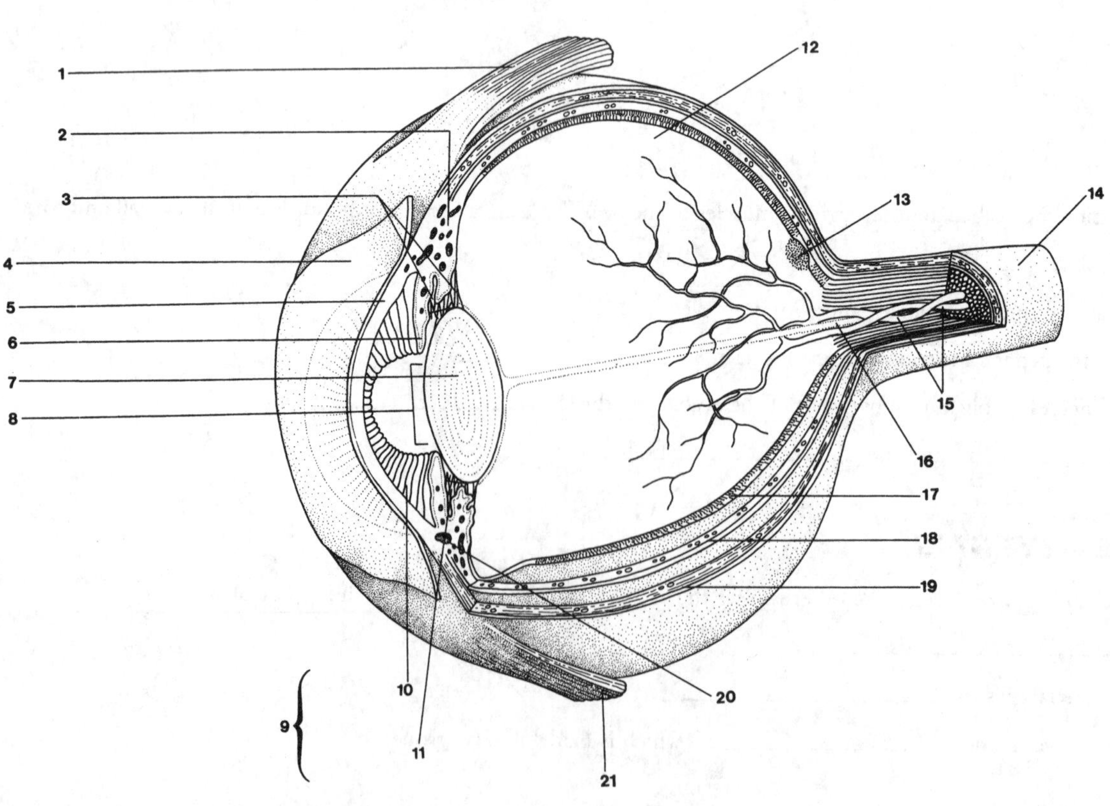

4. Name the following nerves that are involved in vision.
   1) The sensory nerve for sight ____________________
   2) The cranial nerve for constriction of the pupil ____________________
   3) The type of nerves that dilate the pupil ____________________
5. a) The lobes of the cerebral cortex that contain the sensory areas for vision are the ____________________ lobes.
   b) These visual areas put together the slightly different images from each eye to create one image. This is called ____________________ vision.
   c) Since the images on the retinas are upside down, the visual areas also turn the images ____________________.

SUMMARY OF VISION

1. Number the following in the proper sequence as they are involved in the process of vision.

Light rays → _____1_____ cornea

____________________ vitreous humor

____________________ optic nerve

____________________ aqueous humor

____________________ occipital lobe

____________________ lens

____________________ retina

THE EAR

1. The ear contains the receptors for the senses of ____________________ and ____________________.
2. Name the three major portions of the ear. The ____________________, the ____________________, and the ____________________

OUTER EAR

1. a) The auricle, or pinna, is made of skin supported by ____________________.
   b) Does the auricle have a significant function for humans? ____________________ Explain why or why not.
   ____
2. a) The ear canal is a tunnel into the ____________________ bone.
   b) Name the glands that are found in the dermis of the ear canal. ____________________
   c) What do these glands produce? ____________________

MIDDLE EAR

1. The middle ear is a cavity in the ________________ bone that contains ________________ (air or fluid).

2. Match each middle ear structure with its proper function.

   Use each letter once.

   1) Eardrum ________________
   2) Malleus ________________
   3) Eustachian tube ________________
   4) Stapes ________________
   5) Incus ________________

   A. Permits air to enter or leave the middle ear cavity to equalize pressure on both sides of the eardrum
   B. At the end of the ear canal; vibrates when sound waves strike it
   C. Transmits vibrations from the incus to the oval window of the inner ear
   D. Transmits vibrations from the malleus to the stapes
   E. Transmits vibrations from the eardrum to the incus

3. The following diagram depicts the outer, middle, and inner ear.

   Label the parts indicated.

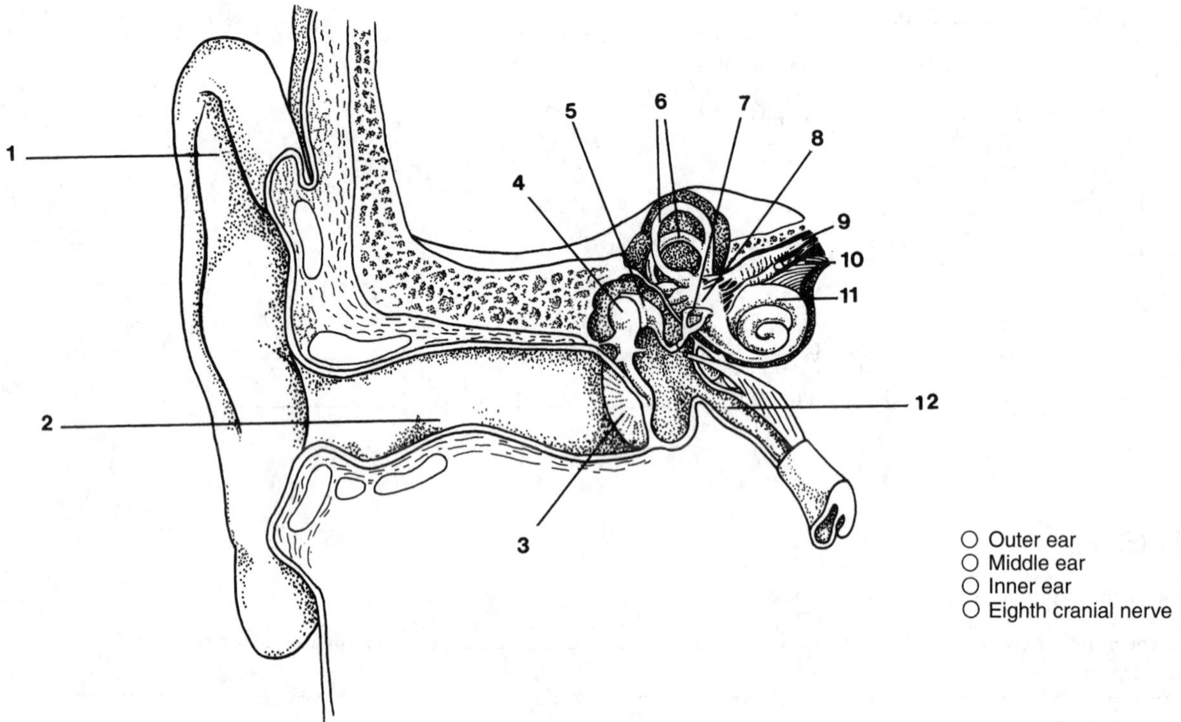

INNER EAR

1. The inner ear is a cavity in the ________________ bone that is lined with ________________ and filled with ________________ (air or fluid).

2. Within the inner ear are the structures that contain the receptors for hearing and for equilibrium. Name these structures.

   1) ________________ contain receptors for the equilibrium of motion
   2) ________________ contain receptors for hearing
   3) ________________ contain receptors for equilibrium of position

3. a) All of these receptors are called hair cells. Each group of hair cells responds to a different type of stimulus, bends, and generates ____________________.
   b) Name the cranial nerve that carries these impulses to the brain. ____________________
4. a) In the cochlea, the hair cells are part of the ____________________.
   b) Vibrations of the ____________________ within the cochlea bend the hair cells, which generate nerve impulses.
   c) The 8th cranial nerve transmits these impulses to the auditory areas in the ____________________ lobes of the cerebrum.
5. a) The utricle and saccule are two membranous sacs located in an area called the ____________________.
   b) Their hair cells are embedded in a gelatinous membrane that contains crystals called ____________________.
   c) Their hair cells bend in response to the pull of ____________________ as the position of the ____________________ changes.
6. a) In each inner ear, there are ____________________ semicircular canals.
   b) Their hair cells bend in response to ____________________.
7. a) The nerve impulses from the utricle and saccule and the semicircular canals are transmitted to these parts of the brain: the ____________________ and the ____________________, which are concerned with the maintenance of equilibrium at a subconscious level.
   b) And to the ____________________, which is concerned with awareness of motion or changes in position.
8. The following diagram depicts the inner ear.

Label the parts indicated.

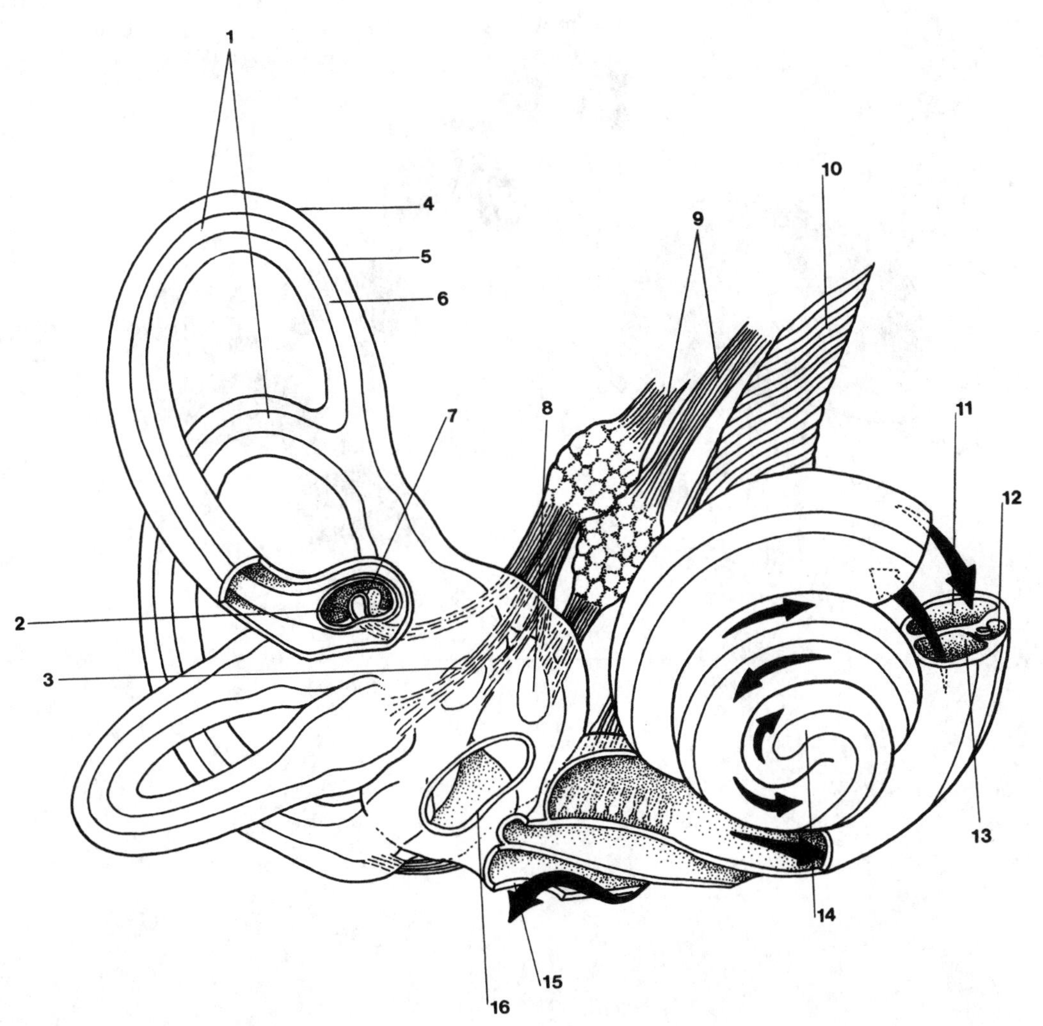

9. The following diagrams depict the parts of the inner ear concerned with equilibrium.

   Label the parts indicated.

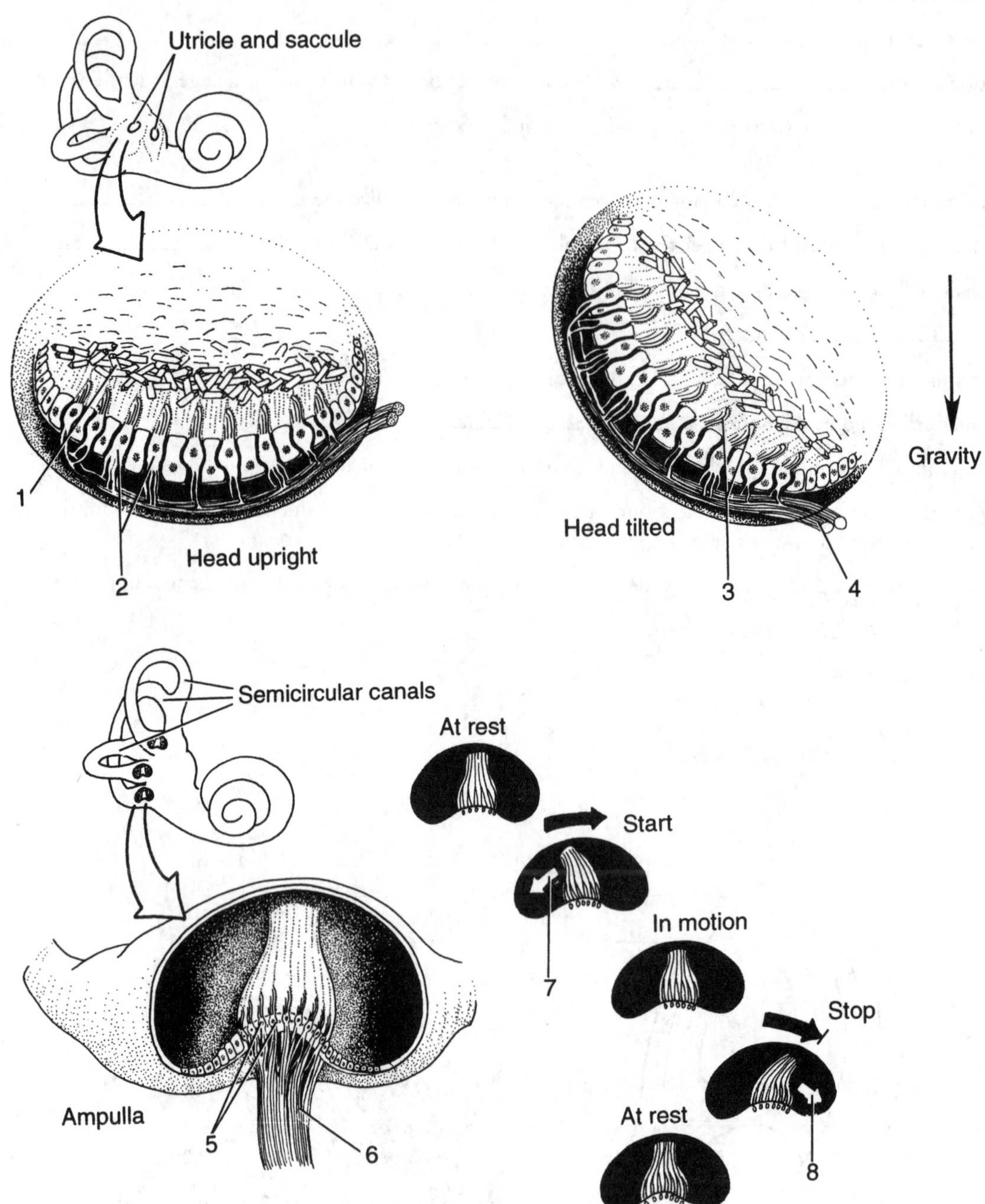

SUMMARY OF HEARING

1. Number the following in the order that they function in the hearing process.

   Sound waves (vibrations) → ____1____ eardrum

   ________ oval window

   ________ incus

   ________ 8th cranial nerve

   ________ malleus

   ________ stapes

   ________ fluid in the cochlea

   ________ hair cells in the organ of Corti

   ________ temporal lobes

2. When the stapes pushes in the oval window, the ________ bulges out to prevent damage to the hair cells.

ARTERIAL RECEPTORS

1. Arterial pressoreceptors detect changes in ________ and are located in the ________ and ________.
2. Arterial chemoreceptors detect changes in ________, ________, and ________ levels in the blood and are located in the ________ and ________.
3. a) When these receptors detect changes and generate impulses, we do not feel sensations but rather this information is used to make changes in ________ or ________.

   b) What part of the brain regulates these vital functions? ________

CROSSWORD PUZZLE

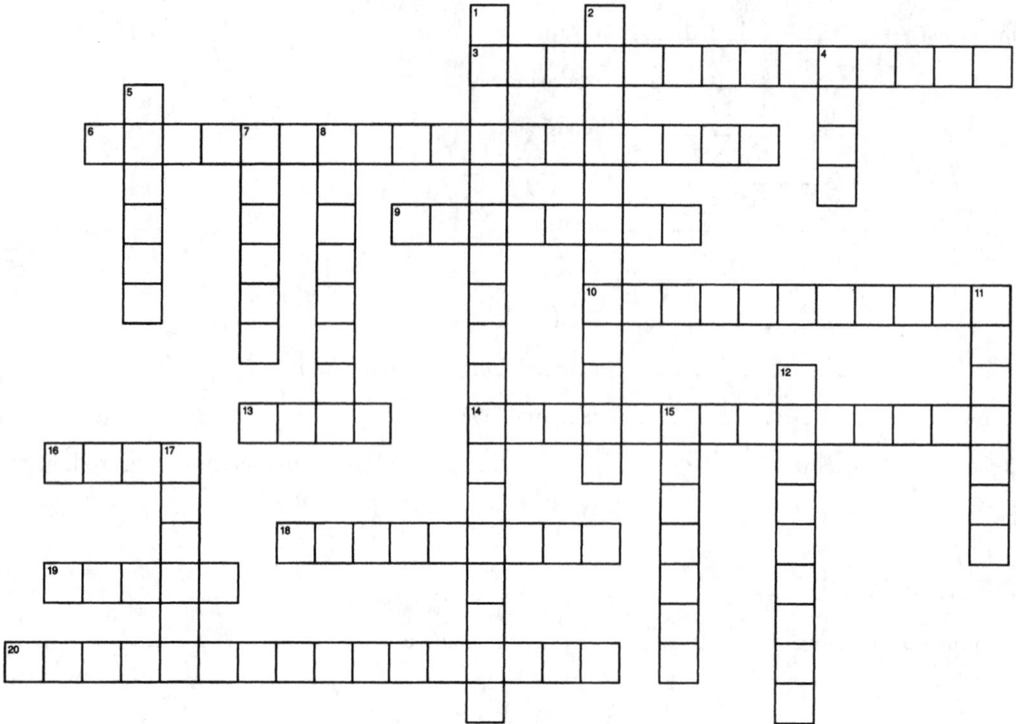

ACROSS

3. Produce tears (two words)
6. Contain receptors that detect motion
9. ________________ humor; keeps the retina in place
10. Lines the eyelids
13. Receptors that detect the presence of light
14. Permits air to enter or leave the middle ear cavity (two words)
16. Colored part of the eye
18. This chemical in rods reacts with light, and its breakdown generates an electrical impulse
19. Receptors that detect colors
20. The eardrum (two words)

DOWN

1. Generate impulses when stimulated by vapor molecules (two words)
2. Contains the receptors for hearing (three words)
4. Enables the eye to focus light
5. Contains the rods and cones
7. Anterior portion of the sclera; transparent
8. Pain that originates in an organ but is felt in the skin (adj.)
11. ________________ humor; nourishes the lens and cornea
12. Receptors for taste are found in these (two words)
15. Contains the organ of Corti; shaped like a snail shell
17. Outer layer of the eyeball

CLINICAL APPLICATIONS

1. a) Mr. G is 58 years old and has come to his eye doctor to get a new prescription for his glasses. The doctor performs another test and tells Mr. G that his intraocular pressure is abnormally high. Mr. G has the condition called ____________________, and the elevated intraocular pressure is caused by poor drainage of ____________________ in the anterior cavity of the eye.

   b) Fortunately, Mr. G's glaucoma can be treated, but if untreated, the higher pressure within the eye would eventually damage the ____________________ and result in blindness.

2. a) Mrs. A is 63 years old and does not hear as well as she used to. Her doctor tells her that a simple hearing aid will improve her hearing. From this, you would suspect that Mrs. A has what type of deafness?

   ____

   b) In this type of deafness, which parts of the hearing pathway may be impaired?

   ____

3. Mrs. T brings her 8-year-old daughter Sara to the doctor because Sara's eyes have suddenly become very red, watery, and itchy. The doctor diagnoses a bacterial infection of the membranes that line the eyelids and cover the whites of the eyes. This membrane is called the ____________________, and the inflammation is called ____________________.

4. a) Mr. S tells his doctor that his vision seems cloudy, even with his glasses. Mr. S thinks he needs new glasses. He is 60 years old and is farsighted, which means that his vision is best for ____________________, but he needs glasses for ____________________.

   b) The doctor determines that new glasses are not needed but that Mr. S does have cataracts. This means that the ____________________ of the eye has become opaque and is causing Mr. S's cloudy vision.

5. Mr. F is red-green color blind, and he and his wife ask their doctor if all their children will also be color blind. There is no history of color blindness in Mrs. F's family. The doctor tells them that Mrs. F probably does not have a gene for color blindness, and that Mr. F has one gene, which he can pass only to his daughters.

   1) Therefore, each son has what chance of being color blind?
      a) 0% b) 25% c) 50% d) 75% e) 100%

   2) Each daughter has what chance of being color blind?
      a) 0% b) 25% c) 50% d) 75% e) 100%

   3) However, if a daughter inherits the gene for color blindness, she may be called a ____________________ of the trait.

MULTIPLE CHOICE TEST #1

Choose the correct answer for each question.

1. The receptors for vision are located in which part of the eye?
   a) cornea b) lens c) iris d) retina

2. The receptors for hearing and equilibrium are located in the:
   a) outer ear b) eardrum c) middle ear d) inner ear

3. The receptors for touch and pressure are:
   a) stretch receptors in the epidermis
   b) free nerve endings in the dermis
   c) encapsulated nerve endings in the dermis
   d) free nerve endings in the epidermis

4. When holding a pencil, you experience the touch of it in the hand. This is an example of:
   a) contrast b) after-image c) projection d) adaptation

5. Bacterial growth on the front of the eyeball is inhibited by:
   a) aqueous humor on the cornea
   b) lysozyme in tears
   c) lysozyme in the lens
   d) aqueous humor in tears

6. The lens and cornea have no capillaries and are nourished by:
   a) tears b) aqueous humor c) vitreous humor d) intracellular fluid

7. The chemoreceptors that detect vaporized chemicals are the:
   a) olfactory receptors in the upper nasal cavities
   b) taste receptors in the taste buds
   c) olfactory receptors in the taste buds
   d) taste receptors in the upper nasal cavities

8. The part of a sensory pathway that detects changes is the:
   a) receptor b) sensory neuron c) sensory tract d) sensory area in the brain

9. The shape of the lens of the eye is changed by the:
   a) retina b) cornea c) iris d) ciliary muscle

10. The pain of a heart attack may be felt in the left shoulder; this is an example of:
    a) visceral pain b) referred pain c) adaptation pain d) after-image pain

11. Which of these cranial nerves is not involved in movement of the eyeball?
    a) optic b) oculomotor c) abducens d) trochlear

12. Becoming unaware of a continuing stimulus is called:
    a) after-image b) adaptation c) contrast d) projection

13. The receptors that detect movement of the body are located in the:
    a) vestibule b) organ of Corti c) semicircular canals d) middle ear

14. The receptors for hearing are part of the ________ within the ________:
    a) stapes/cochlea
    b) cochlea/middle ear
    c) organ of Corti/middle ear
    d) organ of Corti/cochlea

15. The visual receptors that detect colors are the:
    a) rhodopsins b) optic discs c) cones d) rods

16. The first part of the eye that refracts light rays is the:
    a) lens b) retina c) aqueous humor d) cornea

17. The receptors for muscle sense are ________ that detect ________.
    a) free nerve endings/contraction
    b) proprioceptors/stretching
    c) proprioceptors/contraction
    d) free nerve endings/stretching

18. The receptors for thirst are in the ________, and they detect ________:
    a) mouth/changes in saliva
    b) hypothalamus/changes in body water content
    c) hypothalamus/changes in saliva
    d) mouth/changes in body water content

19. The retina is kept in place by the:
    a) sclera b) extrinsic muscles c) vitreous humor d) aqueous humor

20. Keeping dust off the front of the eyeball is a function of the:
    a) tears and lens
    b) conjunctiva and aqueous humor
    c) sclera and eyelids
    d) eyelids and eyelashes

21. The size of the pupil is regulated by the:
    a) iris b) lens c) conjunctiva d) ciliary muscle

22. The adjustable part of the light refraction pathway is the:
    a) lens b) cornea c) iris d) aqueous humor

23. The part of a sensory pathway that feels the sensation is the:
    a) receptor b) cerebral cortex c) sensory tract d) sensory neuron

24. The first part of the ear to vibrate with sound waves is the:
    a) malleus b) pinna c) stapes d) eardrum

25. Otoliths that are pulled by gravity and bend hair cells are found in the:
    a) cochlea b) utricle and saccule c) stapes d) middle ear

26. The eyelids are lined by the:
    a) choroid membrane b) iris c) cornea d) conjunctiva

27. Unconscious muscle sense is used by the ________________ to regulate coordination.
    a) frontal lobes b) medulla c) cerebellum d) temporal lobes

28. The part of the retina that contains only cones is the:
    a) bipolar disc b) papillary layer c) optic disc d) fovea

29. Aqueous humor is reabsorbed into the:
    a) canal of Schlemm b) iris c) lens cavity d) vitreous cavity

30. The optic nerve is made of:
    a) rods b) cones c) rods and cones d) ganglion neurons

MULTIPLE CHOICE TEST #2

Read each question and the four answer choices carefully. When you have made a choice, follow the instructions to complete your answer.

1. Which statement is NOT true of the structure of the ear?
   a) The middle ear and inner ear structures are located within the temporal bone.
   b) The auditory bones are the malleus, incus, and stapes.
   c) The eardrum is at the end of the semicircular canal.
   d) The auricle does not have an important function for people's hearing.

   Reword your choice to make it a correct statement.

   ____

2. Which statement is NOT true of the brain and sensation?
   a) The visual areas are in the occipital lobes.
   b) The auditory areas are in the parietal lobes.
   c) Subconscious equilibrium is integrated by the cerebellum and midbrain.
   d) The olfactory areas are in the temporal lobes.

   Reword your choice to make it a correct statement.

   ____

3. Which statement is NOT true of the physiology of vision?
   a) Light rays are focused on the retina by the cornea, aqueous humor, lens, and vitreous humor.
   b) The visual receptors are the rods and cones in the retina.
   c) If too much light strikes the eye, the iris will dilate the pupil.
   d) The only adjustable part of the focusing mechanism is the lens.

   Reword your choice to make it a correct statement.

   ____

4. Which statement is NOT true of the physiology of hearing?
   a) The stapes transmits vibrations to the oval window of the inner ear.
   b) Vibrations in the fluid in the cochlea cause bending of the hair cells in the organ of Corti.
   c) The eardrum transmits vibrations to the malleus, incus, and round window.
   d) The cranial nerves for hearing are the acoustic, or 8th, cranial nerves.

   Reword your choice to make it a correct statement.

   ____

5. Which statement is NOT true of taste and smell?
   a) Much of what we taste is actually the contribution of the sense of smell.
   b) The receptors for taste are chemoreceptors that detect chemicals dissolved in saliva.
   c) The cranial nerves for the sense of taste are the facial and glossopharyngeal nerves.
   d) The cranial nerves for the sense of smell are the nasal nerves.

   Reword your choice to make it a correct statement.

   ____

6. Which statement is NOT true of the cutaneous senses?
   a) The receptors for pain are free nerve endings in the epidermis.
   b) The sensory tracts include white matter in the spinal cord.
   c) The receptors for pressure and touch are encapsulated nerve endings in the dermis.
   d) The sensory areas are in the parietal lobes of the cerebrum.

   Reword your choice to make it a correct statement.

   ___

7. Which statement is NOT true of the characteristics of sensations?
   a) The degree to which a sensation is felt is called intensity.
   b) An after-image is a sensation that continues after the stimulus stops.
   c) The effect of a previous sensation on a current sensation is called adaptation.
   d) Projection is a function of the brain, and the sensation seems to come from the area of the stimulated receptors.

   Reword your choice to make it a correct statement.

   ___

8. Which statement is NOT true of the structure of the eye?
   a) The vitreous humor in the posterior cavity helps keep the retina in place.
   b) The sclera is the outermost layer of the eyeball.
   c) Aqueous humor is reabsorbed back to the blood at the canal of Schlemm.
   d) The eyeball is moved from side to side by the intrinsic muscles.

   Reword your choice to make it a correct statement.

   ___

9. Which statement is NOT true of equilibrium?
   a) The utricle and saccule contain the receptors that detect changes in the position of the head.
   b) The cerebellum and medulla regulate the reflexes that keep us upright.
   c) The semicircular canals contain the receptors that detect motion.
   d) The nerves for equilibrium are the acoustic, or 8th, cranial nerves.

   Reword your choice to make it a correct statement.

   ___

10. Which statement is NOT true of the internal (visceral) sensations?
    a) Pressoreceptors in the carotid and aortic sinuses detect changes in blood pressure.
    b) The hypothalamus contains the receptors for hunger and thirst.
    c) Adaptation does occur in the sense of hunger but not in the sense of thirst.
    d) Chemoreceptors in the aortic and carotid bodies detect changes in the blood levels of nutrients.

    Reword your choice to make it a correct statement.

    ___

MULTIPLE CHOICE TEST #3

Each question is a series of statements concerning a topic in this chapter. Read each statement carefully and select all of the correct statements.

1. Which of the following statements are true of sensations?
   a) Contrast is a comparison to a previous sensation.
   b) Projection is where a sensation is felt on or in the body.
   c) Receptors are specialized to detect specific kinds of changes.
   d) Sensations are interpreted by the brain, usually the cerebral cortex.
   e) A sensory pathway is: receptor → sensory neuron → sensory tract → sensory area.
   f) Receptors change the energy of a nerve impulse to the energy of a stimulus.

2. Which of the following statements are true of the ear, hearing, and equilibrium?
   a) The hair cells in the organ of Corti detect tilting of the head.
   b) The auditory nerves are the 7th pair of cranial nerves.
   c) The receptors for hearing are in the middle ear, and the receptors for equilibrium are in the inner ear.
   d) The round window prevents pressure damage to the hair cells in the cochlea.
   e) The stapes transmits vibrations to the inner ear at the vestibule.
   f) The Eustachian tube regulates air pressure in the middle ear cavity.
   g) The utricle and saccule contain small crystals called otoliths.
   h) The auditory areas are in the frontal lobes.
   i) The cerebrum is the major regulator of coordination and balance.
   j) The four semicircular canals detect movement in all planes.
   k) The cranial nerves for equilibrium are the 8th pair.
   l) The eardrum is stretched across the end of the Eustachian tube.

3. Which of the following statements are true of the eye and vision?
   a) The conjunctiva cover the eyelids and line the cornea.
   b) The lateral rectus muscle pulls the eyeball toward the nose.
   c) The first structure to refract light rays is the cornea.
   d) The circular muscle fibers of the iris constrict the pupil.
   e) The ciliary muscle changes the shape of the lens.
   f) The choroid layer prevents glare within the eyeball.
   g) Aqueous humor lubricates the eye for blinking.
   h) The blind spot is the site where the optic nerve passes through the eyeball.
   i) The best receptors for color are the rods in the fovea.
   j) The lens adjusts to focus light from different distances.

4. Which of the following statements are true of cutaneous sense and muscle sense?
   a) The receptors for touch and pressure are encapsulated nerve endings.
   b) Conscious muscle sense and the cutaneous senses are interpreted in the parietal lobes.
   c) Stretch receptors in muscles may also be called proprioceptors.
   d) The number of receptors per square inch determines the sensitivity of the skin.
   e) Contrast and adaptation are both important to our perception of the cutaneous senses.
   f) Coordination of voluntary movement requires unconscious muscle sense.

5. Which of the following statements are true of taste and smell?
   a) The cranial nerves for the sense of smell are the facial nerves.
   b) Taste buds in the papillae of the tongue contain the receptors for the chemicals that make up foods.
   c) If a chemical does not vaporize, we cannot smell it.
   d) The sense of taste is very much dependent on the sense of smell.
   e) Olfactory receptors are specific for particular molecular shapes.
   f) The cranial nerves for the sense of taste are the facial and trigeminal.

Chapter 10

The Endocrine System

The endocrine system is one of the regulating systems of the body. This chapter describes the endocrine glands and the functions of their hormones. Each hormone has specific target organs and also specific functions in particular aspects of homeostasis.

ENDOCRINE GLANDS

1. a) The secretions of endocrine glands are called ________________, which enter capillaries and circulate in the ________________.
   b) The cells (organ) on which a hormone exerts its specific effects are called its ________________ cells (organ).
   c) These cells respond to particular hormones because of the presence of ________________ for these hormones, often on the cell membrane.
2. Hormones may be classified in three groups based on their chemical structure. These groups are ________________, ________________, and ________________.
3. a) The pituitary gland is enclosed and protected by the ________________ bone.
   b) The thyroid gland is on the anterior side of the trachea just below the ________________.
   c) The parathyroid glands are located on the posterior sides of the lobes of the ________________.
   d) The pancreas is located in the upper abdominal cavity between the ________________ and the ________________.
   e) The adrenal glands are located on top of the ________________.
   f) The ovaries are located in the pelvic cavity on either side of the ________________.
   g) The testes are located outside the abdominal cavity in the ________________.

4. The following diagram depicts the endocrine glands of the body.

 Label each.

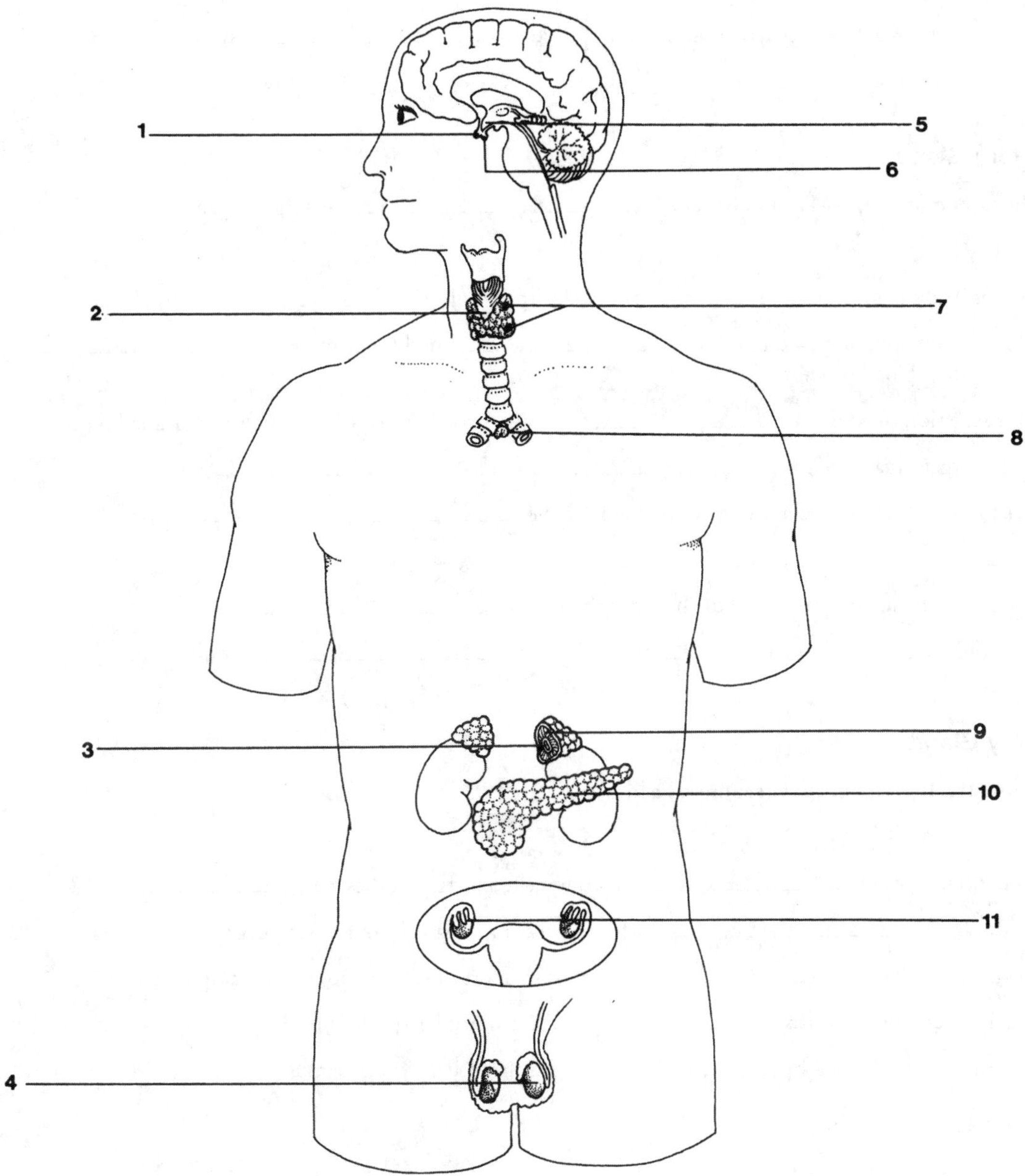

THE PITUITARY GLAND

1. The two parts of the pituitary gland are the ____________ and the ____________.
2. a) The posterior pituitary gland stores two hormones that are actually produced by the ____________.
   b) The anterior pituitary gland secretes its hormones in response to ____________ hormones (factors) from the ____________.

Posterior Pituitary Gland

1. **Antidiuretic hormone (ADH)**—May also be called ____________, and its target organs are the ____________.
2. a) The function of ADH is to ____________ reabsorption of ____________ by the kidneys.
   b) As a result of this function, urinary output ____________, and blood volume ____________, which helps maintain blood pressure.
3. The stimulus for secretion of ADH is ____________ (increased or decreased) water within the body.
4. **Oxytocin**—a) Its target organs are the ____________ and ____________.
   b) With respect to the uterus, oxytocin causes contractions of the ____________ for delivery of the ____________ and the ____________.
   c) With respect to the mammary glands, oxytocin causes the release of ____________.
   d) The stimulus for secretion of oxytocin is nerve impulses from the ____________ during labor or when nursing a baby.

Anterior Pituitary Gland

1. **Growth hormone (GH)**—Has many target organs and tissues.
   a) Functions:
      1) Increases the transport of ____________ into cells and the synthesis of ____________.
      2) Increases the rate of ____________, which results in more cells in growing organs.
      3) Increases the use of ____________ for energy, by increasing its removal from adipose tissue.
   b) The stimulus for secretion of GH is ____________ from the hypothalamus.
   c) The secretion of GH is inhibited by ____________ from the hypothalamus.

d) The following diagram depicts the functions of growth hormone.

Label the parts indicated.

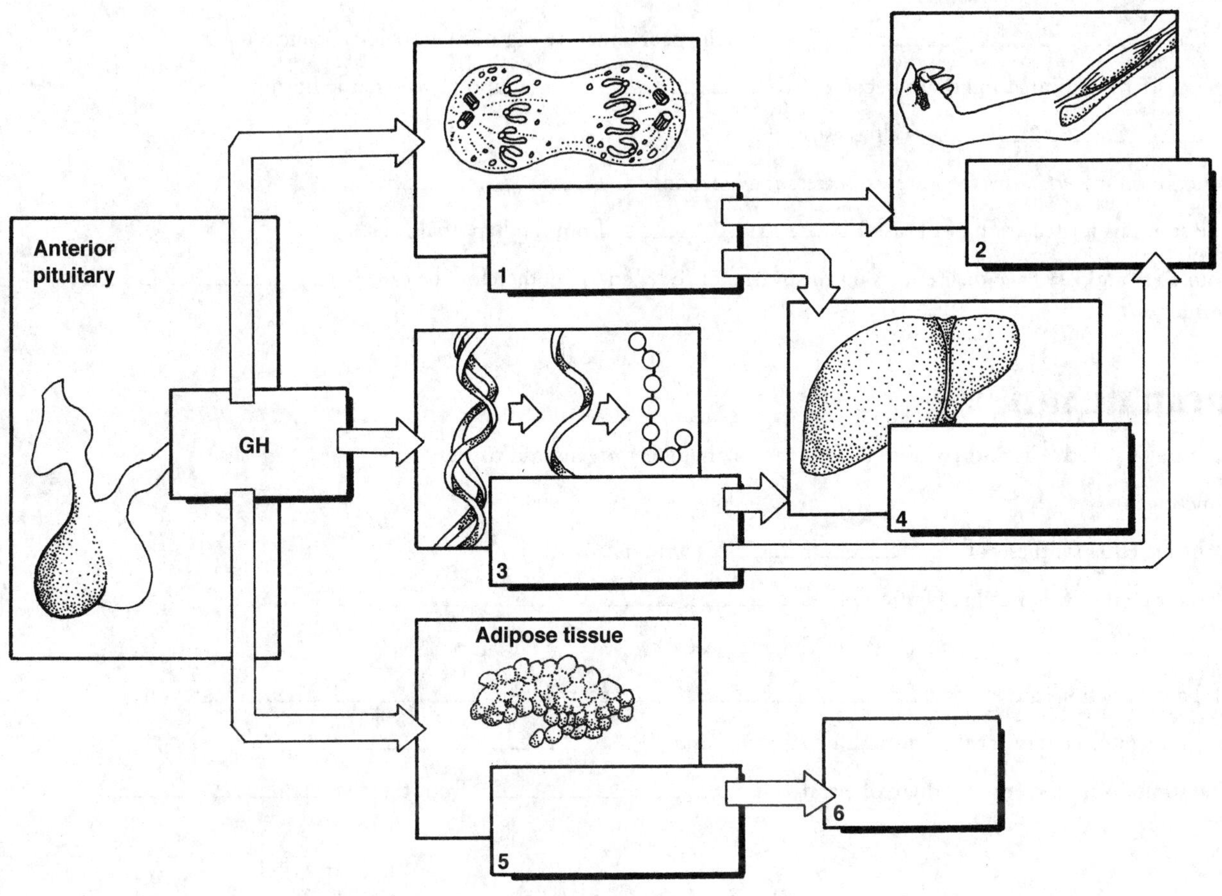

2. **Thyroid-stimulating hormone (TSH)**—a) Its target organ is the ____________________.

   b) Function: Stimulates the thyroid gland to secrete ____________________ and ____________________.

   c) The stimulus for secretion of TSH is ____________________ from the hypothalamus.

3. **Adrenocorticotropic hormone (ACTH)**—a) Its target organs are the ____________________.

   b) Function: Stimulates the adrenal cortex to secrete ____________________.

   c) The stimulus for secretion of ACTH is ____________________ from the hypothalamus.

4. **Prolactin**—a) Its target organs are the ____________________.

   b) Function: Causes production of ____________________ by the mammary glands.

   c) The secretion of prolactin is regulated by PRF and PRIF from the ____________________.

5. **Follicle-stimulating hormone (FSH)**—a) Its target organs in women are the ____________________ and in men are the ____________________.

   b) Functions in women: Initiates the development of ____________________ in ovarian follicles and increases the secretion of the hormone ____________________ by the follicle cells.

   c) Function in men: Initiates the production of ____________________ in the testes.

   d) The stimulus for secretion of FSH is ____________________ from the hypothalamus.

6. **Luteinizing hormone (LH)**—a) Its target organs in women are the ________________ and in men are the ________________.

   b) Functions in women:

      1) Causes ________________, which is the release of a mature egg from an ovarian follicle.

      2) Causes the ruptured follicle to become the ________________ and to secrete the hormone ________________, as well as estrogen.

   c) Function in men: Causes the testes to secrete the hormone ________________.

   d) The stimulus for secretion of LH is ________________ from the hypothalamus.

   e) Both FSH and LH have their effects on the ovaries or testes and may therefore be called ________________ hormones.

THYROID GLAND

1. **Thyroxine (T_4)** and **triiodothyronine (T_3)**—Have many target organs and tissues.

   a) Functions:

      1) Increase the synthesis of ________________ within cells.

      2) Increase the rate of cell respiration of ________________, ________________, and ________________ to produce ATP and heat.

      3) These functions are essential for ________________ growth and ________________ growth.

   b) The mineral necessary for the synthesis of thyroxine and T_3 is ________________.

   c) The stimulus for secretion of thyroxine and T_3 is ________________ from the ________________.

d) The following diagram depicts the functions of thyroxine.

Label the parts indicated.

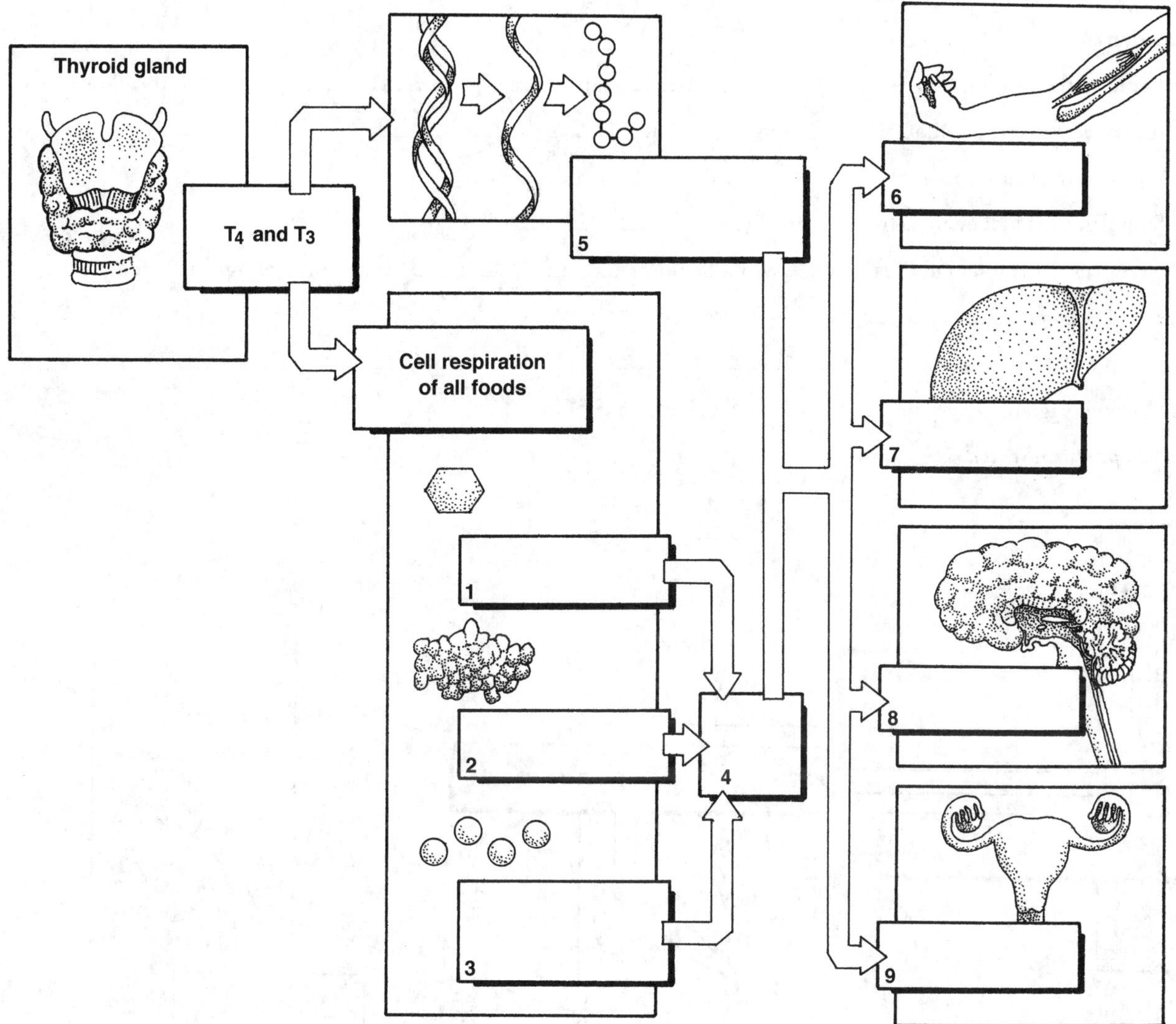

2. **Calcitonin**—a) Its target organs are the ____________________.

 b) Function: Decreases the reabsorption of ____________________ and ____________________ from bones.

 c) As a result of this function, the blood levels of calcium and phosphorus are ____________________.

 d) The stimulus for secretion of calcitonin is ____________________.

PARATHYROID GLANDS

1. **Parathyroid hormone (PTH)**—a) Its target organs are the ____________________, ____________________, and ____________________.

   b) Functions:

      1) Increases the reabsorption of ____________________ and ____________________ from bones to the blood.

      2) Increases the absorption of Ca and P from food in the ____________________.

      3) Increases the reabsorption of ____________________ by the kidneys.

      4) Stimulates the kidneys to activate vitamin ____________________.

   c) As a result of these functions, the blood level of calcium is ____________________, and the phosphate blood level is ____________________.

   d) Secretion of PTH is stimulated by ____________________ and inhibited by ____________________.

2. The following diagram depicts the functioning of PTH and calcitonin.

   Label the parts indicated.

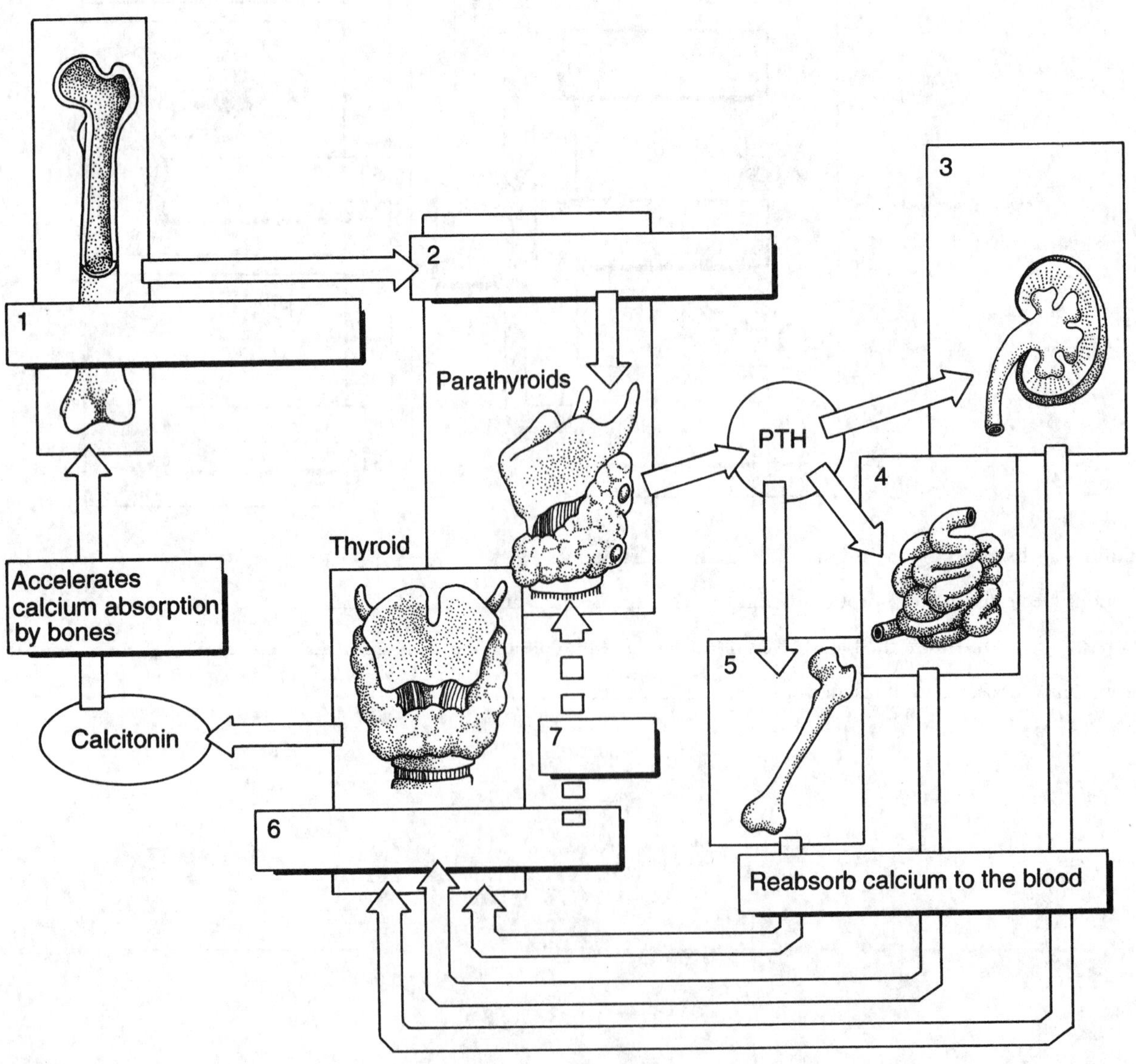

PANCREAS

1. a) The endocrine glands of the pancreas are called ____________________, which contain alpha cells and beta cells.

   b) Alpha cells produce the hormone ____________________, and beta cells produce the hormone ____________________.

2. **Glucagon**—a) Its primary target organ is the ____________________.

   b) Functions:

      1) Causes the liver to convert stored ____________________ to glucose to be used for energy production.

      2) Increases the use of ____________________ and ____________________ for energy production.

   c) As a result of these functions, the blood glucose level is ____________________, and all three ____________________ types may be used to produce ATP.

   d) The stimulus for secretion of glucagon is ____________________.

3. **Insulin**—Has many target organs and tissues.

   a) Functions:

      1) Causes the liver to change ____________________ to glycogen to be stored.

      2) Glycogen is also stored in ____________________ muscles.

      3) Enables other body cells to take in ____________________ from the blood to use for ____________________ production.

      4) Increases the intake of ____________________ and ____________________ by cells, to be used for the synthesis of ____________________ and ____________________.

   b) As a result of these functions, the blood glucose level is ____________________.

   c) The stimulus for secretion of insulin is ____________________.

4. The following diagram depicts the functioning of insulin and glucagon.

Label the parts indicated.

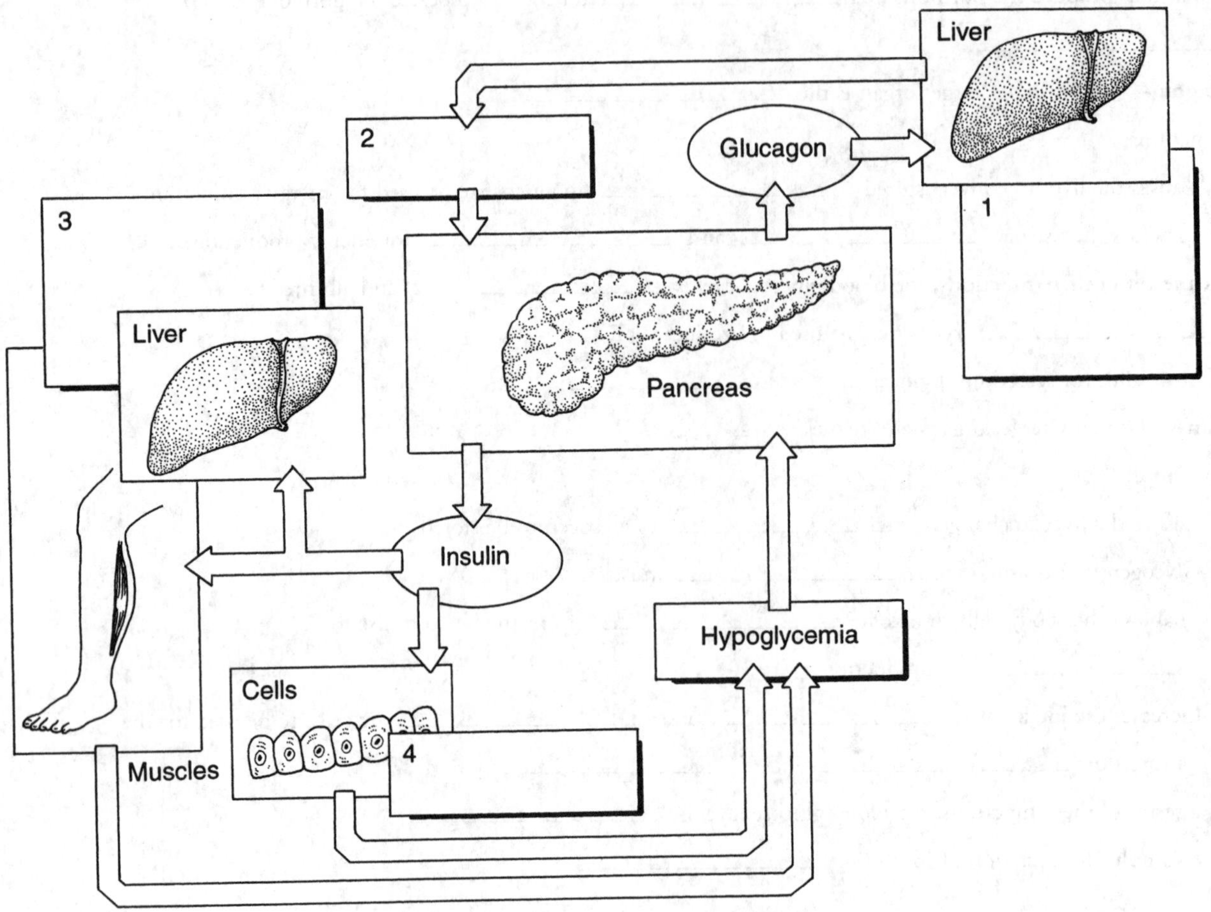

ADRENAL GLANDS

1. The two parts of an adrenal gland are the ____________________ and the ____________________.
2. The term *catecholamines* is a collective term for the hormones ____________________ and ____________________, which are secreted by the adrenal ____________________.
3. The adrenal ____________________ secretes a group of hormones called mineralocorticoids, of which ____________________ is the most important.
4. The adrenal ____________________ secretes a group of hormones called glucocorticoids, of which ____________________ is the most important.

Adrenal Medulla

1. **Norepinephrine**—a) Its primary target organs are ____________________.
   b) Function: Causes ____________________ throughout the body, which raises blood pressure.
2. **Epinephrine**—Has many target organs and tissues.
   a) Functions:
      1) Effect on the heart: ____________________
      2) Effect on blood vessels in skeletal muscle: ____________________
      3) Effect on blood vessels in skin and viscera: ____________________
      4) Effect on intestines: ____________________
      5) Effect on bronchioles: ____________________
      6) Effect on liver: ____________________
      7) Effect on use of fats for energy: ____________________
   b) Many of these functions mimic those of the ____________________ division of the autonomic nervous system.
   c) The stimulus for secretion of both norepinephrine and epinephrine is ____________________ impulses from the hypothalamus during ____________________ situations.

3. The following diagram depicts the functions of epinephrine and norepinephrine.

Label the parts indicated.

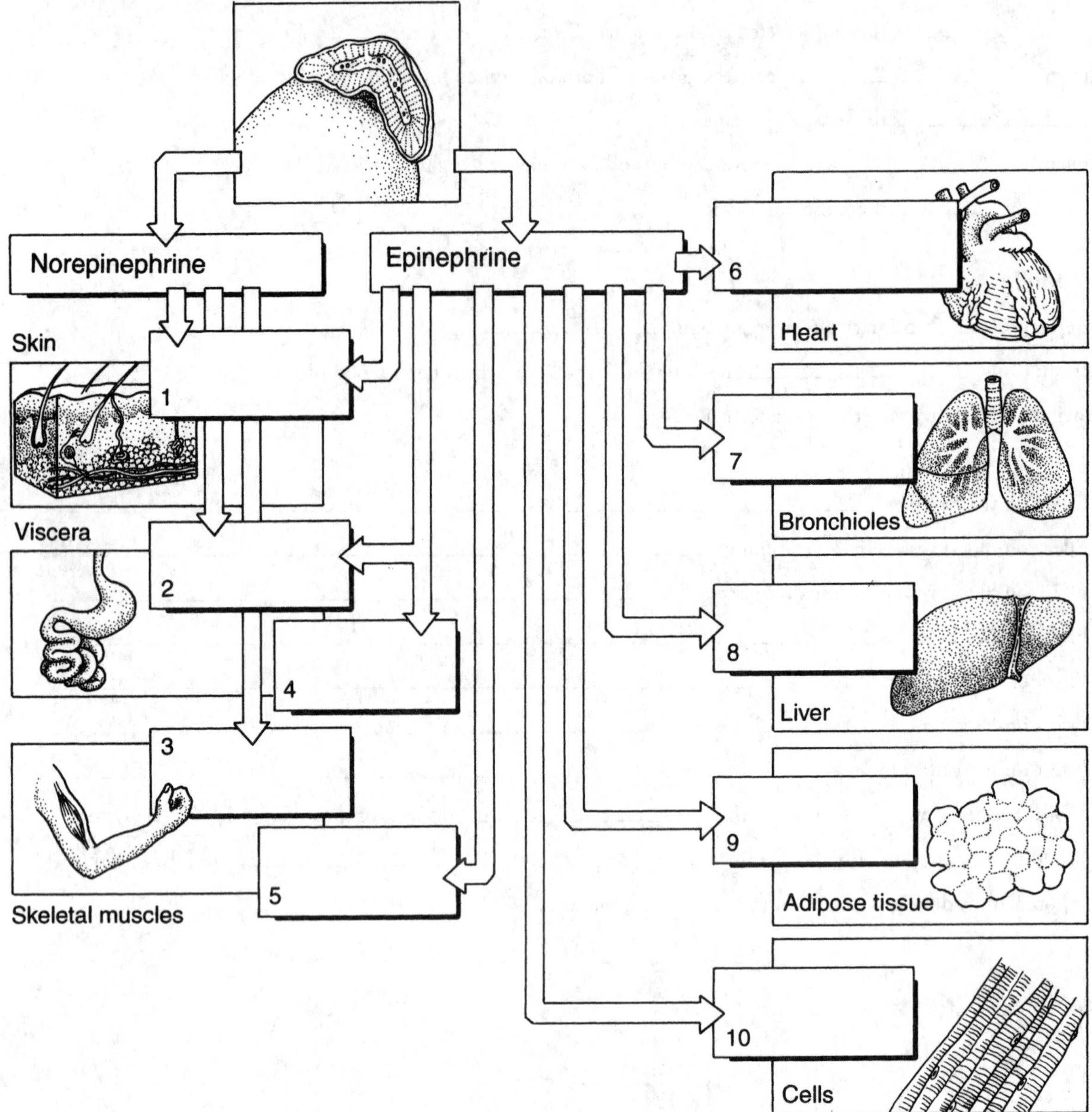

Adrenal Cortex

1. **Aldosterone**—a) Its target organs are the ____________________.

   b) Functions (direct effects): Increases the reabsorption of ____________________ ions and the excretion of ____________________ ions by the kidneys.

   c) As a result of this function (indirect or secondary effects):

   1) ____________________ ions are excreted in urine.

   2) ____________________ ions and ____________________ ions are returned to the blood.

   3) ____________________ is returned to the blood by osmosis following Na^+ ion reabsorption.

d) Therefore, aldosterone helps maintain normal blood volume and ____________________.

e) The stimuli for secretion of aldosterone are decreased blood level of ____________________ ions, or increased blood level of ____________________ ions, or a decrease in ____________________.

f) The following diagram depicts the functions of aldosterone.

Label the parts indicated.

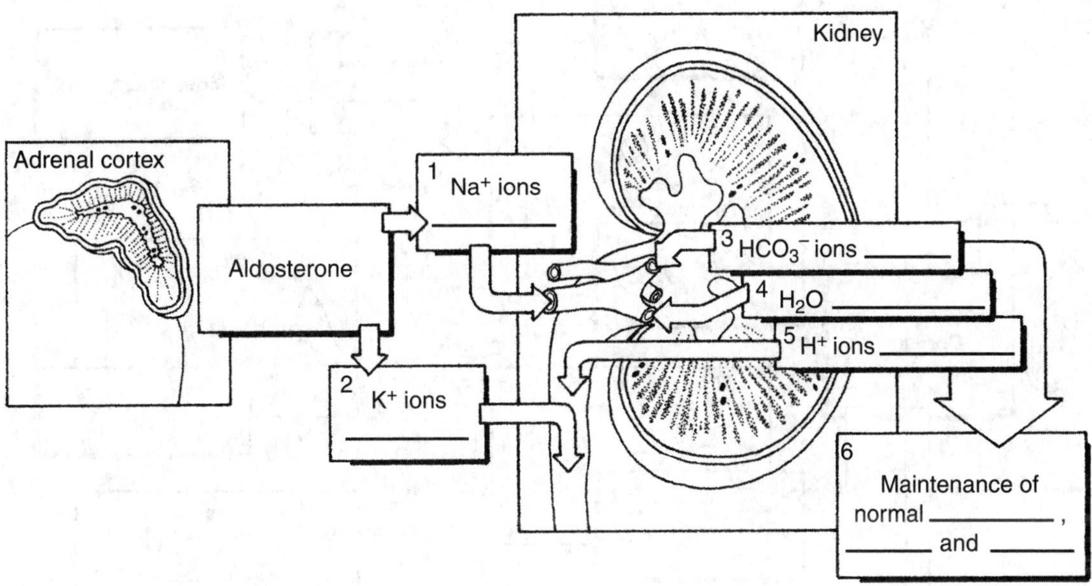

2. **Cortisol**—Has many target organs and tissues.

a) Functions:

1) Increases the use of ____________________ and ____________________ for energy production.

2) Decreases the use of ____________________ for energy so that this energy source is available for use by brain cells.

3) Has an ____________________ effect, which prevents excessive tissue destruction when damage occurs.

b) The stimulus for secretion of cortisol is ____________________ from the ____________________ in situations of physiological stress.

c) The following diagram depicts the functions of cortisol.

Label the parts indicated.

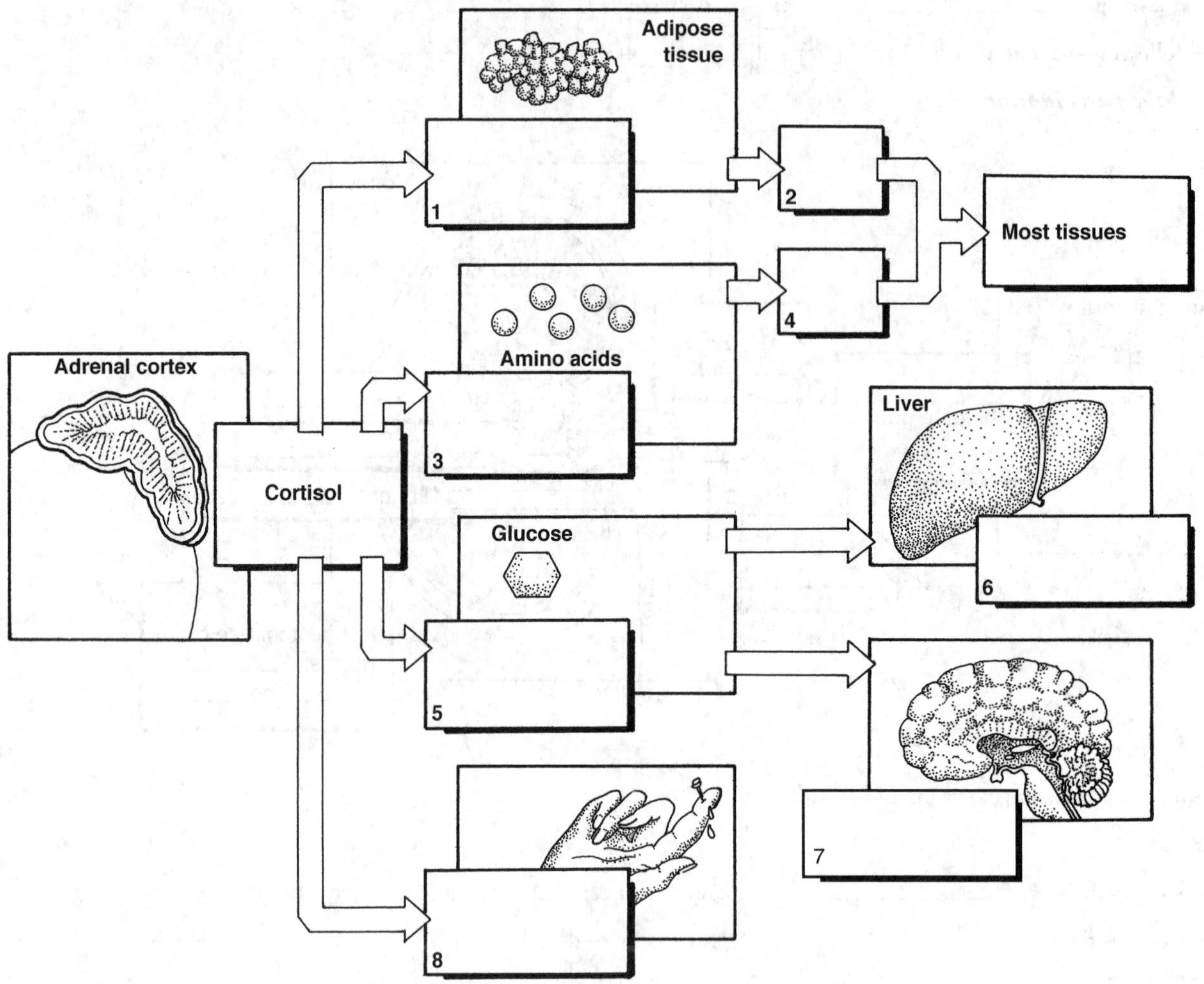

OVARIES

1. **Estrogen**—a) Its target organs include the ____________________ and the ____________________.

   b) Functions:

   1) Promotes maturation of the ____________________ in an ovarian follicle.

   2) Promotes growth of blood vessels in the ____________________ of the uterus, to prepare for a fertilized egg.

   3) Promotes the development of the female secondary sex characteristics, which include

   ____.

   4) Stops growth in height by promoting ____________________ in long bones.

   c) The stimulus for secretion of estrogen is ____________________ from the ____________________.

2. **Progesterone**—a) Its target organs are the ____________________ and ____________________.

   b) Functions:

   1) Increases the growth of ____________________ and the storage of ____________________ in the endometrium of the uterus.

   2) Promotes the growth of the ____________________ of the mammary glands.

   c) The stimulus for secretion of progesterone is ____________________ from the ____________________.

3. **Inhibin**—a) Its target organs are the ____________________ and ____________________.

   b) Function: Decreases the secretion of ____________________ and ____________________.

TESTES

1. **Testosterone**—a) Its target organs include the ____________________ and ____________________.
   b) Functions:
      1) Promotes maturation of ____________________ in the testes.
      2) Promotes the development of the male secondary sex characteristics, which include ____.
      3) Stops growth in height by promoting ____________________ in long bones.
   c) The stimulus for secretion of testosterone is ____________________ from the ____________________.
2. **Inhibin**—a) Its target organ is the ____________________.
   b) Function: Decreases the secretion of ____________________, which helps maintain ____________________ at a constant rate.
   c) The stimulus for secretion of inhibin is ____________________.

OTHER HORMONES

1. a) **Melatonin** is produced by the ____________________.
   b) In people, melatonin brings about the onset of ____________________.
2. a) **Prostaglandins** are hormone-like substances made by cells from the ____________________ of their cell membranes.
   b) In contrast to hormones, whose site of action is often distant from the site of production, prostaglandins usually exert their effects ____________________.
3. There are many different prostaglandins with many functions. State three different functions of prostaglandins.
   1) ____
   2) ____
   3) ____

MECHANISMS OF HORMONE ACTION

1. a) Cells respond to some hormones but not to others because of the presence of ____________________ for certain hormones on the cell membrane or within the cell.
   b) When a cell has receptors for a particular hormone, the cell is said to be a ____________________ cell for that hormone.
2. **The two-messenger mechanism—Protein hormones**
   1) In this mechanism, the first messenger is the ____________________.
   2) The bonding of the hormone to cell receptors stimulates the formation of ____________________ within the cell, and this is the second messenger.
   3) Cyclic AMP then activates the ____________________ within the cell to bring about the cell's response to the hormone.
   4) State two general types of cellular responses to hormones. ____________________ and ____________________
3. **Action of steroid hormones**
   1) Steroid hormones diffuse into cells because they are soluble in the ____________________ of cell membranes.
   2) The steroid hormone combines with a ____________________ in the cytoplasm of the cell and then enters the ____________________ of the cell.

3) Within the nucleus, the steroid protein complex activates specific ________________ to start the process of ________________.

4) The cell's response to the hormone is thus brought about by the ________________ that are produced.

ENDOCRINE DISORDERS

Hypersecretion Disorders

1. Match each disorder with the correct hormone in excess (a letter statement) and the correct description (a number statement).

Use each letter and number once.

1) Giantism ________________

2) Graves' disease ________________

3) Cushing's syndrome ________________

4) Acromegaly ________________

Excess Hormone

A. Growth hormone in childhood
B. Cortisol
C. Growth hormone in adulthood
D. Thyroxine

Description

1. Rapid heart rate, excessive heat production, and weight loss
2. Bones and skin become fragile; fat is deposited in the trunk of the body
3. Excessive growth of long bones
4. Excessive growth of bones of hands, feet, and face

Hyposecretion Disorders

1. Match each disorder with the correct hormone deficiency (a letter statement) and the correct description (a number statement).

Use each letter and number once.

1) Myxedema ________________

2) Cretinism ________________

3) Dwarfism ________________

4) Addison's disease ________________

5) Diabetes mellitus ________________

Hormone Deficiency

A. Growth hormone in childhood
B. Insulin
C. Thyroxine in infancy
D. Thyroxine in adulthood
E. Cortisol and aldosterone

Description

1. Minimal growth of long bones; very short stature
2. Cells cannot utilize glucose for energy; the blood glucose level rises
3. Severe dehydration and hypoglycemia
4. Muscular weakness, slow heart rate, weight gain
5. Severe mental and physical disability

STIMULUS FOR SECRETION—SUMMARY

1. Match each hormone with the proper stimulus for secretion.

Each letter is used once, and two letters are used twice. Each answer line will have only one correct letter.

1) ADH ____________________
2) Aldosterone ____________________
3) Calcitonin ____________________
4) Cortisol ____________________
5) Estrogen ____________________
6) Epinephrine and norepinephrine ____________________
7) FSH ____________________
8) Glucagon ____________________
9) Growth hormone ____________________
10) Insulin ____________________
11) LH ____________________
12) Oxytocin ____________________
13) PTH ____________________
14) Progesterone ____________________
15) T_4 and T_3 ____________________
16) Testosterone ____________________

A. Decreased sodium ion concentration in the blood
B. GHRH
C. Impulses from the hypothalamus during labor or nursing
D. ACTH
E. Decreased water content in the body
F. FSH
G. Sympathetic impulses from the hypothalamus during stressful situations
H. Hyperglycemia
I. Hypoglycemia
J. Hypercalcemia
K. Hypocalcemia
L. LH
M. GnRH
N. TSH

CROSSWORD PUZZLE

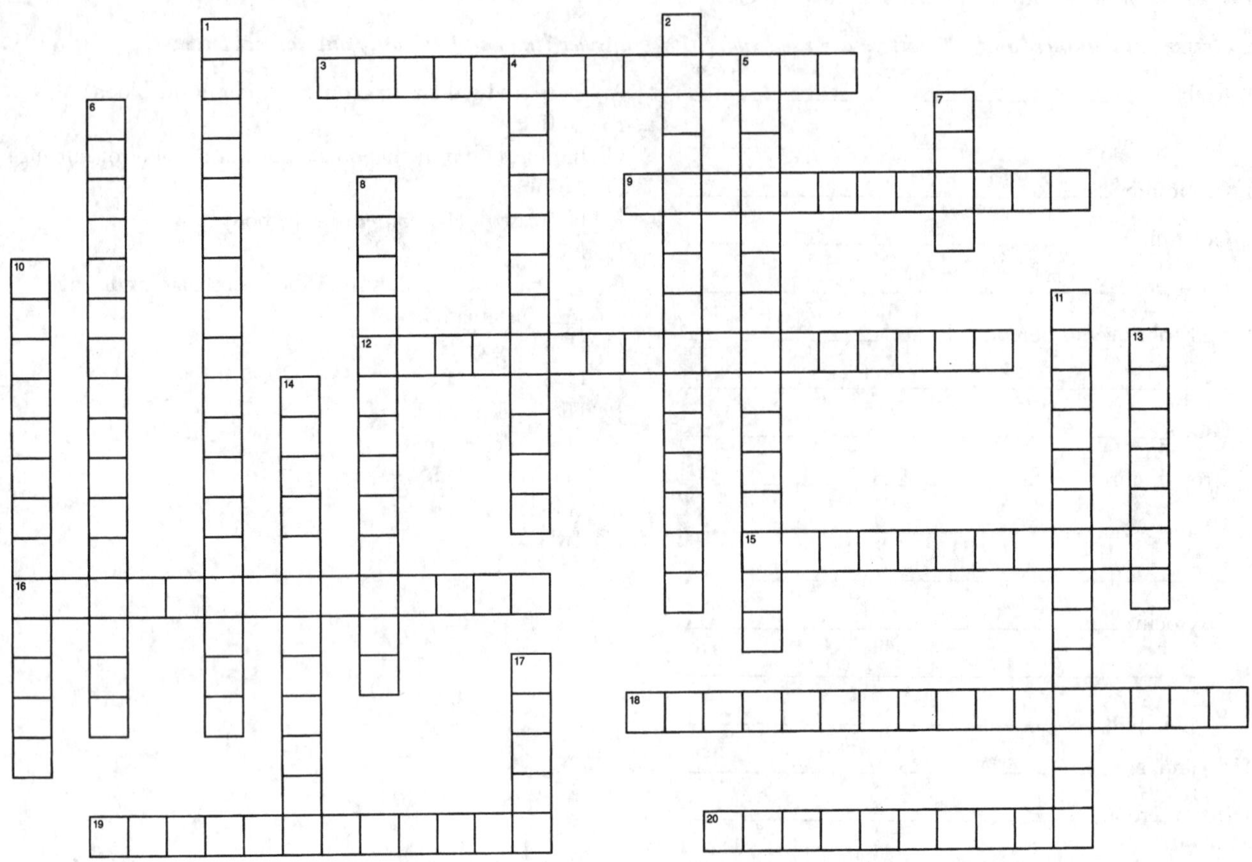

ACROSS

3. The conversion of glycogen to glucose
9. What the ovarian follicle becomes when an ovum is released (two words)
12. ________________ - ________________ mechanism; process culminates in the formation of angiotensin II (hyphenated)
15. The organ that a hormone affects (two words)
16. Epinephrine and norepinephrine, collectively
18. ________________ mechanism; effects of the hormone decrease the secretion of the hormone (two words)
19. Low blood-calcium level
20. The pituitary gland

DOWN

1. Hormone-producing cells of the pancreas (three words)
2. The conversion of amino acids to carbohydrates
4. The conversion of glucose to glycogen
5. "Mimics" the sympathetic division of the ANS
6. Regulate the secretion of hormones of the anterior pituitary gland (two words)
7. ________________ cells; produce insulin
8. High blood glucose level
10. High blood calcium level
11. Produced by cells from the phospholipids of their cell membranes
13. Chemical secreted by an endocrine gland
14. Low blood glucose level
17. ________________ cells; produce glucagon

CLINICAL APPLICATIONS

1. a) Mr. A is 49 years old and 30 pounds overweight. He sees his doctor and says he is "always tired—no energy." He also reports being constantly thirsty and urinating more frequently. Mr. A's doctor asks one question about Mr. A's family medical history (parents, grandparents, siblings). What do you think this question is?

   ____

   b) Further lab work that includes blood and urine tests confirms the doctor's diagnosis that Mr. A has ______________.

   c) Of which type? ______________

   d) Mr. A will be taking medication to enable the insulin his pancreas produces to work effectively. What else should Mr. A do to help control his disease? ______________

2. Mrs. C is 35 years old and reports these symptoms to her doctor: fatigue, weight gain, a constant feeling of being cold. The doctor suspects that Mrs. C has myxedema, which is a hyposecretion of which hormone?

   ____

3. Mrs. M is 29 years old and has fractured her wrist. She tells her doctor that she wasn't doing anything strenuous, her wrist "just broke." One of the first tests the doctor orders is a determination of the blood calcium level, which is found to be abnormally high. The doctor suspects that Mrs. M has a tumor of the ______________ glands that is causing hypersecretion of the hormone ______________, which is rapidly removing calcium from her bones.

4. a) Mr. T is 45 years old and has been diagnosed with Addison's disease. His symptoms are related to hyposecretion of hormones from which gland? ______________

   b) Mr. T's low blood pressure is the result of insufficient ______________, and his muscular weakness is the result of insufficient ______________.

MULTIPLE CHOICE TEST #1

Choose the correct answer for each question.

1. The hormone that lowers blood glucose level by enabling cells to take in glucose is:
   a) glucagon b) cortisol c) insulin d) growth hormone

2. The hormone that increases the rate of cell division is:
   a) thyroxine b) calcitonin c) insulin d) growth hormone

3. The two hormones that regulate blood calcium level are:
   a) insulin and glucagon c) calcitonin and growth hormone
   b) parathyroid hormone and calcitonin d) parathyroid hormone and thyroxine

4. The hormone that initiates egg or sperm production is:
   a) FSH b) LH c) estrogen d) testosterone

5. In men, the hormone necessary for maturation of sperm is:
   a) testosterone b) FSH c) LH d) aldosterone

6. In women, the hormone that causes ovulation is:
   a) FSH b) LH c) estrogen d) progesterone

7. Two hormones that cause the liver to change glycogen to glucose are:
   a) insulin and cortisol c) thyroxine and insulin
   b) glucagon and epinephrine d) insulin and epinephrine

8. The hormone that increases protein synthesis and the use of all three food types for energy is:
   a) insulin b) glucagon c) growth hormone d) thyroxine

9. The hormone that slows peristalsis and dilates the bronchioles is:
   a) glucagon b) cortisol c) epinephrine d) thyroxine

10. The hormone that has an anti-inflammatory effect is:
    a) epinephrine b) aldosterone c) cortisol d) calcitonin
11. The hormone that increases water reabsorption by the kidneys is:
    a) PTH b) oxytocin c) thyroxine d) ADH
12. The hormone that increases calcium reabsorption by the kidneys is:
    a) aldosterone b) ADH c) calcitonin d) PTH
13. The hormone that increases sodium reabsorption by the kidneys is:
    a) aldosterone b) ADH c) PTH d) cortisol
14. In women, the two hormones that promote growth of blood vessels in the endometrium are:
    a) FSH and estrogen c) LH and progesterone
    b) estrogen and progesterone d) FSH and LH
15. In women, the hormone that promotes growth of the corpus luteum is:
    a) progesterone b) estrogen c) LH d) FSH
16. The hormone that stimulates milk production in the mammary glands is:
    a) oxytocin b) estrogen c) progesterone d) prolactin
17. The hormone that causes strong contractions of the uterus during labor is:
    a) estrogen b) prolactin c) oxytocin d) progesterone
18. The hormone that increases the use of fats and excess amino acids for energy while sparing glucose for use by the brain is:
    a) epinephrine b) cortisol c) thyroxine d) insulin
19. Two hormones that help maintain normal blood pressure by maintaining normal blood volume are:
    a) ADH and aldosterone c) insulin and cortisol
    b) aldosterone and thyroxine d) ADH and oxytocin
20. Localized hormones that are synthesized from the phospholipids of cell membranes are called:
    a) steroids b) amines c) prostaglandins d) proteins
21. Steroid hormones are believed to exert their effect by stimulating the synthesis of:
    a) fats b) glycogen c) proteins d) DNA for cell division
22. The two-messenger mechanism of hormone action describes the action of:
    a) steroid hormones b) prostaglandins c) lipid hormones d) protein hormones
23. The hormone produced by the ovaries or testes that inhibits the secretion of FSH is:
    a) estrogen b) inhibin c) testosterone d) progesterone
24. The hormone that brings about sleep is:
    a) melatonin b) epinephrine c) insulin d) cortisol
25. The hormone that increases excretion of potassium by the kidneys is:
    a) ADH b) aldosterone c) cortisol d) PTH
26. The secretion of insulin in response to fluctuating blood glucose levels is a(n):
    a) positive feedback mechanism c) negative feedback mechanism
    b) antifeedback mechanism d) linear feedback mechanism
27. Secretion of the hormones of the anterior pituitary gland is regulated by the:
    a) hypothalamus b) posterior pituitary gland c) cerebrum d) thyroid gland
28. The stimulus for secretion of glucagon is:
    a) hyperglycemia b) hypercalcemia c) hypocalcemia d) hypoglycemia
29. The functions of epinephrine are very similar to the functions of:
    a) thyroxine c) the sympathetic nervous system
    b) the parasympathetic nervous system d) growth hormone
30. The stimulus for secretion of aldosterone is:
    a) low blood sodium level c) high blood sodium level
    b) low blood potassium level d) high blood calcium level

MULTIPLE CHOICE TEST #2

Read each question and the four answer choices carefully. When you have made a choice, follow the instructions to complete your answer.

1. Which statement is NOT true of hormone effects on the liver?
   a) Insulin causes the liver to change glycogen to glucose.
   b) Cortisol causes the liver to store glucose as glycogen.
   c) Glucagon causes the liver to change glycogen to glucose.
   d) Epinephrine causes the liver to change glycogen to glucose.

   Reword your choice to make it a correct statement.

   ____

2. Which statement is NOT true of hormone effects on bones?
   a) GH decreases the rate of mitosis in growing bones.
   b) PTH increases the reabsorption of calcium from bones.
   c) Calcitonin decreases the reabsorption of calcium from bones.
   d) GH and thyroxine increase the rate of protein synthesis in growing bones.

   Reword your choice to make it a correct statement.

   ____

3. Which statement is NOT true of the hormones of the anterior pituitary gland?
   a) TSH stimulates the thyroid gland to secrete thyroxine and T_3.
   b) ACTH stimulates the adrenal cortex to secrete cortisol.
   c) These hormones are secreted in response to releasing hormones from the hypothalamus.
   d) FSH and LH are called gonadotropic hormones because their target organs are the kidneys.

   Reword your choice to make it a correct statement.

   ____

4. Which hormone is NOT paired with its correct stimulus for secretion?
   a) Calcitonin—hypercalcemia
   b) Insulin—hypoglycemia
   c) PTH—hypocalcemia
   d) Glucagon—hypoglycemia

   For your choice, state the correct stimulus for secretion.

   ____

5. Which statement is NOT true of the functions of hormones during stress situations?
   a) Cortisol has an anti-inflammatory effect.
   b) Norepinephrine causes vasoconstriction throughout the body and raises blood pressure.
   c) Epinephrine increases heart rate and decreases peristalsis.
   d) Epinephrine causes the liver to change glucose to glycogen.

   Reword your choice to make it a correct statement.

   ____

6. Which statement is NOT true of hormone effects on the kidneys?
   a) PTH increases reabsorption of calcium ions.
   b) ADH increases reabsorption of water.
   c) Aldosterone decreases the reabsorption of sodium ions.
   d) Aldosterone increases the excretion of potassium ions.

   Reword your choice to make it a correct statement.

   ____

7. Which statement is NOT true of the hormones of the posterior pituitary gland?
   a) ADH is secreted during states of dehydration.
   b) Oxytocin causes release of milk from the mammary glands.
   c) The target organs of ADH are the kidneys.
   d) These hormones are actually produced by the thalamus.

   Reword your choice to make it a correct statement.

   ____

8. Which statement is NOT true of the locations of endocrine glands?
   a) The pancreas is located between the duodenum and the spleen.
   b) The thyroid gland is superior to the larynx on the front of the trachea.
   c) The adrenal glands are superior to the kidneys.
   d) The pituitary gland hangs by a stalk from the hypothalamus.

   Reword your choice to make it a correct statement.

   ____

9. Which statement is NOT true of hormone effects on reproduction?
   a) FSH initiates production of ova or sperm cells.
   b) LH stimulates ovulation in women.
   c) LH stimulates secretion of testosterone in men.
   d) Secondary sex characteristics in men and women are regulated by testosterone and progesterone, respectively.

   Reword your choice to make it a correct statement.

   ____

10. Which statement is NOT true of the mechanisms of hormone action?
    a) The presence of specific receptors determines which hormones a cell will respond to.
    b) Steroid hormones exert their effects by increasing the process of mitosis.
    c) In the two-messenger mechanism, the first messenger is the hormone.
    d) In the two-messenger mechanism, the second messenger is cyclic AMP that activates cellular enzymes.

    Reword your choice to make it a correct statement.

    ____

MULTIPLE CHOICE TEST #3

Each question is a series of statements concerning a topic in this chapter. Read each statement carefully and select all of the correct statements.

1. Which of the following statements are true of the regulation of secretion of hormones?
   a) A negative feedback mechanism is a cycle that shuts itself off.
   b) Secretion of some hormones is regulated by changing blood levels of nutrients.
   c) The positive feedback mechanism that regulates the secretion of thyroxine keeps the body's metabolic rate from dropping too low.
   d) Loss of water or salts will stimulate the release of some hormones.
   e) Hormones that trigger the release of other hormones are TSH, FSH, PTH, LH, and GH.
   f) Some hormones work as an antagonistic pair to regulate an aspect of body chemistry.
   g) Hypoglycemia is the stimulus for secretion of insulin.
   h) The secretion of epinephrine is stimulated by parasympathetic impulses.
   i) GnRH stimulates the secretion of glucagon.
   j) A rising blood sodium level stimulates secretion of aldosterone.

2. Which of the following statements are true of the pituitary gland?
   a) The posterior pituitary is regulated by releasing hormones from the hypothalamus.
   b) GH increases the rate of cell division and decreases the rate of protein synthesis.
   c) ADH decreases urinary output by decreasing the water reabsorbed by the kidneys.
   d) ACTH stimulates the release of cortisol by the adrenal medulla.
   e) TSH increases secretion of T_4 and decreases secretion of T_3.
   f) Prolactin causes the release of milk from the mammary glands.
   g) FSH stimulates the growth of follicles of the thyroid gland.
   h) LH stimulates secretion of oxytocin during birth.

3. Which of the following statements are true of the thyroid or parathyroid glands?
   a) Calcitonin from the thyroid is a synergist for parathyroid hormone.
   b) A molecule of thyroxine contains four atoms of iodine.
   c) PTH increases the absorption of Ca^{+2} from food in the small intestine.
   d) PTH helps regulate the blood calcium level by lowering it.
   e) Thyroxine and T_3 increase energy production from all food types.
   f) Thyroxine and T_3 increase the rate of mitosis, especially in growing bones.
   g) Calcitonin stimulates excretion of excess calcium by the kidneys.
   h) After adolescence, thyroxine is no longer necessary for the proper functioning of the brain.

4. Which of the following statements are true of the pancreas or adrenal glands?
   a) Insulin lowers the blood glucose level by increasing glucose intake by cells.
   b) Cortisol increases the use of amino acids and fats for energy production.
   c) Epinephrine increases heart rate and causes vasodilation in skeletal muscle.
   d) Glucagon increases the conversion of glycogen to glucose by the liver.
   e) Norepinephrine causes vasodilation in skin and viscera.
   f) Aldosterone causes the kidneys to excrete Na^+ ions and reabsorb K^+ ions.
   g) The beta cells of the islets of Langerhans secrete glucagon.
   h) The adrenal cortex provides an anti-inflammatory effect because of its secretion of aldosterone.

Chapter 11

Blood

The blood consists of blood plasma and the blood cells, which are red blood cells, white blood cells, and platelets. As blood circulates, each component carries out specific functions that are essential for homeostasis. This chapter describes the components of blood and their functions.

GENERAL FUNCTIONS OF BLOOD

1. Match each function of blood with the proper examples.

   Use each letter once. Each answer line will have more than one correct letter.

   1) Transportation ______________________
   2) Regulation ______________________
   3) Protection ______________________

   A. Acid–base balance
   B. Blood clotting
   C. Destroys pathogens
   D. Nutrients and waste products
   E. Body temperature
   F. Fluid–electrolyte balance
   G. Gases and hormones

CHARACTERISTICS OF BLOOD

1. The amount of blood within the body varies with the size of the person; this amount is in the range of ______________ liters.
2. a) The blood cells make up ______________ % to ______________ % of the total blood.
   b) The blood plasma makes up ______________ % to ______________ % of the total blood.
3. a) The normal pH range of blood is ______________ to ______________.
   b) This pH range is slightly ______________.
4. a) The viscosity of blood refers to its ______________.
   b) The presence of ______________ and ______________ make blood more viscous than water.

BLOOD PLASMA

1. Blood plasma is approximately ______________ % water.
2. a) The water of plasma is a solvent, which means that substances may ______________ in this water and be transported.
   b) Name two types of substances that are transported in dissolved form in the plasma ______________ and ______________.

3. a) Carbon dioxide is carried in the plasma in the form of ______________ ions.

   b) State the chemical formula of this ion. ______________

4. Match each plasma protein with the proper descriptive statements.

   One letter is used twice, and each other letter is used once. Each answer line will have three correct letters.

   1) Albumin ______________
   2) Clotting factors ______________
   3) Globulins ______________

   A. Synthesized only by the liver
   B. Synthesized by lymphocytes or by the liver
   C. Include fibrinogen and prothrombin
   D. Pulls tissue fluid into capillaries to maintain blood volume
   E. Include antibodies
   F. The most abundant plasma protein
   G. Help prevent blood loss when blood vessels rupture
   H. Include carrier molecules for fats in the blood

HEMOPOIETIC TISSUES

1. The term *hemopoietic tissue* means a tissue in which ______________ are formed.

2. After birth, the primary hemopoietic tissue is ______________, which is found in ______________ and ______________.

3. a) In the red bone marrow, the precursor cell for blood cells is called a ______________.

   b) These cells constantly undergo the process of ______________ to produce new cells.

   c) Name the types of cells formed in red bone marrow. ______________, ______________, and ______________

4. a) Lymphatic tissue is found in lymphatic organs such as the ______________, ______________, and ______________.

   b) The stem cells of lymphatic tissue produce the WBCs called ______________.

5. The following diagram depicts the hemopoietic tissues.

Label each one.

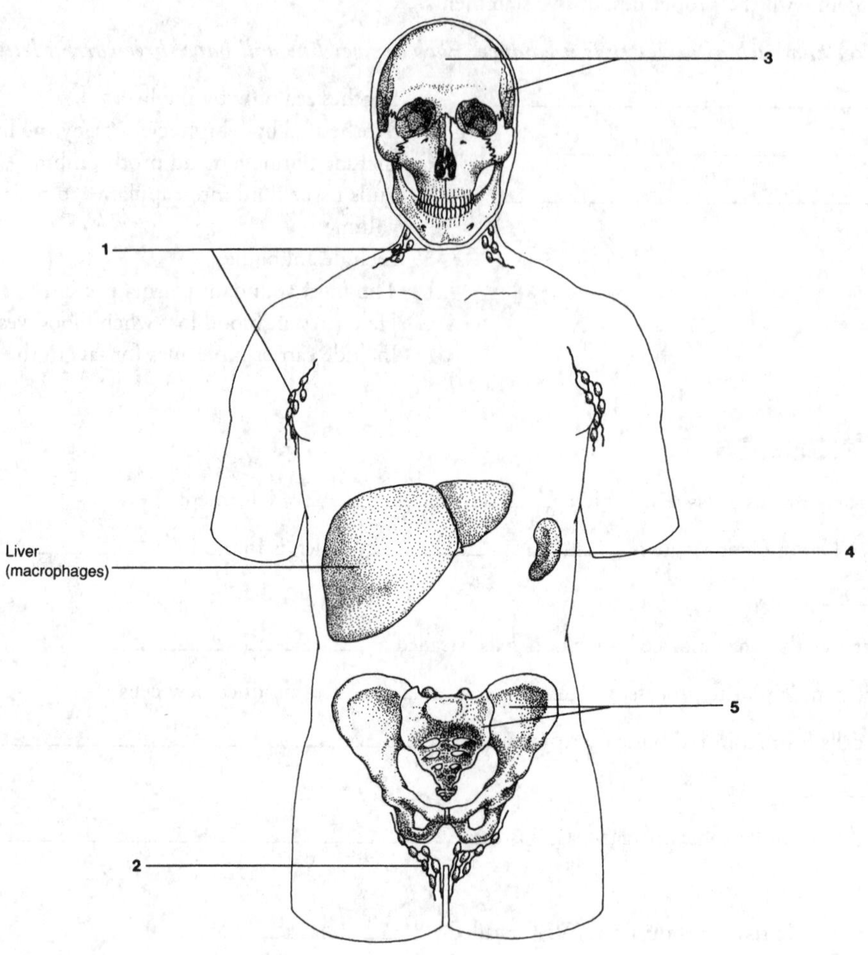

RED BLOOD CELLS

1. a) Red blood cells (RBCs) are also called ________________ and are formed in ________________.
   b) What major cellular structure do mature RBCs lack? ________________
   c) Describe the appearance of RBCs. ________________
2. a) In the embryo, RBCs are first produced by an external membrane called the ________________.
   b) Before the red bone marrow takes over completely, two other fetal organs contribute to RBC production; these are the ________________ and the ________________.
3. a) The oxygen-carrying protein in RBCs is ________________.
   b) The oxygen-carrying mineral in hemoglobin is ________________.
4. a) RBCs pick up oxygen when they circulate through the ________________ capillaries (in the ________________), and this hemoglobin is now called ________________.
   b) RBCs release oxygen in ________________ capillaries, and their hemoglobin is then called ________________.

5. The following diagram depicts the types of blood cells.

   Label the cells indicated.

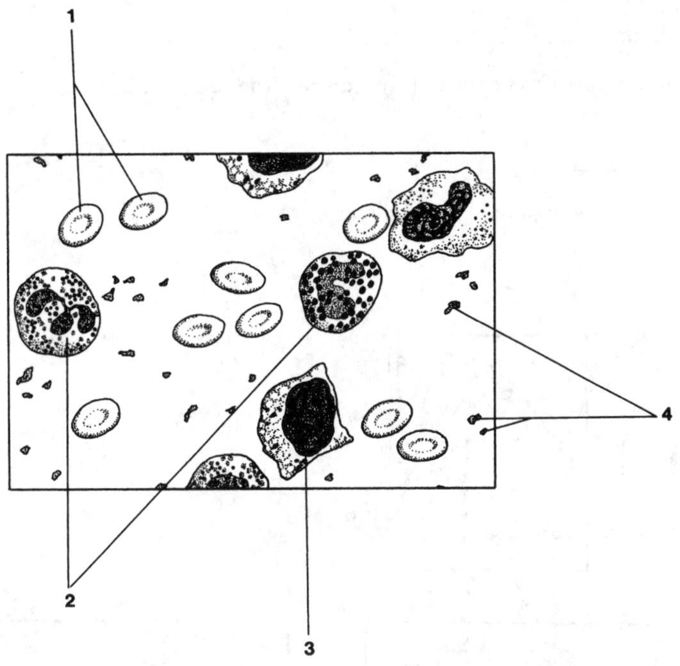

6. a) The major regulating factor for RBC production is the amount of ________________ in the blood and tissues.

   b) The term *hypoxemia* means ____.

   c) The term *hypoxia* means ____.

   d) When hypoxia occurs, the kidneys produce a hormone called ________________, which stimulates the red bone marrow to increase the rate of ________________.

7. a) In RBC formation, the last stage with a nucleus is called a ________________.

   b) The stage in which fragments of the ER are present is called a ________________.

   c) When these immature RBCs are present in large numbers in circulating blood, it means that there are not enough ________________ to transport sufficient ________________ throughout the body.

8. a) The nutrients needed for RBC formation include ________________ and ________________, which will become part of the hemoglobin molecule.

   b) The extrinsic factor is ________________, which is needed for the synthesis of ________________ by the stem cells in the red bone marrow.

   c) The intrinsic factor is produced by the lining of the ________________ (organ).

   d) The function of intrinsic factor is to prevent the digestion of ________________ and promote its absorption in the small intestine.

9. a) The life span of RBCs is approximately ________________ days.

   b) Macrophages (RE cells) that phagocytize old RBCs are found in the ________________, the ________________, and the ________________.

   c) The iron from old RBCs may be stored in the ________________ or transported to the red bone marrow for the synthesis of new ________________.

   d) The globin portion of the hemoglobin is digested to ________________, which may be used in the process of ________________.

10. a) The heme portion of the hemoglobin of old RBCs is converted to ____________________ by RE cells.

    b) Since it has no usefulness, bilirubin is considered a waste product. It is removed from circulation by the ____________________ (organ) and excreted into ____________________.

    c) The colon eliminates bilirubin in ____________________.

11. a) If the blood level of bilirubin rises, perhaps because of liver disease, the ____________________ may appear yellow.

    b) This is called ____________________.

12. The following diagram depicts the life cycle of red blood cells.

    Label the parts indicated.

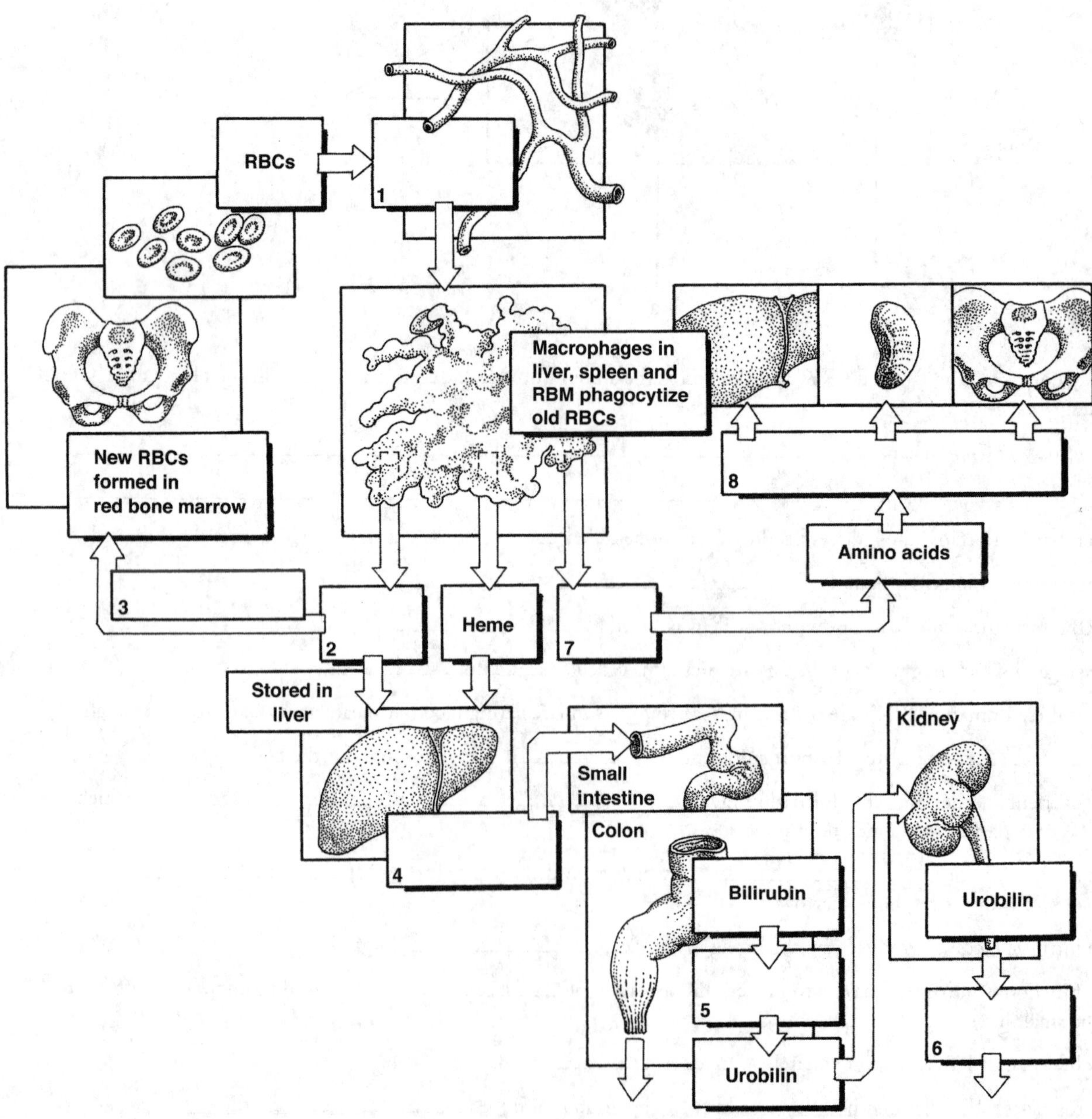

13. CBC values:

    1) The range of a normal RBC count is ____________________ to ____________________ cells/μL.

    2) The range of a normal hematocrit (Hct) is ____________________ % to ____________________ %.

    3) The range of a normal hemoglobin (Hb) level is ____________________ to ____________________ g/100 mL.

RED BLOOD CELL TYPES

1. The two most important RBC types are the ____________________ group and the ____________________ factor.

2. The ABO group contains four blood types, which are ____________________, ____________________, ____________________, and ____________________.

3. Complete the following chart with respect to RBC antigens and plasma antibodies for the four ABO group blood types.

	Type A	Type B	Type AB	Type O
Antigens present on RBCs				
Antibodies present in plasma				

4. The following diagram depicts RBCs representing each of the ABO types.

   Use the letters A or B to label the antigens and antibodies indicated.

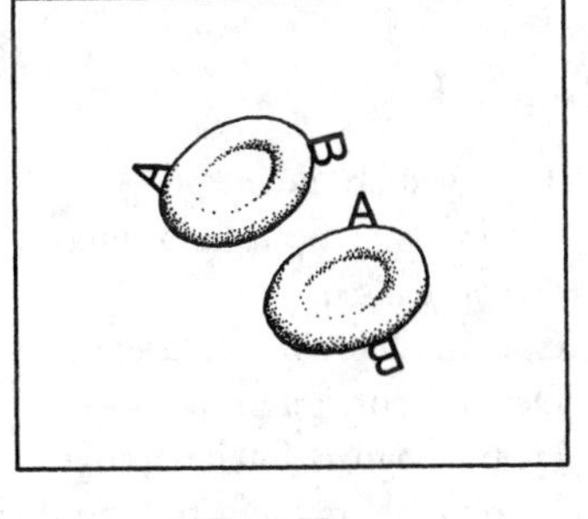

Type AB

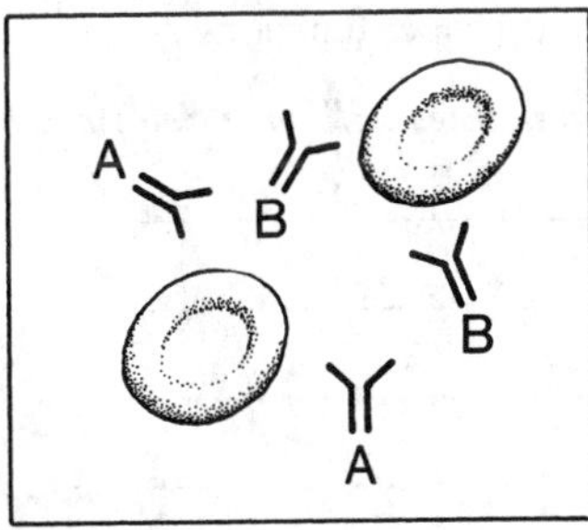

Type O

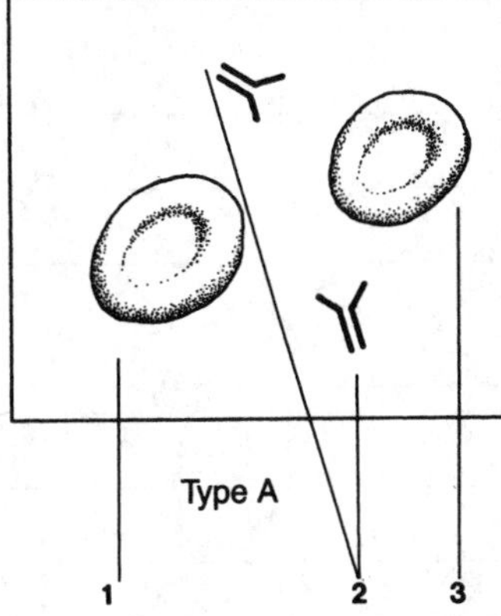

Type A

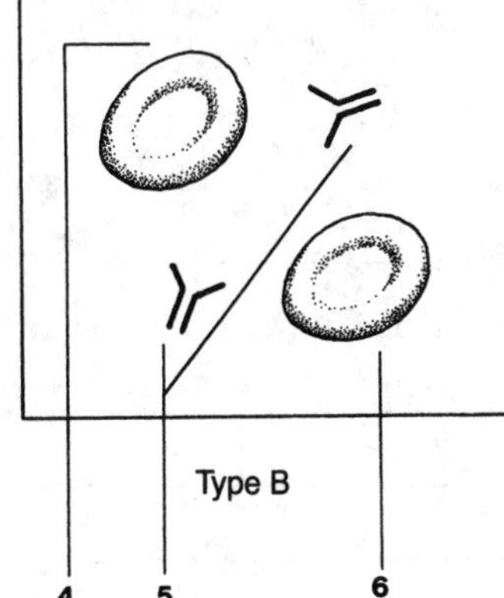

Type B

5. a) The Rh factor is another RBC antigen; it is often called D. A person who is Rh ________________ has this antigen on the RBCs. A person who is Rh ________________ does not have this antigen on the RBCs.

   b) Are anti-Rh antibodies naturally present in the plasma of a person who is Rh negative? ________________

   c) When would such antibodies be formed? ________________

6. a) If a patient receives a transfusion of an incompatible blood type, the donated RBCs will rupture; this is called ________________.

   b) The most serious effects of such a transfusion reaction occur in the ________________ (organ), when the capillaries there become clogged by ________________ from the ruptured RBCs.

WHITE BLOOD CELLS

1. a) White blood cells (WBCs) are also called ________________.

   b) The five kinds of WBCs are in two groups called ________________ and ________________.

2. a) The granular WBCs are the ________________, ________________, and ________________.

   b) What is a "band" cell? ________________

   c) The agranular WBCs are the ________________ and ________________.

3. In appearance, WBCs differ from RBCs in that all WBCs have ________________ present when the cells are mature.

4. The general function of WBCs is to protect the body from ________________ and to provide ________________ to certain infectious diseases.

5. Match each kind of WBC with its proper functions.

   Use each letter once. Two answer lines will have two correct letters.

   1) Neutrophils ________________
   2) Monocytes ________________
   3) T lymphocytes ________________
   4) B lymphocytes ________________
   5) Basophils ________________
   6) Eosinophils ________________

   A. The most abundant phagocytes
   B. Become macrophages to phagocytize pathogens or damaged tissue
   C. Contain heparin to prevent abnormal blood clotting
   D. Detoxify foreign proteins
   E. Help recognize foreign antigens
   F. Become plasma cells that produce antibodies
   G. Contain histamine, which contributes to inflammation
   H. Important in allergic reactions and parasitic infections

6. CBC values:

   1) The range of a normal WBC count is ________________ to ________________ cells/μL.

   2) A high WBC count is called ________________ and often indicates ________________.

   3) A low WBC count is called ________________. State one cause. ________________.

   4) State a normal range for each kind of WBC in a differential count.

      a) Neutrophils ________________
      b) Lymphocytes ________________
      c) Monocytes ________________
      d) Eosinophils ________________
      e) Basophils ________________

7. a) HLA are antigens found on WBCs that represent the antigens found on ____________________.

   b) The normal purpose of HLA is to provide a comparison for the immune system to be able to recognize ____________________ antigens.

   c) The HLA are also important when organs are transplanted. If tissue typing shows that the donated organ has one or more HLA types that match the HLA types of the recipient, there is less chance of ____________________ of the transplanted organ by the immune system of the recipient.

PLATELETS

1. Platelets are also called ____________________ and are formed in ____________________.
2. Platelets are fragments of the large bone marrow cells called ____________________, and ____________________ that is produced by the liver increases the rate of platelet formation.
3. a) The function of platelets is hemostasis, which means ____________________.

   b) The three mechanisms of hemostasis are ____________________, ____________________, and ____________________.
4. Vascular spasm is the mechanism of hemostasis necessary in large vessels that are ruptured or cut.

   a) What tissue in arteries and veins permits them to constrict? ____________________

   b) This tissue contracts in response to ____________________ released by platelets or to the ____________________ caused when the vessel ruptures.

   c) As a result, the opening in the vessel is made ____________________ and may then be covered by a ____________________.
5. a) Platelet plugs are the only effective mechanism of hemostasis for rupture of ____________________ (type of vessel).

   b) The rough surface of a ruptured capillary causes platelets to ____________________ and form a mechanical barrier over the opening.
6. The range of a normal platelet count is ____________________ to ____________________ cells/μL.
7. The term for a low platelet count is ____________________.

CHEMICAL BLOOD CLOTTING

1. The stimulus for chemical clotting (the clotting cascade) is a ____________________ surface within a vessel or a break in a vessel that also creates a ____________________ surface.
2. The clotting factors such as prothrombin and fibrinogen are synthesized by the ____________________ and circulate in the ____________________ until activated in the clotting mechanism.
3. a) The vitamin necessary for prothrombin synthesis is ____________________.

   b) Most of a person's supply of this vitamin is produced by the ____________________ in the person's own ____________________.
4. The mineral necessary for chemical clotting is ____________________, which the body stores in ____________________.

5. a) Stage 1 of clotting involves chemical factors released by ____________________ and other chemicals from ____________________.

   b) The result of stage 1 is the formation of ____________________.

   c) In stage 2, ____________________ converts prothrombin to ____________________.

   d) In stage 3, ____________________ converts fibrinogen to ____________________.

   e) The clot itself is made of ____________________, which forms a mesh over the break in the vessel.

6. a) The process of clot ____________________ pulls the edges of the break in the vessel together, which makes ____________________ of the area easier.

   b) Once a clot has accomplished its function, it is dissolved in a process called ____________________.

7. Abnormal clotting within vessels is prevented in several ways.

   a) The ____________________ epithelium (endothelium) that lines blood vessels is very smooth and repels platelets.

   b) An anticoagulant produced by basophils is ____________________.

   c) Antithrombin is produced by the ____________________ to inactivate excess ____________________.

   d) If excess thrombin is not inactivated, clotting may become a ____________________ cycle of harmful clotting because it is a ____________________ mechanism that requires an external brake.

8. a) The term for an abnormal clot in an intact vessel is ____________________.

   b) The term for a clot that dislodges and travels to another vessel is ____________________.

9. The following diagram depicts the chemical clotting cascade.

Label the parts indicated.

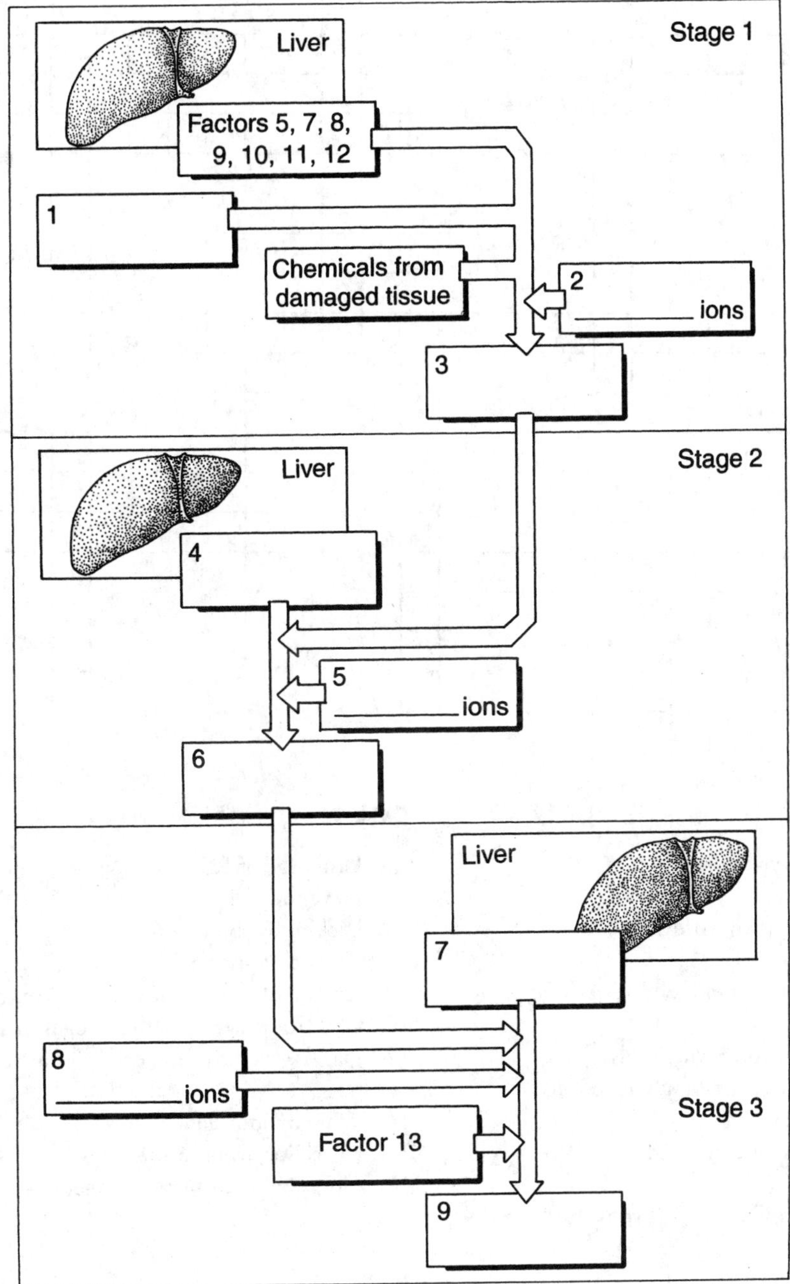

CROSSWORD PUZZLE

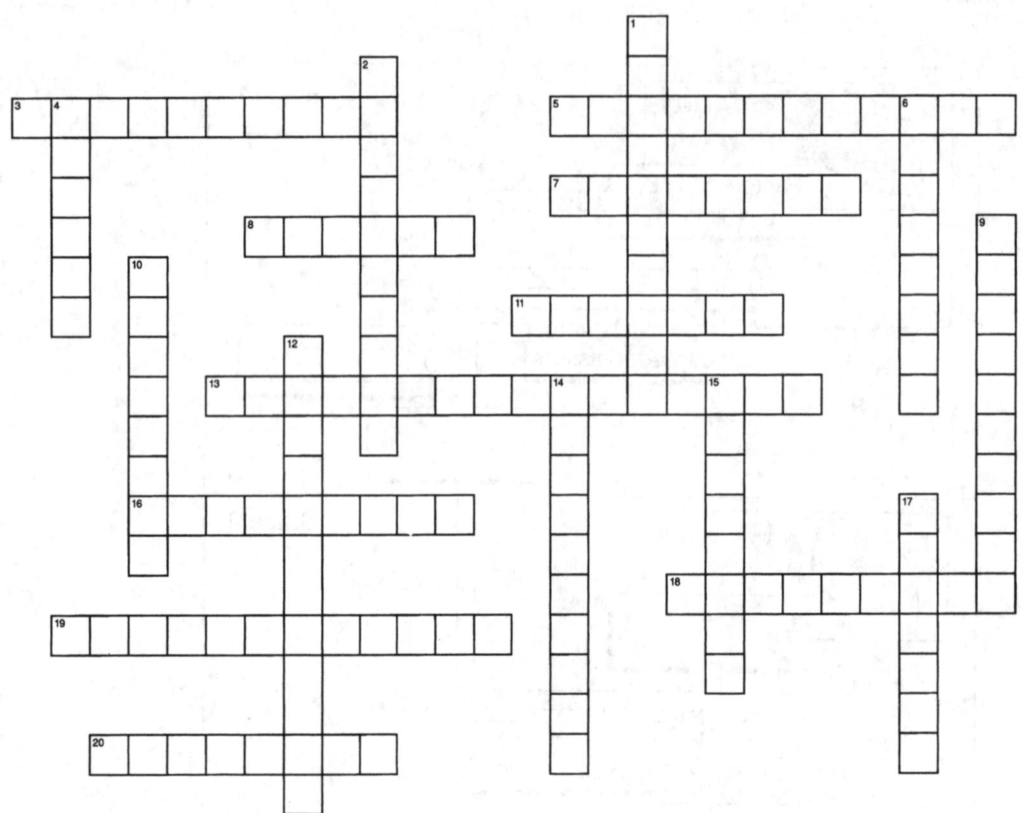

ACROSS

3. The most efficient phagocytic cell
5. Platelets
7. Contains four blood types (two words)
8. Liquid part of the blood
11. Prevents abnormal clotting within blood vessels; produced by basophils
13. Process to stop bleeding; involves chemicals
16. Precursor cells for the production of all blood cells (two words)
18. A bile pigment; what heme is converted into
19. Red blood cells
20. Another antigen on the RBCs, + or – (two words)

DOWN

1. Last stage of RBC development with a nucleus
2. Prevention of blood loss
4. Lack of red blood cells
6. What an abnormal clot is called
9. Gives RBCs the ability to carry oxygen
10. Clot from elsewhere that lodges in and obstructs a vessel
12. Has fragments of the endoplasmic reticulum (an RBC stage)
14. White blood cells
15. Protection from future cases of a disease
17. Most abundant plasma protein

CLINICAL APPLICATIONS

1. The lab results for a particular patient show these CBC values:

   RBCs—4.9 million/µL   WBCs—23,000/µL
   Hct—42%   platelets—190,000/µL
   Hb—14 g/100 mL

   On the basis of these values, you would expect that this patient (choose one answer):
   a) is anemic
   b) has an infection
   c) has bruising and bleeding
   d) is in good health, because all of these values are within normal ranges

2. a) Mr. and Mrs. R's first child has been diagnosed with hemophilia A. The parents have been very worried about the episodes of prolonged bleeding their child has experienced, but they are assured by their doctor that this form of hemophilia can be controlled. What clotting factor is the child lacking? ____________

   b) What will the child now receive? ____________

   c) Is this child a boy or a girl? ____________ Explain your answer.

   ____

3. a) Mrs. C has just given birth to her second child, and the baby is severely jaundiced. She is told that the baby has Rh disease of the newborn, which may also be called ____________.

   b) On the basis of this information, you can now say that Mrs. C is Rh ____________, Mr. C is Rh ____________, Baby C is Rh ____________, and the first child was probably Rh ____________.

   c) If this were Mrs. C's first pregnancy, she could have been given RhoGAM to prevent this from happening again in a later pregnancy. Explain what RhoGAM would have done.

   ____

4. a) Fifteen-year-old Mark needs a kidney transplant, and his six older brothers and sisters agree to be tested to see if they are suitable donors. The simple procedure they undergo is called tissue typing. What are these types called?

   ____

   b) One of Mark's sisters is found to have three HLA types in common with Mark. If she donates a kidney to Mark, what advantage will this provide?

   ____

   c) The search for a suitable kidney donor started with Mark's family because the HLA types are a ____________ characteristic.

MULTIPLE CHOICE TEST #1

Choose the correct answer for each question.

1. A band cell is an:
   a) immature red blood cell
   b) immature white blood cell
   c) immature platelet
   d) immature stem cell

2. A person with type AB blood has:
   a) A and B antigens on the RBCs and neither anti-A nor anti-B antibodies in the plasma
   b) A antigens on the RBCs and anti-B antibodies in the plasma
   c) B antigens on the RBCs and anti-A antibodies in the plasma
   d) A and B antibodies on the RBCs and no antigens in the plasma

3. The plasma protein that helps maintain blood volume by pulling tissue fluid into capillaries is:
   a) prothrombin b) albumin c) gamma globulin d) hemoglobin

4. The WBCs that recognize foreign antigens and produce antibodies are:
   a) monocytes b) neutrophils c) lymphocytes d) eosinophils

5. When old RBCs are destroyed, the waste product ________________ is formed and then excreted by the ________________.
   a) iron/kidneys, in urine
   b) bilirubin/liver, in bile
   c) hemoglobin/liver, in bile
   d) bilirubin/kidneys, in urine

6. The red bone marrow produces:
   a) RBCs only
   b) RBCs and platelets only
   c) RBCs and WBCs only
   d) all the types of blood cells

7. The cells in the hemopoietic tissues that undergo mitosis to produce all the types of blood cells are called:
   a) normoblasts b) stem cells c) reticulocytes d) megakaryocytes

8. Plasma makes up ________________ % of the total blood and is itself ________________ % water.
   a) 52% to 62%/91% b) 91% to 92%/55% c) 65% to 75%/80% d) 50% to 60%/50%

9. The extrinsic factor needed for DNA synthesis in the red bone marrow is:
   a) calcium b) vitamin B_{12} c) vitamin K d) iron

10. The organ that produces erythropoietin during hypoxia is the:
    a) liver b) red bone marrow c) spleen d) kidney

11. The oxygen-carrying protein of RBCs is:
    a) prothrombin b) hemoglobin c) myoglobin d) erythropoietin

12. A blood clot is made of:
    a) thrombin b) fibrinogen c) fibrin d) prothrombin activator

13. The mineral needed for chemical clotting is:
    a) sodium b) calcium c) iron d) potassium

14. The normal pH range of blood is:
    a) 7.15 to 7.25, slightly alkaline
    b) 7.35 to 7.45, slightly alkaline
    c) 7.00 to 7.15, neutral
    d) 7.45 to 7.55, slightly acidic

15. A large artery that is cut can contract in vascular spasm because its wall contains:
    a) smooth muscle
    b) elastic connective tissue
    c) blood
    d) simple squamous epithelium

16. The stimulus for the formation of a platelet plug or a blood clot is:
    a) a rough surface
    b) a smooth surface
    c) a very smooth surface
    d) intact endothelium

17. Lymphatic tissue is found in all of these except the:
    a) lymph nodes b) thymus gland c) liver d) spleen

18. A hematocrit is a measure of the:
    a) percentage of WBCs in the total blood
    b) percentage of blood in the body fluids
    c) percentage of plasma in total blood
    d) percentage of RBCs in the total blood

19. An abnormal clot that forms on a rough surface in an intact vessel is called:
   a) a thrombus b) an embolism c) a platelet plug d) inflammation

20. The Rh factor is an antigen that is found on the:
   a) WBCs of people who are Rh negative
   b) RBCs of people who are Rh negative
   c) WBCs of people who are Rh positive
   d) RBCs of people who are Rh positive

21. The intrinsic factor needed for absorption of the extrinsic factor is produced by cells lining the:
   a) liver b) stomach c) small intestine d) red bone marrow

22. In chemical clotting, fibrinogen is split to fibrin by:
   a) prothrombin activator b) prothrombin c) thrombin d) Factor 8

23. The WBCs that carry out most phagocytosis of pathogens are the:
   a) neutrophils and basophils
   b) basophils and monocytes
   c) monocytes and lymphocytes
   d) monocytes and neutrophils

24. The last immature stage in RBC production is the:
   a) reticulocyte, which is never found in circulating blood
   b) normoblast, which may be found in circulating blood
   c) normoblast, which is never found in circulating blood
   d) reticulocyte, which may be found in circulating blood

25. The function of erythropoietin is to:
   a) decrease RBC production
   b) increase RBC production
   c) decrease all blood cell production
   d) increase all blood cell production

MULTIPLE CHOICE TEST #2

Read each question and the four answer choices carefully. When you have made a choice, follow the instructions to complete your answer.

1. Which of the following is NOT transported by blood plasma?
   a) carbon dioxide
   b) nutrients
   c) hormones
   d) oxygen

   For your choice, state how it is transported in the blood.

   ____

2. Which statement is NOT true of the functions of WBCs?
   a) Lymphocytes produce antibodies.
   b) Neutrophils help detoxify foreign proteins.
   c) Monocytes become macrophages that phagocytize pathogens.
   d) Basophils contain heparin and histamine.

   For your choice, state the type of WBC that does have this function.

   ____

3. Which statement is NOT true of RBC formation?
   a) Red bone marrow is found in flat and irregular bones.
   b) The nutrients iron and protein are needed to become part of hemoglobin.
   c) The intrinsic factor is produced by the liver to prevent digestion of vitamin B_{12}.
   d) Erythropoietin increases the rate of RBC production in the red bone marrow.

   Reword your choice to make it a correct statement.

   ____

4. Which statement is NOT true of the chemical blood clotting cascade?
   a) The mineral necessary is calcium.
   b) Prothrombin, fibrinogen, and other clotting factors are synthesized by the spleen.
   c) The second stage of clotting results in the formation of thrombin.
   d) The clot itself is made of fibrin.

   Reword your choice to make it a correct statement.

   ____

5. Which statement is NOT true of the ABO blood types?
   a) Type O has neither A nor B antigens on the RBCs.
   b) A person has antibodies in the plasma for those antigens NOT present on the person's own RBCs.
   c) Type B has A antigens on the RBCs.
   d) This blood type is a hereditary characteristic.

   Reword your choice to make it a correct statement.

   ____

6. Which of these is NOT a normal value in a CBC?
   a) RBCs—5.1 million/μL
   b) Hb—8 g/100 mL
   c) WBCs—9000/μL
   d) Platelets—210,000/μL

   For your choice, state the correct normal range.

   ____

7. Which statement is NOT true of the destruction of old RBCs?
   a) The normal life span of RBCs is 90 days.
   b) Macrophages that destroy old RBCs are found in the liver, spleen, and red bone marrow.
   c) The iron may be stored in the liver or returned to the red bone marrow.
   d) The bilirubin formed is excreted by the liver into bile.

   Reword your choice to make it a correct statement.

   ____

8. Which statement is NOT true of the plasma proteins?
   a) The presence of albumin in plasma helps maintain blood volume.
   b) Gamma globulins are antibodies produced by basophils.
   c) Alpha and beta globulins are carriers for molecules such as fats.
   d) Fibrinogen and prothrombin are clotting factors synthesized by the liver.

   Reword your choice to make it a correct statement.

   ____

9. Which statement is NOT true of RBC formation?
   a) A normoblast is the last stage with a cell membrane.
   b) Small numbers of reticulocytes are often found in circulating blood.
   c) The precursor cells are stem cells.
   d) Vitamin B_{12} may also be called the extrinsic factor.

   Reword your choice to make it a correct statement.

   ____

10. Which statement is NOT true of the prevention of abnormal clotting?
    a) Antithrombin inactivates excess thrombin.
    b) Heparin is a natural anticoagulant that blocks the chemical clotting cascade.
    c) The simple squamous epithelium that lines blood vessels is smooth to prevent abnormal clotting.
    d) Heparin is produced by monocytes.

    Reword your choice to make it a correct statement.

    ____

11. Which statement is NOT true of the prevention of blood loss in ruptured vessels?
    a) Vascular spasm in a large vessel will make the size of the ruptured area smaller.
    b) Platelet plugs are useful only in medium-sized vessels.
    c) Chemical clotting is stimulated by damage to a vessel.
    d) If a large vessel did not first constrict, the clot that forms would be washed out by the flow of blood.

    Reword your choice to make it a correct statement.

    ____

12. Which term is NOT paired with its correct meaning?
    a) leukocytosis—increased WBC count
    b) thrombus—abnormal blood clot
    c) thrombocytopenia—decreased platelet count
    d) anemia—increased RBC count

    For your choice, state the correct meaning of this term.

    ____

MULTIPLE CHOICE TEST #3

Each question is a series of statements concerning a topic in this chapter. Read each statement carefully and select all of the correct statements.

1. Which of the following statements are true of blood cells?
   a) A megakaryocyte is a large cell that will break up into platelets.
   b) Basophils become macrophages that can phagocytize pathogens.
   c) B lymphocytes become plasma cells that produce blood plasma.
   d) Old RBCs are phagocytized by the macrophages of the colon.
   e) Red bone marrow produces all of the kinds of blood cells.
   f) During an infection, WBCs become more numerous than RBCs.
   g) Reticulocytes are immature neutrophils and are usually found in the red bone marrow.
   h) RBCs carry oxygen that is bonded to the iron in hemoglobin.

2. Which of the following statements are true of hemostasis?
   a) Clotting factors such as fibrinogen are formed by stem cells in the RBM.
   b) Breaks in capillaries are usually blocked by platelet plugs.
   c) Any blood vessel with smooth muscle can respond to damage with vascular spasm.
   d) Platelets will stick to any rough surface in a blood vessel.
   e) Severe damage to a blood vessel will prevent clotting from starting.
   f) The reactions of chemical clotting require the mineral magnesium.
   g) The result of stage 2 of clotting is fibrin.
   h) Stage 1 of clotting results in the formation of thrombin.

3. Which of the following statements are true of blood?
   a) The normal pH of blood varies from 7.5 to 8.5.
   b) The RBC type O (+) is the most common type and also the universal donor.
   c) RBC type AB means that both alpha and beta chains are present in the hemoglobin.
   d) Plasma distributes heat from warm organs such as skeletal muscles to cooler body areas.
   e) The HLA types contribute to the immune system's ability to recognize self and distinguish self from foreign.
   f) Albumin is a plasma protein made by the spleen and RBM.
   g) A normal range of hemoglobin is 38 to 48 g/100 mL.
   h) A type A (–) person cannot donate blood to a type O (+) person.
   i) A normal hematocrit range is 12% to 18%.
   j) Lymphatic tissue produces lymphocytes and is found in the liver.
   k) A WBC count of 6,000/µL is considered high normal.
   l) Both folic acid and vitamin B_{12} are required for RBC production in the RBM.

Chapter 12

The Heart

This chapter describes the heart, which is the pump of the circulatory system. The anatomy of the heart includes the chambers and their vessels and the coronary blood vessels. The physiology of the heart includes the generation and regulation of the heartbeat and the relationship between cardiac output and the functioning of the body as a whole.

CARDIAC MUSCLE TISSUE

1. a) Cardiac muscle cells are branched and may also be called ____________________.

   b) These cells have many mitochondria, the function of which is to produce ____________________.

2. a) The units of contraction of cardiac muscle fibers are ____________________.

   b) The striations of cardiac fibers are the result of the arrangement of the contracting proteins ____________________ and ____________________.

3. a) Electrically, an action potential is ____________________ followed by ____________________.

   b) An action potential involves the entry of ____________________ into the muscle fiber, followed by the exit of ____________________ from the muscle fiber.

   c) Cardiac muscle cells generate their own action potentials, which means that they do not require ____________________ in order to contract.

   d) The electrical impulses of cardiac myocytes spread quickly to adjacent cells because of the presence of ____________________ at the ends of the cells.

   e) The speed of the electrical impulses ensures that in one heartbeat, the two ____________________ will simultaneously contract first, followed by the simultaneous contraction of the two ____________________.

4. a) Cardiac muscle is an endocrine tissue because some cells produce a hormone called ____________________ (abbreviated ____________________); the stimulus for its secretion is ____________________.

   b) The effect of ANP on the kidneys is to decrease the reabsorption of ____________________, which in turn increases the excretion of ____________________.

   c) The effect of ANP on vascular smooth muscle is to cause ____________________.

   d) The combined effect of b) and c) is to ____________________.

   e) ANP also stimulates the conversion of white adipocytes to brown adipocytes so that fats are not stored, but rather their energy is released in the form of ____________________.

FUNCTION AND LOCATION

1. The function of the heart is to ______________________________.
2. a) The heart is located in the mediastinum, which is the area between the ______________ in the ______________ cavity.
   b) Name the organ directly below the heart. ______________

PERICARDIAL MEMBRANES

1. How many layers of pericardial membranes are around the heart? ______________
2. The outermost layer is called the ______________ pericardium and is made of ______________ tissue.
3. a) The serous membrane that lines the fibrous pericardium is called the ______________ pericardium.
   b) The serous membrane that is on the surface of the heart muscle is called the ______________ pericardium, or the ______________.
4. The function of the serous fluid produced by the serous layers is to ______________ as the heart beats.

CHAMBERS OF THE HEART

1. The heart has ______________ chambers: two upper chambers called ______________ and two lower chambers called ______________.
2. Name the tissue that forms the walls of these chambers: ______________, or its other name, ______________.
3. a) The chambers of the heart are lined with simple squamous epithelium called the ______________, which also covers the ______________ of the heart.
   b) The most important physical characteristic of the endocardium is that it is very ______________, which prevents abnormal ______________ in the chambers of the heart.
4. a) Which pair of chambers has the thicker walls? ______________
   b) The wall of myocardium between the two atria is the ______________.
   c) The wall of myocardium between the two ventricles is the ______________.

CHAMBERS—VESSELS AND VALVES

1. Match each chamber of the heart with its proper vessel (a letter) and the function of the vessel (a number).

   Use each letter and number once.

   1) Right atrium ______________
   2) Left atrium ______________
   3) Right ventricle ______________
   4) Left ventricle ______________

   Vessel(s)
   A. Pulmonary veins
   B. Pulmonary artery
   C. Aorta
   D. Superior and inferior caval veins

   Function of Vessel(s)
   1. Return blood from the body
   2. Takes blood to the body
   3. Takes blood to the lungs
   4. Return blood from the lungs

2. The following diagram depicts an anterior view of the exterior of the heart.

Label the parts indicated.

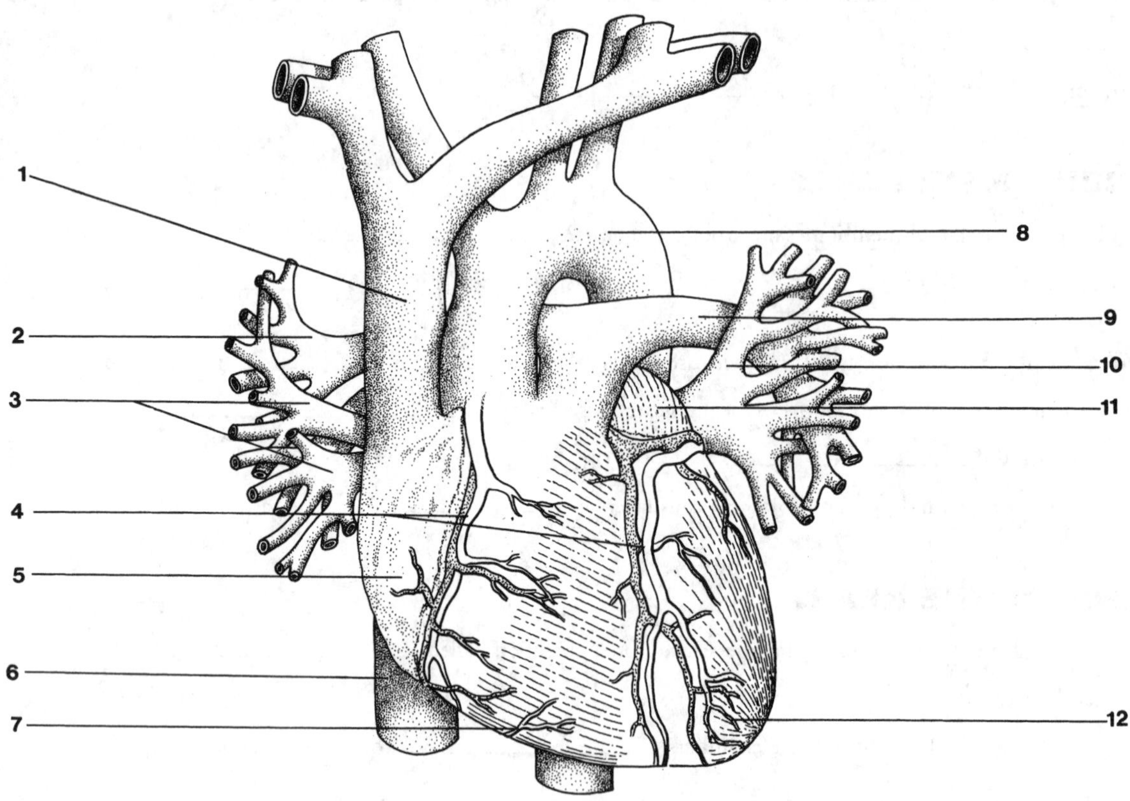

3. Match each heart valve with its proper location (a letter statement) and its function (a number statement).

Use each letter and number once.

1) Tricuspid valve ______________________

2) Mitral valve ______________________

3) Pulmonary semilunar valve ______________________

4) Aortic semilunar valve ______________________

Location

A. At the junction of the right ventricle and the pulmonary artery
B. Between the left atrium and left ventricle
C. Between the right atrium and right ventricle
D. At the junction of the left ventricle and the aorta

Function

1. Prevents backflow of blood to the left ventricle when the ventricle relaxes
2. Prevents backflow of blood to the left atrium when the left ventricle contracts
3. Prevents backflow of blood to the right atrium when the right ventricle contracts
4. Prevents backflow of blood to the right ventricle when the ventricle relaxes

4. The following diagram depicts the interior of the heart (anterior view).

Label the parts indicated.

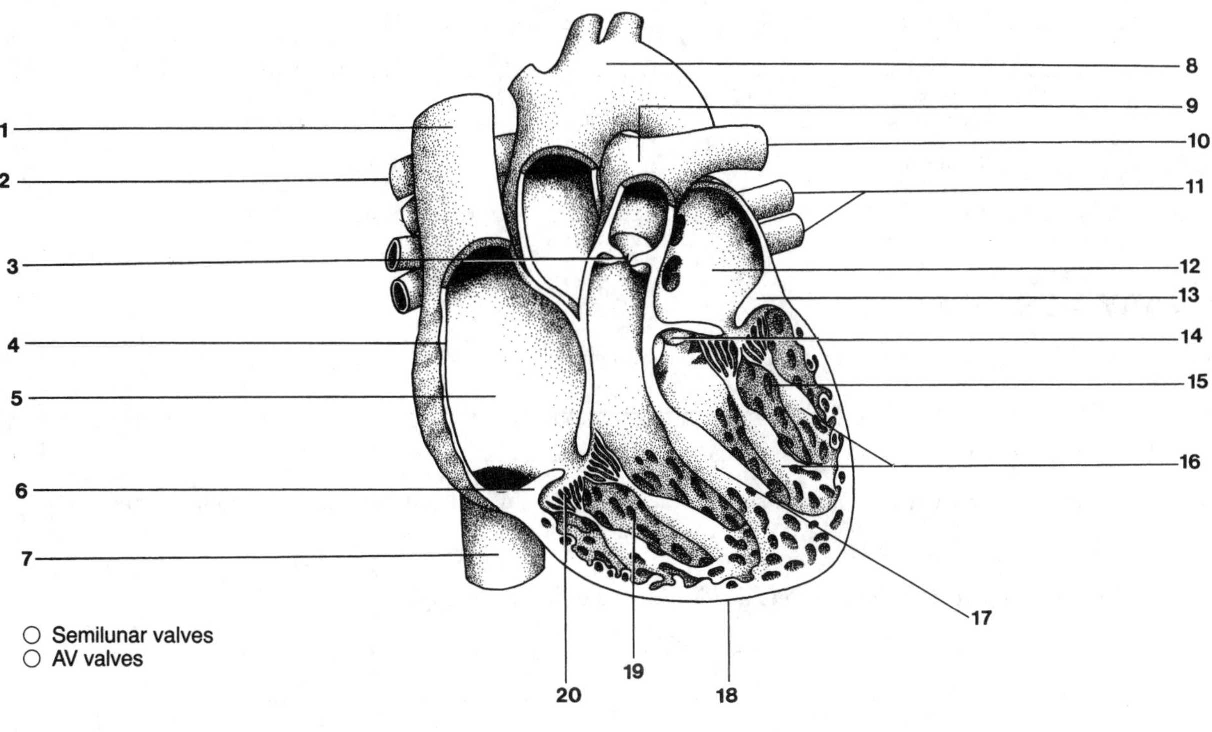

5. a) The papillary muscles are columns of ______________ that project into each ______________ (chamber) of the heart.

   b) The chordae tendineae are strands of ______________ tissue that extend from the papillary muscles to the ______________.

   c) The function of these structures is to prevent inversion of the AV valves when the ventricles ______________. (AV valves is the term for the mitral and tricuspid valves.)

CIRCULATION THROUGH THE HEART

1. a) The heart is actually a double pump, and both pumps work simultaneously. The left side of the heart receives blood from the ______________ and pumps this blood to the ______________.

   b) The right side of the heart receives blood from the ______________ and pumps this blood to the ______________.

2. Number the following in proper sequence with respect to the flow of blood through the heart to and from the body and lungs.

Begin at the left atrium.

1	Left atrium	________	Lungs
________	Left ventricle	________	Pulmonary artery
________	Right atrium	________	Pulmonary veins
________	Right ventricle	________	Aorta
________	Body	________	Superior and inferior caval veins

CORONARY VESSELS

1. a) The purpose of the coronary vessels is to circulate blood throughout the ________.

   b) The most important substance in the blood is ________.

2. a) The right and left coronary arteries are branches of the ________.

   b) The coronary sinus is formed by the union of coronary veins, and it returns blood from the myocardium to the ________ (chamber).

3. a) What will happen to a part of the heart muscle that is deprived of its blood supply?

   ____

   b) State one possible cause of this. ________

CARDIAC CYCLE AND HEART SOUNDS

1. The cardiac cycle is the term for the sequence of events in one ________.
2. a) The term *systole* means ________.

   b) The term *diastole* means ________.

3. In the cardiac cycle, when the atria are in systole, the ventricles are in ________, and when the ventricles are in systole, the atria are in ________.
4. Number the events of one cardiac cycle in the proper sequence. Two events have been properly numbered to get you started.

1	Blood continuously flows into both atria.
________	Two-thirds of the atrial blood flows passively into the ventricles.
________	The right and left AV valves are closed, and the aortic and pulmonary semilunar valves are opened.
________	The pressure of incoming blood opens the right and left AV valves.
________	The atria contract to pump the remaining blood into the ventricles.
________	Ventricular contraction pumps blood into the arteries.
5	The atria relax, and the ventricles begin to contract.
________	The ventricles relax, and the semilunar valves are closed.

5. From atria to ventricles, most blood ________ (flows passively or is pumped).
6. From ventricles to arteries, all blood ________ (flows passively or is pumped).

7. a) The cardiac cycle normally creates ____________________ (how many) heart sounds.
   b) The first sound is caused by the closure of the ____________________ valves.
   c) The second sound is caused by the closure of the ____________________ valves.
   d) An extra sound heard during the cardiac cycle is called a ____________________.

CARDIAC CONDUCTION PATHWAY

1. a) The cardiac conduction pathway is the pathway of electrical ____________________ throughout the heart during each heartbeat.
   b) Must the heart receive nerve impulses to cause contraction? ____________________
   c) Nerve impulses regulate only the ____________________ of contraction.
2. Name the parts of the conduction pathway in order.
   1) ______SA node______ In the wall of the right atrium
   2) ____________________ In the lower interatrial septum
   3) ____________________ In the upper interventricular septum
   4) ____________________ Extend through the interventricular septum
   5) ____________________ To the rest of the myocardium of the ventricles
3. The following diagram depicts the interior of the heart.

   Label the parts of the conduction pathway.

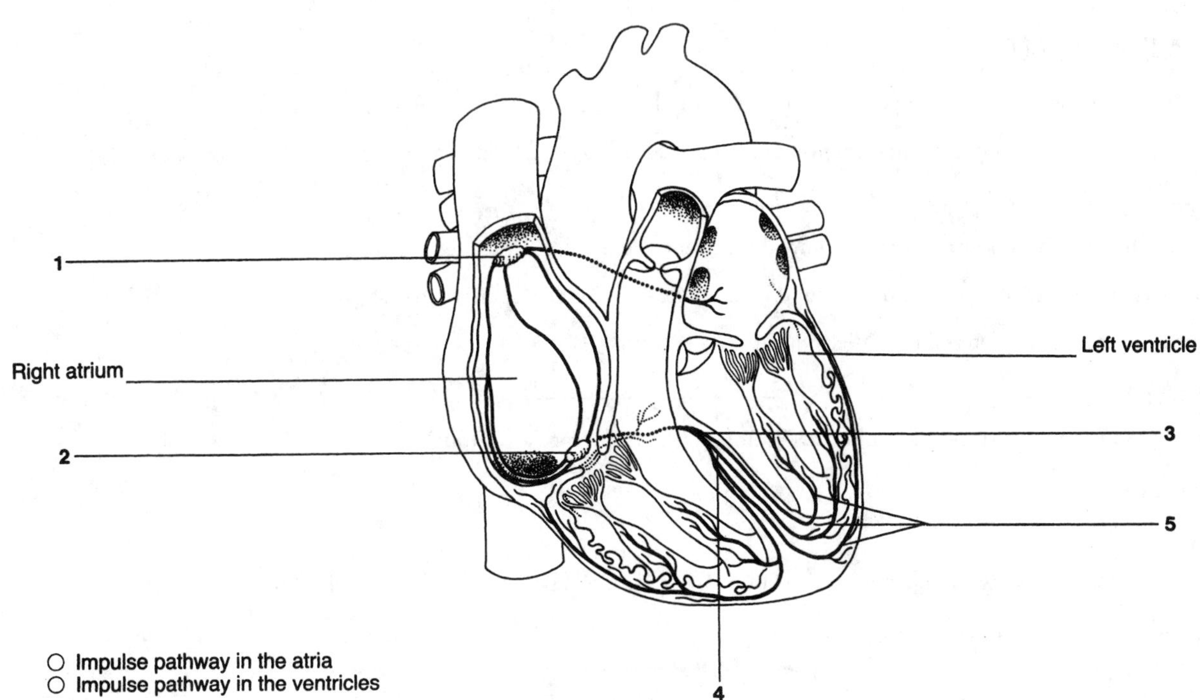

4. a) The SA node initiates each heartbeat because it has the most ______________ rate of contraction.
   b) This is so because the cells of the SA node are more permeable to ______________ ions and therefore ______________ more rapidly than other cardiac muscle cells.
5. Name the parts of the conduction pathway that bring about atrial systole. ______________ and ______________
6. Name the parts of the conduction pathway that bring about ventricular systole. ______________, ______________, and ______________
7. a) The electrical activity of the heart can be recorded and depicted in an ______________.
   b) The term for irregular heartbeats is ______________.
   c) The most serious arrhythmia is ______________, which is very rapid and uncoordinated.

HEART RATE

1. a) State a normal range of resting heart rate for a healthy adult. ______________
   b) Name the vital sign that is a measure of heart rate. ______________
2. Children and infants usually have ______________ (higher or lower) heart rates than do adults, because children are ______________ (younger or smaller) than adults.
3. Athletes in good physical condition often have ______________ (higher or lower) heart rates than do other adults.

CARDIAC OUTPUT

1. Cardiac output is the amount of blood pumped by a ventricle in ______________ (time).
2. Cardiac output is important to maintain normal ______________ and to provide for the transport of ______________ to tissues.
3. a) Stroke volume is the amount of blood pumped by a ventricle in ______________.
   b) State a normal range of resting stroke volume. ______________ mL per beat
4. State the formula used to determine cardiac output:
   cardiac output = ______________ × ______________
5. Problem: Pulse is 60 bpm and stroke volume is 75 mL. What is the cardiac output?
   ____
6. Problem: An athlete's resting cardiac output is 6,000 mL per minute, and her stroke volume is 100 mL per beat. What is her pulse? ______________
7. Problem: During exercise, pulse is 120 bpm and stroke volume is 100 mL. What is the cardiac output?
   ____
8. If resting cardiac output is 6 L, and maximum exercise cardiac output is 16 L, what is the cardiac reserve?
   ____

REGULATION OF HEART RATE

1. a) The cardiac centers are located in the ____________________ of the brain.

   b) Name the two centers: the ____________________ and the ____________________

2. a) Sympathetic nerves to the heart transmit impulses that ____________________ heart rate and force of contraction.

   b) The parasympathetic nerves to the heart are the ____________________ nerves, which transmit impulses that ____________________ heart rate.

3. To bring about changes in heart rate, the medulla must receive sensory information.

   a) The receptors that detect changes in blood pressure are called ____________________, and they are located in the ____________________ sinus and the ____________________ sinus.

   b) The receptors that detect changes in the oxygen level of the blood are called ____________________, and they are located in the ____________________ body and the ____________________ body.

4. a) The sensory nerves for the carotid sinus and body are the ____________________ nerves.

   b) The sensory nerves for the aortic sinus and body are the ____________________ nerves.

5. The regulation of heart rate is a reflex; that is, an automatic motor response stimulated by sensory information. Complete the following description of the reflex arc that is started when blood pressure decreases.

   1) When blood pressure to the brain decreases, the decrease is detected by ____________________ located in the carotid ____________________.

   2) Sensory impulses travel along the ____________________ nerves to the medulla of the brain.

   3) The ____________________ center is stimulated and generates impulses that are carried by ____________________ nerves to the heart.

   4) The effect of these impulses is to ____________________ heart rate and force of contraction to ____________________ blood pressure back to normal.

6. The hormone that increases heart rate and force of contraction in stressful situations is ____________________, secreted by the ____________________.

7. The following diagram depicts the heart and the nervous system.

Label the parts indicated.

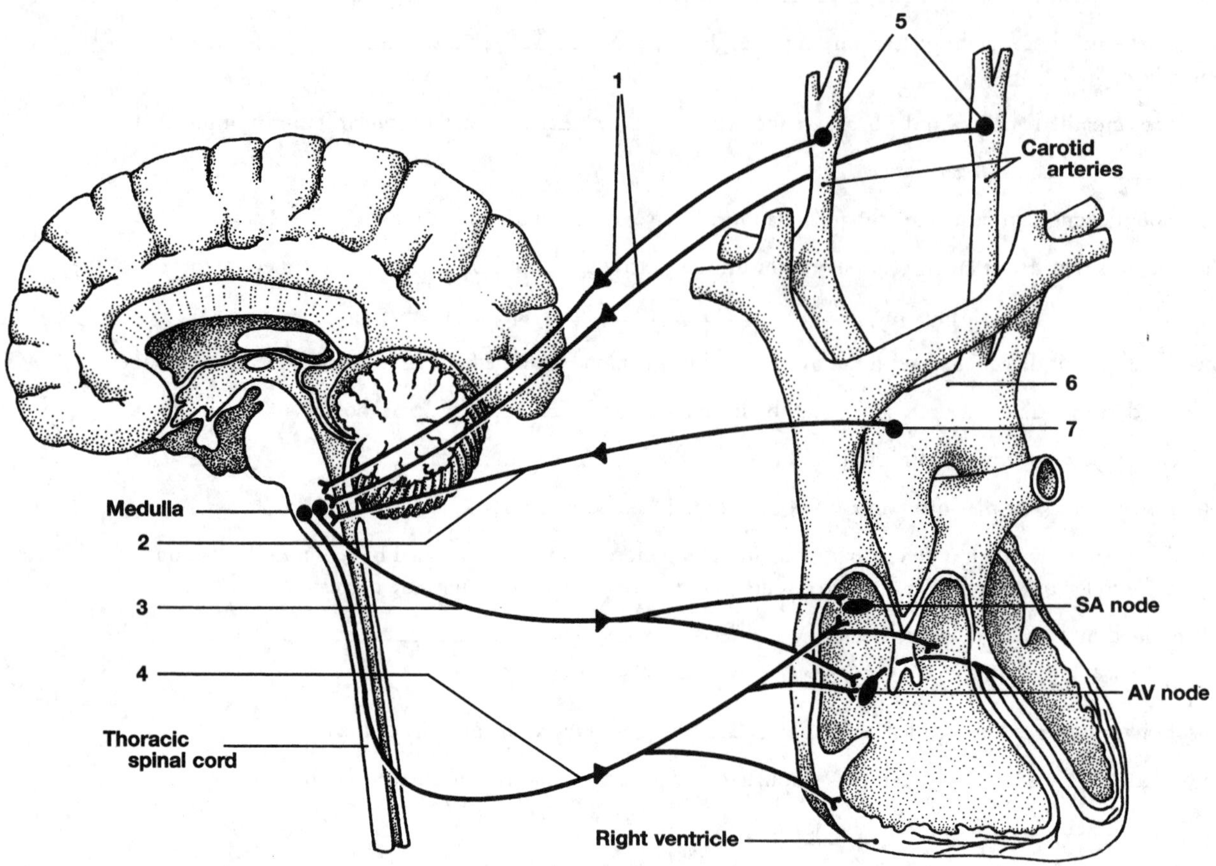

CROSSWORD PUZZLE

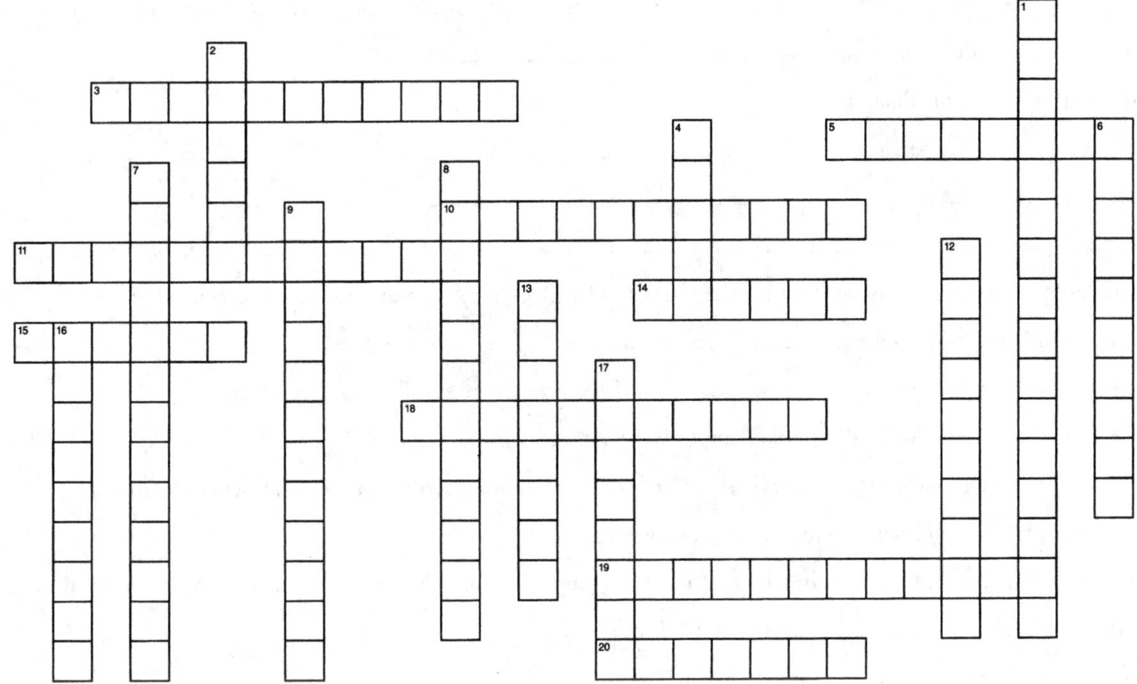

ACROSS

3. Area in the thoracic cavity between the lungs
5. Relaxation of chambers of the heart
10. Heart chambers are lined with this
11. Very rapid and uncoordinated ventricular beat
14. Natural pacemaker of the heart (abbrev.)
15. Upper chamber of the heart
18. Irregular heartbeats
19. Amount of blood pumped by a ventricle per beat (two words)
20. Contraction of chambers of the heart

DOWN

1. First branches of the ascending aorta (two words)
2. ________________ valve; prevents backflow of blood from the left ventricle to the left atrium
4. Largest artery of the body
6. Visceral pericardium
7. Amount of blood pumped by a ventricle in 1 minute (two words)
8. The amount of blood that returns to the heart (two words)
9. Sequence of events in one heartbeat (two words)
12. Cardiac muscle tissue
13. Part of the myocardium is deprived of its blood supply and becomes ________________
16. ________________ valve; prevents backflow of blood from the right ventricle to the right atrium
17. Narrowing of a valve

CLINICAL APPLICATIONS

1. a) Mr. D is 43 years old, and his father died of a heart attack at the age of 40. Mr. D's favorite meal is steak and a baked potato with butter. He also smokes a pack of cigarettes each day and says he doesn't get much chance to exercise. Mr. D. is a person at high risk for a ________________.

   b) List three of the risk factors that Mr. D has.

   1) ________________

   2) ________________

   3) ________________

2. a) Mr. N has been hospitalized following a mild heart attack. His ECG is being monitored to detect possible ventricular fibrillation, which means ________________.

   b) Ventricular fibrillation is a medical emergency because if the ventricles are fibrillating, they are not ________________ blood, and cardiac output will ________________.

3. a) Mr. W is 55 years old and has been hospitalized for surgery to implant an artificial pacemaker for his heart. The heart's own natural pacemaker, which is not functioning properly, is the ________________.

   b) Without an artificial pacemaker to maintain a normal heart rate, the remainder of Mr. W's heart would continue to beat but at a ________________ (faster or slower) rate.

   c) What would then happen to Mr. W's cardiac output? ________________

4. Name the cardiac disorder suggested by each statement.

   1) An abnormal extra sound heard during the cardiac cycle ________________

   2) Death of heart muscle due to lack of oxygen ________________

   3) Lipid deposits within the coronary arteries ________________

MULTIPLE CHOICE TEST #1

Choose the correct answer for each question.

1. During one cardiac cycle:
   a) the atria contract first, followed by contraction of the ventricles
   b) the ventricles contract first, followed by contraction of the atria
   c) the atria and ventricles contract simultaneously
   d) there is no specific pattern of contraction

2. Backflow of blood from the ventricles to the atria is prevented by the:
   a) aortic and pulmonary semilunar valves
   b) tricuspid and pulmonary semilunar valves
   c) mitral and aortic semilunar valves
   d) mitral and tricuspid valves

3. The outermost of the pericardial membranes is the:
   a) visceral pericardium b) fibrous pericardium c) parietal pericardium d) epicardium

4. Each normal heartbeat is initiated by the:
   a) AV node b) bundle of His c) SA node d) ventricles

5. Backflow of blood from the arteries to the ventricles is prevented by the:
   a) aortic and pulmonary semilunar valves
   b) tricuspid and pulmonary semilunar valves
   c) mitral and aortic semilunar valves
   d) mitral and tricuspid valves

6. The vessel into which the left ventricle pumps blood is the:
   a) superior vena cava b) pulmonary artery c) inferior vena cava d) aorta

7. The veins that return blood to the right atrium are the:
   a) thoracic veins
   b) superior and inferior caval veins
   c) pulmonary veins
   d) visceral veins

8. The heart is located:
   a) lateral to the lungs and superior to the diaphragm
   b) medial to the lungs and inferior to the diaphragm
   c) medial to the lungs and superior to the diaphragm
   d) lateral to the lungs and inferior to the diaphragm

9. The function of the serous fluid of the pericardial membranes is to:
   a) prevent friction as the heart beats
   b) prevent abnormal clotting in the chambers of the heart
   c) keep blood flowing through the heart
   d) nourish the myocardium

10. The term *systole* means:
    a) rapid heart rate b) slow heart rate c) relaxation d) contraction

11. The purpose of the coronary vessels is to:
    a) circulate blood through the brain
    b) circulate blood through the heart
    c) circulate blood through the lungs
    d) circulate blood through the body

12. A normal range of heart rate for a healthy adult is ________________ beats per minute.
    a) 90 to 100 b) 70 to 100 c) 60 to 80 d) 50 to 60

13. The chambers of the heart that receive blood from veins are the:
    a) right and left ventricles
    b) right and left atria
    c) right atrium and ventricle
    d) left atrium and ventricle

14. The endocardium lines the chambers of the heart and:
    a) prevents friction when the heart beats
    b) is smooth to prevent abnormal clotting within the heart
    c) prevents backflow of blood
    d) helps pump blood

15. The centers that regulate heart rate are located in the:
    a) medulla b) carotid sinus c) cerebral cortex d) wall of the right atrium

16. The nerves that transmit impulses to decrease the heart rate are the:
    a) sympathetic nerves b) coronary nerves c) glossopharyngeal nerves d) vagus nerves

17. The electrical activity of the heart may be depicted in an:
    a) EEG b) EGG c) ECG d) EEK

18. The amount of blood pumped by a ventricle in 1 minute is called:
    a) stroke volume b) pulse c) coronary blood flow d) cardiac output

19. The normal heart sounds are caused by:
    a) relaxation of the atria
    b) relaxation of the ventricles
    c) opening of the valves
    d) closure of the valves

20. Changes in blood pressure are detected by:
    a) pressoreceptors in the carotid and aortic sinuses
    b) pressoreceptors in the ventricles
    c) pressoreceptors in the medulla
    d) blood vessels in the medulla

21. The hormone ANP increases the loss of ________________ in urine to decrease blood volume and blood pressure.
    a) calcium ions b) sodium ions and water c) potassium ions d) water only

22. A heart rate below 60 bpm is called:
    a) a murmur b) bradycardia c) tachycardia d) galloping

23. The coronary sinus receives blood directly from the:
    a) coronary veins b) coronary arteries c) aorta d) pulmonary veins

24. The first part of the cardiac conduction pathway in the ventricles is the:
    a) bundle of His b) AV node c) bundle branch d) Purkinje fiber

25. The difference between resting cardiac output and maximum exercise cardiac output is called the:
    a) cardiac override b) cardiac limit c) cardiac reserve d) cardiac extra

MULTIPLE CHOICE TEST #2

Read each question and the four answer choices carefully. When you have made a choice, follow the instructions to complete your answer.

1. Which statement is NOT true of the pericardial membranes?
   a) Serous fluid prevents friction between the serous layers as the heart beats.
   b) The fibrous pericardium is the thick, outermost layer.
   c) The parietal pericardium lines the fibrous pericardium.
   d) The visceral pericardium may also be called the endocardium.

   Reword your choice to make it a correct statement.

   ____

2. Which statement is NOT true of the chambers and vessels of the heart?
   a) Both ventricles pump blood into arteries.
   b) The left ventricle pumps blood into the pulmonary artery.
   c) The left atrium receives blood from the pulmonary veins.
   d) The right atrium receives blood from the superior and inferior caval veins.

   Reword your choice to make it a correct statement.

   ____

3. Which statement is NOT true of the valves of the heart?
   a) The mitral valve prevents backflow of blood from the right ventricle to the right atrium.
   b) The right and left AV valves are anchored by papillary muscles and chordae tendineae.
   c) The semilunar valves prevent backflow of blood from the arteries to the ventricles.
   d) The valves are made of endocardium reinforced with connective tissue.

   Reword your choice to make it a correct statement.

   ____

4. Which statement is NOT true of the cardiac cycle?
   a) Most blood flow from the atria to the ventricles flows passively.
   b) All blood flow from ventricles to arteries is actively pumped.
   c) The atria contract first; then the ventricles contract.
   d) When the atria are in systole, the ventricles are in systole.

   Reword your choice to make it a correct statement.

   ____

5. Which statement is NOT true of the cardiac conduction pathway?
   a) The Purkinje fibers carry impulses to the ventricular myocardium.
   b) The bundle of His is located in the upper interventricular septum.
   c) The AV node initiates each heartbeat.
   d) The SA node and AV node promote contraction of the atria.

   Reword your choice to make it a correct statement.

   ____

6. Which statement is NOT true of cardiac output?
   a) Cardiac output is the amount of blood pumped by a ventricle in one beat.
   b) A normal cardiac output is necessary to maintain normal blood pressure.
   c) Cardiac output equals stroke volume times pulse rate.
   d) A normal cardiac output is necessary to maintain oxygenation of tissues.

   Reword your choice to make it a correct statement.

   ___

7. Which statement is NOT true of the regulation of heart rate?
   a) The cardiac centers are located in the medulla.
   b) Sympathetic impulses to the heart decrease the heart rate.
   c) Pressoreceptors detect changes in blood pressure to supply sensory information to the medulla.
   d) The parasympathetic nerves to the heart are the vagus nerves.

   Reword your choice to make it a correct statement.

   ___

8. Which statement is NOT true of the heart's response to exercise?
   a) Starling's law permits the heart to contract more forcefully in response to increased venous return.
   b) The heart rate will increase in response to parasympathetic impulses.
   c) Stroke volume increases during exercise as the heart pumps more forcefully.
   d) Cardiac output increases to supply more oxygen to the body.

   Reword your choice to make it a correct statement.

   ___

9. Which statement is NOT true of coronary circulation?
   a) The two major coronary arteries are branches of the aorta.
   b) The coronary sinus empties blood from the myocardium into the left atrium.
   c) For the myocardium, the most important substance in the blood is oxygen.
   d) Myocardium that is deprived of oxygen becomes ischemic.

   Reword your choice to make it a correct statement.

   ___

10. Which statement is NOT true of heart rate?
   a) The normal range of heart rate for a healthy adult is 60 to 80 beats per minute.
   b) An infant may have a normal heart rate as high as 120 beats per minute.
   c) An athlete's heart rate may range from 40 to 60 beats per minute.
   d) An athlete's heart rate is low because the heart's stroke volume is lower.

   Reword your choice to make it a correct statement.

   ___

MULTIPLE CHOICE TEST #3

Each question is a series of statements concerning a topic in this chapter. Read each statement carefully and select all of the correct statements.

1. Which of the following statements are true of the anatomy of the heart?
   a) The chordae tendineae and the papillary muscles anchor the free edges of the AV valves.
   b) The function of the fibrous skeleton of the heart is to keep the thin-walled atria open at all times.
   c) The epicardium lines the heart and prevents blood clotting within the chambers.
   d) The right atrium receives blood from the lower and upper body.
   e) Blood to the body is pumped by the left ventricle into the aorta.
   f) The six pulmonary veins empty into the left atrium.
   g) The coronary vessels circulate blood throughout the myocardium.
   h) The lungs are lateral to the heart, and the diaphragm is inferior.
   i) The fibrous pericardium is the innermost of the pericardial membranes.
   j) The bundle of His is located in the upper interatrial septum.
   k) Intercalated discs permit the cusps of the AV valves to work smoothly.
   l) The tricuspid valve is between the left atrium and left ventricle.

2. Which of the following statements are true of the physiology of the heart?
   a) Cardiac output equals stroke volume divided by pulse.
   b) The heart is a double pump, and both pumps work simultaneously.
   c) The action potential through the myocardium stimulates the mechanical contraction of the chambers of the heart.
   d) Only half of the blood that flows from the ventricles to the arteries is pumped.
   e) All of the blood that flows from the atria to the ventricles must be pumped.
   f) ANP is secreted by the atria in response to a higher blood volume or pressure.
   g) Impulses along the vagus nerves slow the heart rate.
   h) Ventricular systole closes the arterial semilunar valves.
   i) The cardiac centers of the CNS are located in the hypothalamus.
   j) The AV node has the fastest natural rate of depolarization.
   k) Epinephrine is the hormone that reinforces the effects of the parasympathetic nervous system.
   l) A person's average resting cardiac output is approximately equal to her or his total blood volume.

Chapter 13

The Vascular System

The vascular system consists of the arteries, capillaries, and veins that transport blood throughout the body. This chapter describes these vessels, the pathways of circulation, and the importance and regulation of blood pressure.

BLOOD VESSELS—STRUCTURE AND FUNCTIONS

1. a) The vessels that carry blood from arterioles to venules are ____________________.

   b) The vessels that carry blood from the heart to capillaries are ____________________.

   c) The vessels that carry blood from capillaries to the heart are ____________________.

2. a) The tunica intima is the ____________________ of arteries and veins and is made of ____________________ tissue.

   b) This tissue is very smooth and has what function? ____________________

3. a) The tunica media is the middle layer of arteries and veins and is made of these tissues: ____________________ and ____________________.

   b) What is the function of these tissues? ____________________

4. a) The tunica externa is the ____________________ layer of arteries and veins and is made of this tissue: ____.

   b) What is the function of this tissue? ____

5. Match each type of vessel with the proper descriptive statements.

   Use each letter once. Each answer line will have three correct letters.

   1) Arteries ____________________

   2) Veins ____________________

   A. The lining is not folded into valves.
   B. The outer layer is thin because blood pressure is low.
   C. The lining is folded into valves to prevent back flow of blood.
   D. The middle layer is thin because these vessels are not as important in the maintenance of blood pressure.
   E. The middle layer is thick because these vessels are important in the maintenance of blood pressure.
   F. The outer layer is thick, to prevent rupture by the high blood pressure in these vessels.

6. a) Direct connections between arteries or between veins are called ____________________.

   b) Their purpose is to provide ____________________ for the flow of blood if one vessel becomes blocked or obstructed.

7. a) Capillaries are made of what type of tissue? ____________________

   b) Besides being smooth to prevent abnormal clotting, this tissue is thin to permit ____________________ of materials between the blood and surrounding cells.

8. a) Blood flow through capillary networks is regulated by smooth muscle cells called ____________________ at the beginning of each network.

   b) Precapillary sphincters are not regulated by nerve impulses but by tissue needs. In an active tissue, precapillary sphincters will ____________________ (constrict or dilate) to ____________________ (increase or decrease) blood flow to supply the tissue with more ____________________.

9. a) Large, very permeable capillaries are called ____________________.

   b) Proteins and blood cells can enter or leave sinusoids, which are found in these organs (tissues): ____________________ and ____________________.

10. The following diagram depicts the three types of blood vessels.

    Label the parts indicated.

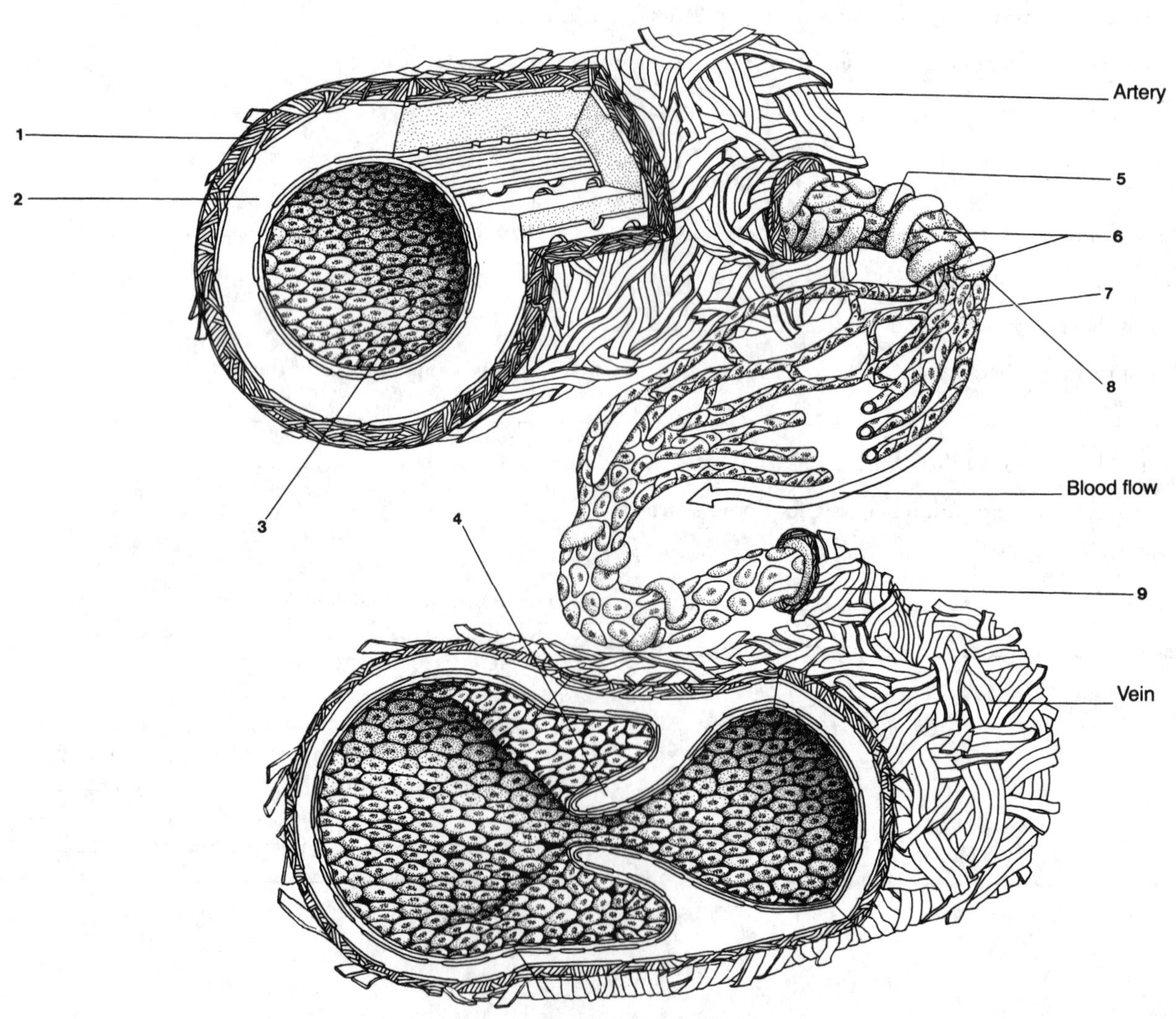

EXCHANGES IN CAPILLARIES

1. Match each process that occurs in capillaries with the proper descriptive statements.

Use each letter once. Each answer line will have more than one correct letter.

1) Diffusion ______________________

2) Filtration ______________________

3) Osmosis ______________________

A. Molecules move from an area of greater concentration to an area of lesser concentration.
B. The albumin in the blood creates a colloid osmotic pressure that pulls water and dissolved materials.
C. The high blood pressure at the arterial end of capillary networks forces plasma out of capillaries.
D. Nutrients move from the blood into tissue fluid.
E. CO_2 moves from tissue fluid into the blood.
F. Waste products move from tissue fluid into the blood.
G. Oxygen moves from the blood to tissue fluid.

2. The following diagram depicts a capillary in a body tissue.

Label the substances that move in the direction of the arrows by the means of the processes indicated.

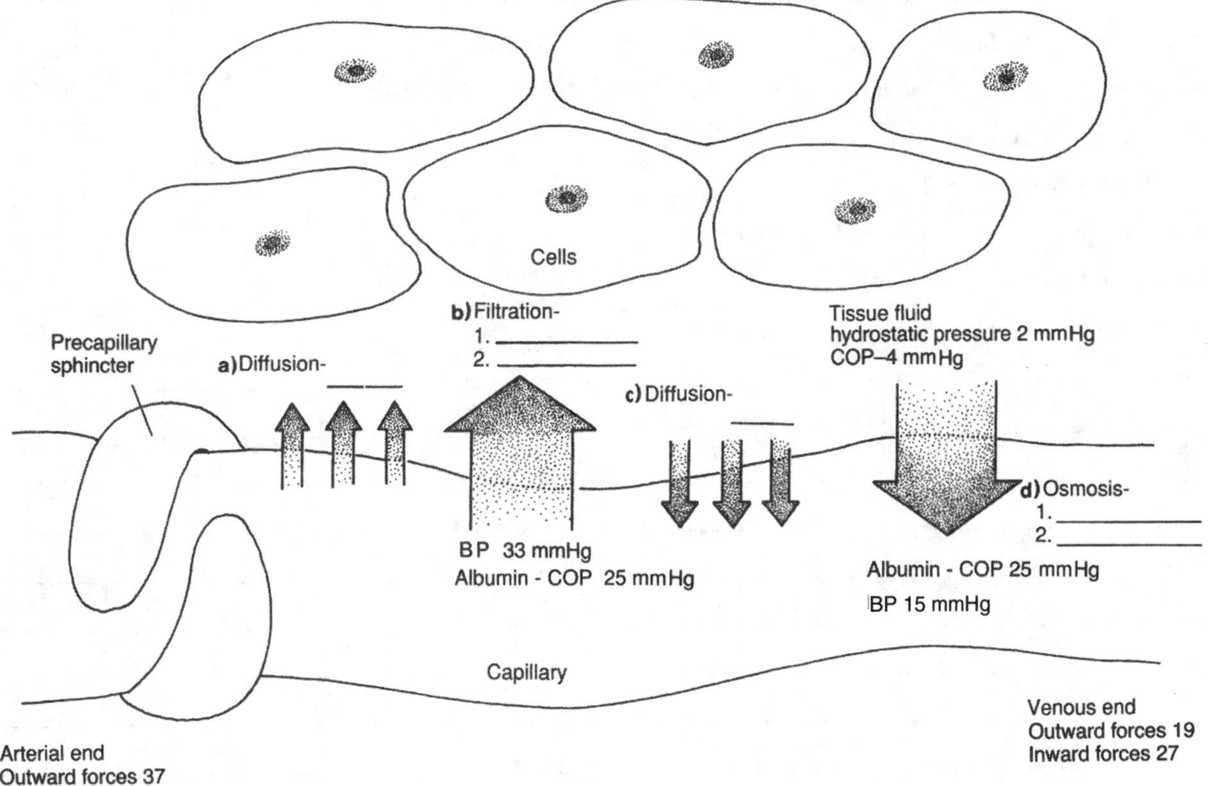

PATHWAYS OF CIRCULATION

Pathway—Pulmonary Circulation

1. a) This pathway begins at the ____________ ventricle, which pumps blood through the pulmonary artery to the ____________.

   b) Blood in the pulmonary veins returns to the ____________ atrium.

2. The purpose of this pathway of circulation is to exchange ____________ and ____________ between the blood in the pulmonary ____________ and the air in the ____________ of the lungs.

Pathway—Systemic Circulation

1. a) This pathway begins at the ____________ ventricle, which pumps blood through the aorta to the ____________.

   b) Blood in the superior and inferior caval veins returns to the ____________ atrium.

2. The purpose of this pathway of circulation is to exchange materials between the blood in the systemic ____________ and the ____________ throughout the body.

3. The parts of the aorta are named according to their locations. Name each part described:

   1) Below the level of the diaphragm ____________
   2) Emerges from the left ventricle ____________
   3) Passes through the chest cavity to the level of the diaphragm ____________
   4) Curves over the top of the heart ____________

4. Name the part of the body supplied by each of these arteries.

   1) Coronary ____________
   2) Femoral ____________
   3) Brachial ____________
   4) Bronchial ____________
   5) Internal carotid ____________
   6) Renal ____________
   7) Hepatic ____________
   8) Intercostal ____________
   9) Anterior tibial ____________
   10) Subclavian ____________
   11) Radial ____________
   12) Superior mesenteric ____________
   13) Esophageal ____________
   14) Vertebral ____________
   15) Celiac ____________
   16) Popliteal ____________
   17) Common iliac ____________
   18) Plantar arches ____________

5. The following diagram depicts the body in anterior view.

Label the arteries indicated.

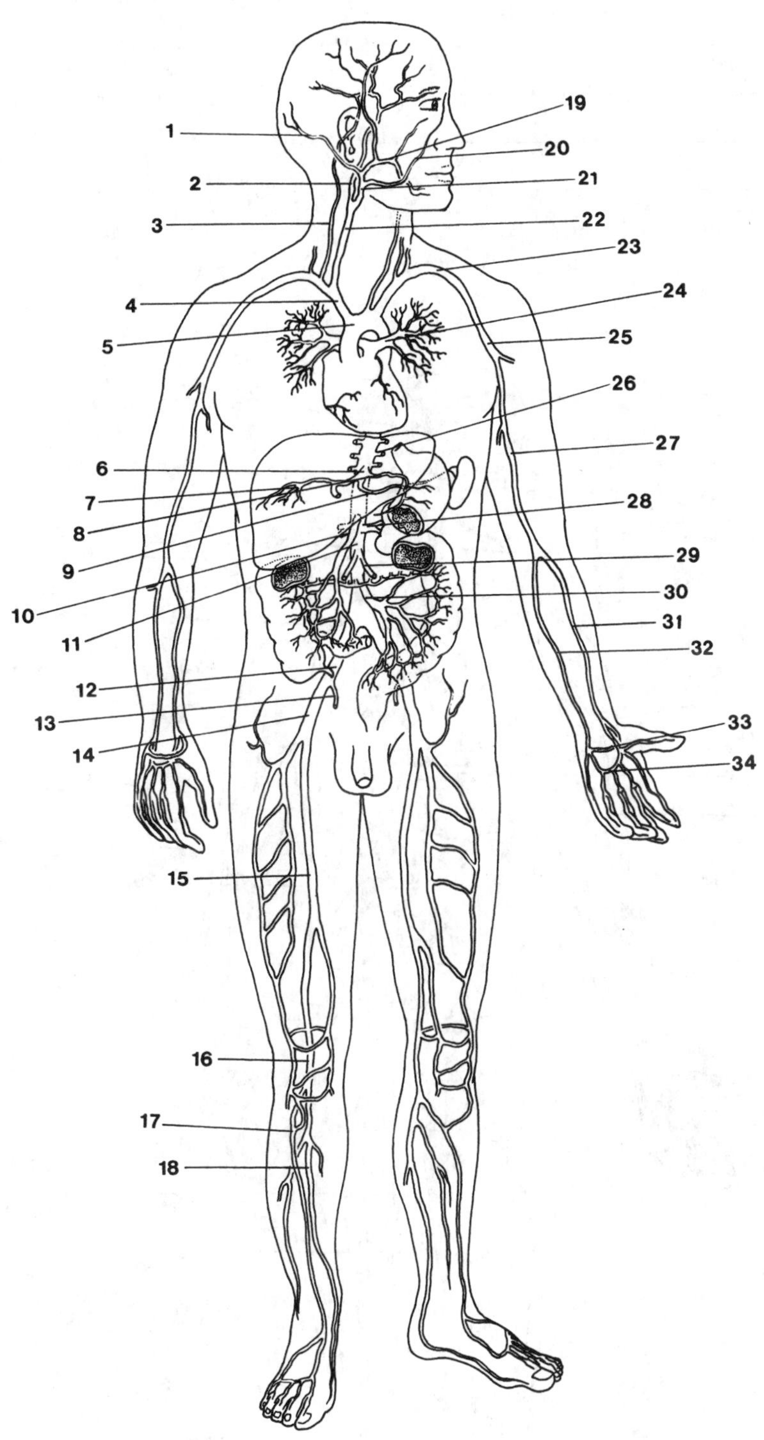

6. Name the part of the body drained of blood by each of these veins.

   1) External jugular ______________________
   2) Axillary ______________________
   3) Great saphenous ______________________
   4) Inferior vena cava ______________________
   5) Common iliac ______________________
   6) Ulnar ______________________
   7) Cranial venous sinuses ______________________
   8) Superior vena cava ______________________
   9) Renal ______________________
   10) Subclavian ______________________
   11) Brachial ______________________
   12) Femoral ______________________

7. The following diagram depicts the body in anterior view.

   Label the veins indicated.

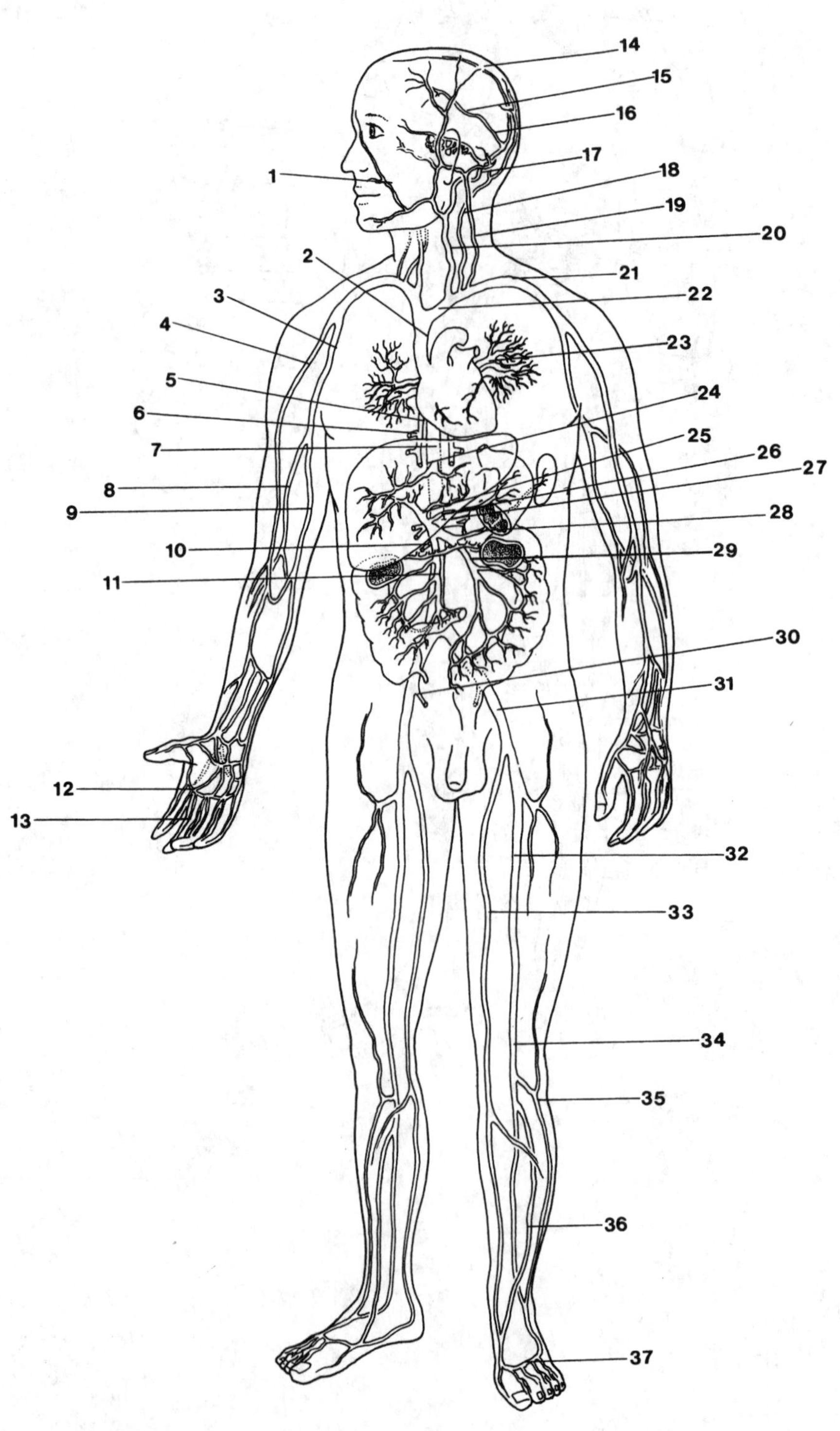

Hepatic Portal Circulation

1. In this specialized pathway of systemic circulation, blood from the ________________ and ________________ circulates through the liver before returning to the heart.
2. The portal vein is formed by the union of two large veins, the ________________ vein and the ________________ vein.
3. From the portal vein, blood flows into the ________________ in the liver and then to the hepatic veins to the ________________, which returns blood to the right atrium.
4. a) This pathway differs from the rest of systemic circulation in that there are two sets of ________________ in which exchanges take place.
   b) The first set of capillaries is in the ________________, and the second set of capillaries is the sinusoids in the ________________.
5. Use a specific example to describe the purpose of portal circulation:
   1) Substance in the blood: ________________
   2) Absorbed or produced by which organ: ________________
   3) Function of the liver with respect to this substance: ________________

Fetal Circulation

1. The site of exchanges of materials between fetal blood and maternal blood is the ________________.
2. a) Name the kinds of materials exchanged: ________________, ________________, and ________________.
   b) Name the mechanisms (processes) of exchange: ________________ and ________________.
3. The fetus is connected to the placenta by the umbilical cord, which contains two ________________ and one ________________.
4. a) The umbilical arteries carry blood from the ________________ to the ________________.
   b) In this blood are ________________ and ________________, which will be eliminated in the placenta.
5. a) The umbilical vein carries blood from the ________________ to the ________________.
   b) In this blood are ________________ and ________________ obtained in the placenta.
6. Within the body of the fetus, the umbilical vein branches into two vessels. One branch takes blood to the fetal liver. The other branch is called the ________________ and takes blood to the ________________, which returns blood to the right atrium of the fetal heart.
7. a) Within the fetal heart, the ________________ is an opening in the interatrial septum that permits some blood to flow from the ________________ atrium to the ________________ atrium.
   b) This blood, therefore, bypasses the fetal ________________, which are still deflated and non-functional.
8. a) Just outside the fetal heart, the ductus arteriosus is a short vessel that permits blood to flow from the ________________ to the ________________.
   b) State the purpose of the ductus arteriosus. ________________
9. a) After birth, the ductus venosus ________________ and becomes nonfunctional.
   b) The foramen ovale is ________________ by a flap on the left side, and the ductus arteriosus ________________.
   c) This ensures that normal ________________ circulation will be established.

10. The following diagram depicts fetal circulation.

Label the parts indicated.

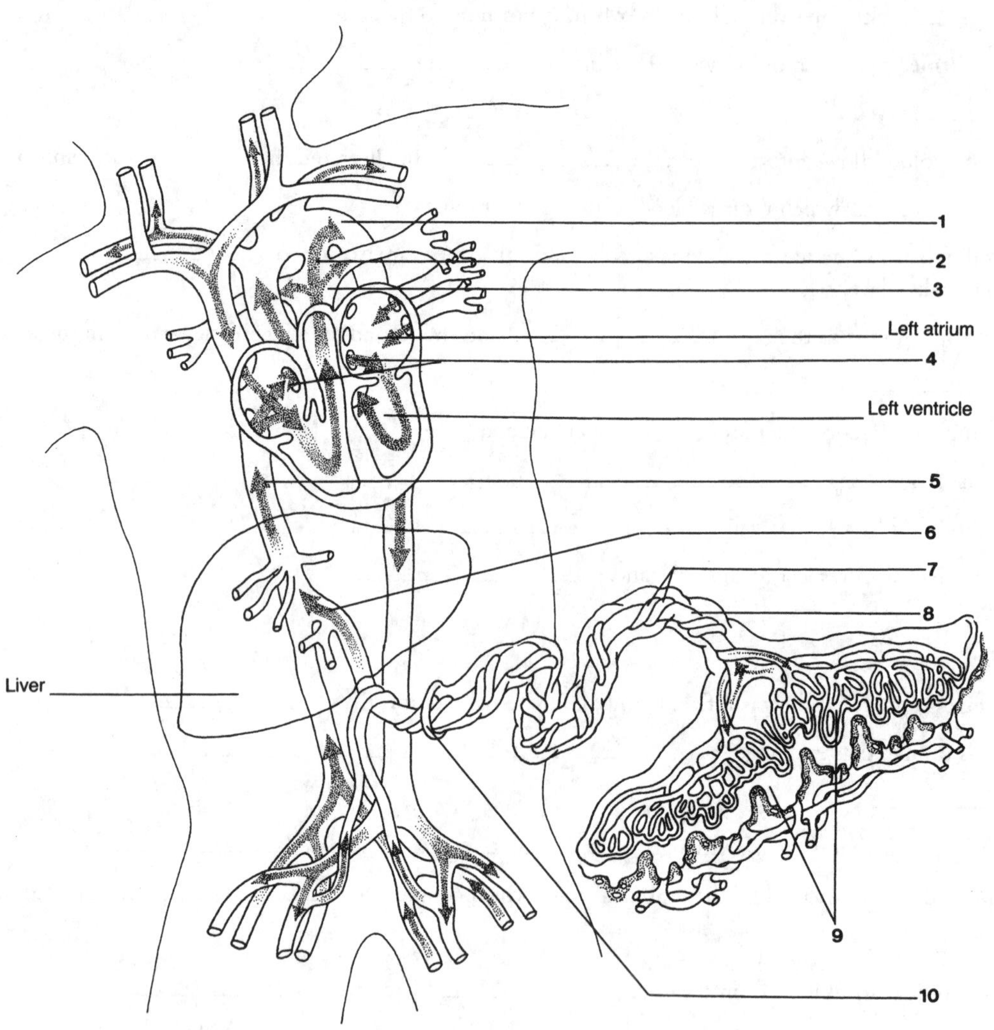

VELOCITY OF BLOOD FLOW

1. When the cross-sectional area of the vascular system increases (as when blood flows from arteries to arterioles), the velocity of blood flow ________________.
2. When blood flows from venules to veins, the cross-sectional area ________________, and the velocity ________________.
3. a) The segment of the vascular system with the greatest cross-sectional area is the ________________.

   b) Velocity of blood flow here is ________________.

   c) Explain the importance of this velocity here.

   ____

BLOOD PRESSURE

1. Blood pressure (BP) is the force exerted by the blood against the walls of the ______________.
2. BP is measured in units called ______________.
3. a) Systemic BP is created by the pumping of the ______________ of the heart.
   b) Systolic BP is the pressure when the left ventricle is ______________.
   c) Diastolic BP is the pressure when the left ventricle is ______________.
4. In systemic circulation, BP is highest in the ______________, then decreases in the arterioles and capillaries, and is lowest in the ______________.
5. a) In systemic capillaries, BP is 30 to 35 mm Hg at the ______________ end and decreases to 12 to 15 mm Hg ______________ at the end.
   b) This pressure is high enough to permit the process of ______________, but it is low enough to prevent ______________ of the capillaries.
   c) The process of filtration in capillaries is important to bring ______________ to tissues.
6. a) Pulmonary BP is always very low because the ______________ pumps less forcefully than does the left ventricle.
   b) The low BP in the pulmonary capillaries is important to ______________ filtration, to prevent the accumulation of tissue fluid in the alveoli.

Maintenance of Systemic Blood Pressure

1. a) Venous return is the amount of blood that returns to the ______________.
   b) Venous return is essential to maintain cardiac output, and if venous return decreases, cardiac output will ______________ and BP will ______________.
   c) 1) Constriction of the veins helps increase venous return. The tissue in the wall of a vein that is capable of contraction is ______________.
      2) The skeletal muscle pump contributes to venous return in veins in which part of the body? ______________ Contractions of skeletal muscles ______________ these veins and force blood toward the heart.
      3) The respiratory pump contributes to venous return in veins in which part of the body? ______________ The pressure changes during ______________ alternately compress and expand these veins and force blood toward the ______________.
2. If heart rate and force increase, BP will ______________.
3. a) Peripheral resistance is the resistance of the blood vessels to the flow of blood. Normal diastolic BP is maintained by slight ______________ of arteries and veins.
   b) Greater vasoconstriction will ______________ BP.
   c) Vasodilation will ______________ BP.
4. a) The elastic walls of the large arteries are stretched during ventricular ______________ and recoil during ventricular ______________.
   b) Therefore, normal elasticity ______________ (increases or decreases) systolic BP and ______________ (increases or decreases) diastolic BP.

5. a) The viscosity of blood depends upon the presence of ____________________ and ____________________.
   b) Decreased blood viscosity will ____________________ BP.
   c) This may occur when there is a decrease in RBCs, called ____________________, or when the ____________________ does not produce sufficient ____________________.
6. a) Following severe hemorrhage, BP will ____________________.
   b) Following a small loss of blood, certain compensating mechanisms will prevent a sharp decrease in BP. State one of these compensating mechanisms. ____________________
7. a) Several hormones have effects on BP.
   1) Norepinephrine raises BP because it stimulates ____________________.
   2) Epinephrine raises BP because it stimulates ____________________ and ____________________.
   3) ADH raises BP by increasing the reabsorption of water by the ____________________, which increases blood ____________________.
   4) Aldosterone raises BP by increasing the reabsorption of Na^+ ions by the ____________________, which is followed by the reabsorption of ____________________ to increase blood volume.
   5) ANP lowers BP by increasing urinary excretion of ____________________, which ____________________ blood volume.
   b) The following diagram depicts the hormones that affect blood pressure.

   Label the parts indicated.

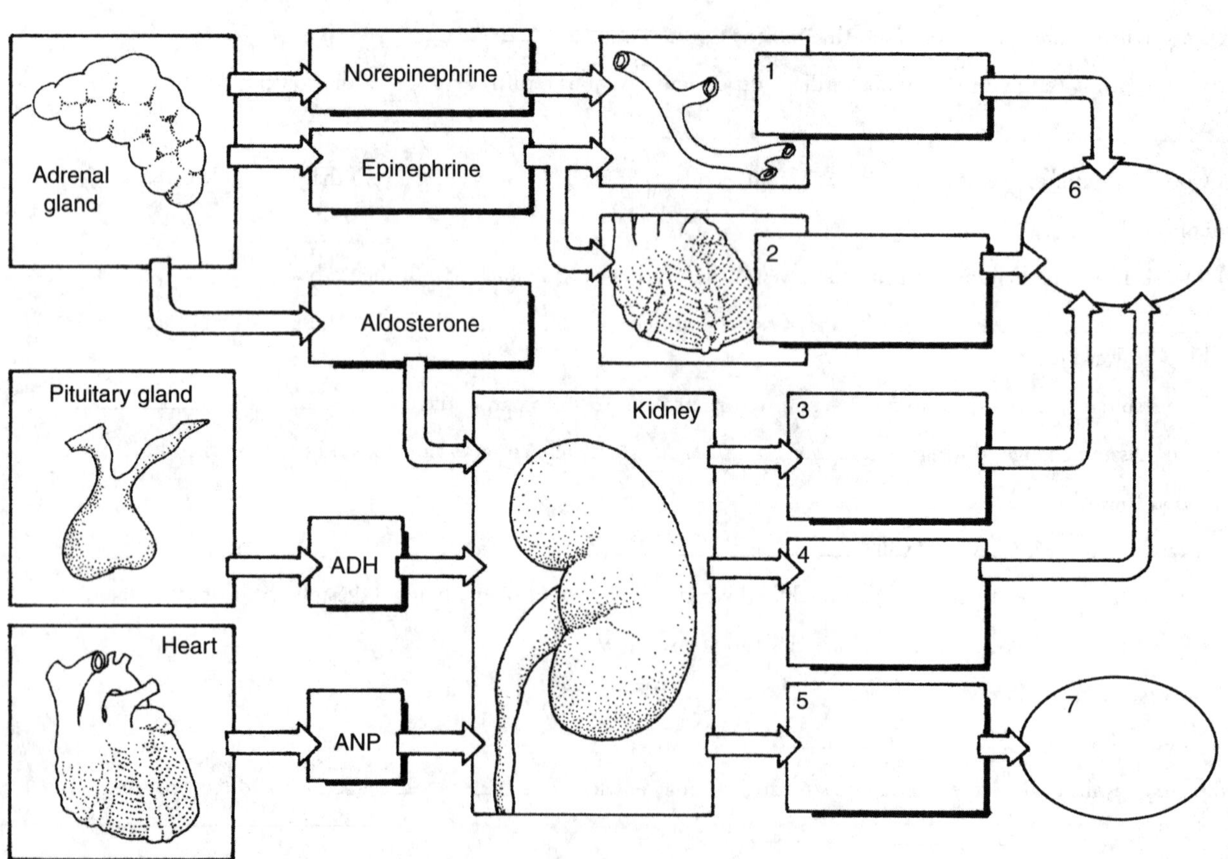

Regulation of Blood Pressure

1. Intrinsic mechanisms:

   1) The heart responds to increased venous return by pumping more ____________________. This will ____________________ cardiac output and BP and depends on the characteristic of cardiac muscle called ____________________ of the heart.

   2) Decreased blood flow to the kidneys will result in decreased ____________________ (process), which will decrease urinary output to maintain blood volume.

   3) The following diagram depicts the renin-angiotensin mechanism.

   Label the parts indicated.

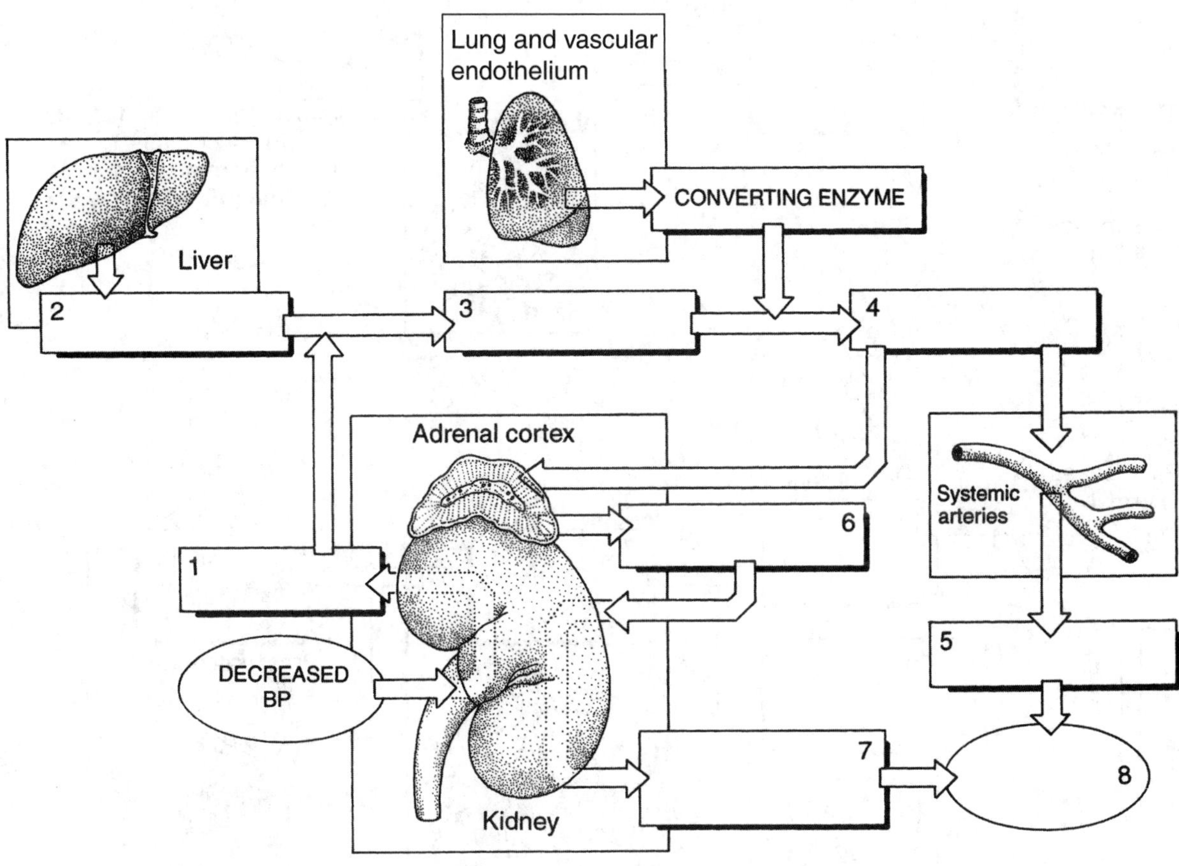

2. Nervous mechanisms for peripheral resistance:

   1) The vasomotor center is located in the ____________________ of the brain.

      a) This center consists of two areas: the ____________________ area and the ____________________ area.

      b) The medulla receives sensory information about the need for changes in vessel diameter from the ____________________ in the carotid and aortic sinuses.

   2) The division of the autonomic nervous system that regulates the diameter of arteries and veins is the ____________________ division.

      a) Several impulses per second maintain normal vasoconstriction of arteries and veins. More vasoconstriction is brought about by ____________________ impulses per second and will ____________________ BP.

      b) Vasodilation is brought about by ____________________ impulses per second and will ____________________ BP.

3. The following diagram depicts the nervous mechanisms for the regulation of blood pressure.

Label the parts indicated.

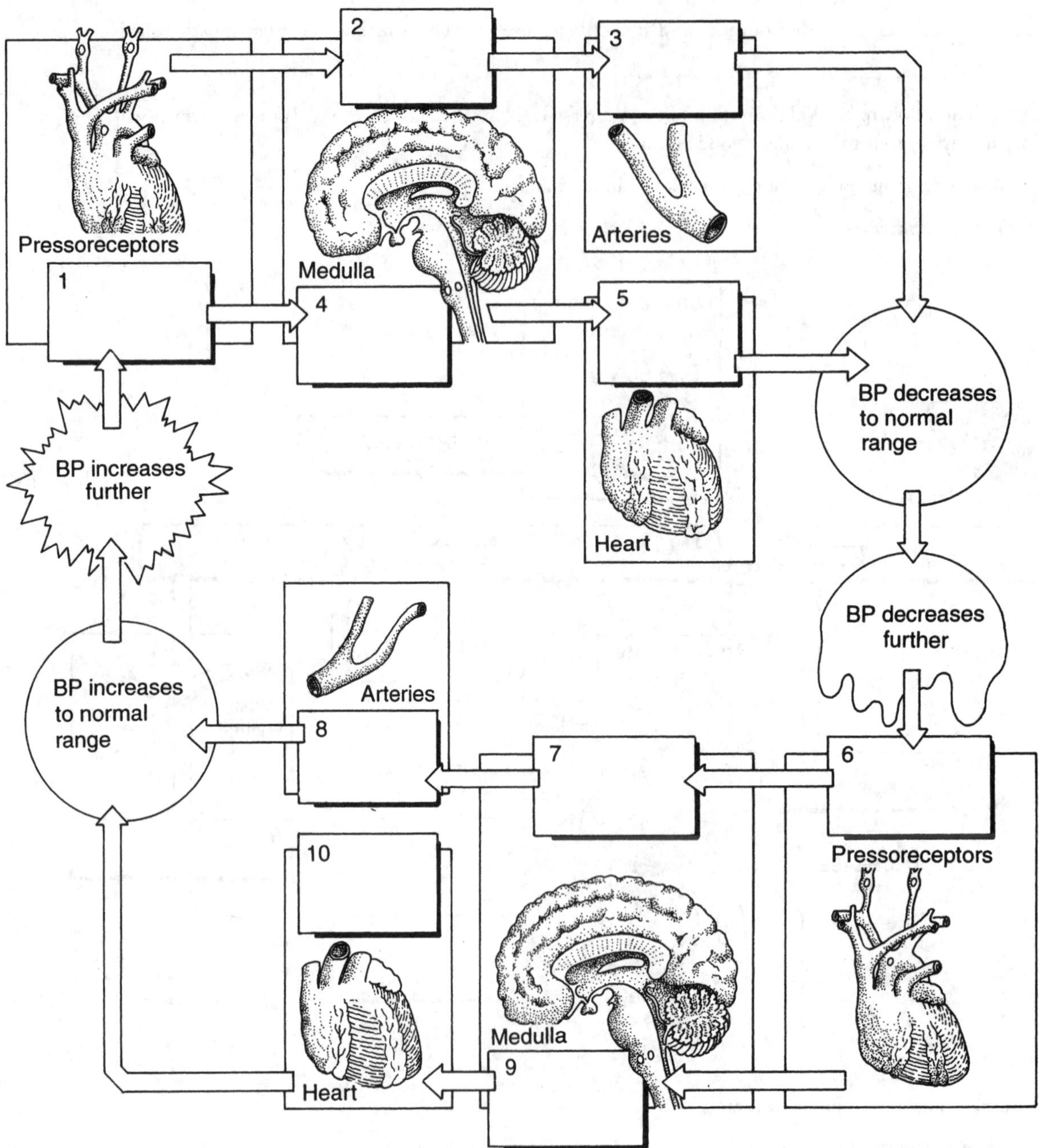

Hypertension

1. Hypertension is systemic BP that consistently has a systolic reading above ____________ mm Hg and a diastolic reading above ____________ mm Hg.
2. a) Chronic hypertension weakens the walls of arteries and contributes to the deterioration called ____________.

   b) Weakened arteries may rupture, which often occurs in these organs: the ____________ and the ____________.
3. Chronic hypertension causes the left ventricle to work harder to pump blood against the higher pressure in the arteries. As a result, the left ventricle may enlarge; this is called left ventricular ____________.
4. a) The coronary capillaries may not be sufficient to supply the abnormally enlarged myocardium, and exercise may bring on chest pains called ____________.

   b) Such pain is the result of lack of ____________ to the myocardium.

CROSSWORD PUZZLE

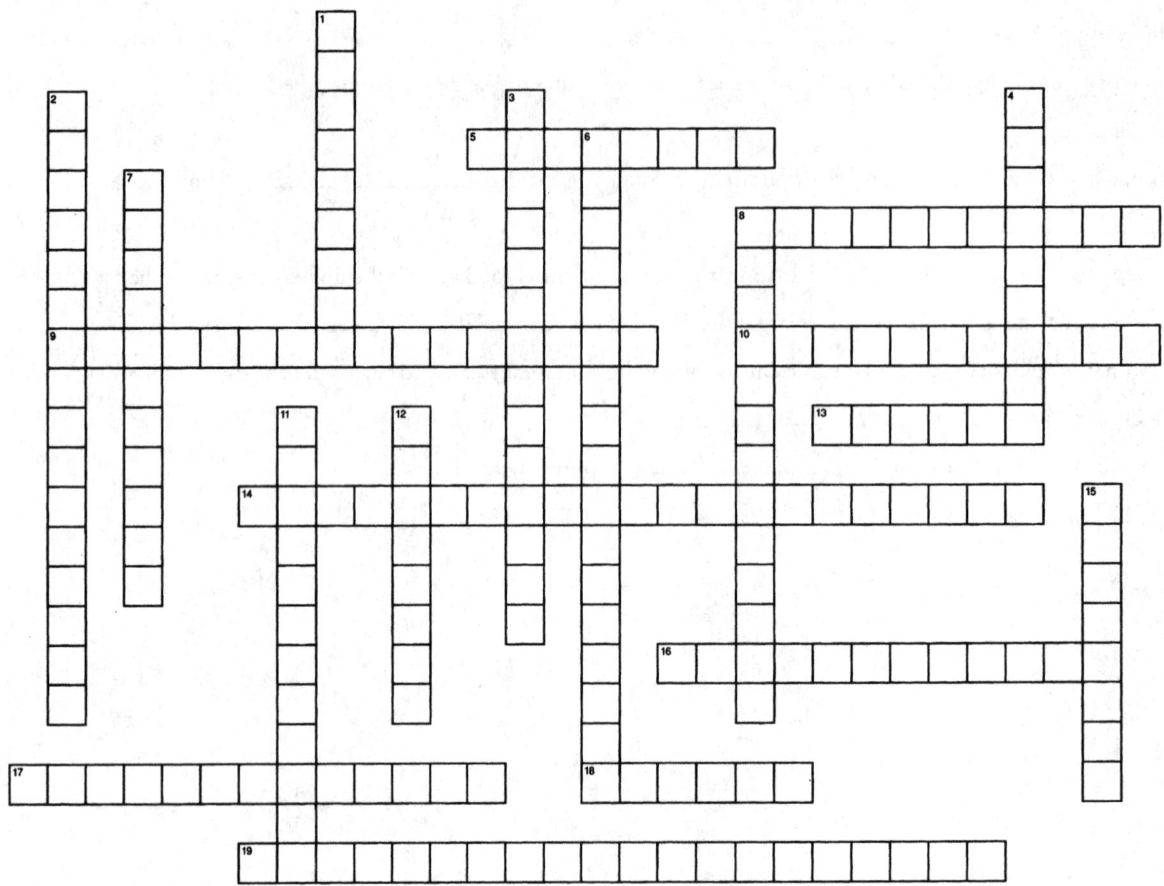

ACROSS

5. A large, permeable capillary
8. Shock resulting from decreased blood volume
9. Arteries lose their elasticity
10. A connection or joining of vessels
13. Smaller vein
14. Regulates blood flow in capillary networks (two words)
16. Massive allergic reaction: ____________ shock
17. Carries oxygenated blood from the placenta to the fetus (two words)
18. ____________ shock; results from the presence of bacteria in the blood
19. Refers to the resistance the vessels offer to the flow of blood (two words)

DOWN

1. Inflammation of veins
2. Fetal vessel that diverts blood from the pulmonary artery to the aorta
3. Arterial anastomosis to supply the brain with blood
4. Smaller artery
6. Carry blood from the fetus to the placenta (two words)
7. The tissue that lines arteries and veins
8. Pathway of circulation in which blood from the digestive organs flows through the liver (two words)
11. Opening in the interatrial septum; permits blood to bypass the fetal lungs (two words)
12. Site of exchanges between fetus and mother
15. ____________ veins; swollen and distended

CLINICAL APPLICATIONS

1. a) Mr. T has just been diagnosed with hypertension. This means that his systolic BP is consistently above

   ________________ mm Hg, and his diastolic BP is consistently above ________________ mm Hg.

   b) The doctor asks Mr. T if he ever experiences chest pain when climbing stairs or playing ball with his son. The doctor's question is an attempt to determine whether Mr. T has a potentially serious consequence of hypertension, which is ________________ of the left ventricle.

   c) Chest pain during exertion is called ________________ and is caused by the lack of ________________ to the heart.

2. Name the vascular disorder suggested by each of these descriptions.

   1) Swollen and painful superficial veins of the leg: ________________

   2) A weakened area in the wall of an artery bulges or bubbles outward: ________________

   3) The walls of arteries deteriorate and lose their elasticity as a person ages: ________________

   4) Deposits of cholesterol accumulate in the lining of arteries: ________________

3. 1) Following a car accident, Mr. C is brought to the emergency department. The nurse sees that he has only a few small abrasions and then takes Mr. C's vital signs. She immediately tells a doctor that Mr. C may be bleeding internally. Which set of vital signs prompted the nurse to make the assessment that Mr. C might be in circulatory shock?

   a) Pulse—86 bpm; BP—120/70 mm Hg; respirations—12 per minute

   b) Pulse—140 bpm; BP—100/70 mm Hg; respirations—30 per minute

   c) Pulse—70 bpm; BP—140/80 mm Hg; respirations—20 per minute

   2) The type of shock Mr. C has is called:

   a) cardiac shock b) hypovolemic shock c) anaphylactic shock

   3) How may Mr. C's respiratory rate be beneficial? ________________

MULTIPLE CHOICE TEST #1

Choose the correct answer for each question.

1. The fetal blood vessel that carries blood from the pulmonary artery to the aorta is the:
   a) ductus arteriosus b) ductus venosus c) pulmonary duct d) ductus pulmonarus

2. The fetal blood vessel that carries blood from the placenta to the fetus is the:
   a) umbilical vein b) ductus venosus c) ductus arteriosus d) umbilical artery

3. The layer of the walls of arteries and veins that is smooth to prevent abnormal clotting is the:
   a) middle layer, made of smooth muscle
   b) lining, made of smooth muscle
   c) lining, made of simple squamous epithelium
   d) outer layer, made of simple squamous epithelium

4. The layer of the walls of arteries that helps maintain diastolic blood pressure is made of:
   a) fibrous connective tissue and elastic connective tissue
   b) smooth muscle and fibrous connective tissue
   c) smooth muscle and elastic connective tissue
   d) fibrous connective tissue and simple squamous epithelium

5. Backflow of blood within veins is prevented by:
   a) smooth muscle b) valves c) precapillary sphincters d) the middle layer

6. In capillaries, nutrients are transported from the blood to tissues by the process of:
   a) osmosis b) diffusion c) active transport d) filtration

7. In capillaries, O_2 and CO_2 are exchanged between the blood and tissue fluid by the process of:
a) osmosis b) filtration c) active transport d) diffusion

8. A systemic blood pressure reading always consists of two numbers, which are called:
a) systemic/systolic b) systolic/diastolic c) diastolic/systolic d) diastolic/systemic

9. Pulmonary blood pressure is always low and thereby prevents:
a) osmosis of tissue fluid into pulmonary capillaries
b) filtration of tissue fluid into the alveoli
c) filtration of tissue fluid into pulmonary capillaries
d) osmosis of blood plasma into the alveoli

10. Systemic circulation begins at the:
a) right atrium b) right ventricle c) left ventricle d) left foot

11. In hepatic portal circulation, blood from the digestive organs and spleen circulates through the ________ before returning to the ________.
a) brain/liver b) liver/heart c) liver/brain d) heart/liver

12. Venous return in veins that pass through the thoracic cavity is increased by the:
a) skeletal muscle pump c) pumping of the left ventricle
b) pumping of the right ventricle d) respiratory pump

13. Venous return in the deep veins in the legs is increased by the:
a) low pressure in the right atrium c) low pressure in the left atrium
b) respiratory pump d) skeletal muscle pump

14. Following a large loss of blood, as in severe hemorrhage, blood pressure will:
a) increase b) decrease c) remain the same d) increase, then decrease

15. Epinephrine increases blood pressure because it:
a) causes vasodilation c) increases heart rate and force
b) decreases heart rate and force d) increases water reabsorption by the kidneys

16. The vasomotor center is located in which part of the brain?
a) medulla b) hypothalamus c) frontal lobes d) cerebellum

17. ADH increases blood pressure because it:
a) causes vasodilation c) decreases water reabsorption by the kidneys
b) increases heart rate and force d) increases water reabsorption by the kidneys

18. The nerves that carry impulses to regulate the diameter of arteries and veins are:
a) sympathetic nerves c) somatic motor nerves
b) parasympathetic nerves d) visceral sensory nerves

19. Connections between arteries or between veins that provide alternate pathways for blood flow are called:
a) capillary networks b) venous sinuses c) sinusoids d) anastomoses

20. Large, very permeable capillaries that permit cells or proteins to enter or leave the blood are called:
a) sinusoids b) venous sinuses c) anastomoses d) precapillary sphincters

21. In the fetus, blood flows from the right atrium to the left atrium through the:
a) foramen magnum b) foramen pulmonis c) foramen atrius d) foramen ovale

22. Norepinephrine increases blood pressure because it:
a) increases blood volume c) causes vasoconstriction
b) decreases blood volume d) causes vasodilation

23. Precapillary sphincters will dilate if the surrounding tissue is:
a) low in oxygen c) low in carbon dioxide
b) high in oxygen d) high in oxygen and low in carbon dioxide

24. When blood pressure decreases, the kidneys secrete:
a) renin b) more water c) erythropoietin d) all of these

25. Angiotensin II causes:
a) vasoconstriction b) increased secretion of aldosterone c) an increase in BP d) all of these

MULTIPLE CHOICE TEST #2

Read each question and the four answer choices carefully. When you have made a choice, follow the instructions to complete your answer.

1. Which statement is NOT true of hepatic portal circulation?
   a) Blood from the digestive organs flows through the liver before returning to the heart.
   b) The liver regulates the blood levels of nutrients such as glucose.
   c) The vein that takes blood into the liver is the hepatic vein.
   d) Blood from the spleen circulates through the liver first, before returning to the heart.

   Reword your choice to make it a correct statement.

   ____

2. Which statement is NOT true of fetal circulation?
   a) The foramen ovale permits blood to flow from the left atrium to the right atrium to bypass the fetal lungs.
   b) The umbilical arteries carry blood from the fetus to the placenta.
   c) The site of exchanges between maternal blood and fetal blood is the placenta.
   d) The fetus depends upon the mother for oxygen and nutrients.

   Reword your choice to make it a correct statement.

   ____

3. Which statement is NOT true of the structure of arteries and veins?
   a) Veins have valves to prevent the backflow of blood.
   b) The lining of both arteries and veins is very smooth to permit exchanges of materials between the blood and tissues.
   c) The outer layer of arteries is thick and made of fibrous connective tissue to prevent rupture.
   d) The smooth muscle of arteries and veins permits vasoconstriction or vasodilation.

   Reword your choice to make it a correct statement.

   ____

4. Which statement is NOT true of venous return?
   a) The constriction of veins helps increase venous return.
   b) The skeletal muscle pump is especially important for deep veins of the legs.
   c) The respiratory pump depends on pressure changes during breathing.
   d) Decreased venous return will result in increased cardiac output.

   Reword your choice to make it a correct statement.

   ____

5. Which artery is NOT paired with the part of the body it supplies with blood?
   a) common carotid artery—neck and head
   b) renal artery—liver
   c) femoral artery—thigh
   d) brachial artery—arm

   For the artery of your choice, state its correct location.

   ____

6. Which statement is NOT true of exchanges in capillaries?
   a) Oxygen diffuses from the blood to tissues.
   b) Waste products dissolved in the tissue fluid are returned to the blood by osmosis.
   c) Carbon dioxide diffuses from the blood to tissues.
   d) By the process of filtration, tissue fluid is formed, and nutrients are brought to tissues.

   Reword your choice to make it a correct statement.

   ____

7. Which statement is NOT true of the pulse and pulse sites?
   a) In healthy people, the pulse rate equals the heart rate.
   b) The radial pulse is felt on the thumb side of the wrist.
   c) The pedal pulse is felt on the top of the foot.
   d) The carotid pulse is felt at the elbow.

   Reword your choice to make it a correct statement.

   ___

8. Which statement is NOT true of the regulation of BP?
   a) The vasomotor center is located in the medulla.
   b) Vasoconstrictor nerve fibers are part of the sympathetic division of the ANS.
   c) Fewer sympathetic impulses to an artery result in vasodilation.
   d) The tissue in arteries and veins that constricts or dilates is fibrous connective tissue.

   Reword your choice to make it a correct statement.

   ___

9. Which statement is NOT true of the effects of hormones on BP?
   a) Epinephrine increases cardiac output, which raises BP.
   b) Norepinephrine causes vasoconstriction, which lowers BP.
   c) ADH increases the reabsorption of water by the kidneys, which raises blood volume and BP.
   d) Aldosterone increases the reabsorption of sodium ions by the kidneys, and water follows to maintain (or raise) blood volume and BP.

   Reword your choice to make it a correct statement.

   ___

10. Which statement is NOT true of systemic circulation?
    a) The left ventricle pumps blood throughout the body.
    b) The superior and inferior caval veins return blood to the right atrium.
    c) BP is highest in the aorta and lowest in the caval veins.
    d) BP in the veins is higher than BP in the capillaries.

    Reword your choice to make it a correct statement.

    ___

MULTIPLE CHOICE TEST #3

Each question is a series of statements concerning a topic in this chapter. Read each statement carefully and select all of the correct statements.

1. Which of the following statements are true of the structure and function of blood vessels?
   a) Only veins and capillaries have valves because BP is low within them.
   b) The smooth muscle layer of arteries is thicker than that of veins.
   c) The fibrous layer of arteries is thinner than that of veins.
   d) Capillaries are the continuation of the endothelium of arteries and become the endothelium of veins.
   e) Blood flow into capillary networks is regulated by precapillary sphincters and their sympathetic nerves.
   f) Sinusoids are located where capillaries need strong walls to prevent leaks.
   g) An anastomosis is a connection between vessels of the same type that provides an alternate pathway for blood flow.
   h) Exchanges of materials between the blood and tissues take place in capillary networks.

2. Which of the following statements are true of the circulation of blood?
   a) Pulmonary circulation begins at the left atrium and ends at the right ventricle.
   b) The systemic arteries begin as branches of the aorta.
   c) The amount of blood that circulates through an organ increases as the organ becomes more metabolically active.
   d) Systemic circulation begins at the left ventricle and ends at the right atrium.
   e) Hepatic portal circulation enables the liver to store or modify substances in the blood coming from the digestive organs and spleen.
   f) The two caval veins drain blood from the body, except for the head.
   g) Blood flow slows in capillaries but speeds up in veins.
   h) Blood in veins is kept moving by external forces, as well as constriction of the veins themselves.

3. Which of the following statements are true of blood pressure (BP)?
   a) If BP decreases, filtration in the kidneys will increase to compensate.
   b) Normal systolic pressure created by the right ventricle is about 120 mm Hg.
   c) Blood pressure in capillaries is lower than in the arterioles and increases in the venules.
   d) The outer layer of the walls of arteries can contract or relax to change BP.
   e) BP is regulated by the vasomotor center in the medulla.
   f) The carotid and aortic sinuses contain pressoreceptors that detect changes in BP.
   g) The arteries provide peripheral resistance, which depends on parasympathetic impulses.
   h) If BP decreases as a result of hemorrhage, heart rate will increase and more vasoconstriction will occur.
   i) ADH and aldosterone both help maintain or raise BP by causing vasoconstriction.
   j) The renin-angiotensin mechanism prevents kidney damage by lowering BP if it becomes too high.

Chapter 14

The Lymphatic System and Immunity

The lymphatic system has two very different functions. It returns tissue fluid to the blood to maintain normal blood volume, and it protects the body from microorganisms and other foreign material that might cause disease. This chapter describes both of these functions, which are essential for homeostasis, the continued proper functioning of the body.

LYMPH AND LYMPH VESSELS

1. The fluid found in lymph vessels is called ____________________.
2. a) In (blood) capillaries, the process of ____________________ forces some plasma out into tissue spaces, and this fluid is now called ____________________.

   b) Tissue fluid becomes lymph when it enters ____________________.
3. Match each lymph vessel or structure with its proper function.

   Use each letter once.

   1) Lymph capillaries ____________________
   2) Valves ____________________
   3) Smooth muscle layer ____________________
   4) Cisterna chyli ____________________
   5) Thoracic duct ____________________
   6) Right lymphatic duct ____________________

   A. Contracts to keep lymph moving through larger lymph vessels
   B. Empties lymph from the lower body and upper left quadrant into the left subclavian vein
   C. Empties lymph from the upper right quadrant into the right subclavian vein
   D. Collect tissue fluid from intercellular spaces
   E. The vessel formed by the union of lymph vessels from the lower body
   F. Prevent backflow of lymph in larger lymph vessels

LYMPH NODES AND NODULES

1. Lymph nodes and nodules are both made of ____________ tissue, which consists primarily of the WBCs called ____________, which have differentiated from ____________.

2. a) The fixed cells in lymph nodes and nodules that produce antibodies are called ____________.

   b) The fixed cells that phagocytize pathogens are called ____________.

3. Match the lymph nodes and nodules with their proper locations (letter statements) and functions (number statements).

   Use each letter and number once. Each answer line will have two correct letters and two correct numbers.

   1) Lymph nodes ____________

   2) Lymph nodules ____________

Location

A. Below the epithelium of all mucous membranes
B. Along the pathways of lymph vessels
C. The major paired groups are cervical, axillary, and inguinal
D. In the pharynx, they are called tonsils

Function

1. Destroy pathogens in the lymph from the extremities before the lymph is returned to the blood
2. Plasma cells produce antibodies that will enter the blood
3. Destroy pathogens that penetrate mucous membranes
4. Plasma cells produce antibodies that act locally in mucous membranes

4. The following diagram depicts the lymphatic system.

Label the parts indicated.

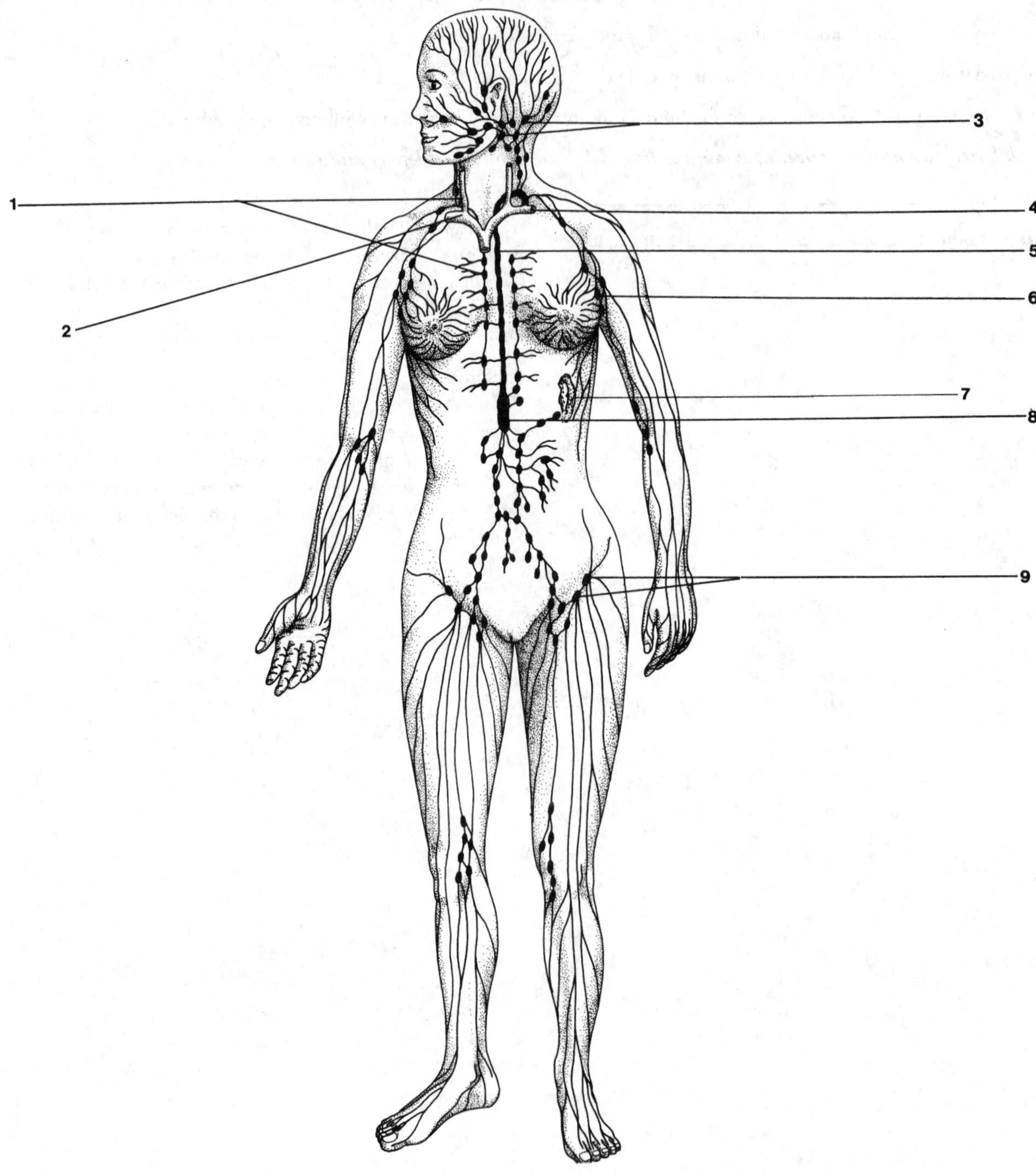

SPLEEN

1. a) The spleen is located in the ____________ cavity below the ____________ and behind the ____________.

   b) The spleen is protected from mechanical injury by which bones? ____________

2. Which type of blood cell is both stored by the spleen and, when damaged, removed from circulation by the spleen?

____

3. The spleen contains fixed macrophages that ____________________ pathogens in the blood and fixed plasma cells that produce ____________________ to foreign antigens.

4. The macrophages of the spleen may also be called reticuloendothelial cells because they phagocytize old ____________________ and form ____________________, which will be excreted by the liver.

5. a) If the spleen must be removed, other organs will compensate for its functions. The ____________________ will have fixed plasma cells.

   b) The ____________________ and ____________________ will remove old RBCs from circulation.

6. The fetal spleen has a temporary function, which is the production of ____________________.

THYMUS

1. The thymus is large and most active in which age group? ____________________

2. The thymus is located ____________________ the thyroid gland.

3. *T cells* is the name for the ____________________ that are produced by or that mature in the thymus.

4. a) Thymic hormones enable the T cells to participate in the recognition of foreign ____________________ and to provide ____________________ for certain diseases.

   b) This capability of the T cells may be called immunological ____________________.

5. The T cells must "learn" the components of the body that belong—that are "self"—especially cells and proteins.

   a) The ability to distinguish cells that belong in the body from those that do not is called ____________________.

   b) The ability *not* to react to the proteins produced by the body's cells is called ____________________.

IMMUNITY

1. Immunity includes the ability to destroy ____________________ and to ____________________ future cases of certain infectious diseases.

2. a) Antigens are chemical markers, and there are two general types: ____________________ antigens, which are found in the body's own cells, and ____________________ antigens, which may also be called "non-self."

   b) Name three different kinds of foreign antigens. ____________________, ____________________, and ____________________

3. The two major components of immunity are innate immunity and adaptive immunity. Match each component with its proper characteristics.

   Use each letter once. One answer line will have four correct answers and the other will have six.

   1) Innate immunity ____________________

   2) Adaptive immunity ____________________

   A. Involves T and B lymphocytes
   B. Is specific as to antigen
   C. Creates memory
   D. May become more efficient with repeated exposure
   E. Does not involve antibodies
   F. Does involve the process of inflammation
   G. Is not specific as to antigen
   H. Does not create memory
   I. Involves the lymphocytes called natural killer cells
   J. Response remains the same despite repeated exposures

INNATE IMMUNITY

1. The three functional aspects of innate immunity are ________________, ________________, and ________________.

2. a) Barriers include the ________________ of the skin, the ________________ tissue below the dermis, and the ________________ membranes of the digestive, respiratory, urinary, and reproductive tracts.

   b) The respiratory mucosa is able to sweep pathogens out because of the presence of ________________ epithelium.

   c) The mucosa of the stomach produces ________________, which destroys most microorganisms in gastric juice.

   d) The enzyme ________________ in tears and saliva inhibits bacterial growth.

   e) The fatty acids in ________________ help inhibit bacterial growth on the skin surface.

   f) The subcutaneous tissue contains many ________________ cells in areolar connective tissue.

3. Match the following defensive cells of innate immunity with their characteristics.

   Use each letter once. Two answer lines will have two correct answers.

   1) Macrophages ________________
   2) Langerhans cells ________________
   3) Natural killer cells ________________
   4) Basophils and mast cells ________________
   5) Neutrophils ________________

   A. Produce histamine and leukotrienes
   B. Use perforins to rupture the membranes of pathogens
   C. The most abundant phagocytic WBCs
   D. Pick up foreign material and take it to the nearest lymph node
   E. Phagocytic cells that also activate the lymphocytes of adaptive immunity
   F. A type of lymphocyte found in the blood, bone marrow, spleen, and lymph nodes
   G. Also called dendritic cells, and found in the epidermis

4. Match the following chemicals of innate immunity with the proper statements.

   Use each letter once. One answer line will have two letters.

   1) Interferons ________________
   2) Complement ________________
   3) Histamine and leukotrienes ________________
   4) Cytokines ________________

   A. Prevent reproduction of viruses within cells
   B. Chemical signals to attract cells
   C. Makes capillaries more permeable and causes vasodilation
   D. Produced by cells infected with viruses
   E. Forms an enzymatic ring to punch a hole in a cellular antigen

5. a) Inflammation is the body's response to any type of ________________.

   b) Inflammation is triggered by the chemicals ________________ and ________________, which are released from mast cells and basophils.

   c) The four characteristics of inflammation are ________________, ________________, ________________, and ________________.

   d) The purpose of inflammation is to try to prevent ________________ of the damage and to permit ________________ of tissue to begin.

6. The following diagram depicts the functional aspects of innate immunity.

   Label the parts indicated.

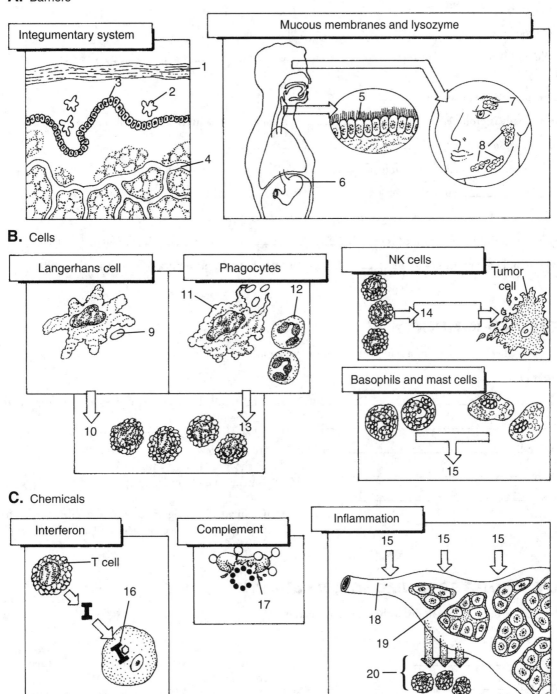

ADAPTIVE IMMUNITY

1. a) Adaptive immunity involves the types of lymphocytes that become specific for a foreign antigen; these are called ________________ and ________________.

   b) In the embryo, B cells are produced in the ________________.

   c) T cells are produced in the ________________, as well as in the ________________.

   d) Both T cells and B cells migrate to the ________________ and ________________, where they function.

2. a) Antibodies are molecules made of ________________.

   b) When an antibody is produced, it is ________________ for a foreign antigen and will label it for destruction.

   c) Antibodies may also be called ________________ or ________________.

3. a) The two mechanisms of adaptive immunity are called ________________ immunity and ________________ immunity.

   b) The first step in both mechanisms is the recognition of an antigen as ________________.

   c) One way in which this is accomplished is by ________________, which phagocytize the foreign antigen and present it and self-antigens to ________________, to activate them specifically for this particular foreign antigen.

CELL-MEDIATED IMMUNITY

1. Match each aspect of cell-mediated immunity with its proper function.

   Use each letter once.

   1) Helper T cells ________________
   2) Cytotoxic T cells ________________
   3) Memory T cells ________________
   4) Macrophages ________________
   5) Cytokines ________________

   A. Chemically destroy foreign antigens by disrupting their cell membranes
   B. Chemicals produced by cytotoxic T cells that attract macrophages
   C. Phagocytize foreign antigens
   D. Initiate a rapid immune response if the antigen enters the body again
   E. Compare the foreign antigen to the self-antigens on the macrophage and become activated and antigen–specific

ANTIBODY–MEDIATED (HUMORAL) IMMUNITY

1. Match each aspect of humoral immunity with its proper function.

   Use each letter once.

   1) Helper T cells ________________
   2) Memory B cells ________________
   3) Plasma cells ________________
   4) Macrophages ________________
   5) Antibodies ________________
   6) Complement ________________

   A. Initiate rapid antibody production if the antigen enters the body again
   B. Proteins produced by plasma cells that bind to a specific foreign antigen
   C. Compare the foreign antigen with self-antigens on the macrophages; are antigen specific and strongly activate B cells
   D. Plasma proteins that are activated by antigen–antibody complexes and lyse cellular antigens
   E. Produce antibodies specific for one foreign antigen
   F. Phagocytize antigen–antibody complexes

2. Part A of the following diagram depicts the events of cell-mediated immunity, and part B depicts antibody-mediated immunity.

Label the parts indicated.

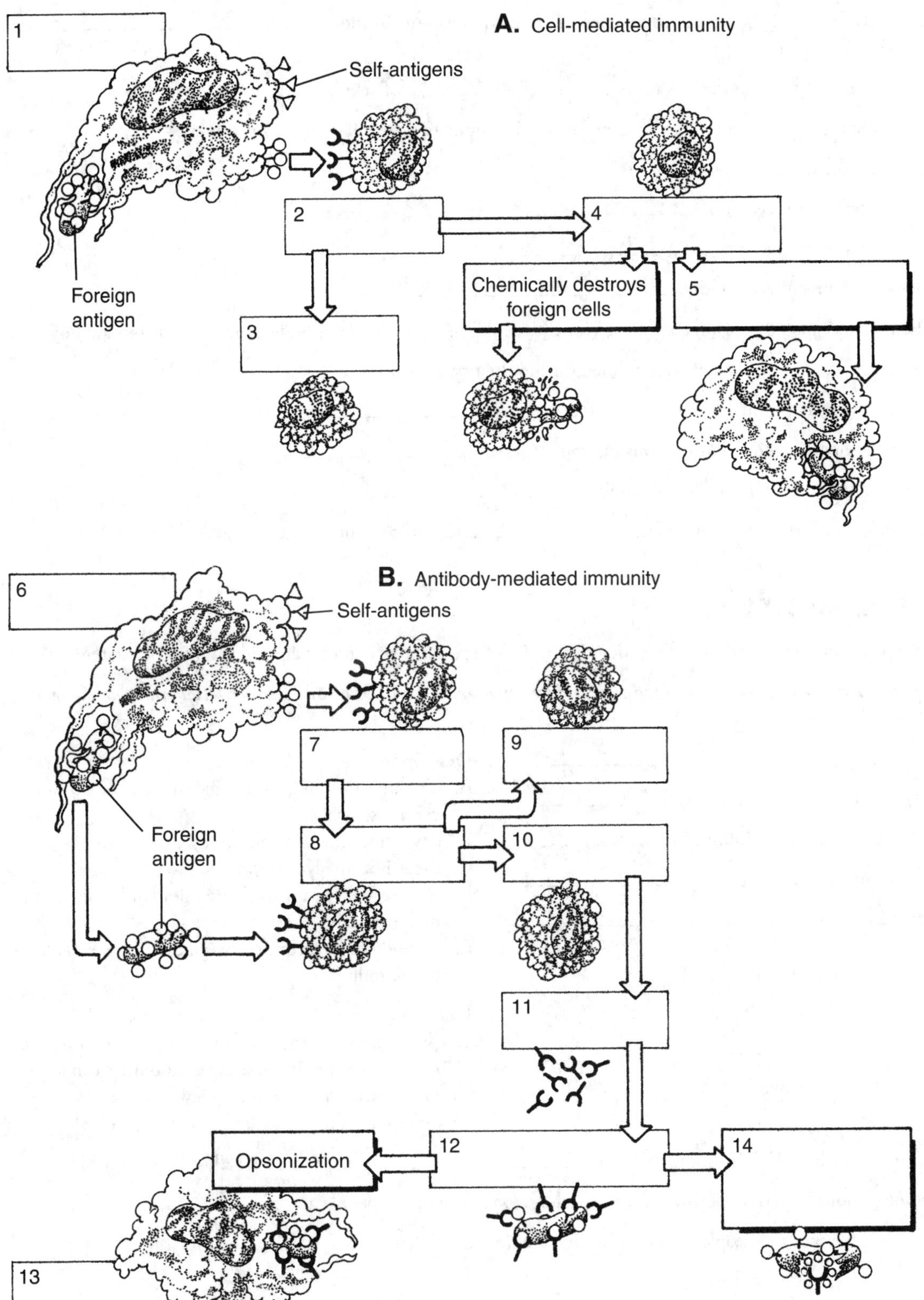

3. a) The term *opsonization* literally means "to buy food." As part of humoral immunity, opsonization is accomplished by ________________ that bond to a specific foreign antigen and label it as "food."

   b) Opsonization attracts ________________ that will phagocytize the antigen–antibody complex.

4. a) Complement fixation may be partial or complete. Partial complement fixation attracts ________________ to phagocytize the foreign antigen.

   b) Complete complement fixation causes the ________________ of cellular antigens.

5. a) When the body is exposed to a foreign antigen for the first time, is there any antibody production?

   ____

   b) Describe it in terms of speed. ________________

   c) Describe it in terms of amount. ________________

   d) What may then happen to the individual? ________________

6. a) When the body is exposed to a foreign antigen for the second or third time, antibody production is initiated by

   ________________ cells that were formed during the first exposure.

   b) Describe this antibody production in terms of speed. ________________

   c) Describe this antibody production in terms of amount. ________________

   d) Because of this, what is often true for the individual? ________________

   e) This knowledge is utilized in our use of ________________ to prevent certain diseases.

TYPES OF IMMUNITY

1. Match each type of immunity with its proper description (a letter statement) and its duration (a number statement).

   Use each letter once. Two numbers are used twice. Each answer line will have one correct letter and one correct number.

   1) Genetic immunity ________________
   2) Naturally acquired passive immunity ________________
   3) Artificially acquired passive immunity ________________
   4) Naturally acquired active immunity ________________
   5) Artificially acquired active immunity ________________

   Description

   A. The injection of immune globulins after exposure to certain diseases
   B. A vaccine stimulates antibody production
   C. Does not involve antibodies
   D. Recovery from a disease provides antibodies and memory cells
   E. Antibodies are acquired by placental transmission or from breast milk

   Duration

   1. Always lasts a lifetime because it is programmed in DNA
   2. Is always temporary because antibodies from another source break down within a few months
   3. Varies with the particular disease or vaccine

2. a) Herd immunity is the collective immunity of a ________________ that may be small or large, in which nonimmune persons are protected from a disease when a majority of the people are ________________.

   b) This is possible because the people who are immune are no longer ________________ for the pathogen.

CROSSWORD PUZZLE

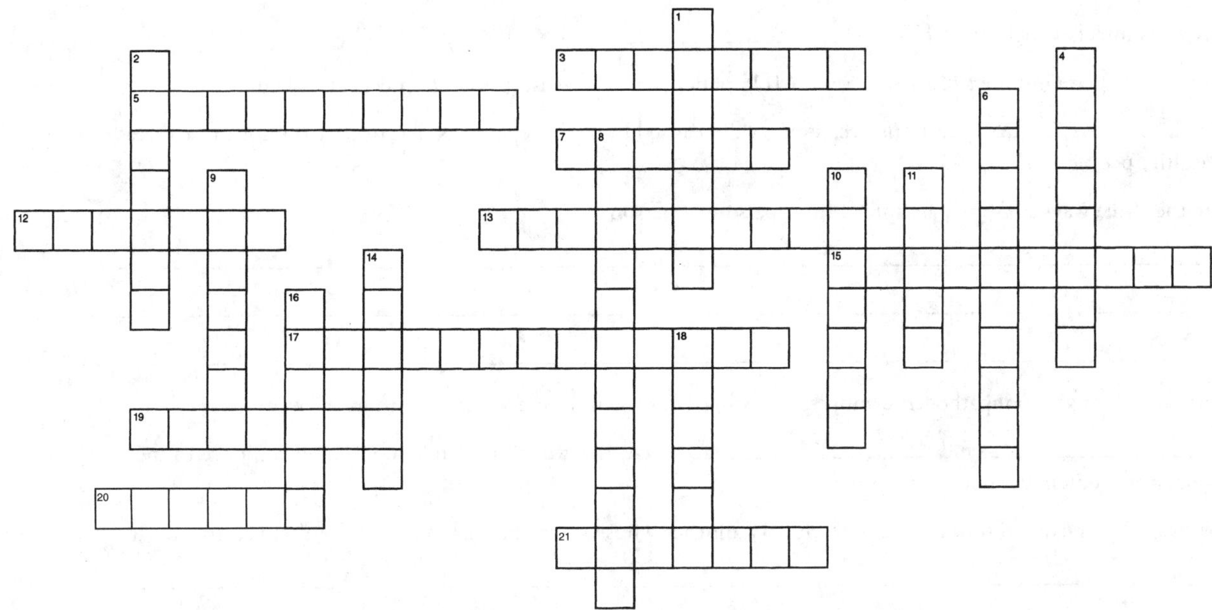

ACROSS

3. ________________ immunity may be passive or active
5. Weakened pathogen; used in a vaccine
7. Inactivated bacterial toxin; used in a vaccine
12. Chemical marker that identifies a cell
13. Produces antibodies to pathogens (two words)
15. Mass of lymphatic tissue along lymph vessels (two words)
17. Immune mechanism that includes cytotoxic T cells (hyphenated)
19. An artificial substance that stimulates production of antibodies and memory cells
20. ________________ immunity; does not create memory
21. Lymph nodules of the pharynx

DOWN

1. Alternate name for antibody-mediated immunity
2. ________________ immunity; antibodies are from another source
4. A protein molecule produced by a plasma cell
6. May be T or B
8. An antigen is "labeled" for phagocytosis
9. Immunity that does not involve antibodies; is programmed in DNA
10. Hypersensitivity to a particular foreign antigen
11. Tissue fluid that enters lymph capillaries
14. Loss of this organ increases susceptibility to pneumonia or meningitis
16. ________________ immunity; person produces her or his own antibodies
18. Produces T cells in the fetus

CLINICAL APPLICATIONS

1. a) AIDS is caused by a ________________ called ________________.

   b) This virus infects which type of T cells? ________________

   c) Without adequate numbers of these T cells, AIDS patients are very susceptible to infections called ________________ infections because they are caused by microorganisms that usually do not cause disease in healthy people.

   d) State the three ways HIV may be spread from person to person.

      1) ________________

      2) ________________

      3) ________________

2. a) A vaccine contains a nonpathogenic antigen that stimulates the immune system to produce specific ________________ and ________________ cells that will stimulate a more rapid response to the pathogen if it enters the body.

   b) How would you explain how a vaccine works to a mother whose infant has just received the first vaccination for measles? ________________

3. a) Six-year-old Kenisha has a strep throat, and her mother tells the doctor that her daughter has "swollen glands" in her neck. These "glands" are the ________________, which have become enlarged.

   b) Within these lymph nodes, ________________ are attempting to phagocytize the bacteria that have entered the lymph from the pharynx.

4. a) Mrs. J is allergic to certain plant pollens, and every summer she develops symptoms that include sneezing, runny nose, and watery eyes. Mrs. J usually takes a medication that contains an antihistamine. The purpose of this is to counteract the effects of ________________ released during the allergic response.

   b) Substances such as pollens that cause allergies are called ________________.

   c) Are these substances pathogenic? ________________

   d) Is the immune (allergic) response to an allergen helpful or protective in any way? ________________

   e) State one possible serious consequence of a severe allergic reaction. ________________

MULTIPLE CHOICE TEST #1

Choose the correct answer for each question.

1. The lymphatic vessel that returns lymph to the left subclavian vein is the:
   a) right lymphatic duct b) cisterna chyli c) left cisterna d) thoracic duct

2. Tissue fluid is called lymph when it enters:
   a) the thoracic duct b) lymph capillaries c) the cisterna chyli d) veins

3. In larger lymph vessels, backflow of lymph is prevented by:
   a) smooth muscle b) dilation c) valves d) expansion

4. An important function of the lymphatic system is to return tissue fluid to the:
   a) cells b) liver c) spleen d) blood

5. The masses of lymphatic tissue located below the epithelium of mucous membranes are called:
   a) lymph nodules b) lymph capillaries c) lymph nodes d) lymph axillaries

6. The lymph nodes that remove pathogens in the lymph coming from the legs are called:
   a) thoracic nodes b) cervical nodes c) axillary nodes d) inguinal nodes

7. Lymph nodes and nodules and the spleen are made primarily of WBCs called:
   a) neutrophils b) basophils c) lymphocytes d) eosinophils

8. The fixed macrophages of the spleen phagocytize old RBCs and form:
   a) more hemoglobin b) bilirubin c) bile d) myoglobin

9. The spleen is located below which organ?
   a) the liver b) the stomach c) the colon d) the diaphragm

10. The functioning of the thymus is most important in which age group?
    a) old age b) adolescence c) childhood d) middle age

11. In the embryo, both T cells and B cells are produced in the:
    a) thymus b) red bone marrow c) liver d) kidney

12. Antibodies are ________________ molecules that may also be called ________________.
    a) inorganic/immune globulins c) lipid/immune globulins
    b) carbohydrate/enzymes d) protein/gamma globulins

13. The two general kinds of antigens, from the perspective of the immune system, are:
    a) self-antigens and foreign antigens c) A antigens and B antigens
    b) T antigens and B antigens d) gamma antigens and beta antigens

14. The return of tissue fluid to the blood is important to maintain normal:
    a) blood volume b) blood pressure c) both of these d) neither of these

15. Plant viruses do not cause disease in people, which is an example of:
    a) passive acquired immunity c) herd immunity
    b) natural acquired immunity d) genetic immunity

16. In adaptive immunity, the cells that remember a foreign antigen and initiate its rapid destruction upon a second exposure are:
    a) plasma cells b) memory cells c) helper cells d) macrophages

17. In adaptive immunity, the cells that participate in the recognition of foreign antigens are:
    a) helper T cells and cytotoxic T cells c) B cells and macrophages
    b) B cells and cytotoxic T cells d) macrophages and helper T cells

18. Recovery from a disease may provide the type of immunity called:
    a) naturally acquired passive immunity c) naturally acquired active immunity
    b) artificially acquired passive immunity d) artificially acquired active immunity

19. Which of these organs does NOT compensate for any function of the spleen if the spleen must be removed?
    a) liver b) thoracic duct c) lymph nodes d) red bone marrow

20. A baby is born temporarily immune to the diseases its mother is immune to; this is an example of:
    a) naturally acquired passive immunity c) naturally acquired active immunity
    b) artificially acquired passive immunity d) artificially acquired active immunity

21. An antibody can best be described as:
    a) a killer b) a defender c) a label d) an enzyme to punch a hole

22. Natural killer cells are believed to eliminate foreign cells by damaging their:
    a) nuclei b) mitochondria c) chromosomes d) cell membranes

23. The first antibody response to a foreign antigen is usually:
    a) fast, with a small amount c) slow, with a large amount
    b) slow, with a small amount d) fast, with a large amount

24. Following a drop of water in the body, which sequence is correct?
    a) plasma → tissue fluid → lymph → plasma
    b) tissue fluid → plasma → lymph → tissue fluid
    c) lymph → tissue fluid → plasma → lymph
    d) plasma → lymph → tissue fluid → plasma

25. In innate immunity, the best anatomic barrier to pathogens is probably the:
    a) stratum corneum b) gastric cilia c) areolar connective tissue d) subcutaneous tissue

26. A chemical involved in innate immunity is:
    a) lysozyme b) histamine c) leukotriene d) all of these

27. An immune mechanism that does not create memory is:
    a) humoral b) innate c) cell mediated d) adaptive

28. Inflammation is the body's response to:
    a) a cut in the skin
    b) brain damage from lack of oxygen
    c) ulcer bacteria in the stomach
    d) all of these

29. The signs of inflammation include all of these except:
    a) pain b) paleness c) warmth d) swelling

30. The cells involved in innate immunity do all of these except:
    a) produce antibodies
    b) phagocytize pathogens
    c) produce histamine
    d) activate the lymphocytes of adaptive immunity

MULTIPLE CHOICE TEST #2

Read each question and the four answer choices carefully. When you have made a choice, follow the instructions to complete your answer.

1. Which statement is NOT true of the spleen?
   a) The spleen stores platelets.
   b) Antibodies are produced in the spleen by cells called fixed monocytes.
   c) The spleen is located in the upper left abdominal quadrant.
   d) Fixed macrophages in the spleen phagocytize pathogens in the blood.

   Reword your choice to make it a correct statement.

   ____

2. Which statement is NOT true of lymph nodes?
   a) They are located along the pathway of lymph vessels.
   b) During a serious infection they may become swollen.
   c) They contain fixed macrophages and plasma cells.
   d) The three major paired groups are the cervical, axillary, and plantar nodes.

   Reword your choice to make it a correct statement.

   ____

3. Which statement is NOT true of lymph nodules?
   a) They are found below the epithelium of serous membranes.
   b) They contain lymphocytes and macrophages.
   c) They destroy pathogens that penetrate mucous membranes.
   d) They are found in the digestive and respiratory tracts.

   Reword your choice to make it a correct statement.

   ____

4. Which statement is NOT true of lymph?
   a) Lymph is tissue fluid that has entered lymph capillaries.
   b) Lymph flows through lymph nodes as it travels back to the blood.
   c) Lymph contains the antibodies produced by the fixed macrophages.
   d) Lymph may contain pathogens that have entered breaks in the skin.

   Reword your choice to make it a correct statement.

   ____

5. Which statement is NOT true of lymph vessels?
   a) The right lymphatic duct returns lymph to the inferior vena cava.
   b) The valves in larger lymph vessels prevent the backflow of lymph.
   c) The return of lymph to the blood is assisted by the respiratory pump and the skeletal muscle pump.
   d) The thoracic duct returns lymph to the left subclavian vein.

   Reword your choice to make it a correct statement.

   ____

6. Which statement is NOT true of the types of immunity?
   a) In passive immunity, antibodies come from another person or source.
   b) Genetic immunity is the result of the DNA makeup of a species.
   c) In active immunity, a person produces his or her own antibodies.
   d) A vaccine provides naturally acquired passive immunity.

   Reword your choice to make it a correct statement.

   ____

7. Which statement is NOT true of the thymus and immunity?
   a) The thymus is most active in the fetus and child.
   b) The thymus is located below the thyroid gland.
   c) T cells are the monocytes produced by the thymus.
   d) The thymus is necessary for T cells to become immunologically competent.

   Reword your choice to make it a correct statement.

   ____

8. Which statement is NOT true of antigens and antibodies?
   a) When an antibody is produced, it will bond to the nearest foreign antigen.
   b) Self-antigens are found on all the cells of an individual.
   c) Foreign antigens stimulate antibody production.
   d) Antibodies are protein molecules.

   Reword your choice to make it a correct statement.

   ____

9. Which statement is NOT true of innate immunity?
   a) The ciliated epithelium of the respiratory tract is a good barrier to pathogens.
   b) Langerhans cells pick up foreign antigens and transport them to lymph nodes.
   c) Mast cells and basophils produce leukotrienes and antibodies.
   d) Neutrophils and macrophages phagocytize pathogens.

   Reword your choice to make it a correct statement.

   ____

10. Which statement is NOT true of adaptive immunity?
    a) Each response is specific for one foreign antigen.
    b) Repeated responses become less efficient.
    c) Its responses create memory.
    d) Antibodies may be involved.

    Reword your choice to make it a correct statement.

    ____

11. Which statement is NOT true of cell-mediated immunity?
   a) Cytotoxic T cells chemically disrupt the cell membranes of foreign antigens.
   b) The foreign antigen is recognized by macrophages and helper B cells.
   c) Tumor cells or cells infected with viruses can be destroyed.
   d) Memory T cells remember a specific foreign antigen.

   Reword your choice to make it a correct statement.

   ___

12. Which statement is NOT true of humoral (antibody-mediated) immunity?
   a) Plasma cells come from activated B cells and produce antibodies.
   b) Opsonization is the labeling of foreign antigens by self-antigens.
   c) Memory B cells remember a specific foreign antigen.
   d) Macrophages phagocytize antigen–antibody complexes.

   Reword your choice to make it a correct statement.

   ___

13. Which statement is NOT true of immunity?
   a) Innate immunity is not specific.
   b) The macrophages of innate immunity activate the lymphocytes of adaptive immunity.
   c) The memory cells of innate immunity are lymphocytes.
   d) Adaptive immunity depends on lymphocytes from the thymus gland.

   Reword your choice to make it a correct statement.

   ___

14. Which statement is NOT true of inflammation?
   a) It is a response by the body to any type of damage.
   b) Its purpose is to contain damage and permit repair.
   c) It is a positive feedback mechanism and may create more damage.
   d) Histamine makes capillaries less permeable.

   Reword your choice to make it a correct statement.

   ___

MULTIPLE CHOICE TEST #3

Each question is a series of statements concerning a topic in this chapter. Read each statement carefully and select all of the correct statements.

1. Which of the following statements are true of the lymphatic system?
   a) Lymph nodes are most abundant along the blood vessels of the hands and feet.
   b) Lymphatic tissue contains stem cells that differentiate into lymphocytes.
   c) Lymph from the lower body flows into the cisterna chyli and then into the thoracic duct.
   d) The thoracic duct empties lymph into the right subclavian vein.
   e) The walls of large lymph vessels have smooth muscle that contracts to promote the flow of lymph.
   f) Lymph vessels have valves to prevent the backflow of lymph.
   g) The spleen contains lymphocytes that become plasma cells that produce antibodies.
   h) If an adult's spleen must be removed, the red bone marrow will compensate for the loss of erythrocyte production.
   i) Most T cells are produced in the red bone marrow.
   j) In the fetus, T cells learn self-recognition, which means working with other T cells to provide immunity.

2. Which of the following statements are true of innate immunity?
   a) The antibacterial enzyme lysozyme is found in tears and saliva.
   b) Macrophages have receptors for common pathogens.
   c) The respiratory ciliated epithelium and other mucous membranes are good barriers to pathogens.
   d) Innate immune responses do not produce memory for pathogens.
   e) Inflammation includes vasodilation, which may produce the symptom of heat in the affected area.
   f) Histamine and leukotrienes are released by mast cells.
   g) Innate immune responses do not become more efficient with repeated exposures to a pathogen.
   h) The process of inflammation in response to tissue damage is always the same.

3. Which of the following statements are true of adaptive immunity?
   a) Antibody-mediated immunity produces memory, but cell-mediated immunity does not.
   b) Antibodies are produced by plasma cells that differentiate from helper T cells.
   c) Adaptive immune responses are not specific as to pathogen.
   d) Self-antigens are not important for adaptive immune responses.
   e) Recognition of foreign antigens is very efficient when macrophages work with helper T cells.
   f) Cytotoxic T cells secrete chemicals to destroy foreign cells.
   g) Antibodies are a label for foreign antigens, to attract macrophages.
   h) The process of complement fixation is activated by an antigen–antibody complex.

4. Which of the following statements are true of immunity?
   a) On the second exposure to a pathogen, antibody production is faster and greater than with a first exposure.
   b) Passive immunity means that immune cells wait for a pathogen to flow by in the blood.
   c) Genetic immunity includes all the genes we have for antibody production.
   d) Foreign antigens for humans include fungi, bacteria, and nonhuman proteins.
   e) IgG antibodies are large enough to cross the placenta to provide active immunity for the newborn.
   f) IgD antibodies are found in breast milk.
   g) The duration of active immunity varies, but passive immunity is always lifelong.
   h) The purpose of a vaccine is to take the place of the first exposure to a pathogen; this provides artificially acquired active immunity.

Chapter 15

The Respiratory System

This chapter describes the structure and functions of the respiratory system. The lungs provide a site for the exchange of oxygen and carbon dioxide between the air and the blood. The other organs of the respiratory system contribute to the movement of air into and out of the lungs. The two divisions of the respiratory system are the upper respiratory tract and the lower respiratory tract.

NOSE, NASAL CAVITIES, AND PHARYNX

1. The nose is the usual passageway for air into and out of the respiratory tract. The ____________ just inside the nostrils help block the entry of dust.

2. a) Within the skull are the two nasal cavities, which are separated by the ____________.

   b) The nasal mucosa (lining) is made of ____________ epithelium.

   c) State the three functions of the nasal mucosa.

      1) ____________

      2) ____________

      3) ____________

   d) In the upper nasal cavities are the receptors for the sense of ____________.

3. a) The paranasal sinuses are air cavities that open into the ____________.

   b) Name two of the four bones that contain paranasal sinuses.

      1) ____________

      2) ____________

   c) State one function of these sinuses. ____________

4. Match each part of the pharynx with the proper descriptive statements.

   Use each letter once. One answer line will have five correct letters; each other answer line will have two correct letters.

   1) Nasopharynx ____________

   2) Oropharynx ____________

   3) Laryngopharynx ____________

   A. An air and food passage that opens into the larynx and esophagus
   B. Covered by the soft palate during swallowing
   C. An air passage only
   D. An air and food passage behind the mouth
   E. The palatine tonsils are on the lateral walls
   F. The adenoid is on the posterior wall
   G. The eustachian tubes open into this part
   H. The swallowing reflex involves contraction of the oropharynx and this part
   I. The only part lined with ciliated epithelium

5. From the pharynx, incoming air enters the ________________ and then the trachea, both of which are part of the upper respiratory tract.

LARYNX

1. a) The larynx is an air passage between the ________________ and the ________________.

   b) The other function of the larynx is ________________.

2. Match each part of the larynx with its proper description.

   Use each letter once. One answer line will have two correct answers.

   1) Epiglottis ________________
   2) Vocal cords ________________
   3) Ciliated epithelium ________________
   4) Thyroid cartilage ________________
   5) Glottis ________________

   A. Two folds on either side of the glottis
   B. Sweeps mucus and pathogens upward
   C. The air passage between the vocal cords
   D. The largest, most anterior cartilage of the larynx
   E. The cartilage that covers the larynx during swallowing
   F. Vibrate to produce speech sounds

3. The cranial nerves that are the motor nerves to the larynx are the ________________ and ________________ nerves.

4. Speech sounds are produced when the intrinsic muscles of the larynx pull the vocal cords ________________ (together or apart) across the glottis. The vocal cords are then vibrated by ________________ air.

5. The following diagrams depict an anterior view (left) and midsagittal section (right) of the larynx.

   Label the parts indicated.

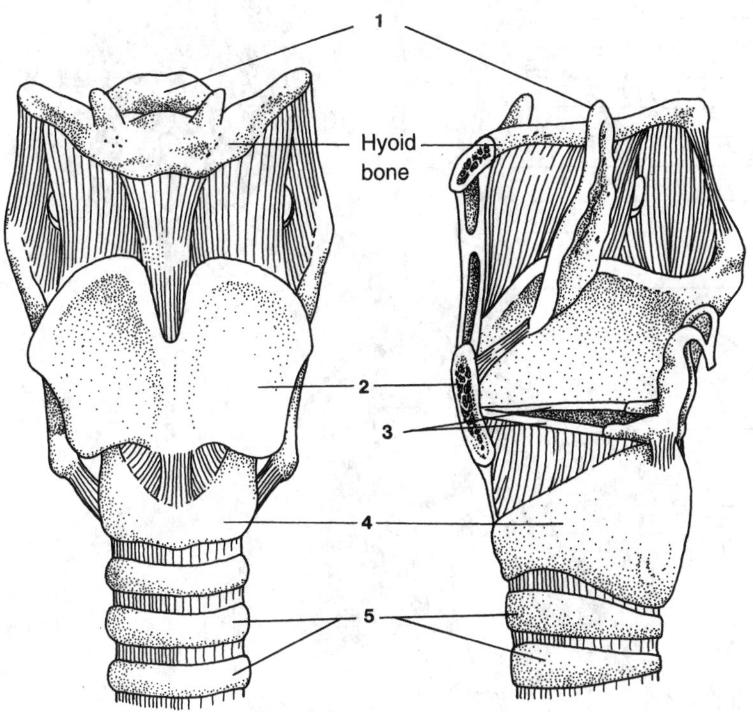

TRACHEA AND BRONCHIAL TREE

1. The trachea is an air passage that extends from the ____________________ to the ____________________.
2. a) The tissue that forms C-shaped rings in the wall of the trachea is ____________________.
   b) State the function of these incomplete rings. ____________________
   c) The tissue of the tracheal mucosa that sweeps mucus and pathogens upward is ____________________.
3. a) The right and left primary bronchi are branches of the ____________________.
   b) The secondary bronchi are within the lungs. There are ____________________ in the left lung and ____________________ in the right lung.
4. a) The smaller branches of the bronchial tree are called ____________________, and they differ in structure from the bronchi in that there is no ____________________ in their walls to keep them open.
   b) The smallest bronchioles end in the clusters of ____________________ in the lungs.
5. The following diagram depicts the respiratory system.

   Label the parts indicated.

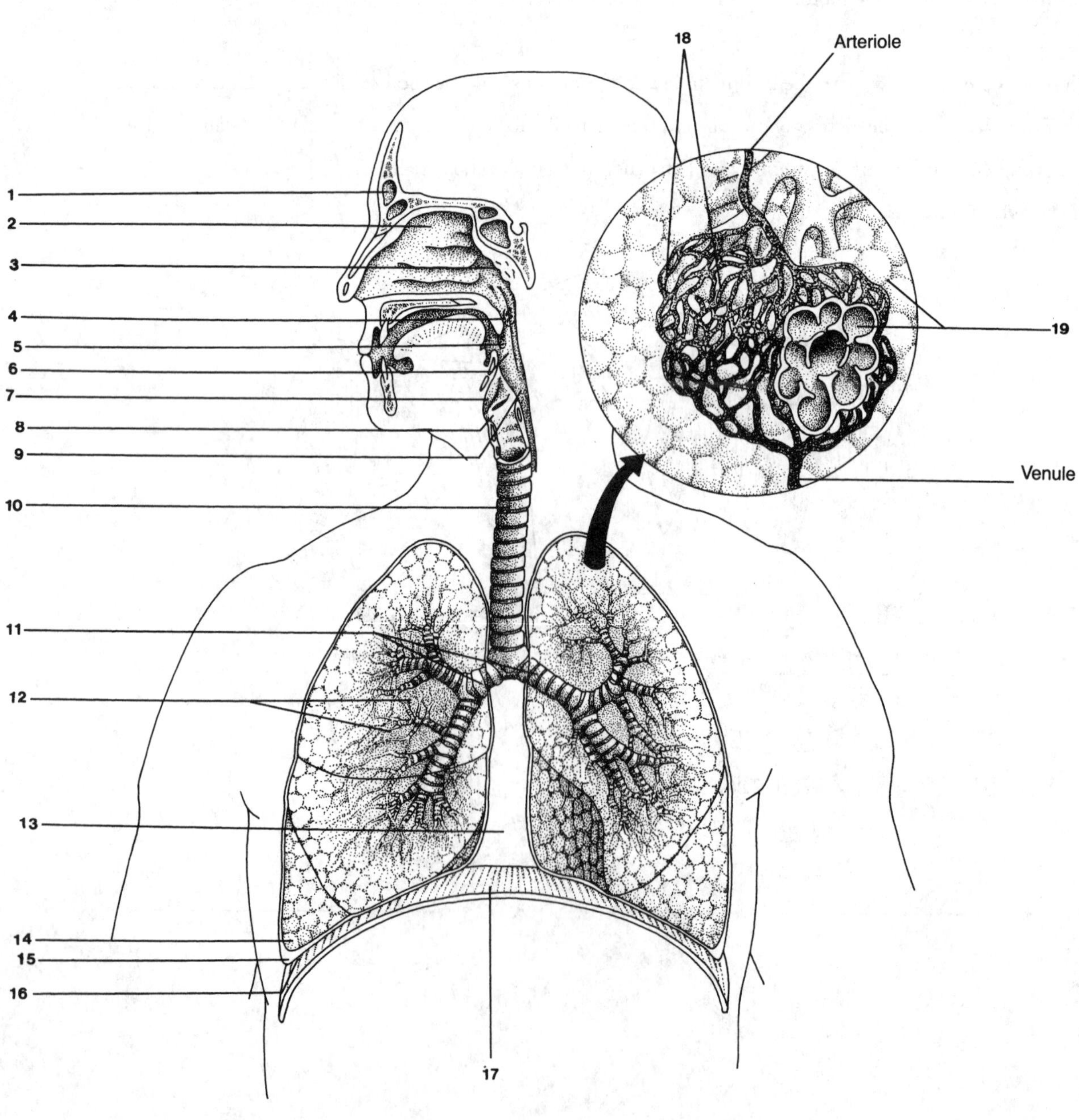

THE PATHWAY OF AIR

1. With respect to the flow of inhaled air, number the following in the proper sequence. The first and last have been indicated to get you started.

1	nose	______	bronchioles
______	larynx	______	nasopharynx
______	primary bronchi	______	trachea
______	laryngopharynx	______	secondary bronchi
______	nasal cavities	11	alveoli
______	oropharynx		

2. Which of the above structures are part of the lower respiratory tract? (Do not include the trachea.)

   1) ______________ 3) ______________

   2) ______________ 4) ______________

PLEURAL MEMBRANES

1. a) The serous membrane that is on the surface of the lungs is the ______________ pleura.

   b) The serous membrane that lines the thoracic cavity is the ______________ pleura.

2. a) The serous fluid ______________ between the pleural membranes as the lungs expand and recoil.

   b) The serous fluid also keeps the pleural membranes ______________ (together or apart) during breathing.

LUNGS

1. a) The lungs are within the thoracic cavity and are protected from mechanical injury by the

   ______________.

   b) Medial to the lungs is the area called the ______________, which contains the ______________ (organ).

   c) Inferior to the lungs is the ______________, one of the respiratory muscles.

2. a) The indentation on the medial side of each lung is called the ______________.

   b) At this site, the ______________ and the ______________ enter the lung.

3. a) The alveoli of the lungs are made of alveolar type I cells, which are what type of tissue?

   ____

   b) The pulmonary capillaries around the alveoli are made of what type of tissue?

   ____

   c) The important characteristic of this tissue is that it is ______________ to permit

   ______________.

   d) The tissue in the spaces between the alveoli that is important for normal exhalation is ______________.

4. a) Each alveolus is lined with a thin layer of tissue fluid that is important to permit ______________.

   b) This tissue fluid is mixed with ______________ that decreases the surface tension of the tissue fluid and permits ______________ of the alveolus.

   c) Surfactant is produced by ______________ cells.

5. The following diagram depicts a cluster of alveoli and pulmonary capillaries.

Label the parts indicated.

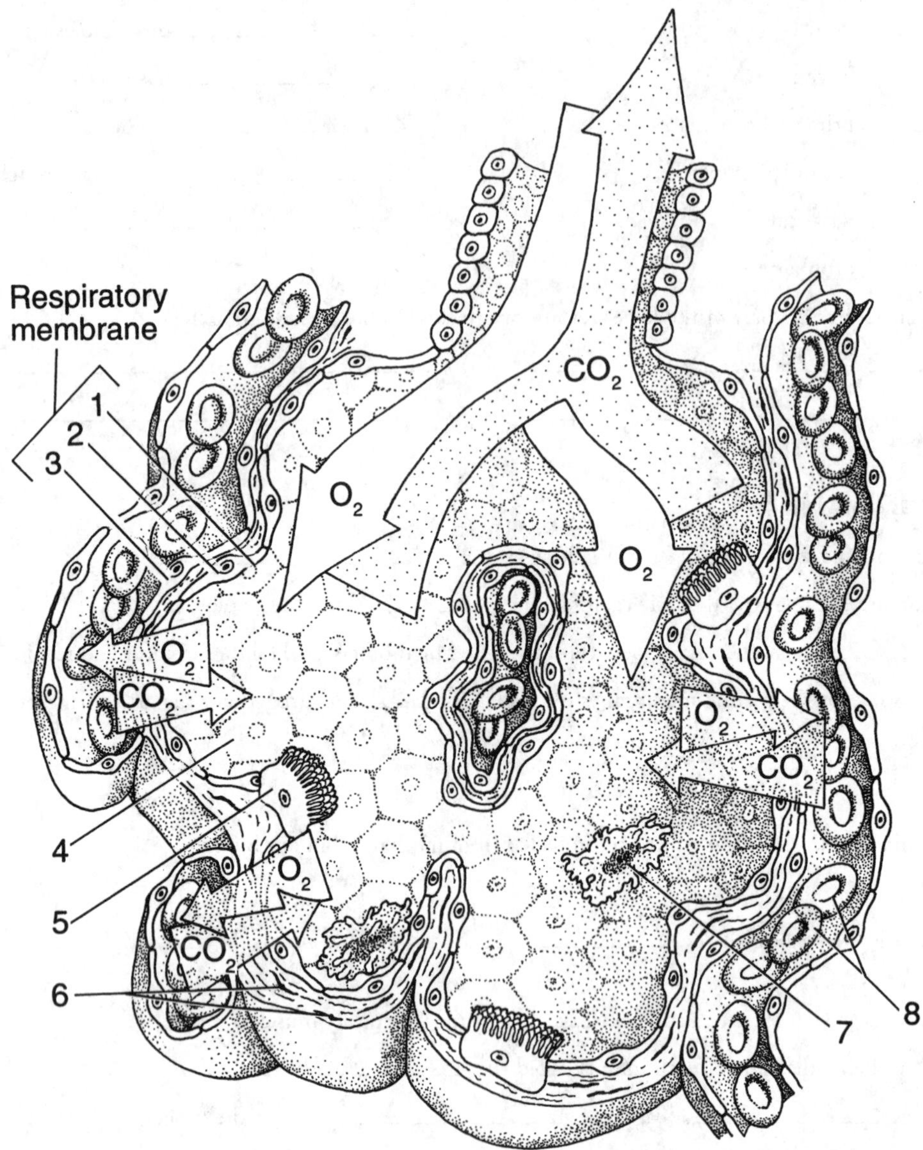

MECHANISM OF BREATHING

1. a) The movement of air into and out of the lungs is called ________________.

   b) The two phases of this movement are ________________ and ________________.

2. The respiratory centers are located in the brain, in the ________________ and ________________.

3. The respiratory muscles are:

   1) the external and internal ________________ muscles, which are supplied by the ________________ nerves, and

   2) the ________________, which is supplied by the ________________ nerves.

4. Contractions of the respiratory muscles produce changes in ________________ within the bronchial tree and alveoli to bring about ventilation.

5. Match each air pressure with the statements that apply to each.

Use each letter once. Each answer line will have two correct letters.

1) Atmospheric pressure ______________

2) Intrapleural pressure ______________

3) Intrapulmonic pressure ______________

A. The pressure in the bronchial tree and alveoli
B. The pressure of the air around us
C. The pressure within the potential pleural space
D. 760 mm Hg at sea level
E. Fluctuates below and above atmospheric pressure during breathing
F. Always slightly below atmospheric pressure

INHALATION (INSPIRATION)

1. With respect to normal inhalation, number these events in proper sequence.

______1______ The medulla generates motor impulses.

______________ The chest cavity is enlarged in all directions.

______________ The diaphragm and external intercostal muscles contract.

______________ Intrapulmonic pressure decreases.

______________ Motor impulses travel along the phrenic and intercostal nerves.

______________ The chest wall expands the parietal pleura, which expands the visceral pleura, which in turn expands the lungs.

______________ Air enters the lungs until intrapulmonic pressure equals atmospheric pressure.

2. A deep breath (more than normal) requires a more forceful ______________ of the respiratory muscles, which in turn would bring about greater expansion of the ______________.

EXHALATION (EXPIRATION)

1. With respect to normal exhalation, number these events in proper sequence.

______1______ Motor impulses from the medulla decrease.

______________ The chest cavity becomes smaller, and the elastic connective tissue around the alveoli recoils.

______________ Intrapulmonic pressure rises above atmospheric pressure.

______________ The lungs are compressed.

______________ The diaphragm and external intercostal muscles relax.

______________ Air is forced out of the lungs until intrapulmonic pressure equals atmospheric pressure.

2. Normal exhalation is considered a passive process because it does not require the ______________ of the respiratory muscles.

3. A forced exhalation requires contraction of the ______________ muscles to pull the ribs ______________, or contraction of the ______________ muscles to compress the abdominal organs and push the diaphragm ______________.

PULMONARY VOLUMES

1. Match each pulmonary volume with its proper definition.

   Use each letter once.

   1) Tidal volume ____________
   2) Vital capacity ____________
   3) Inspiratory reserve ____________
   4) Expiratory reserve ____________
   5) Minute respiratory volume ____________
   6) Residual air ____________

   The Volume of Air

   A. Inhaled and exhaled in 1 minute
   B. Beyond tidal, in the most forceful exhalation
   C. Involved in the deepest inhalation followed by the most forceful exhalation
   D. Remaining in the lungs after the most forceful exhalation
   E. Beyond tidal, in the deepest inhalation
   F. In one normal inhalation and exhalation

2. The lungs need to have residual air in them at all times so that ____________ is a continuous process.

3. a) Alveolar ventilation is the volume of air in each inhalation that reaches the ____________ and participates in ____________.

   b) At the end of an inhalation, the volume of air that is in the respiratory passages is called ____________ dead space.

   c) In a disease such as pneumonia, the fluid-filled alveoli are called ____________ dead space.

4. a) The expansibility of the lungs and thoracic wall is called ____________.

   b) State two possible causes of decreased pulmonary compliance.

   1) ____________ 2) ____________

   c) State two possible causes of decreased thoracic compliance.

   1) ____________ 2) ____________

   d) Normal compliance is necessary for normal ____________, and anything that decreases compliance increases ____________ dead space.

EXCHANGE OF GASES

1. a) External respiration is the exchange of gases between the air in the ____________ and the ____________ in the pulmonary capillaries.

   b) Internal respiration is the exchange of gases between the blood in the ____________ and the tissue fluid (cells).

2. The two respiratory gases are ____________ and ____________.

3. a) Inhaled air (the atmosphere) is approximately ____________ % oxygen and ____________ % CO_2.

   b) Exhaled air is approximately ____________ % oxygen and ____________ % CO_2.

4. The value that is used to express the concentration of O_2 and CO_2 in the air or in body fluids is called ____________ and is abbreviated ____________.

5. The following diagram depicts one alveolus and a pulmonary capillary. The presence of each respiratory gas in each site is shown, indicated as PO_2 or PCO_2.

   a) For each gas in each site, indicate whether the partial pressure is "high" or "low," and then use an arrow to show the direction of diffusion of each gas.

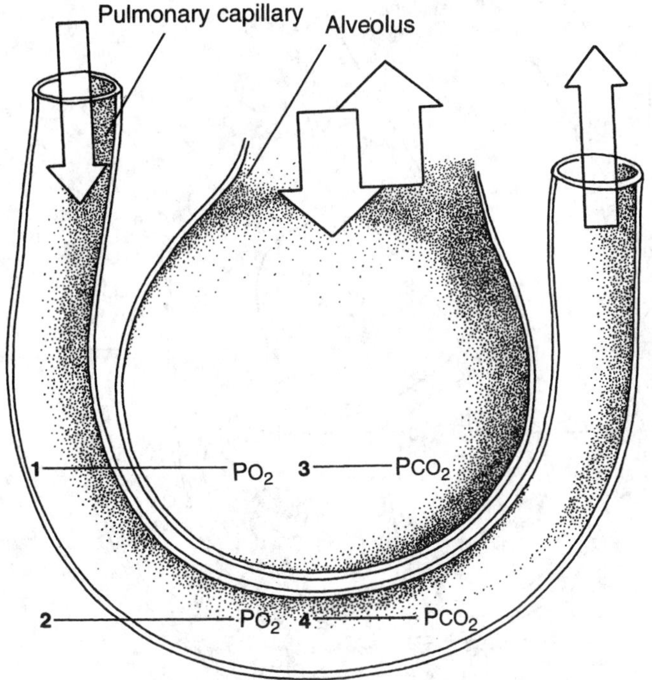

   b) This exchange of gases is called ____________________ respiration.

   c) The blood that leaves the pulmonary capillaries will return to the ____________________ and then be pumped by the ____________________ ventricle to the ____________________.

6. The following diagram depicts a systemic capillary around cells in a body tissue. The presence of each respiratory gas in each site is shown, indicated as PO_2 or PCO_2.

   a) For each gas in each site, indicate whether the partial pressure is "high" or "low," and then use an arrow to show the direction of diffusion of each gas.

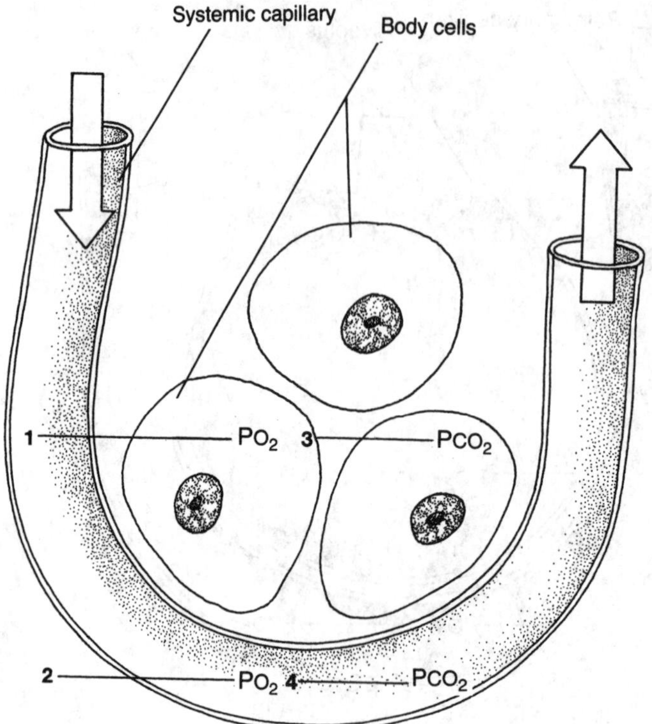

   b) This exchange of gases is called ________________ respiration.

   c) The blood that leaves the systemic capillaries will return to the ________________ and then be pumped by the ________________ ventricle to the ________________.

TRANSPORT OF GASES IN THE BLOOD

1. a) Oxygen is carried in the blood by which cells? ________________

   b) Within these cells, the oxygen is bonded to the mineral ________________ in the protein ________________.

   c) The oxygen-hemoglobin bond is formed when blood circulates through the ________________, where the PO_2 in the alveoli is high.

2. As blood circulates through the systemic capillaries, the O_2-Hb bond tends to break. State the three factors that will increase the release of oxygen from hemoglobin.

   1) ________________

   2) ________________

   3) ________________

3. Most CO_2 is carried in the blood in the form of ________________ ions in the ________________ (part of blood).

4. The function of hemoglobin with respect to CO_2 transport is to act as a buffer for ________________ ions in the RBCs.

NERVOUS REGULATION OF RESPIRATION

1. Within the brain, the respiratory centers are located in the ________________ and ________________.

2. a) In the medulla, the ________________ center generates impulses that travel to the respiratory muscles, causing them to contract.

   b) The expiration center is activated to promote a ________________.

3. In the pons, the ________________ center prolongs inhalation, and the ________________ center interrupts the apneustic center to help bring about ________________.

4. The normal rate of respirations ranges from ________________ to ________________ breaths per minute.

5. The Hering-Breuer inflation reflex also contributes to a normal rate and rhythm of breathing. In this reflex, ________________ in the lungs detect the stretching during inflation and generate impulses that depress the ________________ center in the medulla to help bring about exhalation.

6. a) The hypothalamus may influence the rate or rhythm of breathing during ________________ situations.

   b) The cerebral cortex may bring about ________________ changes in breathing.

   c) State one example. ________________

7. a) With respect to the cough reflex, the cough center is located in the ________________ of the brain.

   b) The stimulus for a cough is irritation of the mucosa of the ________________, ________________, or ________________.

   c) The response is an ________________ exhalation out the ________________ to remove the irritation.

8. a) With respect to the sneeze reflex, the sneeze center is located in the ________________ of the brain.

   b) The stimulus for a sneeze is irritation of the mucosa of the ________________.

   c) The response is an ________________ exhalation out the ________________ to remove the irritation.

9. The following diagram depicts the nervous regulation of respiration.

   Label the parts indicated.

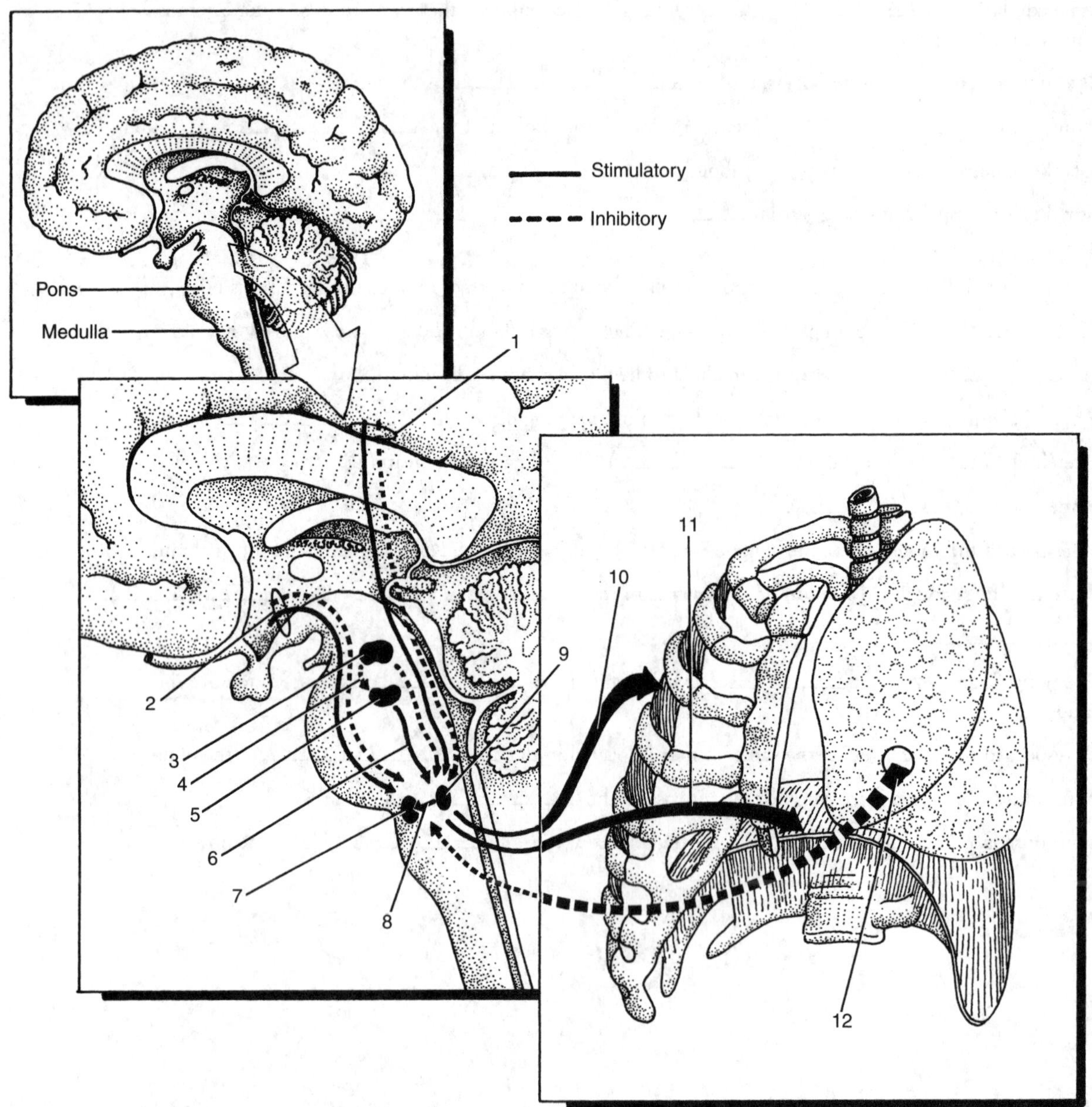

CHEMICAL REGULATION OF RESPIRATION

1. a) Changes in the blood levels of the gases ____________________ and ____________________ may have an effect on the rate of breathing.

   b) A change in the ____________________ of the blood may also have an effect on respiration.

2. Match each change in blood-gas level with the proper descriptive statements.

   Use each letter once. Each answer line will have two correct letters.

   1) Decreased blood O_2 ____________________

   2) Increased blood CO_2 ____________________

   A. Detected by chemoreceptors in the carotid and aortic bodies
   B. The stimulus to increase respirations to inhale more oxygen
   C. Detected by chemoreceptors in the medulla
   D. The stimulus to increase respirations to exhale more carbon dioxide

3. a) Which of the two respiratory gases is the major regulator of respiration? ____________

   b) This is so because a decrease in respirations will permit this gas to accumulate in the body and lower the ____________ of the blood and other body fluids.

   c) In contrast, decreased respirations will not have a great effect on the blood level of oxygen. This is so because of the ____________ air that remains in the lungs after exhalation and because most of the oxygen in inhaled air ____________ (does or does not) enter the blood but is available to do so if needed.

4. The following diagram depicts the sequence of events in chemical regulation of respiration.

   Label the parts indicated.

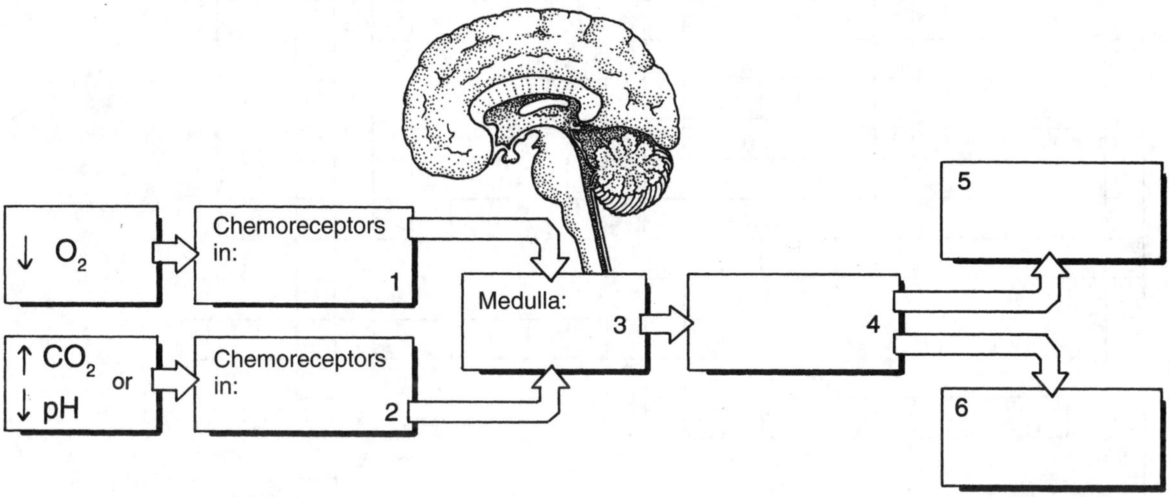

↓ O_2 → Chemoreceptors in: 1

↑ CO_2 or ↓ pH → Chemoreceptors in: 2

Medulla: 3

4

5

6

RESPIRATION AND ACID-BASE BALANCE

1. The respiratory system affects the pH of the blood and other body fluids because it regulates the amount of ____________ present in these fluids.

2. Match each respiratory pH imbalance with proper cause and effects.

   Use each letter once. Each answer line will have three correct letters.

   1) Respiratory alkalosis ____________

   2) Respiratory acidosis ____________

   A. Caused by a decrease in the rate or efficiency of respiration
   B. Fewer H^+ ions are formed
   C. CO_2 accumulates in the body
   D. CO_2 is exhaled more rapidly
   E. Caused by an increase in the rate of respiration
   F. More H^+ ions are formed

3. State one specific cause of respiratory acidosis. ____________

4. State one specific cause of respiratory alkalosis. ____________

5. a) State one possible cause of metabolic acidosis. ____________

   b) To compensate, respirations will ____________ in order to ____________ (exhale or retain) CO_2.

6. a) State one possible cause of metabolic alkalosis. ____________

   b) To compensate, respirations will ____________ in order to ____________ (exhale or retain) CO_2.

CROSSWORD PUZZLE

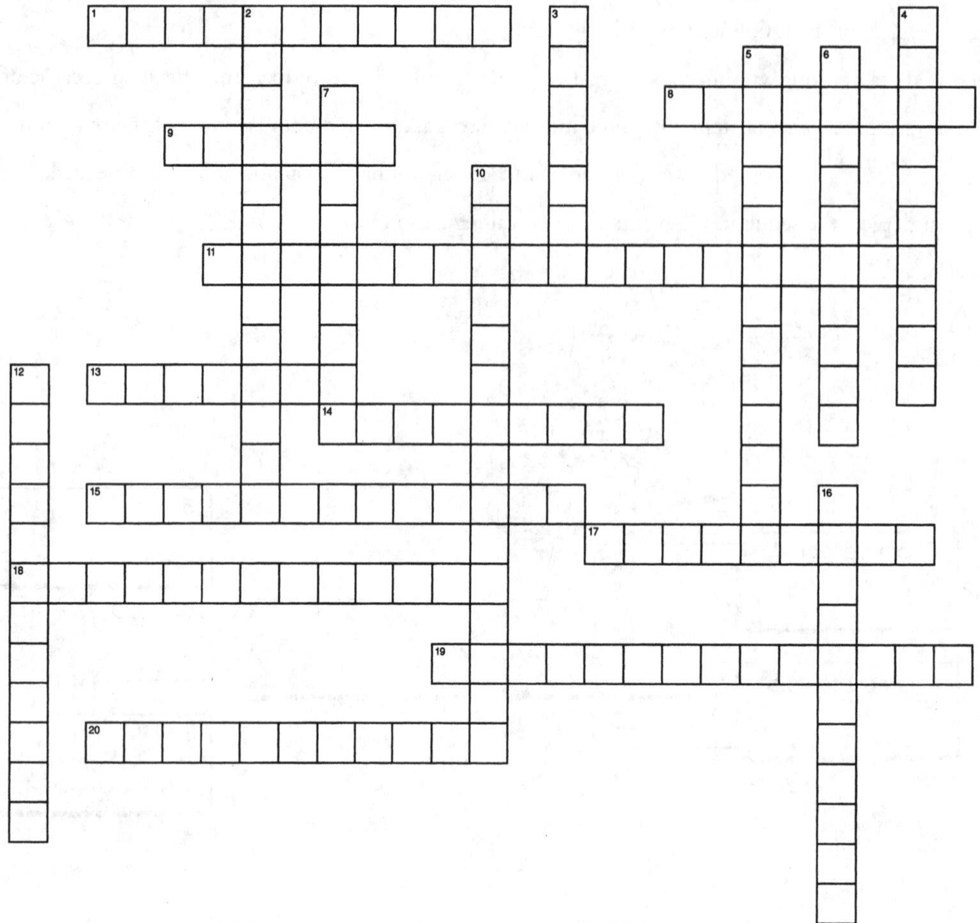

ACROSS

1. Movement of air to and from the alveoli
8. Respiratory __________; excess CO_2 increases H^+ ion formation
9. Voice box
11. Permits inflation of the alveoli (two words)
13. Air sacs of the lungs
14. Respiratory __________; less CO_2 decreases H^+ ion formation
15. The trachea and all of the bronchial tubes (two words)
17. Disease in which alveoli lose their elasticity
18. Motor impulses from the medulla to the diaphragm travel along these to initiate inhalation (two words)
19. Accumulation of fluid in the alveoli (two words)
20. Amount of air involved in one normal inhalation and exhalation (two words)

DOWN

2. __________ pressure; within the bronchial tree and the alveoli
3. Opening between the vocal cords
4. Prevents entry of food into the larynx
5. Sum of tidal volume, inspiratory reserve, and expiratory reserve (two words)
6. Prevents entry of food into the nasopharynx (two words)
7. Bacterial infection of the lungs
10. Value used to express the concentration of a gas in a particular site (two words)
12. __________ pressure; within the potential pleural space between the parietal pleura and the visceral pleura
16. Amount of air left in the lungs after the most forceful exhalation (two words)

CLINICAL APPLICATIONS

1. a) Mr. D has mild emphysema and tells his doctor that he "can't seem to get enough air in." As Mr. D will learn, his problem is not directly with inhalation but rather with exhalation. The deterioration of the ____________ tissue around the alveoli has changed exhalation from a ____________ process to an active process.

   b) Mr. D will have to expend energy to ____________ in order to empty his lungs sufficiently to be able to inhale.

2. a) A baby born prematurely at 8 months of gestation is monitored closely by the nursing staff for any signs of respiratory diseases. Such premature infants may have difficulty breathing if their immature lungs have not yet produced sufficient quantities of ____________.

   b) Without this substance, the alveoli of the lungs are not easily ____________ and collapse after each exhalation.

3. a) Mr. H was in a car accident in which a piece of metal pierced the right side of his chest. In the emergency room, he is conscious and able to say that he is having difficulty breathing. A possible cause of this is that the puncture wound in the chest wall has allowed ____________ to enter the ____________ space on that side.

   b) Because this air is at atmospheric pressure, it will cause the right lung to ____________.

   c) This is called ____________.

4. a) Mrs. M is 68 years old. One year ago, she had a myocardial infarction that weakened her left ventricle. She is now having difficulty with breathing. Mrs. M's respiratory problem may be related to her previous heart attack because if the left ventricle does not pump blood efficiently, blood will back up into ____________ circulation.

   b) This will increase blood pressure in the ____________ capillaries and cause filtration of tissue fluid into the alveoli.

   c) This condition is called ____________.

MULTIPLE CHOICE TEST #1

Choose the correct answer for each question.

1. The upper respiratory tract includes all of these except the:
   a) nasal cavities b) larynx c) primary bronchi d) nasopharynx

2. During swallowing, the larynx is covered by the:
   a) soft palate b) epiglottis c) vocal cords d) thyroid cartilage

3. During swallowing, the nasopharynx is covered by the:
   a) hard palate b) oropharynx c) epiglottis d) soft palate

4. The trachea is kept open by which tissue?
   a) fibrous connective tissue b) cartilage c) ciliated epithelium d) elastic connective tissue

5. In the nasal cavities and trachea, mucus and pathogens are swept to the pharynx by:
   a) ciliated epithelium c) cartilage
   b) simple squamous epithelium d) elastic connective tissue

6. The part of the pharynx that is a passageway for air only is the:
   a) nasopharynx b) oropharynx c) laryngopharynx d) tracheopharynx

7. Inhaled air passes from the trachea to the:
   a) primary bronchi b) larynx c) pharynx d) secondary bronchi

8. Cartilage supports and keeps open all of these structures except the:
   a) secondary bronchi b) bronchioles c) primary bronchi d) larynx

9. The serous fluid between the pleural membranes keeps the membranes together and:
   a) exchanges gases b) creates friction c) destroys pathogens d) prevents friction

10. The primary bronchi and the pulmonary blood vessels enter the lung at the:
    a) apex b) hilus on the lateral side c) hilus on the medial side d) base

11. The tissue fluid that lines the alveoli is important to:
    a) prevent diffusion of gases c) trap pathogens
    b) prevent friction d) permit diffusion of gases

12. Within the alveoli, surface tension is decreased and inflation is possible because of the presence of:
    a) tissue fluid b) pulmonary blood c) pulmonary surfactant d) mucus

13. The respiratory centers in the brain are located in the:
    a) pons and cerebral cortex c) hypothalamus and pons
    b) medulla and cerebral cortex d) medulla and pons

14. During inhalation, the thoracic cavity is enlarged from top to bottom by contraction of the:
    a) external intercostal muscles, which move down c) diaphragm, which moves down
    b) diaphragm, which moves up d) internal intercostal muscles, which move up

15. In the alveoli, the partial pressures of oxygen and carbon dioxide are:
    a) low Po_2 and high Pco_2 c) high Po_2 and high Pco_2
    b) high Po_2 and low Pco_2 d) low Po_2 and low Pco_2

16. Intrapulmonic pressure is the air pressure within the:
    a) bronchial tree and alveoli c) mouth and nose
    b) intrapleural space d) rib cage and diaphragm

17. Irritants on the mucosa of the larynx are removed by:
    a) a deep breath b) yawning c) the sneeze reflex d) the cough reflex

18. Most oxygen is transported in the blood:
    a) on RBC membranes c) bonded to hemoglobin in RBCs
    b) in blood plasma as free oxygen d) bonded to hemoglobin in blood plasma

19. Most carbon dioxide is transported in the blood:
    a) as hydrogen ions in the RBCs c) as CO_2 in the plasma
    b) as bicarbonate ions in the plasma d) as part of hemoglobin in RBCs

20. Internal respiration is the exchange of gases between the:
    a) systemic capillaries and body tissues c) alveoli and systemic capillaries
    b) pulmonary capillaries and alveoli d) pulmonary capillaries and body tissues

21. The factors that increase the release of oxygen from hemoglobin in systemic capillaries include all of these except:
    a) high temperature b) low Po_2 c) high Pco_2 d) low temperature

22. The air that remains in the lungs after the most forceful exhalation is called:
    a) vital capacity b) tidal volume c) residual air d) leftover expiration

23. The gas that is the most important chemical regulator of respiration is:
    a) O_2, because if present in excess it lowers the pH of body fluids
    b) CO_2, because if present in excess it lowers the pH of body fluids
    c) O_2, because if present in excess it raises the pH of body fluids
    d) CO_2, because if present in excess it raises the pH of body fluids

24. The receptors that detect a decrease in the oxygen level of the blood are located in the:
    a) carotid and aortic bodies b) medulla c) pulmonary artery d) hypothalamus

25. If pneumonia decreases the exchange of gases in the lungs, the resulting pH imbalance is called:
    a) metabolic acidosis b) metabolic alkalosis c) respiratory acidosis d) respiratory alkalosis

26. The nasal mucosa has all of these functions except:
   a) warming incoming air
   b) moistening incoming air
   c) increasing the oxygen content of the air
   d) sweeping mucus and pathogens to the pharynx

27. All of these will increase physiological dead space except:
   a) fractured ribs
   b) asthma
   c) stuffed-up sinuses
   d) tuberculosis

28. Anatomic dead space includes all of these except:
   a) bronchioles
   b) larynx
   c) potential pleural space
   d) nasal cavities

29. The expansibility of the lungs and chest wall is called the:
   a) stretchiness
   b) inflation potential
   c) inhalation potential
   d) compliance

30. To compensate for metabolic acidosis, respirations will:
   a) increase to exhale more CO_2
   b) decrease to exhale more CO_2
   c) increase to retain more CO_2
   d) decrease to retain more CO_2

MULTIPLE CHOICE TEST #2

Read each question and the four answer choices carefully. When you have made a choice, follow the instructions to complete your answer.

1. Which statement is NOT true of the bronchial tree?
   a) The trachea branches into the right and left primary bronchi.
   b) Air is brought to the alveoli by the secondary bronchi.
   c) The bronchioles have no cartilage in their walls.
   d) Ciliated epithelium in the trachea sweeps mucus and pathogens upward.

   Reword your choice to make it a correct statement.

   ___

2. Which statement is NOT true of the larynx?
   a) It is an air passageway between the pharynx and the trachea.
   b) The vocal cords are pulled together across the glottis when speaking.
   c) The vagus and accessory nerves are the motor nerves for speech.
   d) The epiglottis covers the top of the larynx during breathing.

   Reword your choice to make it a correct statement.

   ___

3. Which statement is NOT true of gas exchange?
   a) In external respiration, oxygen diffuses from the alveoli to the pulmonary capillaries.
   b) In internal respiration, carbon dioxide diffuses from the tissues to the systemic capillaries.
   c) The blood that enters pulmonary capillaries has a low PCO_2.
   d) The blood that enters systemic capillaries has a high PO_2.

   Reword your choice to make it a correct statement.

   ___

4. Which statement is NOT true of the respiratory muscles?
   a) The external intercostal muscles pull the ribs up and out during inhalation.
   b) The motor nerves to the diaphragm are called the diaphragm nerves.
   c) The diaphragm contracts and moves down during inhalation.
   d) The internal intercostal muscles pull the ribs down and in during a forced exhalation.

   Reword your choice to make it a correct statement.

   ___

5. Which statement is NOT true of the nervous regulation of respiration?
   a) The apneustic center in the pons helps bring about exhalation.
   b) The medulla contains an inspiration center and an expiration center.
   c) The inspiration center in the medulla generates motor impulses that cause contraction of the diaphragm and external intercostal muscles.
   d) Sensory impulses from the baroreceptors in the lungs to the medulla help prevent overinflation of the lungs.

   Reword your choice to make it a correct statement.

   ____

6. Which statement is NOT true of the chemical regulation of respiration?
   a) An increased blood level of CO_2 is detected by chemoreceptors in the medulla.
   b) An increased blood level of CO_2 lowers the pH of the blood.
   c) Hypoxia is detected by the pulmonary and aortic chemoreceptors.
   d) Changes in the blood oxygen level do not affect pH.

   Reword your choice to make it a correct statement.

   ____

7. Which statement is NOT true of the transport of gases in the blood?
   a) Most CO_2 is carried as bicarbonate ions in the plasma.
   b) Hemoglobin in RBCs is able to transport oxygen because it contains copper.
   c) Oxygen is released from hemoglobin in tissues that have a low PO_2 and a high PCO_2.
   d) The oxygen-hemoglobin bond is formed as blood passes through the pulmonary capillaries.

   Reword your choice to make it a correct statement.

   ____

8. Which statement is NOT true of pulmonary volumes?
   a) Tidal volume is the amount of air in one normal inhalation and exhalation.
   b) Residual air is important to provide for continuous gas exchange in the alveoli.
   c) Vital capacity is the deepest inhalation followed by a normal exhalation.
   d) Two factors that determine expected vital capacity are age and height.

   Reword your choice to make it a correct statement.

   ____

9. Which statement is NOT true of inhalation and exhalation?
   a) The normal range of respirations per minute is 20 to 30.
   b) Normal exhalation is a passive process.
   c) Normal inhalation is an active process.
   d) The elasticity of the lungs contributes to normal exhalation.

   Reword your choice to make it a correct statement.

   ____

10. Which statement is NOT true of the respiratory system and acid-base balance?
   a) Hyperventilating may bring about a mild respiratory alkalosis.
   b) The respiratory system helps compensate for metabolic acidosis by decreasing the respiratory rate.
   c) The respiratory system affects the pH because it regulates the amount of CO_2 in body fluids.
   d) A pulmonary disease that interferes with gas exchange results in respiratory acidosis.

   Reword your choice to make it a correct statement.

   ____

MULTIPLE CHOICE TEST #3

Each question is a series of statements concerning a topic in this chapter. Read each statement carefully and select all of the correct statements.

1. Which of the following statements are true of the upper respiratory tract?
   a) The lymph nodules of the larynx are called tonsils.
   b) The epiglottis covers the pharynx during swallowing.
   c) The most important function of the larynx is to be an airway.
   d) Inhaled air is warmed and moistened by the nasal mucosa.
   e) The nasal cavities are lined with ciliated epithelium, which has goblet cells that produce mucus.
   f) The oropharynx and laryngopharynx are passageways, at different times, for both air and food.

2. Which of the following statements are true of the anatomy of the lower respiratory tract?
   a) The trachea is the "trunk" of the bronchial tree.
   b) The parietal pleura lines the lungs.
   c) The external intercostal muscles pull the ribs up and down for exhalation.
   d) When the diaphragm relaxes, it moves down and air rushes into the lungs.
   e) The trachea and primary bronchi are kept open by C-shaped pieces of cartilage.
   f) The alveoli of the lungs are surrounded by elastic connective tissue, which contributes to normal exhalation.
   g) The bronchioles do not have smooth muscle in their walls, only epithelium.
   h) The alveoli are made of simple squamous epithelium.

3. Which of the following statements are true of the mechanism of breathing?
   a) Contraction and relaxation of the respiratory muscles creates changes in intrapulmonic pressure.
   b) The respiratory muscles are skeletal (striated) muscle tissue and must receive nerve impulses to contract.
   c) As the chest cavity is enlarged, air rushes out to equalize the pressure inside and out.
   d) Serous fluid keeps the pleural membranes together during breathing.
   e) The diaphragm increases the front-to-back dimensions of the chest cavity.
   f) A normal inhalation is passive but a normal exhalation is active.

4. Which of the following statements are true of gas exchange and transport?
   a) Most oxygen is carried in the blood dissolved in blood plasma.
   b) Pulmonary surfactant lines each alveolus and contributes to easy inflation.
   c) Most CO_2 is carried in the blood bonded to hemoglobin.
   d) In the lungs, oxygen and CO_2 move in opposite directions.
   e) In the tissues, oxygen and CO_2 move in the same direction.
   f) As temperature rises in tissues, less oxygen is released from hemoglobin.

5. Which of the following statements are true of the regulation of respiration?
   a) Both the medulla and pons contribute to the normal breathing rate and rhythm.
   b) If the pH of the blood and tissue fluid decreases, respirations will decrease to compensate.
   c) Breathing is a reflex.
   d) Hypoxemia will stimulate a decrease in respiratory rate and depth.
   e) The strongest stimulus to increase rate and depth of breathing is an increase in CO_2 in the blood.
   f) The chemoreceptors that detect changes in the CO_2 in the blood are located in the hypothalamus.

Chapter 16

The Digestive System

The digestive system consists of the organs that contribute to the physical and chemical breakdown of complex food molecules into simpler molecules. These simpler molecules are then absorbed into the blood and lymph to be transported to cells throughout the body. This chapter describes the digestive organs and the specific functions of each that contribute to digestion and absorption.

DIVISIONS OF THE DIGESTIVE SYSTEM

1. Match each division of the digestive system with the statements that apply to each.

   Use each letter once. One answer line will have four correct letters, and the other will have three correct letters.

   1) Alimentary tube ______________________________

   2) Accessory organs ______________________________

   A. Begins at the mouth
   B. Include the teeth, tongue, and salivary glands
   C. Includes the stomach, small intestine, and large intestine
   D. Include the liver, gallbladder, and pancreas
   E. No digestion takes place here
   F. Digestion does take place in some organs
   G. Ends at the anus

2. Name the parts of the alimentary tube in which digestion takes place. ____________________,

   ____________________, and ____________________

3. Name the part of the alimentary tube in which most absorption of nutrients takes place. ____________________

4. The following diagram depicts the digestive system.

Label the parts indicated.

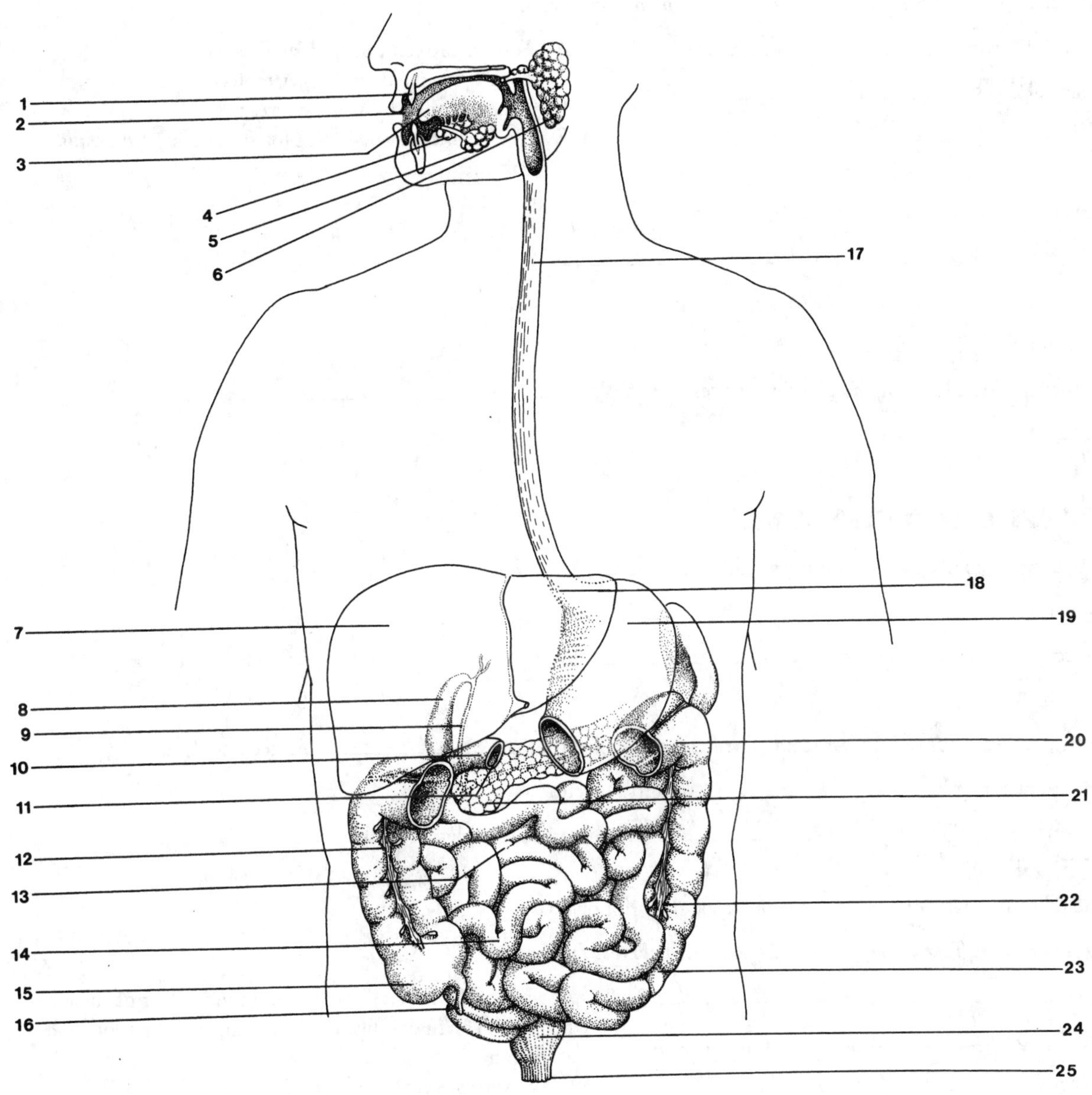

TYPES OF DIGESTION AND END PRODUCTS OF DIGESTION

1. Match each type of digestion with the proper descriptive statements.

   Use each letter once. Each answer line will have two correct letters.

   1) Chemical digestion ____________________

   2) Mechanical digestion ____________________

   A. Food is broken down to smaller pieces
   B. Food is changed to simpler molecules
   C. Accomplished by specific enzymes
   D. Creates more surface area for the action of digestive enzymes

2. Name the end products of digestion of:

   1) Fats: ______________ and ______________

   2) Proteins: ______________

   3) Carbohydrates: ______________

3. Other nutrients released during the digestive process are ______________, ______________, and ______________.

ORAL CAVITY AND PHARYNX

1. a) The opening for food into the oral cavity is the ______________.

   b) The superior boundary of the oral cavity is the ______________.

2. Both the teeth and the tongue contribute to ______________ digestion in the oral cavity by what we call ______________.

3. a) An individual develops two sets of teeth. The first set is called the ______________ teeth and, if complete, consists of ______________ (number) teeth.

   b) The second set is called the ______________ teeth and, if complete, consists of ______________ (number) teeth.

4. Match the following parts of a tooth with the proper descriptions.

   Use each letter once. Two answer lines will have two correct letters.

   1) Enamel ____________________

   2) Dentin ____________________

   3) Pulp cavity ____________________

   4) Periodontal membrane ____________________

   A. Forms the roots of a tooth and the interior of the crown
   B. Produces a bone-like cement to anchor the roots of a tooth
   C. Covers the crown of a tooth
   D. Lines the tooth sockets in the mandible and maxillae
   E. Forms a hard chewing surface
   F. Contains blood vessels and nerves

5. a) The tongue is important for the sense of ______________.

   b) As the first step in swallowing, the tongue is elevated to push food toward the ______________.

6. Name the pairs of salivary glands with these locations:

   1) Below the floor of the mouth ______________

   2) In front of the ears ______________

   3) At the posterior corners of the mandible ______________

7. The salivary glands are exocrine glands because they have ____________ to take saliva to the ____________.

8. a) The digestive enzyme in saliva is ____________, which digests starch to ____________.
   b) The water of saliva is important to dissolve food so that it may be ____________ and to moisten a mass of food so that it may be ____________.
   c) The enzyme in saliva that inhibits the overgrowth of bacteria in the oral cavity is ____________.

9. The only eating-related function of the pharynx is ____________, which is a reflex regulated by a center in the ____________ of the brain.

ESOPHAGUS

1. The esophagus takes food from the ____________ to the ____________.

2. a) At the junction of the esophagus and the stomach is a circular smooth muscle called the ____________.
   b) Contraction of this sphincter prevents the backup of ____________ into the ____________.

TYPICAL STRUCTURE OF THE ALIMENTARY TUBE

1. Match each layer of the alimentary tube with the proper descriptive statements.

   Use each letter once. Two answer lines will have three correct letters, and two will have two correct letters.

   1) Mucosa ____________
   2) Submucosa ____________
   3) External muscle layer ____________
   4) Serosa ____________

   A. Includes the epithelial tissue that lines the organ
   B. Above the diaphragm; made of fibrous connective tissue
   C. Made of areolar connective tissue
   D. Secretes mucus and digestive enzymes
   E. Made of two layers (typically) of smooth muscle
   F. Contains lymph nodules to destroy pathogens
   G. Contains Meissner's plexus, enteric nerves that regulate secretions of the mucosa
   H. Contains Auerbach's plexus, enteric nerves that regulate peristalsis
   I. Below the diaphragm; is the mesentery
   J. Provides mechanical digestion and peristalsis

2. The following diagram depicts the typical layers of the alimentary tube as found in the small intestine.

Label the parts indicated.

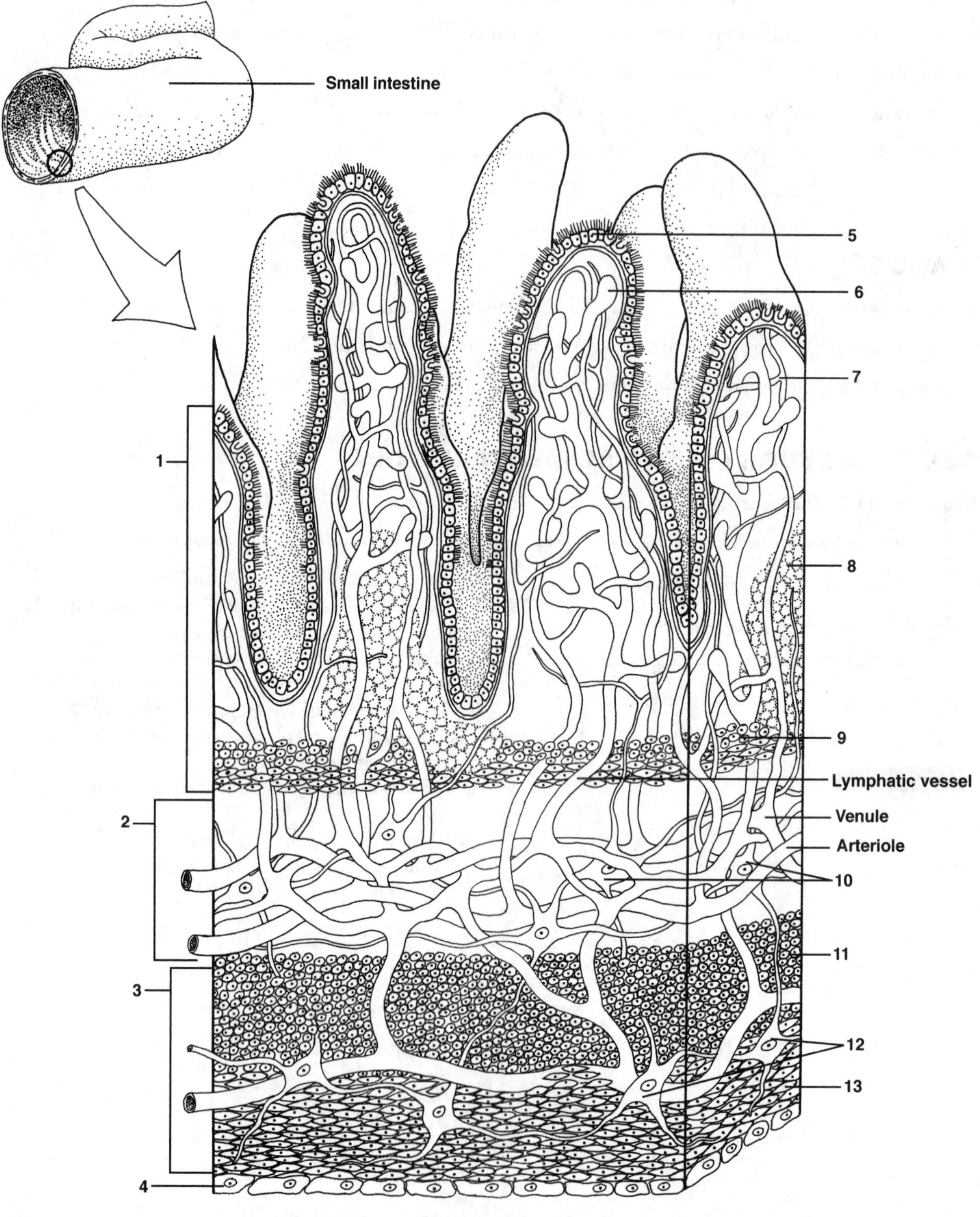

STOMACH

1. The stomach is the saclike portion of the alimentary tube that extends from the ____________ to the ____________.

2. Some digestion does take place in the stomach, which also serves as a ____________ for food so that digestion takes place gradually.

3. a) The folds of the gastric mucosa that are present when the stomach is empty are called ____________.

   b) The glands of the stomach are called ____________, and their secretion is called ____________.

4. Match the cells of the gastric pits with their proper secretion.

   Use each letter once.

   1) Mucous cells ____________
   2) Chief cells ____________
   3) Parietal cells ____________
   4) G cells ____________

   A. Secrete pepsinogen, an inactive form of the enzyme pepsin
   B. Secrete hydrochloric acid, which activates pepsin
   C. Secrete mucus, which helps protect the gastric mucosa
   D. Secrete gastrin when food enters the stomach

5. a) The part of gastric juice that kills most microorganisms that enter the stomach is ____________.

   b) The part of gastric juice that digests proteins to polypeptides is the enzyme ____________.

6. a) Secretion of gastric juice may begin with the sight or smell of food. This is a ____________ (sympathetic or parasympathetic) response.

   b) When food actually reaches the stomach, the ____________ cells of the gastric pits secrete the hormone ____________, which stimulates the secretion of more gastric juice.

7. a) Mechanical digestion in the stomach is a function of the ____________ layer of the wall of the stomach.

   b) The pyloric sphincter is located at the junction of the pylorus of the ____________ and the ____________ of the small intestine.

   c) When the pyloric sphincter contracts, it prevents the backup of food from the ____________ to the ____________.

8. The following diagram depicts a partial longitudinal section of the stomach.

Label the parts indicated.

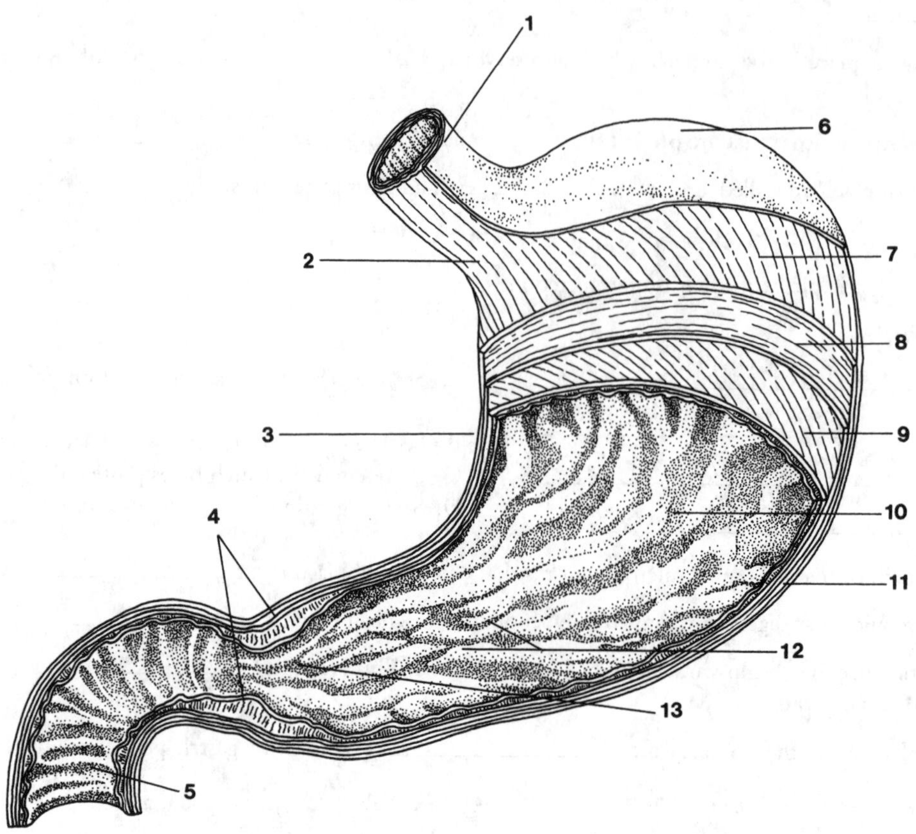

LIVER AND GALLBLADDER

1. a) The liver is located just below the __________________ in the upper right and center of the __________________ cavity.

   b) The functional unit of the liver is called a liver __________________, which is made of liver cells and the large capillaries called __________________.

   c) Between liver lobules are branches of the __________________ artery, the __________________ vein, and __________________ ducts.

2. a) The digestive function of the liver is the production of __________________, which contains bile salts that emulsify __________________.

   b) This is an aspect of __________________ (mechanical or chemical) digestion.

3. Bile leaves the liver through the __________________ duct, which joins the __________________ duct of the gallbladder to form the __________________ duct, which carries bile to the __________________ of the small intestine.

4. Bile also has an excretory function because __________________ and __________________ are transported to the intestines to be eliminated in __________________.

5. a) The gallbladder is located on the undersurface of the __________________.

   b) State the two functions of the gallbladder. __________________ and __________________

6. a) The production of bile and the contraction of the gallbladder are regulated by hormones produced by the
______________________ cells of the mucosa of the ______________________.
   b) The hormone ______________________ stimulates production of bile by the liver.
   c) The hormone ______________________ stimulates contraction of the gallbladder.

7. The following diagram depicts the liver, gallbladder, and pancreas.
Label the parts indicated.

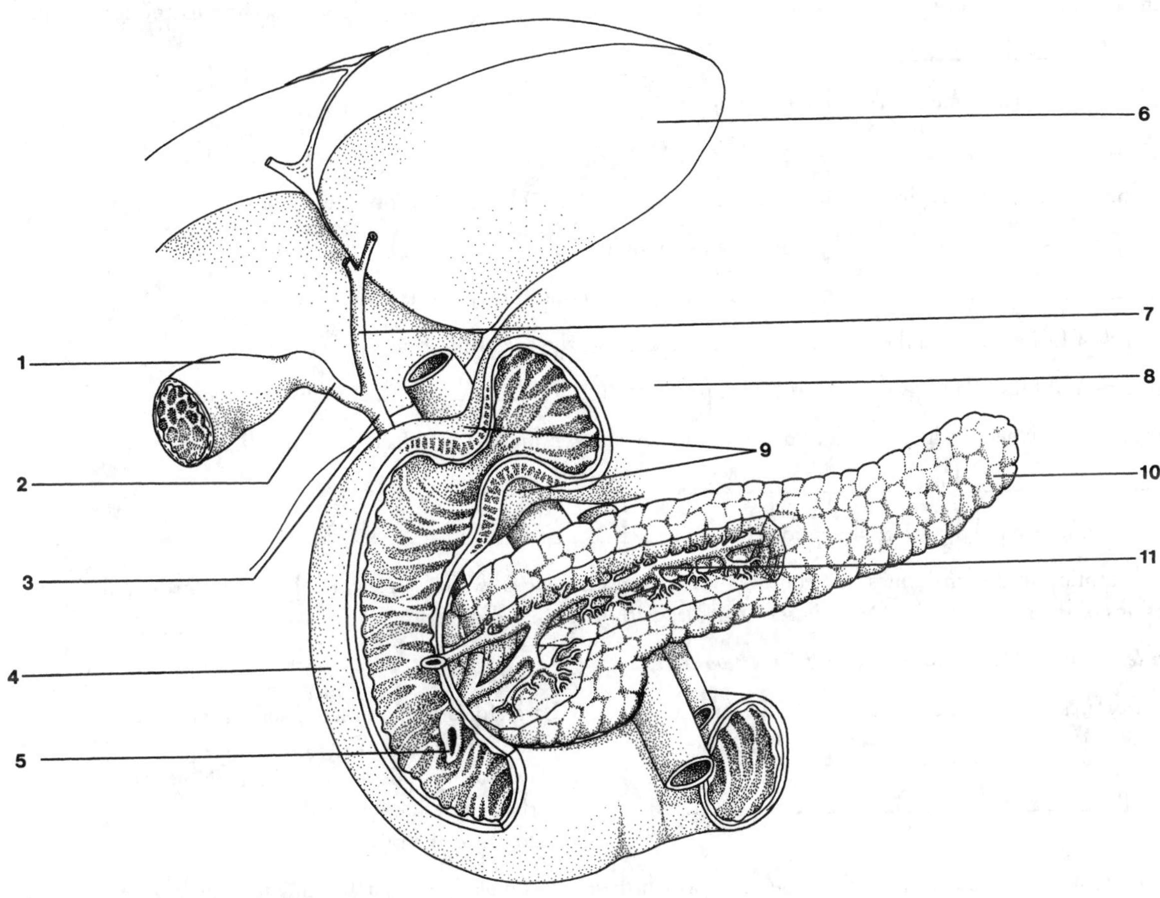

PANCREAS

1. The pancreas is located in the upper abdominal cavity between the ______________________ and
______________________.

2. Pancreatic juices are carried by the main pancreatic duct to the ______________________ duct to the duodenum, which is their site of action.

3. a) Enzyme pancreatic juice contains several enzymes. The enzyme ______________________ digests polypeptides to shorter chains of amino acids.
   b) The enzyme ______________________ digests emulsified fats to ______________________ and
______________________.
   c) The enzyme ______________________ digests starch to ______________________.

4. Bicarbonate pancreatic juice contains sodium bicarbonate that neutralizes the hydrochloric acid from the
______________________ in the ______________________ (site of action).

SMALL INTESTINE

1. a) The small intestine is coiled within the abdominal cavity and is encircled by the ____________.

   b) The small intestine carries food from the ____________ to the ____________.

2. a) The three parts of the small intestine, in order, are the ____________, ____________, and ____________.

   b) The common bile duct enters which of these parts? ____________

3. a) Which layer of the wall of the small intestine is responsible for mixing chyme with digestive secretions and for peristalsis? ____________

   b) What is the collective name for all of the nerve fibers and networks of the alimentary tube?

   ____

   c) The name Peyer's patches is given to the ____________ of the small intestine.

4. Digestion of food is completed in the small intestine and requires ____________ from the liver, ____________ from the pancreas, and the enzymes produced by the small intestine itself.

5. a) The crypts of Lieberkühn are the ____________ of the small intestine.

   b) The secretion of these glands is stimulated by the presence of ____________ in the duodenum.

6. a) The digestive enzymes produced by the small intestine are sucrase, maltase, and lactase, which digest ____________ to ____________.

   b) Peptidases complete the digestion of peptide chains to ____________.

7. Efficient absorption in the small intestine requires a very large surface area, which is provided by several structural modifications of the small intestine. Match each structure with its proper description.

   Use each letter once. One answer line will have two letters.

   1) Plica circulares ____________
   2) Villi ____________
   3) Microvilli ____________

   A. Large folds of the mucosa and submucosa
   B. Small projections of the mucosa
   C. Microscopic folds of the cell membrane on the free surface of each columnar cell
   D. Also called the brush border

8. The villi of the small intestine contain the vessels into which the end products (nutrients) of digestion are absorbed. Each villus contains a ____________, which absorbs the water-soluble nutrients, and a ____________, which absorbs the fat-soluble nutrients.

9. Match each of the following nutrients with the vessel in the villi into which it is absorbed (a letter statement) and the mechanism of absorption (a number statement).

   Each answer line will have one letter and one number.

   1) Amino acids ____________
   2) Fat-soluble vitamins ____________
   3) Water-soluble vitamins ____________
   4) Fatty acids and glycerol ____________
   5) Monosaccharides ____________
   6) Positive ions ____________
   7) Negative ions ____________
   8) Water ____________

   Vessel
   A. Capillary network
   B. Lacteal

   Mechanism
   1. Active transport
   2. Passive transport
   3. Osmosis
   4. Requires bile salts

10. a) The absorption of vitamin ______________ requires the ______________ factor produced by the stomach lining.

 b) The absorption of calcium ions requires vitamin ______________ and ______________ hormone.

11. a) Blood from the capillary networks in the small intestine travels through the portal vein to the ______________ before returning to the heart.

 b) Lymph from the lacteals in the small intestine enters the blood in the ______________ vein.

12. The following diagram depicts a villus of the small intestine.

Label the parts indicated.

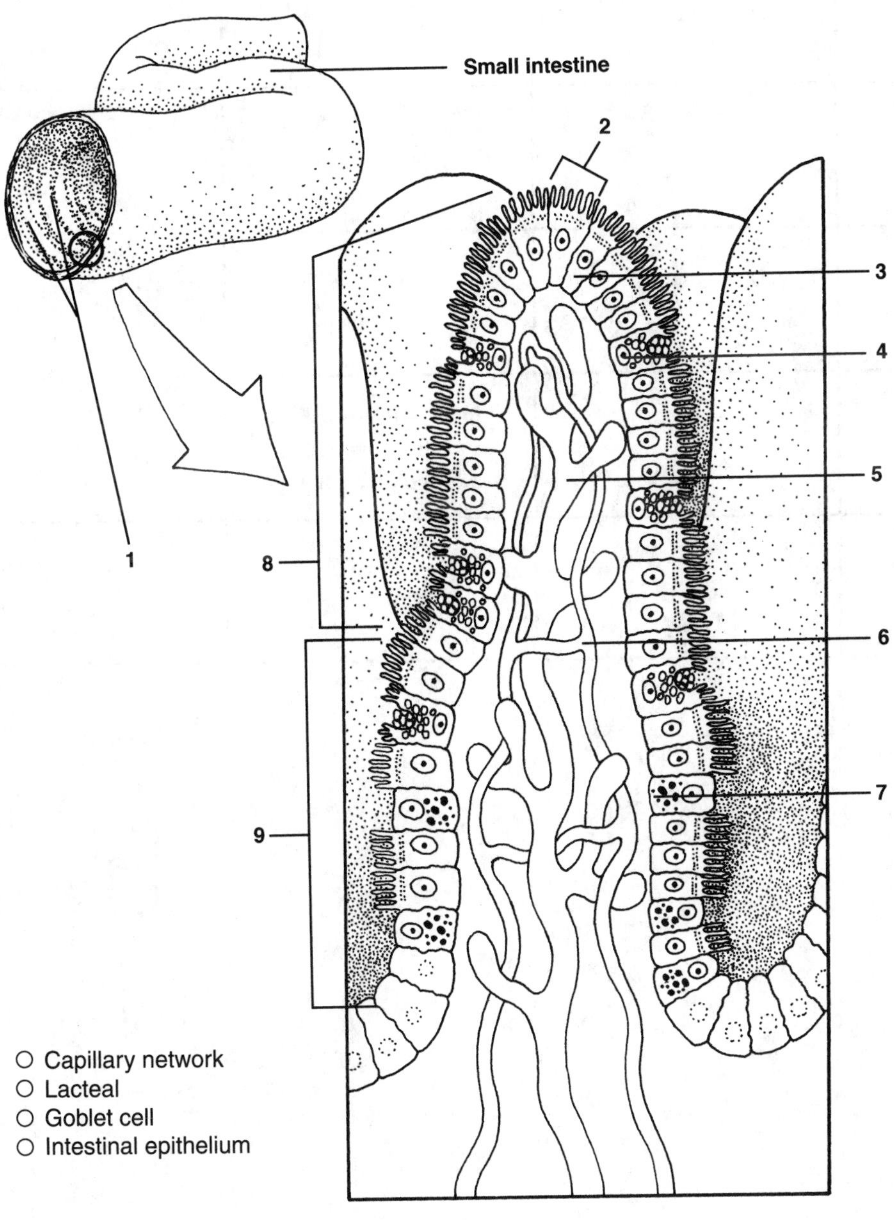

REVIEW OF DIGESTION

1. Complete the following chart by naming the digestive secretion and its action for each organ listed in the column under the proper food type. If an organ does not contribute to the digestion of a food type, write "none." Several boxes have been completed to get you started.

FOOD TYPE

Organ	Carbohydrates	Fats	Proteins
Salivary glands	Amylase—digests starch to maltose		
Stomach		None	
Liver			None
Pancreas		Lipase—digests emulsified fats to fatty acids and glycerol	
Small intestine		None	

2. The following diagram depicts digestion.

Label the parts indicated.

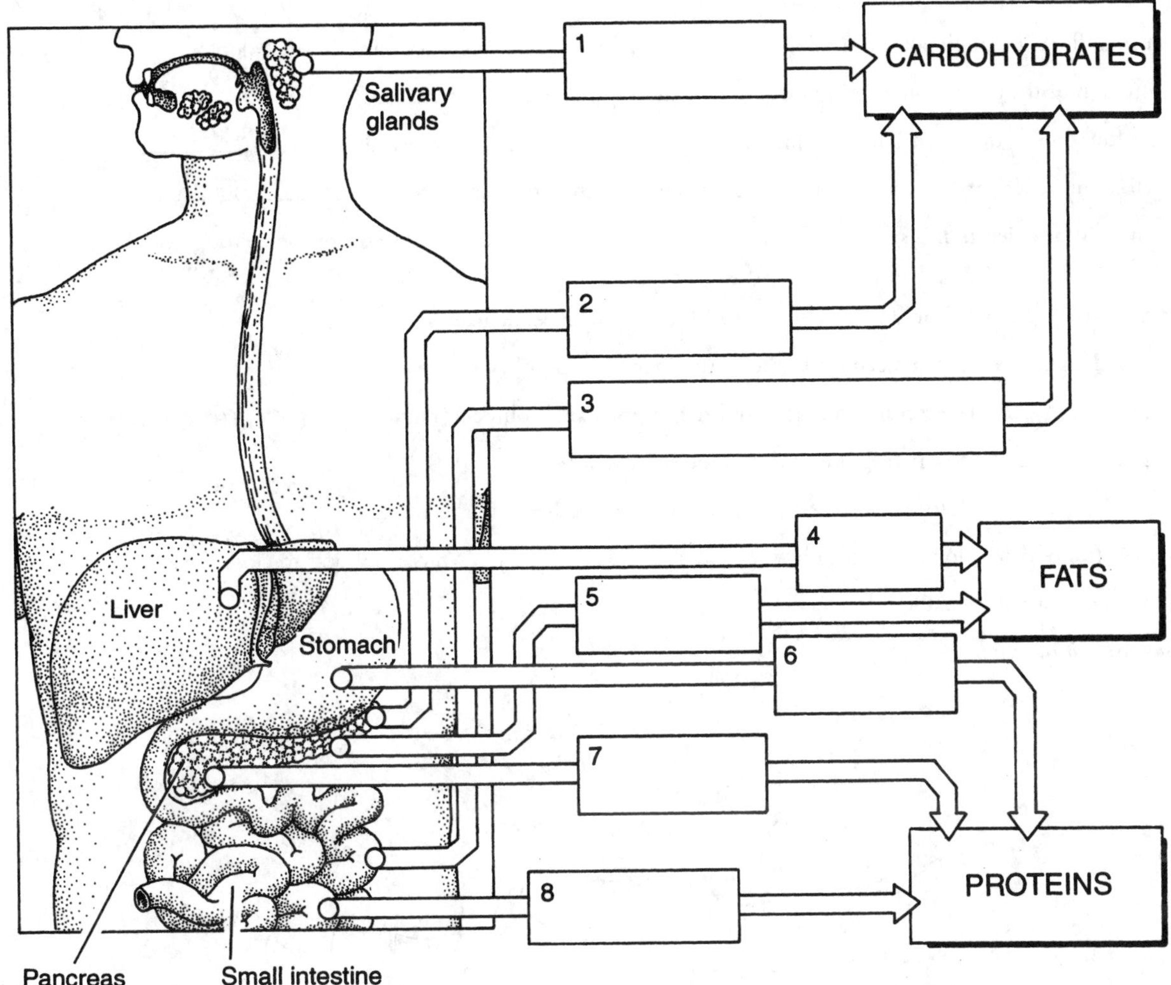

LARGE INTESTINE

1. The large intestine is also called the ____________ and extends from the ____________ of the small intestine to the ____________.
2. The first part of the colon is the ____________, to which the appendix is attached.
3. Number the parts of the large intestine in order, beginning with the cecum.

_____1_____ cecum

____________ rectum

____________ ascending colon

____________ sigmoid colon

____________ descending colon

____________ anal canal

____________ transverse colon

4. The functions of the colon are:

   1) The absorption of ____________________, ____________________, and ____________________.

   2) The elimination of ____________________.

5. a) The normal flora (microbiota) of the colon are the ____________________ that live in the colon.

   b) One function of the colon flora is to inhibit the growth of ____________________.

   c) Another function of the colon flora is to produce ____________________, especially ____________________.

6. a) The defecation reflex for the elimination of feces involves which part of the CNS? ____________________

   b) The stimulus for defecation is stretching of the ____________________ as peristalsis of the colon pushes feces into it.

7. Number the parts of the reflex arc for the defecation reflex in proper sequence.

   ________1________ Stretch receptors in the rectum generate impulses.

   ____________________ The rectum contracts, and the internal anal sphincter relaxes.

   ____________________ Sensory impulses travel to the spinal cord.

   ____________________ Motor impulses return to the smooth muscle of the rectum.

8. Voluntary control of defecation is provided by the ____________________, which contracts to close the anus.

9. The following diagram depicts the colon.

   Label the parts indicated.

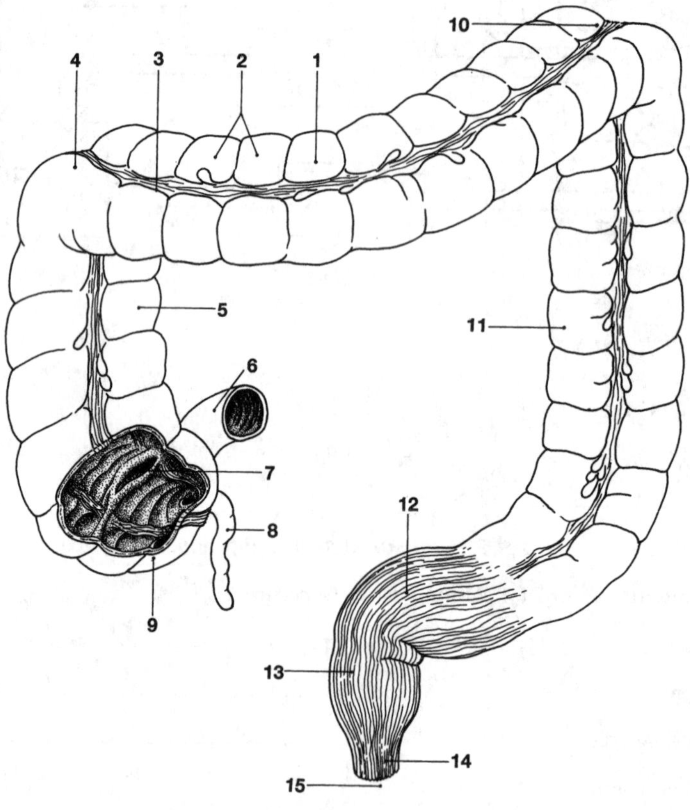

LIVER—OTHER FUNCTIONS

1. With respect to liver functions, match each of the following with the proper descriptive statement.

Use each letter once.

1) Transamination ________________

2) Glycogen ________________

3) Synthesis of clotting factors ________________

4) Beta-oxidation ________________

5) A, D, E, K, and B_{12} ________________

6) Fructose and galactose ________________

7) Synthesis of albumin ________________

8) Synthesis of enzymes ________________

9) Deamination ________________

10) Synthesis of lipoproteins ________________

11) Synthesis of bilirubin ________________

12) Cholesterol ________________

13) Iron and copper ________________

14) Kupffer cells ________________

A. To transport fats in the blood
B. Monosaccharides that are changed to glucose
C. To help maintain blood volume
D. Formed from old RBCs and eliminated in feces
E. The process in which the amino group is removed from an amino acid
F. A steroid that is synthesized by the liver; excess is excreted in bile
G. The process by which a fatty acid molecule is converted into two-carbon molecules
H. Includes fibrinogen and prothrombin
I. The form in which excess glucose is stored
J. The process by which the nonessential amino acids are synthesized
K. The macrophages of the liver that phagocytize pathogens
L. To change harmful substances such as alcohol to less harmful ones
M. The vitamins stored by the liver
N. The minerals stored by the liver

CROSSWORD PUZZLE

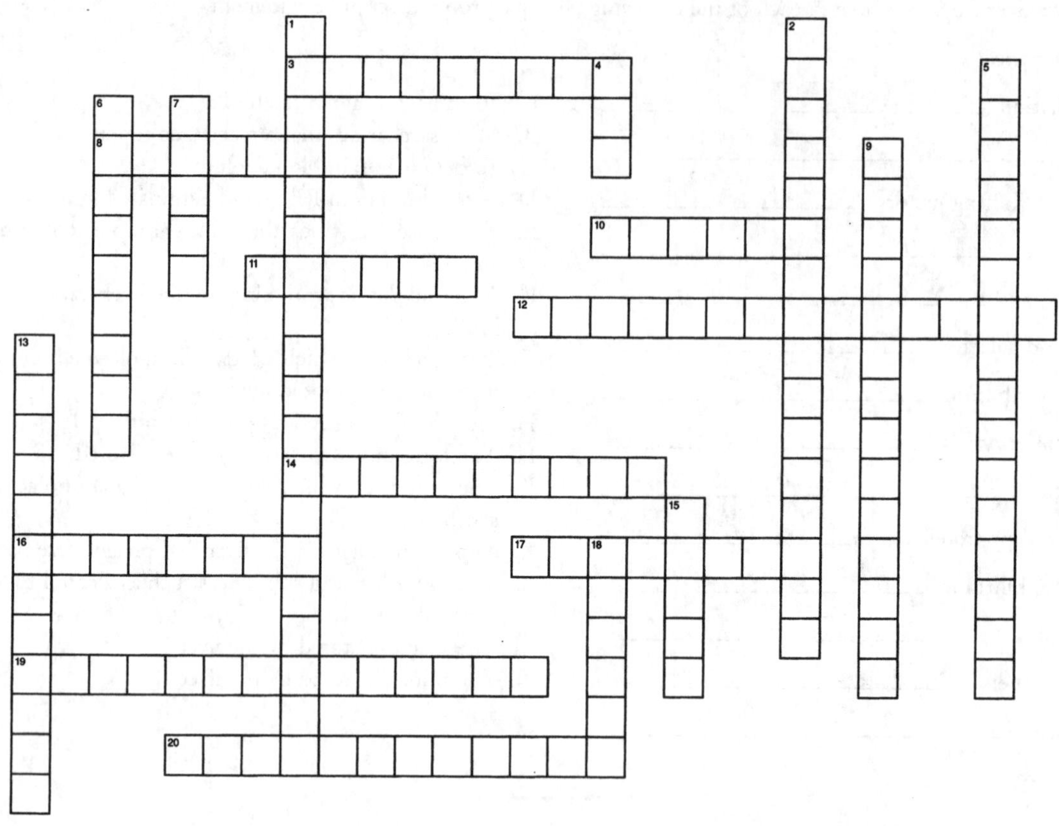

ACROSS

3. __________ amino acids; cannot be synthesized by the liver
8. Bile salts do this to large fat globules
10. Secretion of the parotid glands
11. __________ flora; the bacteria that live in the colon
12. Takes bile and pancreatic juice to the duodenum
14. __________ digestion; the physical breakdown of food
16. The first 10 inches of the small intestine
17. __________ digestion; accomplished by digestive enzymes
19. Prevents backup of fecal material into the ileum (two words)
20. __________ amino acids; can be synthesized by the liver

DOWN

1. Anchors a tooth in its socket (two words)
2. Involuntary response for elimination of feces (two words)
4. Prevents backup of stomach contents (initials)
5. Prevents backup of intestinal contents (two words)
6. Inflammation of the liver; often caused by viruses
7. Folds of the mucosa when the stomach is empty
9. Extends from the mouth to the anus (two words)
13. Inflammation of the appendix
15. Folds of the mucosa of the small intestine
18. Provides a hard chewing surface

CLINICAL APPLICATIONS

1. a) Mrs. L is 50 years old and has been feeling pain in the upper right abdominal quadrant after eating. Her doctor suspects that Mrs. L has stones in which accessory organ of digestion? ____________________
   b) Such stones are usually made of what substance? ____________________
   c) If gallstones are producing frequent and severe pain, the most effective procedure is to ____________________.
2. a) Mr. A is 67 years old and has been hospitalized with diverticulitis. Diverticula are small outpouchings of the weakened wall of the colon, and they may become inflamed if ____________________ become trapped within them.
   b) Mr. A is being treated with ____________________ to prevent further infection and the need for surgery.
3. a) Mr. F has been diagnosed with hepatitis A. He feels very fatigued, has no appetite, and the whites of his eyes appear ____________________.
   b) This color is caused by excess ____________________ in the blood, which Mr. F's damaged ____________________ (organ) cannot excrete rapidly. Mr. F asks his doctor what causes hepatitis and how he got it.
   c) The doctor tells him that hepatitis is caused by a ____________________, and that hepatitis A is usually spread from person to person by ____________________.
4. Name the digestive disorder suggested by each of these descriptions.
   1) A 6-month-old baby boy often vomits after his feeding. ____________________
   2) A 40-year-old woman feels abdominal discomfort after eating foods that contain milk.
   ____
   3) A very serious infection that may follow a ruptured appendix or a perforated ulcer.
   ____

MULTIPLE CHOICE TEST #1

Choose the correct answer for each question.

1. The vitamin produced by the normal flora of the colon in amounts sufficient to meet a person's daily need is:
   a) K b) A c) D d) niacin
2. Backup of food from the small intestine to the stomach is prevented by the:
   a) lower esophageal sphincter c) internal anal sphincter
   b) ileocecal valve d) pyloric sphincter
3. All of the following are accessory organs of digestion except the:
   a) liver b) salivary glands c) stomach d) pancreas
4. Mechanical digestion includes all of the following except:
   a) chewing c) emulsification of fat by bile salts
   b) conversion of starch to maltose d) contractions of the stomach
5. The hard chewing surface of a tooth is formed by:
   a) dentin b) gingiva c) bone d) enamel
6. The only voluntary aspect of swallowing is:
   a) contraction of the pharynx c) elevation of the soft palate
   b) peristalsis of the esophagus d) elevation of the tongue

7. The functions of saliva include all of these except:
a) taste b) swallowing c) starch digestion d) protein digestion

8. A tooth is anchored in its socket in the jaw by the:
a) root b) pulp cavity c) periodontal membrane d) dentin

9. The layer of the alimentary tube that is responsible for peristalsis is the:
a) serosa b) external muscle layer c) submucosa d) mucosa

10. In the gastric mucosa, the parietal cells secrete:
a) hydrochloric acid b) pepsin c) pepsinogen d) complete gastric juice

11. The liver synthesizes all of these except:
a) hemoglobin b) fibrinogen c) albumin d) lipoproteins

12. When food reaches the stomach, secretion of gastric juice is stimulated by the hormone:
a) epinephrine b) secretin c) gastrin d) cholecystokinin

13. Bile is stored by the:
a) common bile duct b) gallbladder c) liver cells d) hepatic duct

14. Bile and pancreatic juices are carried to the duodenum by the:
a) pancreatic duct b) hepatic duct c) cystic duct d) common bile duct

15. Bicarbonate pancreatic juice is important to neutralize hydrochloric acid that enters the:
a) pancreas b) stomach c) duodenum d) esophagus

16. The digestion of protein involves all of these except:
a) pepsin from the stomach c) peptidases from the small intestine
b) trypsin from the pancreas d) amylase from the pancreas

17. The capillary networks and lacteals in the villi of the small intestine are important for:
a) absorption of nutrients c) increasing surface area
b) chemical digestion d) mechanical digestion

18. The absorption of amino acids and glucose into the capillary networks of the villi is accomplished by the process of:
a) diffusion b) active transport c) passive transport d) osmosis

19. The functions of the large intestine include all of these except:
a) digestion of starch c) absorption of water
b) elimination of undigested material d) absorption of vitamins and minerals

20. By the process of transamination, the liver synthesizes the:
a) enzymes for detoxification c) essential amino acids
b) nonessential amino acids d) clotting factors

21. The stimulus for the defecation reflex is:
a) contraction of the rectum c) stretching of the rectum
b) contraction of the internal anal sphincter d) relaxation of the internal anal sphincter

22. The liver stores all of these except:
a) iron b) glycogen c) vitamins A and D d) fat

23. The liver is able to detoxify potentially harmful substances by means of the synthesis of specific:
a) plasma proteins b) lipoproteins c) steroids d) enzymes

24. Greater surface area in the small intestine is provided by all of these except:
a) villi b) microvilli c) rugae d) plica circulares

25. The "brush border" refers to the:
a) microvilli of the small intestine c) rugae of the stomach
b) papillae of the tongue d) haustra of the colon

26. The function of the enteroendocrine cells of the stomach is the production of:
a) pepsin b) mucus c) HCl d) gastrin

27. The enteric nervous system is found in all of these organs except the:
a) stomach b) liver c) small intestine d) large intestine

28. In the small intestine, the Peyer's patches are:
    a) all of the villi collectively
    b) the glands between the villi
    c) the lymph nodules
    d) the epithelium except the goblet cells

29. Contraction of the gallbladder is stimulated by:
    a) gastrin
    b) secretin
    c) cholecystokinin
    d) epinephrine

30. The cells of the liver that phagocytize pathogens are:
    a) macrophages
    b) Kupffer cells
    c) both a and b are correct
    d) neither a nor b is correct

MULTIPLE CHOICE TEST #2

Read each question and the four answer choices carefully. When you have made a choice, follow the instructions to complete your answer.

1. Which statement is NOT true of the locations of digestive organs?
    a) The parotid glands are below the floor of the mouth.
    b) The pancreas extends from the duodenum to the spleen.
    c) The gallbladder is on the undersurface of the right lobe of the liver.
    d) The part of the colon that follows the descending colon is the sigmoid colon.

    Reword your choice to make it a correct statement.

    ____

2. Which statement is NOT true of the functions of the liver?
    a) Fructose and galactose are changed to glucose.
    b) Bilirubin is synthesized and excreted into bile.
    c) Deamination is the first step in the use of excess amino acids for energy.
    d) The process of beta-oxidation permits the use of starches for energy production.

    Reword your choice to make it a correct statement.

    ____

3. Which statement is NOT true of digestive enzymes?
    a) Amylase from the pancreas digests starch to maltose.
    b) Lipase from the pancreas digests fats to amino acids.
    c) Sucrase from the small intestine digests sucrose to monosaccharides.
    d) Pepsin from the stomach digests proteins to polypeptides.

    Reword your choice to make it a correct statement.

    ____

4. Which statement is NOT true of the regulation of digestive secretions?
    a) Gastrin stimulates the secretion of gastric juice.
    b) Increased secretion of saliva is a sympathetic response.
    c) Cholecystokinin stimulates contraction of the gallbladder.
    d) Secretin stimulates production of bile by the liver.

    Reword your choice to make it a correct statement.

    ____

5. Which statement is NOT true of the digestive process?
    a) The end product of starch digestion is glucose.
    b) Mechanical digestion increases the surface area of food particles.
    c) Each digestive enzyme digests two types of foods.
    d) Emulsification of fats by bile salts is a form of mechanical digestion.

    Reword your choice to make it a correct statement.

    ____

6. Which statement is NOT true of the teeth and tongue?
   a) Both the teeth and the tongue are involved in chewing.
   b) Blood vessels and nerves are found in the enamel of a tooth.
   c) Chewing is an example of mechanical digestion.
   d) The tongue is important for taste and swallowing.

   Reword your choice to make it a correct statement.

   ____

7. Which statement is NOT true of the passage of food through the alimentary tube?
   a) The pyloric sphincter prevents the backup of small intestinal contents.
   b) The ileocecal valve prevents the backup of feces from the colon.
   c) Swallowing propels food from the oral cavity through the pharynx and esophagus to the stomach.
   d) The lower esophageal sphincter prevents the backup of esophageal contents.

   Reword your choice to make it a correct statement.

   ____

8. Which statement is NOT true of the stomach?
   a) Hydrochloric acid destroys most microorganisms that enter the stomach because it has a pH of 7 to 8.
   b) Rugae are folds of the mucosa that permit expansion of the stomach.
   c) Pepsinogen is converted to pepsin by hydrochloric acid.
   d) The smooth muscle layers of the stomach contract to mix food with gastric juice.

   Reword your choice to make it a correct statement.

   ____

9. Which statement is NOT true of the villi of the small intestine?
   a) Capillary networks absorb water-soluble nutrients.
   b) Blood from the capillaries of the villi flows through the portal vein to the liver.
   c) Villi are folds of the submucosa of the small intestine.
   d) Lacteals absorb the fat-soluble nutrients.

   Reword your choice to make it a correct statement.

   ____

10. Which statement is NOT true of the liver and gallbladder?
    a) The common bile duct is formed by the hepatic duct and the accessory pancreatic duct.
    b) The hepatic duct takes bile out of the liver.
    c) The gallbladder stores bile and concentrates bile by absorbing water.
    d) The cystic duct takes bile into and out of the gallbladder.

    Reword your choice to make it a correct statement.

    ____

11. Which statement is NOT true of the small intestine?
    a) The small intestine is part of the alimentary tube.
    b) The small intestine extends from the stomach to the colon.
    c) Enzymes are produced to complete the digestion of proteins and fats.
    d) Most absorption of nutrients takes place in the small intestine.

    Reword your choice to make it a correct statement.

    ____

12. Which statement is NOT true of the colon?
    a) The internal and external anal sphincters surround the anus.
    b) The ileum of the small intestine opens into the ascending colon.
    c) The normal flora of the colon produces vitamins and inhibits the growth of pathogens.
    d) The colon absorbs water, minerals, and vitamins.

    Reword your choice to make it a correct statement.

    ____

MULTIPLE CHOICE TEST #3

Each question is a series of statements concerning a topic in this chapter. Read each statement carefully and select all of the correct statements.

1. Which of the following statements are true of the structure of the digestive organs?
   a) The sigmoid flexure is the last part of the small intestine.
   b) The functional units of the liver are liver lobules, which are made of rows of hepatocytes separated by sinusoids.
   c) The pyloric sphincter prevents fecal backup into the small intestine.
   d) The lower esophageal sphincter contracts to push food into the stomach.
   e) Peyer's patches are lymphatic tissue found in the submucosa of the stomach.
   f) Enteric nerves are those in the wall of the alimentary tube.
   g) The external anal sphincter is the smooth muscle layer of the rectum.
   h) The trigeminal nerves are sensory for teeth and for the taste buds on the tongue.
   i) The pancreas is between the duodenum and the spleen in the upper right abdominal quadrant.
   j) The gallbladder secretes bile into the cystic duct, which joins the hepatic duct to form the common bile duct.

2. Which of the following statements are true of the process of digestion?
   a) Lipase from the pancreas and bile from the liver are needed for the digestion of fats.
   b) Mechanical digestion changes complex organic molecules into simpler organic molecules.
   c) Cholecystokinin and secretin are gastric hormones that have the gallbladder and pancreas as their target organs.
   d) Parietal cells of the gastric mucosa secrete HCl and the extrinsic factor.
   e) Trypsin from the pancreas continues the work of pepsin from the stomach.
   f) The digestion of carbohydrates requires only amylase from the pancreas or salivary glands.
   g) The end products of protein digestion are amino acids.
   h) Monosaccharides and water are absorbed into the lacteals of the villi of the small intestine.
   i) The intrinsic factor produced by the duodenum prevents the digestion of vitamin B_{12}.
   j) A digestive enzyme digests only one food type and may require a specific pH in order to function properly.

3. Which of the following statements are true of the liver?
   a) The process of beta-oxidation changes glycerol to the essential fatty acids.
   b) Deamination is necessary to get an amino group to make an essential amino acid.
   c) The fixed macrophages of the liver are Kupffer cells, and they phagocytize old RBCs as well as pathogens.
   d) Albumen is synthesized for the blood-clotting mechanism.
   e) Bilirubin from old WBCs is excreted into bile.
   f) Iron, copper, and vitamins A and D are stored by the liver.
   g) The detoxification of potential poisons requires the liver to synthesize specific enzymes.
   h) The liver synthesizes cholesterol and excretes excess cholesterol into bile.
   i) Glycogen is changed to glucose when the blood glucose level decreases, and the hormone insulin speeds up this process.
   j) The clotting factors fibrin and thrombin are synthesized.

Chapter 17

Body Temperature and Metabolism

This chapter describes the mechanisms of heat production and heat loss and the integration of these mechanisms to maintain a constant body temperature. Also in this chapter are simple descriptions of the metabolic pathways involved in the use of foods to produce adenosine triphosphate (ATP) and a discussion of metabolic rate and the factors that affect energy and heat production.

BODY TEMPERATURE

1. a) State the normal range of human body temperature in °F: ____________ to ____________,

   and in °C: ____________ to ____________.

   b) The average body temperature is considered to be ____________ °F or ____________ °C.

2. For which two age groups is temperature regulation not as precise as it is at other times during life?

   ____________ and ____________

HEAT PRODUCTION

1. a) Heat is produced as one of the energy products of the process of ____________.

   b) The other energy product of this process is ____________.

2. a) The hormone that is the most important regulator of energy production is ____________, produced by the ____________ gland.

   b) This hormone increases heat production by increasing the rate of ____________ within cells.

3. The hormone that increases the rate of cell respiration in stressful situations is ____________, produced by the ____________ (gland).

4. a) Active organs produce significant amounts of heat because they are constantly producing ____________ for their metabolic activities.

   b) The ____________ produce about 25% of the total body heat at rest because they are usually in a state of slight contraction called ____________.

   c) Because it has so many important and continuous functions, the ____________ produces about 20% of the total body heat at rest.

   d) Active organs do not "overheat" because the ____________ that circulates through them carries heat to cooler parts of the body.

5. When food is consumed, more heat is produced as the digestive organs produce more ATP for ____________________ (name a process).

6. When a person has a fever, heat production ____________________ (increases or decreases) because metabolic rate is faster at higher temperatures.

HEAT LOSS

1. a) The major pathway of heat loss from the body is by way of the ____________________ because it covers the body surface.

   b) A secondary pathway of heat loss is by way of the ____________________ tract as heat is lost in water vapor in exhaled air.

   c) Minor pathways of heat loss are by way of the ____________________ and ____________________ tracts because urine and feces are at body temperature when excreted.

2. a) The temperature of the skin influences how much heat will be lost by the related processes of ____________________, ____________________, and ____________________.

   b) Of these three processes, the one that involves air currents moving warmer air away from the skin is ____________________.

   c) The processes of ____________________ and ____________________ both involve loss of heat to air or objects that touch the skin.

   d) For radiation and conduction to be effective heat loss mechanisms, the external environment must be ____________________ than the body temperature.

3. a) The temperature of the skin is determined by the flow of ____________________ through the skin.

   b) In the dermis, ____________________ will increase blood flow and heat loss, while ____________________ will decrease blood flow and help conserve heat.

4. a) In the process of sweating, excess body heat is lost as this heat ____________________ the sweat on the skin surface.

   b) Sweat is secreted by ____________________ sweat glands.

   c) Sweating is not efficient when the atmospheric humidity is ____________________ but is efficient when the atmospheric humidity is ____________________.

   d) State the potential disadvantage of excessive sweating. ____________________

REGULATION OF BODY TEMPERATURE

1. The part of the brain that may be likened to a thermostat and that regulates body temperature is the ____________________.

2. a) Specialized neurons in the hypothalamus detect changes in the temperature of the ____________________ that circulates through the brain.

   b) The hypothalamus receives sensory information about the environmental temperature from the ____________________ in the skin.

3. Match each environmental situation with the proper responses brought about by the hypothalamus.

   Use each letter once. Each answer line will have three correct letters.

   1) Cold environment ______________________

   2) Warm environment ______________________

   A. Sweating increases
   B. Vasodilation in the dermis
   C. Muscle tone decreases to produce less heat
   D. Vasoconstriction in the dermis
   E. Shivering may occur to produce more heat
   F. Sweating decreases

FEVER

1. A fever is an abnormally ______________ body temperature and is caused by substances called ______________ that raise the setting of the hypothalamic thermostat.
2. a) Pyrogens that come from outside the body include ______________ and ______________.
   b) Endogenous pyrogens are chemicals released during ______________.
3. a) When a pyrogen first affects the hypothalamus, the hypothalamic thermostat is reset ______________ (higher or lower), and the person feels ______________.
   b) To raise the body temperature to the setting of the hypothalamic thermostat, ______________ will occur.
4. a) When the pyrogen has been destroyed, the hypothalamic thermostat is reset ______________ (higher or lower), and the person feels ______________.
   b) At this time ______________ will occur to lower the body temperature.
5. A fever may be beneficial because the metabolism of some ______________ is inhibited by higher temperatures, and the activity of the person's own ______________ is increased.
6. a) A very high fever may cause enzymes within cells to become denatured, which means they ______________, and as a result, cells may die.
   b) This has the most serious consequences in the ______________ (organ) because ______________ (cells) cannot reproduce to replace lost cells.

METABOLISM

1. The term for all of the reactions that take place within the body is ______________.
2. a) The term for breakdown (or decomposition) reactions is ______________.
   b) The term for synthesis (or building) reactions is ______________.
   c) Which of these types of reactions requires energy (ATP) to form bonds? ______________.
   d) Which of these types of reactions often releases energy that is used to synthesize ATP? ______________.
   e) The catalysts for many reactions in the body are specific proteins called ______________.
3. Complete the summary reaction of cell respiration:

   $C_6H_{12}O_6$ (glucose) + ______________ → ______________ + H_2O + ______________ + ______________
4. The three stages of the cell respiration of a molecule of glucose are called ______________, ______________, and ______________.

5. Match each statement about cell respiration with the proper stage.

   Use each letter as many times as it is correct (some lines will have two letters) for a total of 18 letters.

   A. Glycolysis B. Krebs (citric acid) cycle C. Cytochrome (electron) transport system

   1) ________________ Takes place in the cytoplasm
   2) ________________ Takes place in the mitochondria
   3) ________________ Does not require oxygen
   4) ________________ Requires oxygen
   5) ________________ Glucose is broken down to pyruvic acid.
   6) ________________ A pyruvic acid molecule is broken down, and CO_2 is formed.
   7) ________________ Metabolic water is produced.
   8) ________________ A small amount of ATP is produced.
   9) ________________ Most ATP is produced.
   10) ________________ Energy is released by the flow of hydrogen ions through ATP synthase.
   11) ________________ Pairs of hydrogens are removed by NAD.
   12) ________________ A pair of hydrogens is removed by FAD.
   13) ________________ Requires energy of activation in the form of ATP to start the reaction
   14) ________________ Lactic acid will be formed if no oxygen is present.

6. 1) The vitamin that is part of the hydrogen carrier molecule NAD is ________________.
   2) The vitamin that is part of the hydrogen carrier molecule FAD is ________________.
   3) The vitamin necessary to remove CO_2 from pyruvic acid is ________________.
   4) The minerals that are part of some of the cytochromes are ________________ and ________________.

PROTEINS AND FATS AS ENERGY SOURCES

1. a) When excess amino acids are used for energy production, the first step is deamination in which the ________________ group is removed.
   b) The remaining carbon chain is converted to a molecule that can enter the ________________.
   c) Such molecules may be a two-carbon ________________ group or a three-carbon ________________.
2. a) The end products of fat digestion are glycerol and fatty acids. To be used for energy production, a three-carbon glycerol molecule is converted to ________________, which will enter the Krebs cycle.
   b) To be used for energy production, a fatty acid is broken down in the process of ________________ into two-carbon ________________ groups, which also enter the Krebs cycle.
3. a) Both amino acids and fatty acids may be converted by the liver into two- or four-carbon molecules called ________________ to be used for energy production.
   b) The only noncarbohydrate molecules that can be converted to glucose to supply the brain are the ________________.

4. The following diagram depicts the cell respiration of carbohydrates, proteins, and fats.

Label the parts indicated.

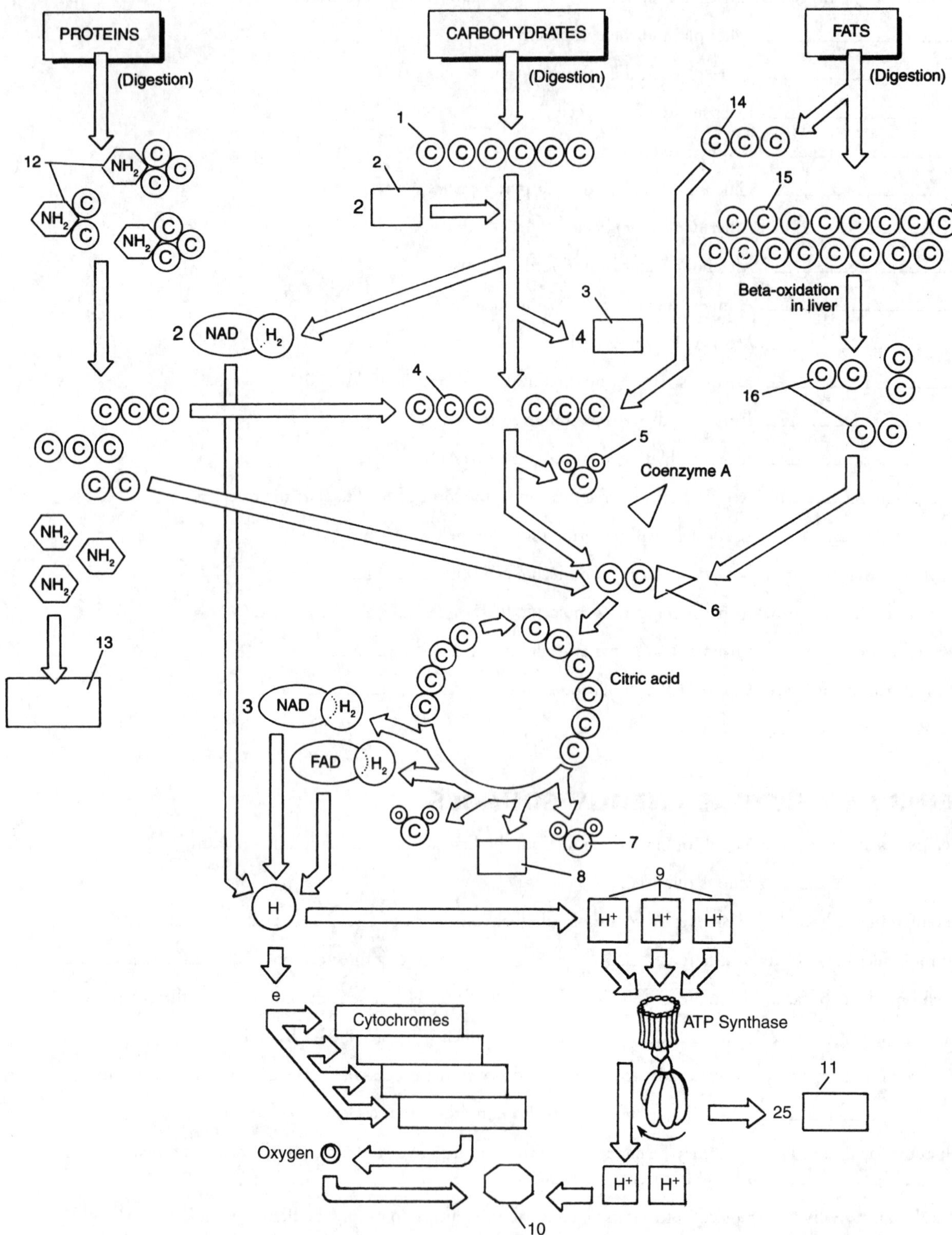

ENERGY AVAILABLE FROM FOOD

1. The potential energy in food is measured in units called ________________ or ________________.
2. How many kilocalories are available in 1 gram of the following? Carbohydrates: ________________; fat: ________________; and protein: ________________.

SYNTHESIS USES OF FOODS

1. a) Excess glucose is stored as ________________ in the liver and ________________, to be used for energy at a later time.
   b) Glucose is converted to ________________ (monosaccharides), which are part of the nucleic acids ________________ and ________________.
2. The primary uses for amino acids are the synthesis of the ________________ amino acids and the synthesis of ________________.
3. Name three specific proteins and the cells (or organs) in which they are produced.
   1) ________________ produced by ________________
   2) ________________ produced by ________________
   3) ________________ produced by ________________
4. a) Fatty acids and glycerol may be used to synthesize ________________, which are then stored in adipose tissue.
   b) Fatty acids and glycerol may also be used to synthesize the ________________ that are part of cell membranes.
   c) The acetyl groups obtained from fatty acids may be used by the liver to synthesize the steroid called ________________.
5. Name two other steroids that are synthesized from cholesterol, and name the organ that produces each.
   1) ________________ produced by ________________
   2) ________________ produced by ________________

METABOLIC RATE

1. a) Although metabolic rate means energy production, it is usually expressed as the production of which form of energy? ________________
   b) The units used to measure heat production are called ________________.
2. The energy production required simply to maintain life is called the ________________.
3. With respect to age, metabolic rate is highest in ________________ and lowest in ________________.
4. Men usually have higher metabolic rates than do women because the hormone ________________ increases the metabolic rate to a greater extent than does the hormone ________________.
5. The metabolic rate increases during stressful situations because of the greater activity of the ________________ nervous system and the hormones ________________ and ________________.
6. Exercise increases the metabolic rate because of the greater activity of the ________________ (organs).

CROSSWORD PUZZLE

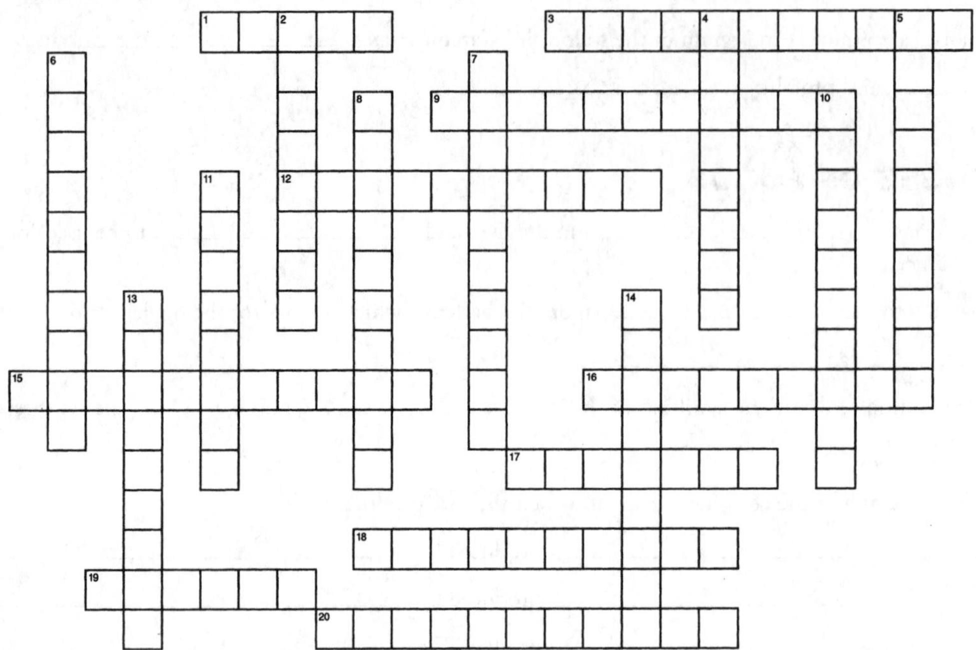

ACROSS

1. Abnormally high body temperature
3. Proteins in the mitochondria that contain either iron or copper
9. Lowers a fever; aspirin is one
12. Total of all reactions that take place in the body
15. 1 gram of carbohydrate yields about four of these (singular)
16. Heat from the body is lost to cooler air
17. Substance that may cause a fever
18. Air currents move warmer air away from skin
19. Heat ________; results when extreme water loss causes the body to stop sweating
20. Abnormally low body temperature

DOWN

2. Organic molecules needed in small amounts for normal body functioning
4. Necessary for the function of certain enzymes
5. Heat ________; caused by excessive sweating with loss of H_2O and salts
6. Anaerobic stage in the breakdown of glucose
7. ________ pyrogens; chemicals released during inflammation
8. Decomposition or "breakdown" reactions
10. Loss of heat to cooler objects
11. Inorganic substances needed for fluid-electrolyte balance
13. Freezing of part of the body
14. Synthesis reactions

CLINICAL APPLICATIONS

1. a) After spending the morning making a snowman, 8-year-old Michael asks his mother to unzip his jacket because he cannot seem to grasp the zipper. His mother immediately sees that Michael's fingertips are bluish-white.

   Michael has ________________ of his fingertips.

   b) If Michael had stayed outside in the cold, permanent tissue damage might have resulted. When water in capillaries freezes, it expands and ________________ the capillaries, leading to a lack of ________________ in the surrounding tissues.

2. a) Mrs. K is elderly and lives alone. On a very hot afternoon, her daughter comes to visit and finds Mrs. K unconscious. Mrs. K's skin is dry with no perspiration. This serious condition is called ____________________.

   b) In this condition, sweating stops to prevent a decrease in blood ____________________, but without sweating as a heat loss mechanism, body temperature ____________________.

3. Mr. B has been told by his doctor to lose 10 pounds. Mr. B's usual caloric intake is 3,000 calories per day. If Mr. B consumes only 2,000 calories per day, how long will it take him to lose 10 pounds? (one pound = 3,500 calories)

   a) Calories in 10 pounds: $10 \times 3{,}500$ = ____________________ calories

   b) Mr. B's diet: ____________________ fewer calories per day

   c) Time needed: calories in 10 pounds/decreased calories per day = ____________________

   = ____________________ days

   d) What else can Mr. B do to help himself lose weight? ____________________

MULTIPLE CHOICE TEST #1

Choose the correct answer for each question.

1. The hormone that is most important in the daily regulation of cell respiration is:
   a) insulin b) thyroxine c) epinephrine d) growth hormone

2. Heat is distributed from active organs to cooler parts of the body by:
   a) the liver b) blood c) lymph d) the kidneys

3. The average human body temperature in °F and °C is:
   a) 98.6°F / 37°C b) 97.6°F / 35°C c) 99.6°F / 39°C d) 97°F / 36.6°C

4. In a cold environment, the arterioles in the dermis will:
   a) dilate to conserve heat c) constrict to conserve heat
   b) dilate to lose heat d) constrict to lose heat

5. When sweating takes place, excess body heat is lost in the:
   a) osmosis of sweat b) secretion of sweat c) evaporation of sweat d) filtration of sweat

6. When the body is at rest, the skeletal muscles produce a significant amount of heat because of:
   a) oxygen debt b) forceful contractions c) lactic acid production d) muscle tone

7. From the respiratory tract, a small amount of heat is lost in:
   a) exhaled water vapor b) inhaled water vapor c) inhaled oxygen d) exhaled carbon dioxide

8. The movement of air across the skin results in heat loss by the process of:
   a) conduction b) convection c) radiation d) evaporation

9. The part of the brain that regulates body temperature is the:
   a) cerebrum b) thalamus c) hypothalamus d) medulla

10. The body's response to a warm environment includes all of these except:
    a) increased sweating c) vasodilation in the dermis
    b) decreased muscle tone d) increased muscle tone

11. Chemicals produced during inflammation that cause fevers are called:
    a) endogenous pyrogens c) exogenous pyrogens
    b) antibacterial pyrogens d) pyromaniac pyrogens

12. A fever may be beneficial because:
    a) WBCs may be inhibited c) RBCs are activated
    b) the growth of some pathogens may be inhibited d) RBCs are inhibited

13. The term for synthesis reactions is:
a) metabolism b) catabolism c) anabolism d) metabolic rate

14. The Krebs cycle and the cytochrome (electron) transport system take place in which part of the cell?
a) cytoplasm b) ribosomes c) nucleus d) mitochondria

15. The most important synthesis uses for glucose are:
a) pentose sugars and glycogen
b) glycogen and polysaccharides
c) fructose and pentose sugars
d) fructose and glycogen

16. The most important synthesis uses for amino acids are:
a) essential amino acids and enzymes
b) proteins and nonessential amino acids
c) proteins and essential amino acids
d) enzymes and phospholipids

17. Most of the ATP produced during cell respiration is produced during which stage?
a) glycolysis b) Krebs cycle c) cytochrome (electron) transport system d) transamination

18. The carbon dioxide produced in cell respiration is produced in which stage?
a) cytochrome (electron) transport system b) deamination c) glycolysis d) Krebs cycle

19. All of these vitamins are necessary for cell respiration except:
a) niacin b) vitamin D c) riboflavin d) thiamine

20. The basal metabolic rate is the term for the body's heat production when the body is:
a) in a stressful situation b) at rest c) exercising strenuously d) performing light activity

21. In order to be used for energy production, amino acids may be changed to all of these except:
a) acetyl groups b) glucose c) ketones d) fatty acids

22. In order to be used for energy production, fatty acids and glycerol may be changed to all of these except:
a) ketones b) pyruvic acid c) acetyl groups d) glucose

23. In the cytochrome (electron) transport system, acidosis is prevented by the formation of:
a) acetyl CoA b) CO_2 c) oxygen d) water

24. A meal that consists of 20 grams of starch, 20 grams of protein, and 10 grams of fat has __________ calories.
a) 220 b) 230 c) 240 d) 250

25. Vitamins can best be described as:
a) building blocks of new tissue
b) sources of energy
c) a chemical form of stored energy
d) chemicals often necessary for the functioning of enzymes

MULTIPLE CHOICE TEST #2

Read each question and the four answer choices carefully. When you have made a choice, follow the instructions to complete your answer.

1. Which statement is NOT true of heat production?
a) In stressful situations, the hormone epinephrine increases heat production.
b) Eating increases the activity of the digestive organs and increases heat production.
c) Heat is an energy product of cell respiration, which is regulated primarily by thyroxine.
d) Almost half the body's total heat at rest is produced by the liver and heart.

Reword your choice to make it a correct statement.

____

2. Which statement is NOT true of heat loss?
   a) Radiation is an effective heat loss mechanism only when the environment is warmer than body temperature.
   b) Most heat is lost from the body by way of the skin.
   c) Excessive sweating may result in serious dehydration.
   d) Sweating is not an effective heat loss mechanism when the atmospheric humidity is high.

   Reword your choice to make it a correct statement.

   ___

3. Which statement is NOT true of the body's responses in a warm environment?
   a) The hypothalamus initiates these responses as body temperature rises.
   b) More heat will be lost as a result of vasoconstriction in the dermis.
   c) Sweating increases so that excess body heat will evaporate sweat on the skin surface.
   d) Muscle tone decreases so that less heat is produced.

   Reword your choice to make it a correct statement.

   ___

4. Which statement is NOT true of the body's responses in a cold environment?
   a) Putting on a coat is a behavioral response.
   b) Vasoconstriction in the dermis helps conserve heat.
   c) Sweating will decrease to help conserve heat.
   d) Shivering may occur as muscle tone decreases.

   Reword your choice to make it a correct statement.

   ___

5. Which statement is NOT true of a fever?
   a) Pyrogens activate the heat-conserving mechanisms of the hypothalamus.
   b) High fevers may cause brain damage because carbohydrates become denatured.
   c) Pyrogens include bacteria, viruses, and chemicals released during inflammation.
   d) A low fever may be beneficial because the activity of WBCs may be increased.

   Reword your choice to make it a correct statement.

   ___

6. Which statement is NOT true of metabolic rate?
   a) Heat production by the body at rest is called basal metabolic rate.
   b) Heat production is measured in kilocalories.
   c) Exercise and stressful situations increase metabolic rate.
   d) Men and young children usually have lower metabolic rates than do women and elderly people.

   Reword your choice to make it a correct statement.

   ___

7. Which statement is NOT true of carbohydrate metabolism?
   a) An important function of glucose is the synthesis of the pentose sugars.
   b) The complete breakdown of glucose in cell respiration does not require oxygen.
   c) The stages of cell respiration for glucose are glycolysis, the Krebs cycle, and the cytochrome (electron) transport system.
   d) One gram of carbohydrate yields 4 kilocalories of energy.

   Reword your choice to make it a correct statement.

   ___

8. Which statement is NOT true of fat metabolism?
   a) One gram of fat yields 9 kilocalories of energy.
   b) To be used for energy production, fatty acids are converted to acetyl groups.
   c) Fatty acids and glycerol are used to synthesize true fats to be stored in the liver.
   d) Fatty acids and glycerol are used to synthesize phospholipids for cell membranes.

   Reword your choice to make it a correct statement.

   ____

9. Which statement is NOT true of amino acid metabolism?
   a) One gram of protein yields 7 kilocalories of energy.
   b) The liver synthesizes the nonessential amino acids.
   c) Most amino acids are used to synthesize the body's own proteins.
   d) To be used for energy production, excess amino acids may be converted to pyruvic acid or to acetyl groups.

   Reword your choice to make it a correct statement.

   ____

10. Which statement is NOT true of the stages of cell respiration?
   a) Glycolysis takes place in the cytoplasm and does not require oxygen.
   b) The Krebs cycle takes place in the mitochondria, and CO_2 is produced.
   c) The cytochrome (electron) transport system takes place in the mitochondria, and metabolic water is produced.
   d) Most of the ATP produced in cell respiration is produced in the Krebs cycle.

   Reword your choice to make it a correct statement.

   ____

MULTIPLE CHOICE TEST #3

Each question is a series of statements concerning a topic in this chapter. Read each statement carefully and select all of the correct statements.

1. Which of the following statements are true of body temperature and its regulation?
   a) The skeletal muscles produce almost 25% of the total body heat when the body is at rest.
   b) A pyrogen is a substance that resets the hypothalamic thermostat upward.
   c) The single organ that produces about 20% of the total body heat is the skin.
   d) Vasoconstriction in the dermis will conserve heat by keeping warm blood in the skin.
   e) The higher the humidity, the more efficiently sweating works to lose heat.
   f) In a cold environment, muscle tone decreases to conserve heat.
   g) In the process of convection, air is cooled as it passes over the warm skin surface.
   h) An individual's body temperature may vary 1° to 2°F in 24 hours.
   i) Thyroxine is the hormone most responsible for regulating heat production.
   j) Holding a warm puppy (or kitten) will warm you by the process of conduction.

2. Which of the following statements are true of metabolism, energy production, and synthesis?
   a) In cell respiration, most of the ATP is produced in the cytochrome (electron) transport system.
   b) Women tend to have higher metabolic rates than do men.
   c) Catabolic reactions include the stages of cell respiration.
   d) A gram of fat yields half as much energy as a gram of protein.
   e) Pyruvic acid, to enter the Krebs cycle, can be formed from glucose, glycerol, or certain amino acids.
   f) In cell respiration, CO_2 is produced in glycolysis.
   g) In cell respiration, metabolic water is produced in the Krebs citric acid cycle.
   h) Basal metabolic rate is the energy required to simply remain alive—that is, with the body at rest.
   i) Excess fatty acids can be used to synthesize amino acids and ketones.
   j) Excess glucose is converted to glycogen or to fat, both of which are stored in the liver.
   k) Urea is a waste product of fatty acid use for energy production.
   l) The liver synthesizes the nonessential amino acids.

Chapter 18

The Urinary System

The urinary system consists of the kidneys, ureters, urinary bladder, and urethra. This chapter describes the work of the kidneys in the formation of urine and the elimination of urine by the other organs of the urinary system. Of great importance is the formation of urine as it is related to many aspects of the homeostasis of body fluids.

FUNCTIONS OF THE KIDNEYS

1. The kidneys form urine from blood plasma and excrete ____________________, such as urea.
2. The kidneys regulate the electrolyte content of the blood by excreting or conserving ____________________.
3. The kidneys regulate the pH of the blood by excreting or conserving ions such as ____________________ and ____________________.
4. The kidneys regulate the volume of blood by excreting or conserving ____________________.
5. By regulating all of these aspects of the blood, the kidneys also regulate these same aspects of ____________________.

KIDNEYS—LOCATION AND EXTERNAL ANATOMY

1. The kidneys are located on either side of the ____________________ column in the upper abdominal cavity, ____________________ the peritoneum.
2. The upper part of the kidneys is protected by which bones? ____________________
3. a) Surrounding the kidneys is ____________________ tissue that acts as a cushion.
   b) The ____________________ is the fibrous connective tissue membrane that covers the adipose tissue and helps hold the kidneys in place.
4. a) On the medial side of each kidney is an indentation called the ____________________.
   b) At this site, the ____________________ takes blood from the abdominal aorta into the kidney, and the ____________________ returns blood from the kidney to the inferior vena cava.

5. The following diagram depicts the urinary system (left—male; right—female).

Label the parts indicated.

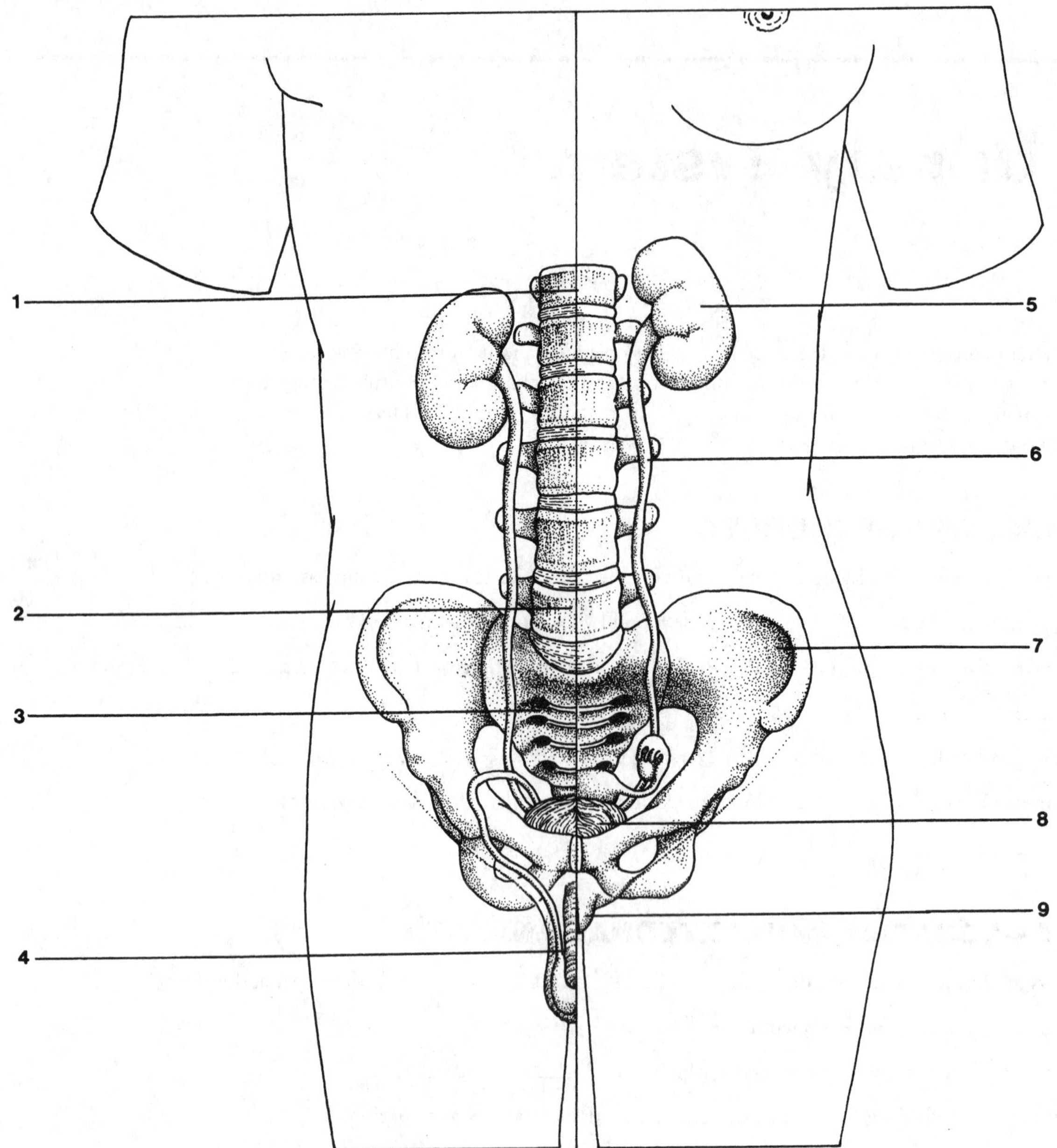

KIDNEYS—INTERNAL STRUCTURE

1. Match each area of the kidney with the proper descriptive statements.

Use each letter once. Two answer lines will have three correct letters, and the other will have two correct letters.

1) Renal cortex ____________________

2) Renal medulla ____________________

3) Renal pelvis ____________________

A. A cavity formed by the expansion of the ureter within the kidney at the hilus
B. The outer layer of kidney tissue
C. The inner layer of kidney tissue
D. The calyces are funnel-shaped extensions of this part
E. Urine enters from the papillae of the pyramids
F. Consists of wedge-shaped pieces called renal pyramids
G. Contains the loops of Henle and the collecting tubes of nephrons
H. Contains the renal corpuscles and convoluted tubules of nephrons

2. The following diagrams depict the frontal section of a kidney and a magnified wedge of kidney tissue.

Label the parts indicated.

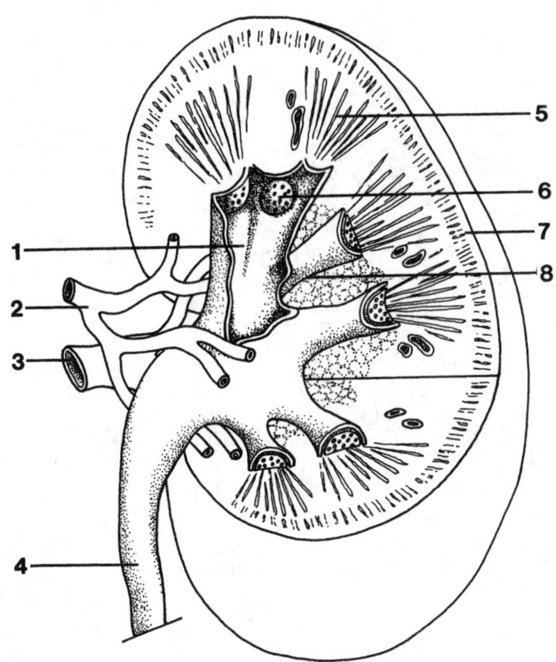

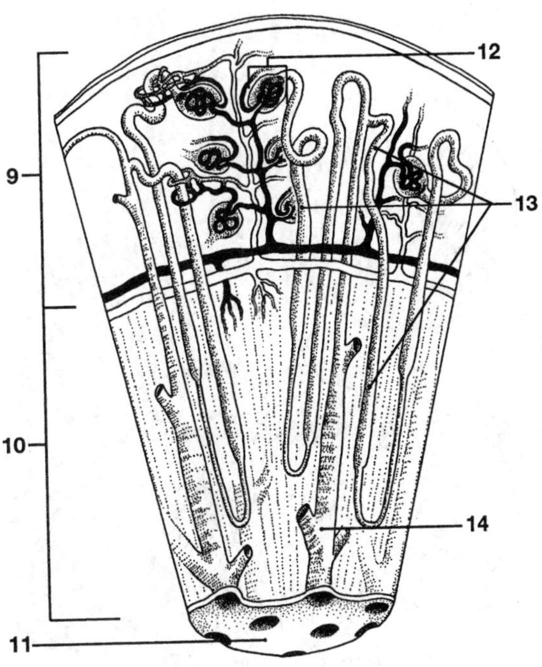

THE NEPHRON

1. a) Nephrons are the structural and ____________________ units of the kidneys.

   b) Each nephron consists of two major parts, called the ____________________ and the ____________________.

2. A renal corpuscle consists of a capillary network called the ____________________, which is surrounded by ____________________, the expanded end of a renal tubule.

3. a) Blood enters the glomerulus from an ____________________ arteriole and leaves the glomerulus by way of an ____________________ arteriole.

   b) Which of these arterioles has the smaller diameter? ____________________

4. a) The inner layer of Bowman's capsule is made of cells called podocytes and is very permeable because it has ____________________.

   b) The outer layer of Bowman's capsule has no pores and is not ____________________.

   c) The fluid that enters Bowman's capsule from the glomerulus is called ____________________.

5. Number the parts of the renal tubule in the order that renal filtrate flows through them.

   ____________________ distal convoluted tubule

   ____________________ loop of Henle

   ____________________ collecting tubule

   ____________________ proximal convoluted tubule

6. Collecting tubules unite to form a papillary duct that empties urine into a calyx of the ____________________.

7. The blood vessels that surround all of the parts of a renal tubule are called the ____________________, and they carry blood that has come from an ____________________ arteriole.

8. The following diagram depicts a nephron and some sections of its parts.

Label the parts indicated.

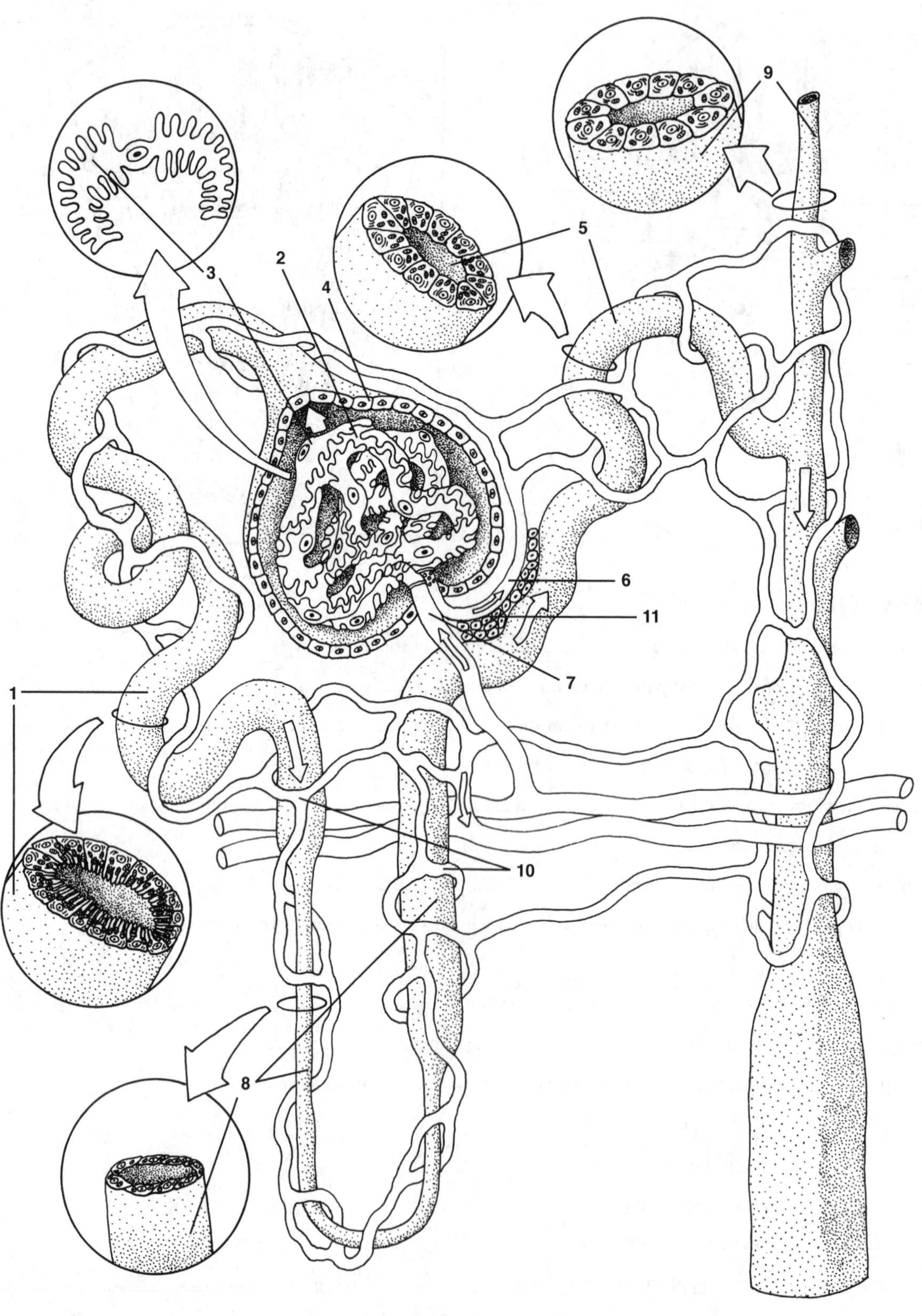

BLOOD VESSELS OF THE KIDNEY

1. a) The renal artery is a branch of the ____________________.

   b) The renal vein takes blood to the ____________________.

2. a) Number these blood vessels of the kidney in the order that blood flows through them. Begin at the renal artery and end at the renal vein.

   ____1____ renal artery

   ____________________ small veins in the kidneys (interlobular, arcuate, interlobar)

   ____________________ afferent arterioles

   ____________________ efferent arterioles

   ____________________ smaller arteries in the kidney (interlobar, arcuate, interlobular)

   ____________________ glomeruli

   ____________________ peritubular capillaries

   ____8____ renal vein

   b) In this pathway of blood flow, there are two sets of ____________________ in which exchanges take place between the blood and surrounding kidney tissue.

FORMATION OF URINE—GLOMERULAR FILTRATION

1. Glomerular filtration takes place in which major part of the nephron? ____________________

2. a) In glomerular filtration, blood pressure forces plasma, dissolved substances, and small proteins out of the ____________________ and into ____________________.

   b) The fluid in Bowman's capsule is now called ____________________.

   c) The components of blood that remain in the blood are ____________________ and ____________________ because they are too ____________________ to be forced out of the glomerulus.

3. a) Are useful materials such as nutrients and minerals present in renal filtrate? ____________________

   b) Are waste products present in renal filtrate? ____________________

   c) Therefore, glomerular filtration is selective in terms of (choose one answer) ____________________.

      1) usefulness—only waste materials enter the filtrate

      2) size—anything small or dissolved in blood plasma may enter the filtrate

      3) both of these

   d) Renal filtrate, therefore, is very similar to the ____________________ from which it is made, except that there is little protein and no ____________________ cells in filtrate.

4. a) The glomerular filtration rate (GFR) is the amount of renal filtrate formed by the kidneys in ____________________ (time) and is about 100 to 125 mL per minute.

   b) If blood flow through the kidneys decreases, the GFR will ____________________.

   c) If blood flow through the kidneys increases, the GFR will ____________________.

FORMATION OF URINE—TUBULAR REABSORPTION AND TUBULAR SECRETION

1. In the process of tubular reabsorption, useful materials are transported from the filtrate in the

   ________________ to the blood in the ________________.

2. a) How much of the renal filtrate is reabsorbed back into the blood? ________________ %

   b) The filtrate that is not reabsorbed enters the renal pelvis and is then called ________________.

3. Name the mechanism by which each of these substances is reabsorbed from the filtrate back to the blood.

   1) Water ________________
   2) Glucose ________________
   3) Amino acids ________________
   4) Negative ions ________________
   5) Small proteins ________________
   6) Positive ions ________________

4. a) There is a threshold level of reabsorption for glucose, amino acids, and vitamins. This means that there is a

   ________________ to how much of each can be reabsorbed.

   b) For example, the threshold level for the reabsorption of glucose will be exceeded when the blood glucose level is

   too ________________ (high or low) and results in a ________________ (high or low) filtrate level of glucose.

   c) If the blood glucose level is normal, ________________ (some, most, or all) of the glucose present in the filtrate will be reabsorbed.

5. a) The hormone that increases the reabsorption of sodium ions and the excretion of potassium ions is

   ________________.

   b) The hormone that increases the reabsorption of calcium ions is ________________.

6. a) In the process of tubular secretion, substances are transported from the blood to the ________________ in the renal tubules.

   b) Waste products that may be secreted into the filtrate are ________________ and ________________,

   as well as the metabolic products of ________________.

   c) To help maintain the normal pH of the blood, ________________ ions may also be secreted into the renal filtrate.

7. a) The hormone aldosterone, by increasing the reabsorption of ________________ ions, also increases the

   reabsorption of ________________ by osmosis.

   b) This is very important to maintain blood ________________ and blood ________________.

8. a) Antidiuretic hormone (ADH) directly increases the reabsorption of ________________ by the distal convoluted tubules and collecting tubules.

   b) This reabsorption produces a urine that is more ________________ (dilute or concentrated) than body fluids.

   c) When ADH secretion decreases, (more or less) water is reabsorbed and the urine will be more

   ________________ (dilute or concentrated).

9. a) The hormone ANP increases the excretion of ________________ ions and ________________.

   b) This will have what effect on blood volume? ________________ On blood pressure?

   ____

10. Number the following in the order in which filtrate/urine passes through them. Begin at the glomerulus and end at the urethra.

1	glomerulus	______	distal convoluted tubule
______	loop of Henle	______	Bowman's capsule
______	ureter	______	urinary bladder
______	calyx	______	collecting tubule
______	renal pelvis	11	urethra
______	proximal convoluted tubule		

11. The following diagram depicts the processes that take place within a nephron.

Name the substances reabsorbed or secreted by the mechanisms indicated.

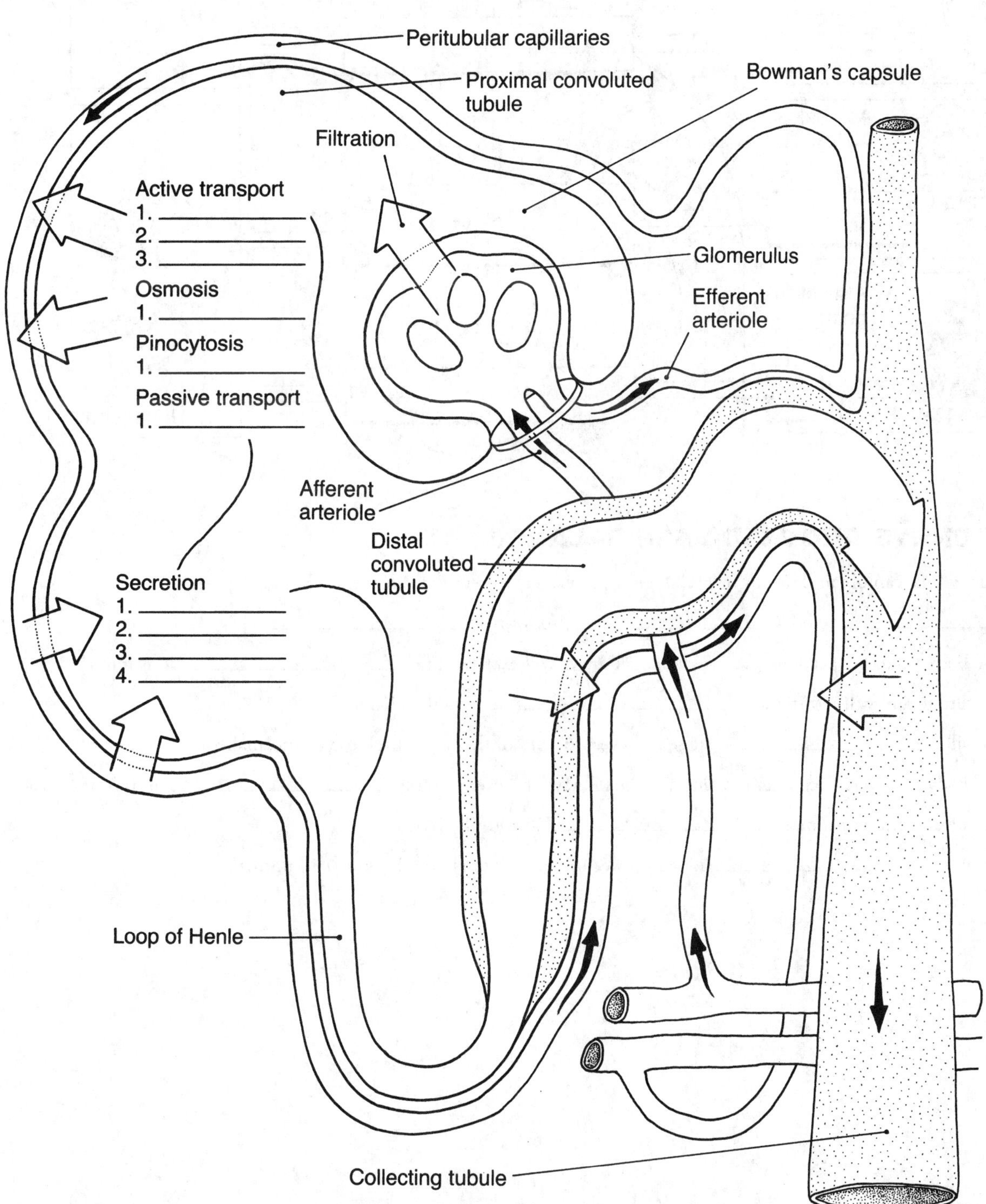

12. The following diagram depicts the effects of hormones on the kidneys.

Label the parts indicated.

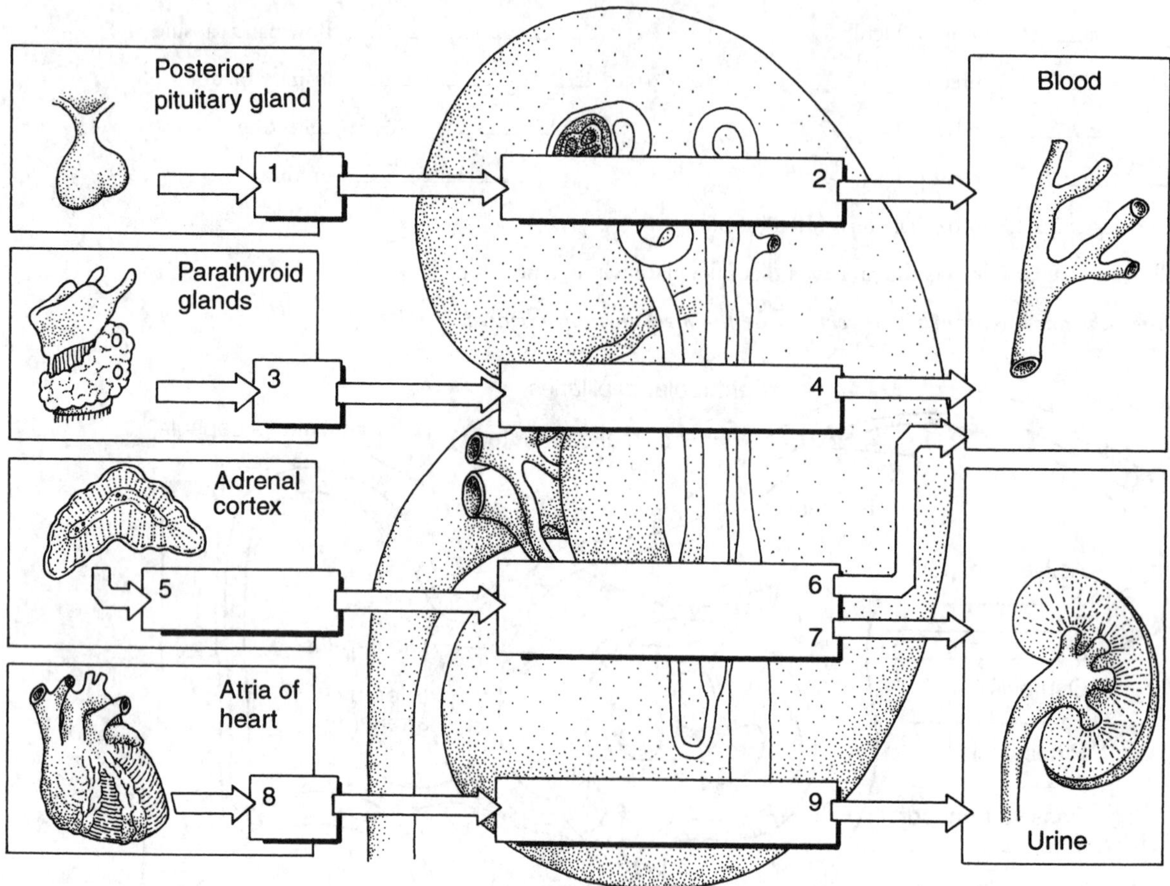

THE KIDNEYS AND ACID-BASE BALANCE

1. The kidney cells can obtain H^+ ions or HCO_3^- ions, to excrete or retain, from the reaction of ________________ and ________________ to form ________________.

2. a) If the body fluids are becoming too acidic, the kidneys will excrete more ________________ ions into the renal filtrate and will return more ________________ ions to the blood.

   b) This will ________________ (raise or lower) the pH of the blood back toward normal.

3. a) If the body fluids are becoming too alkaline, the kidneys will excrete more ________________ ions in the renal filtrate and return more ________________ ions to the blood.

   b) This will ________________ (raise or lower) the pH of the blood back toward normal.

OTHER FUNCTIONS OF THE KIDNEYS

1. a) The kidneys secrete the enzyme renin when blood pressure ____________.
   b) The cells that secrete renin are called ____________ cells, which are located in the wall of the ____________ arteriole.
   c) Renin starts the renin-angiotensin mechanism that results in the formation of ____________.
   d) State the two functions of angiotensin II. ____________ and ____________
   e) Both of these functions will ____________ (raise or lower) blood pressure.
2. a) The kidneys secrete the hormone ____________ when body tissues are in a state of hypoxia, which means ____________.
   b) This hormone then stimulates the ____________ to increase the rate of production of ____________ to increase the oxygen-carrying capacity of the blood.
3. The kidneys change the inactive forms of vitamin ____________ to the most active form, which increases the absorption of ____________ and ____________ in the small intestine.

ELIMINATION OF URINE

1. a) The two ureters are located ____________ the peritoneum of the dorsal abdominal cavity.
   b) Each ureter extends from the ____________ of a kidney to the ____________ side of the ____________.
   c) Peristalsis of the ureter propels urine into the bladder. Name the tissue in the wall of the ureter that is ____________ responsible for peristalsis.
2. a) The urinary bladder is located behind the ____________ bones.
   b) In men, the bladder is superior to the ____________ gland.
   c) In women, the bladder is inferior to the ____________.
3. The functions of the urinary bladder are to serve as a ____________ for accumulating urine and to ____________ to eliminate urine.
4. a) When the bladder is empty, the mucosa has folds called ____________.
   b) The epithelium of the bladder is ____________ epithelium.
   c) Both of these permit ____________ of the bladder without tearing the lining.
   d) The triangular area on the floor of the bladder that has no rugae is called the ____________.
   e) The boundaries of this area are formed by the openings of the two ____________ and the ____________.
5. a) The smooth muscle layer of the wall of the bladder is called the ____________.
   b) Around the opening of the urethra, the detrusor muscle forms the ____________ sphincter, which is ____________ (voluntary or involuntary).

6. The following diagram depicts both female and male urinary bladders and urethras.

Label the parts indicated.

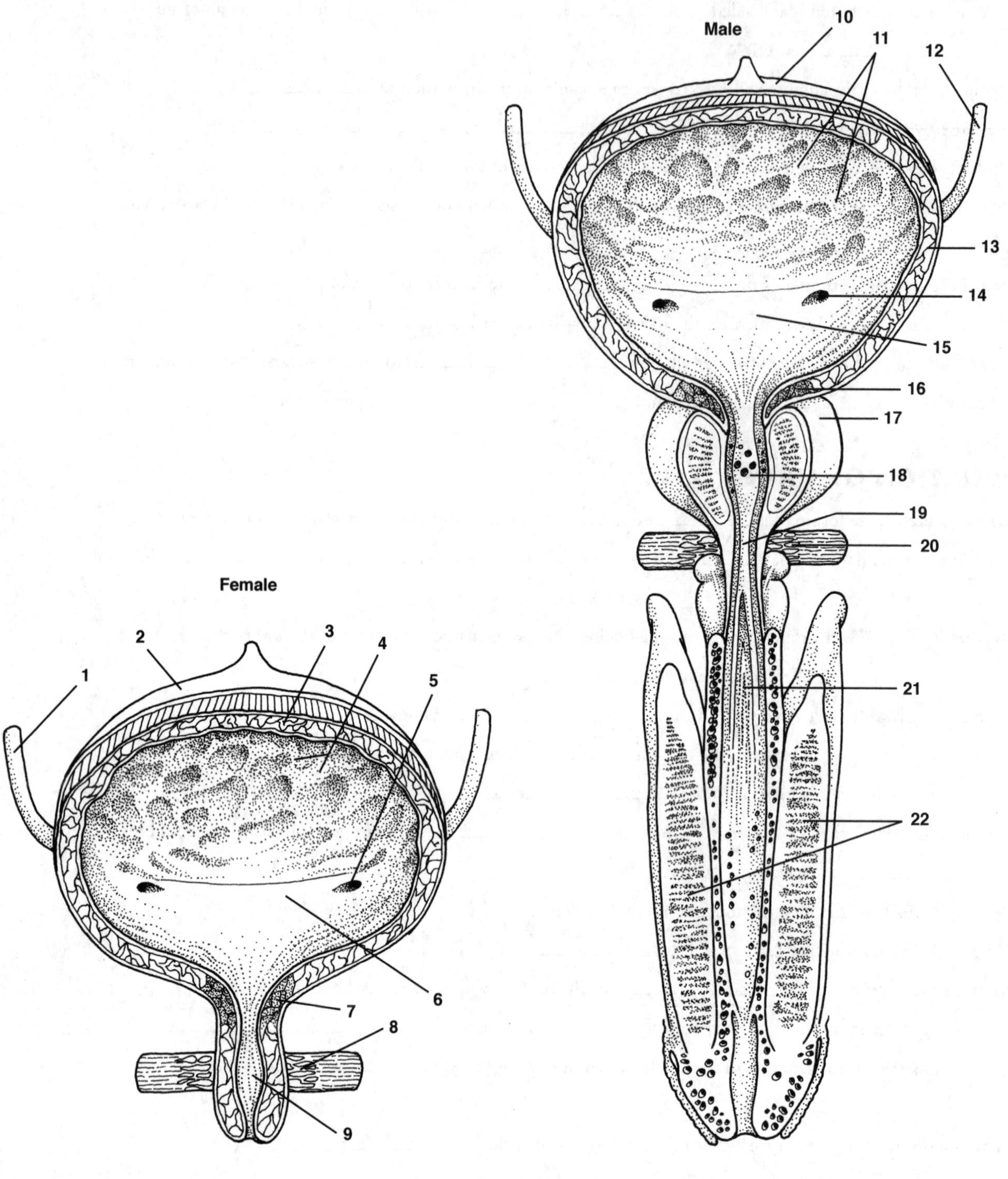

7. a) The urethra carries urine from the ______________ to the ______________.
   b) In women, the urethra is ______________ to the vagina.
   c) In men, the urethra extends through the ______________ gland and the ______________ and carries ______________, as well as urine.
   d) Surrounding the wall of the urethra is the ______________ sphincter, which is ______________ (voluntary or involuntary).

8. a) The term *micturition reflex* is another name for the ______________ reflex.
   b) Which part of the CNS is directly involved in this reflex? ______________

9. a) Number the following events of the urination reflex in the proper sequence.

   ______1______ stretching of the detrusor muscle by accumulating urine
   ______________ the detrusor muscle contracts, and the internal urethral sphincter relaxes
   ______________ motor impulses along parasympathetic nerves return to the detrusor muscle
   ______________ sensory impulses travel to the sacral spinal cord
   ______________ stretch receptors in the detrusor muscle generate impulses
   ______6______ the bladder is emptied

   b) Urination may be prevented by voluntary contraction of the ______________.

CHARACTERISTICS OF URINE

1. a) The normal range of urinary output in 24 hours is ______________.
   b) State one factor that may decrease urinary output. ______________
   c) State one factor that may increase urinary output. ______________

2. a) The yellow color of urine is often called ______________ or ______________.
   b) Urine is usually ______________ (clear or cloudy) in appearance.

3. a) The normal range of specific gravity of urine is ______________ to ______________.
   b) Specific gravity is a measure of the ______________ in urine and is an indicator of the ______________ ability of the kidneys.
   c) Urine with a specific gravity of 1.023 is more ______________ (dilute or concentrated) than urine with a specific gravity of 1.015.
   d) State one factor that will result in the formation of a more concentrated urine.

   ____

4. The average pH of urine is ______________, but a pH range of ______________ to ______________ is considered normal.

5. Urine is approximately 95% ______________, which is the ______________ for nitrogenous wastes and excess minerals.

6. Match each nitrogenous waste with its proper source.

   Use each letter once.

   1) Creatinine ______________________
   2) Uric acid ______________________
   3) Urea ______________________

   A. From the deamination of excess amino acids that are used for energy production
   B. From the use of creatine phosphate as an energy source in muscles
   C. From the decomposition of nucleic acids

7. If the blood levels of nitrogenous wastes are higher than normal, the kidneys are (choose one answer):

   a) functioning properly
   b) not functioning properly
   c) working overtime

ABNORMAL CONSTITUENTS OF URINE

Match each abnormal constituent of urine with the proper reason for it.

Use each letter once.

1) Protein ______________________
2) Blood ______________________
3) Glucose ______________________
4) Ketones ______________________
5) Bacteria ______________________

A. Infection somewhere in the urinary tract
B. Increased use of fats and proteins for energy
C. A higher than normal blood glucose level
D. Bleeding somewhere in the urinary tract
E. The glomeruli have become too permeable, allowing large molecules through

CROSSWORD PUZZLE

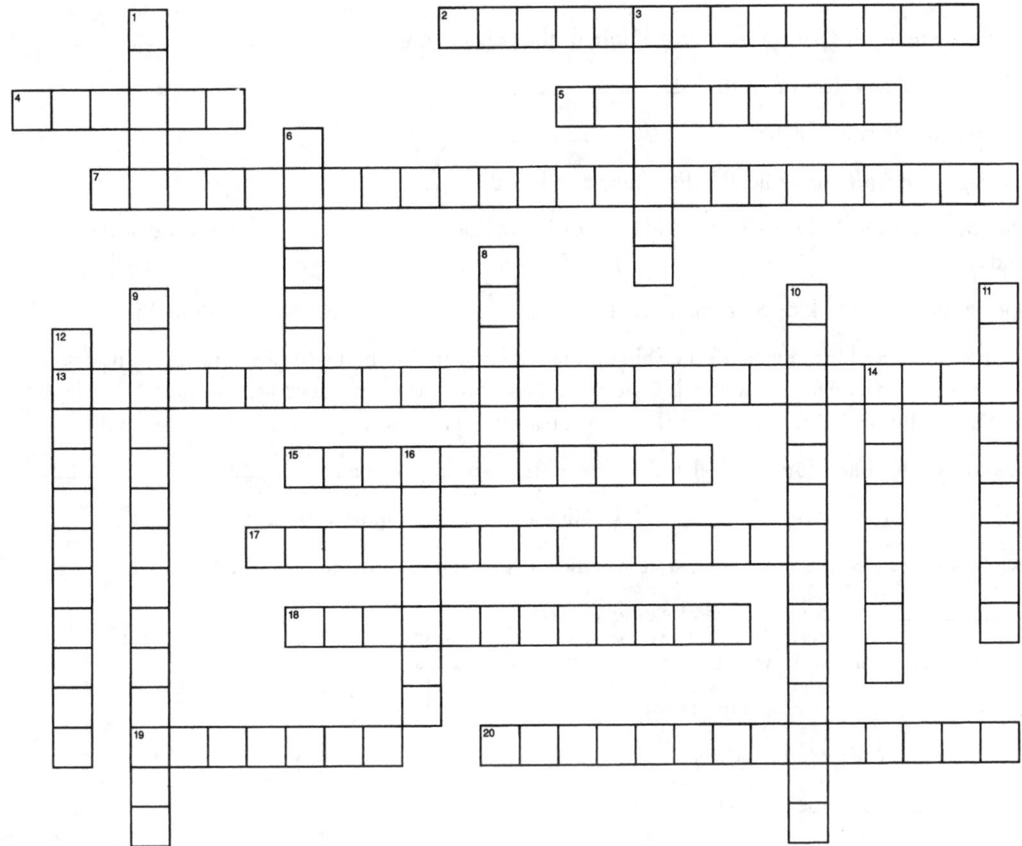

ACROSS

2. Receives filtrate from a glomerulus (two words)
4. Organ responsible for the formation of urine
5. Inflammation of the kidneys
7. Initials are GFR
13. Sphincter of the bladder (three words)
15. The urination reflex (synonym)
17. Measure of the dissolved materials in urine (two words)
18. Use of an artificial kidney machine
19. Renal ________; also called kidney stones
20. The limit to how much glucose the renal tubules can reabsorb (two words)

DOWN

1. ________ failure; inability of the kidneys to function properly
3. Structural and functional unit of the kidney
6. Carries urine from the bladder to the exterior
8. Carries urine from a kidney to the bladder
9. Spherical muscle layer in the wall of the bladder (two words)
10. Reservoir for accumulating urine (two words)
11. Renal ________; consists of a glomerulus surrounded by a Bowman's capsule
12. ________ wastes, such as urea
14. Inflammation of the urinary bladder
16. Triangular area on the floor of the bladder

CLINICAL APPLICATIONS

1. a) Urinary tract infections are usually caused by what type of microorganism? ____________
   b) These microorganisms are usually part of the normal flora of the person's own ____________.
   c) The term *nephritis* refers to an infection of the ____________.
   d) The term *cystitis* refers to infection of the ____________.
2. a) Mrs. R has been diagnosed with renal calculi, which may also be called ____________.
   b) To prevent further formation of stones, Mrs. R will be advised to ____________ (increase or decrease) her intake of fluids.
   c) This will be important to have her kidneys form a ____________ (concentrated or dilute) urine.
3. a) Mr. G was in a car accident and lost a great deal of blood. He has been in the intensive care unit for 24 hours, and his condition is stable. His doctor, however, is concerned about possible kidney damage and finds these reports about Mr. G: urinary output in 24 hours is 500 mL, blood urea nitrogen is elevated, blood creatinine level is elevated. On the basis of this information, are Mr. G's kidneys functioning normally? ____________
   b) Is Mr. G in total renal failure? ____________ Give a reason to support your answer.
4. a) For each of these possible causes of renal failure, state whether it is prerenal, intrinsic renal, or postrenal.
      1) ____________ a kidney stone is blocking a ureter
      2) ____________ an extensive bacterial infection of the kidneys
      3) ____________ a ureter becomes twisted
      4) ____________ severe hemorrhage
      5) ____________ a serious side effect of some antibiotics
   b) For someone with chronic renal failure, the procedure called ____________ may be lifesaving.
   c) In this procedure, the patient's blood is passed through a hemodialysis machine to remove ____________ and ____________.

MULTIPLE CHOICE TEST #1

Choose the correct answer for each question.

1. A renal corpuscle consists of:
   a) a glomerulus and peritubular capillaries
   b) a renal tubule and Bowman's capsule
   c) a glomerulus and Bowman's capsule
   d) a renal tubule and peritubular capillaries
2. The kidneys are located behind the:
   a) renal artery b) peritoneum c) spinal column d) renal vein
3. All of these are found at the hilus of a kidney except the:
   a) renal artery b) urethra c) ureter d) renal vein
4. The cavity within the kidney that collects urine is the:
   a) renal pelvis b) urinary bladder c) ureter d) renal vein
5. The renal vein takes blood from the kidney to the:
   a) abdominal aorta b) renal artery c) superior vena cava d) inferior vena cava
6. All of these are parts of the renal tubule except the:
   a) loop of Henle
   b) glomerulus
   c) proximal convoluted tubule
   d) distal convoluted tubule

7. If body fluids are becoming too acidic, the kidneys will excrete more of these ions in urine:
   a) sodium ions b) bicarbonate ions c) hydrogen ions d) potassium ions

8. The hormone that directly increases the reabsorption of water by the kidneys is:
   a) ADH b) aldosterone c) PTH d) ANP

9. The process of glomerular filtration takes place from the:
   a) glomerulus to Bowman's capsule
   b) renal tubule to peritubular capillaries
   c) Bowman's capsule to glomerulus
   d) peritubular capillaries to renal tubule

10. The process of tubular reabsorption takes place from the:
    a) glomerulus to afferent arteriole
    b) peritubular capillaries to renal tubule
    c) efferent arteriole to glomerulus
    d) renal tubule to peritubular capillaries

11. The renal pyramids make up the:
    a) renal cortex b) renal pelvis c) renal medulla d) renal fascia

12. Renal filtrate differs from blood plasma in this way:
    a) only waste products are present in filtrate
    b) there are no nutrients in filtrate
    c) there is more protein in filtrate
    d) there are no blood cells in filtrate

13. When the blood level of oxygen decreases, the kidneys secrete:
    a) renin b) aldosterone c) angiotensin II d) erythropoietin

14. When blood pressure decreases, the kidneys secrete:
    a) angiotensin II b) renin c) aldosterone d) epinephrine

15. The part of the urinary bladder that actually eliminates the urine is the:
    a) detrusor muscle b) trigone c) rugae d) transitional epithelium

16. Voluntary control of urination is provided by the:
    a) detrusor muscle
    b) external urethral sphincter
    c) spinal cord
    d) internal urethral sphincter

17. In tubular reabsorption, glucose and amino acids are reabsorbed by the process of:
    a) active transport b) passive transport c) pinocytosis d) osmosis

18. The glomerular filtration rate will decrease if:
    a) the amount of waste products in the blood increases
    b) blood flow through the kidneys decreases
    c) blood flow through the kidneys increases
    d) the amount of waste products in the blood decreases

19. Urea is a nitrogenous waste product that comes from the metabolism of:
    a) nucleic acids b) muscle tissue c) amino acids d) carbohydrates

20. Normal values for daily urinary output and pH of urine might be:
    a) 2.5 liters / 3.0 b) 4.0 liters / 6.5 c) 2.0 liters / 9.0 d) 1.5 liters / 6.0

21. Urine is propelled through a ureter by:
    a) striated muscle b) ciliated epithelium c) skeletal muscle d) smooth muscle

22. The kidneys are protected from mechanical injury by the:
    a) pelvic bone and smooth muscle
    b) smooth muscle and rib cage
    c) adipose tissue and pelvic bone
    d) rib cage and adipose tissue

23. If body fluids are becoming too alkaline, the kidneys will excrete more of these ions in urine:
    a) calcium ions b) bicarbonate ions c) hydrogen ions d) potassium ions

24. Creatinine is a nitrogenous waste product that comes from energy metabolism in:
    a) the brain b) the liver c) the lungs d) the muscles

25. Atrial natriuretic peptide (ANP) is produced by the ________________ and causes the kidneys to excrete ________________.
    a) liver, potassium ions b) heart, potassium ions c) liver, sodium ions d) heart, sodium ions

MULTIPLE CHOICE TEST #2

Read each question and the four answer choices carefully. When you have made a choice, follow the instructions to complete your answer.

1. Which statement is NOT true of tubular reabsorption and tubular secretion?
   a) Water is reabsorbed by osmosis.
   b) Reabsorbed materials enter the glomerular capillaries.
   c) Creatinine and the metabolic products of medications may be secreted into the renal filtrate.
   d) Positive ions are reabsorbed by active transport, and negative ions and water follow.

   Reword your choice to make it a correct statement.

   ___

2. Which statement is NOT true of hormone effects on the kidneys?
   a) ADH increases the reabsorption of water.
   b) PTH increases the reabsorption of calcium ions.
   c) Aldosterone decreases the reabsorption of sodium ions.
   d) ANP increases the excretion of sodium ions.

   Reword your choice to make it a correct statement.

   ___

3. Which statement is NOT true of the urination reflex?
   a) The stimulus is stretching of the rugae of the urinary bladder by accumulating urine.
   b) From the bladder, sensory impulses travel to the sacral spinal cord.
   c) The detrusor muscle contracts to eliminate the urine.
   d) The internal urethral sphincter relaxes to permit urination.

   Reword your choice to make it a correct statement.

   ___

4. Which statement is NOT true of the location of the kidneys?
   a) The kidneys are posterior to the peritoneum.
   b) The kidneys are medial to the spinal column.
   c) Adipose tissue surrounds the kidneys and cushions them.
   d) The renal fascia is a fibrous connective tissue membrane that helps keep the kidneys in place.

   Reword your choice to make it a correct statement.

   ___

5. Which statement is NOT true of a nephron?
   a) A glomerulus is surrounded by Bowman's capsule.
   b) The proximal and distal convoluted tubules are found in the renal cortex.
   c) The loops of Henle and collecting tubules are found in the renal pelvis.
   d) All of the parts of a renal tubule are surrounded by peritubular capillaries.

   Reword your choice to make it a correct statement.

   ___

6. Which statement is NOT true of glomerular filtration?
   a) Blood pressure in the glomeruli is high because the efferent arterioles are smaller in diameter than the afferent arterioles.
   b) Blood pressure in the glomeruli provides the force for filtration.
   c) Blood cells and large proteins remain in the blood in Bowman's capsule.
   d) Glomerular filtration increases if blood flow through the kidneys increases.

   Reword your choice to make it a correct statement.

   ___

7. Which statement is NOT true of the characteristics of normal urine?
   a) Specific gravity is a measure of the dissolved materials in urine.
   b) Cloudy urine may be an indicator of a urinary tract infection.
   c) The volume of urinary output may be changed by factors such as excessive sweating or diarrhea.
   d) Glucose is present in urine only if the blood glucose level is too low.

   Reword your choice to make it a correct statement.

   ____

8. Which statement is NOT true of the elimination of urine?
   a) A ureter carries urine from the renal pelvis to the urinary bladder.
   b) The micturition reflex is a spinal cord reflex.
   c) The urethra carries urine from the urinary bladder to the exterior.
   d) The voluntary external urethral sphincter can close the ureter.

   Reword your choice to make it a correct statement.

   ____

9. Which statement is NOT true of the kidneys?
   a) The kidneys change inactive forms of vitamin C to the active form.
   b) When blood pressure decreases, the kidneys secrete renin to start the formation of angiotensin II.
   c) In a state of hypoxia, the kidneys secrete erythropoietin to increase the rate of RBC production.
   d) The kidneys help maintain normal blood volume and blood pressure.

   Reword your choice to make it a correct statement.

   ____

10. Which statement is NOT true of the kidneys and regulation of the homeostasis of body fluids?
    a) The kidneys regulate blood volume by excreting or conserving water.
    b) By regulating the composition of blood, the kidneys also regulate the composition of tissue fluid.
    c) The kidneys excrete or conserve minerals to regulate the electrolyte balance of the blood.
    d) The kidneys respond to a decreasing pH of the blood by excreting fewer hydrogen ions.

    Reword your choice to make it a correct statement.

    ____

MULTIPLE CHOICE TEST #3

Each question is a series of statements concerning a topic in this chapter. Read each statement carefully and select all of the correct statements.

1. Which of the following statements are true of the kidneys?
   a) The work of the kidneys takes place in the two sets of capillaries in and around each nephron.
   b) The renal pyramids make up the renal medulla and contain loops of Henle and renal corpuscles.
   c) The hormone ADH increases the reabsorption of Na^+ ions in the distal convoluted tubule and collecting tubule.
   d) The calyces are extensions of the renal pelvis, which opens into the ureter.
   e) The blood pressure in the glomeruli is relatively high, which promotes efficient filtration.
   f) A Bowman's capsule is the expanded end of a distal convoluted tubule.
   g) Juxtaglomerular cells secrete renin when blood pressure decreases.
   h) Hydrogen ions and medications may be actively secreted into renal filtrate.
   i) A threshold level for reabsorption means that all of a useful substance in the renal filtrate will be returned to the blood.
   j) Podocytes increase the surface area for reabsorption in the proximal convoluted tubule.
   k) To compensate for acidosis, the kidneys will store any excess H^+ ions.
   l) The hormone that increases the excretion of K^+ ions and the reabsorption of Na^+ ions is aldosterone produced by the adrenal cortex.

2. Which of the following statements are true of the ureters, urethra, and urinary bladder?
   a) The ureters are posterior to the peritoneum and enter the upper anterior side of the urinary bladder.
   b) The bladder is lined with simple squamous epithelium that stretches as the bladder fills.
   c) Urine flows from a kidney to the bladder only because of gravity.
   d) In women, the urethra is anterior to the vagina.
   e) Both the internal and external urethral sphincters are made of smooth muscle.
   f) Parasympathetic impulses cause the detrusor muscle to contract and the internal urethral sphincter to relax.
   g) Urination is a spinal cord reflex, and the stimulus is contraction of the detrusor muscle.
   h) In men, the urethra is also part of the reproductive system.

3. Which of the following statements are true of urine and its constituents?
   a) Urine with a high specific gravity will often be an amber color rather than straw.
   b) The metabolism of excess amino acids for energy production results in the formation of the nitrogenous waste product urea.
   c) If fats become a primary energy source for the body, a consequence may be ketonuria.
   d) The average pH of urine is about 6, slightly alkaline.
   e) The average person in a normal state of hydration will void 1 to 2 liters of urine per 24 hours.
   f) Proteinuria is a consequence of the glomeruli becoming too permeable.
   g) Bacteruria always indicates an infection of the bladder.
   h) An increase in the blood level of creatinine may indicate a decrease in the functioning of the kidneys.

Chapter 19

Fluid-Electrolyte and Acid-Base Balance

The regulation of the water and mineral content of the body is the fluid-electrolyte balance, and the regulation of the pH of the body fluids is the acid-base balance. Some of this material has been part of previous chapters and is included in this chapter to provide a complete discussion of these important aspects of homeostasis.

WATER COMPARTMENTS

1. The water found within cells is called ______________ fluid and is about ______________ (fraction) of the total body water.

2. a) All the water found outside cells is called ______________ fluid.

   b) This fluid may be given more specific names when found in specific locations. Name each of these fluids.

      1) ______________ in tissue spaces between cells

      2) ______________ in blood vessels

      3) ______________ in lymph vessels

3. Match each of these specialized fluids with its proper location.

   Use each letter once.

   1) Cerebrospinal fluid ______________

   2) Aqueous humor ______________

   3) Synovial fluid ______________

   4) Serous fluid ______________

   A. Within the eyeball
   B. Between membranes such as the pleural membranes
   C. Around and within the CNS
   D. Within joint cavities

4. The two processes by which water moves from one compartment to another within the body are ______________ and ______________.

WATER INTAKE AND OUTPUT

1. a) The body's major source of water is ______________.

   b) A secondary source of water is the water in ______________.

   c) A small amount of water is produced within the body as a product of the process of ______________.

2. a) The major pathway of water loss from the body is by way of the ________________ system in the form of ________________.

   b) A secondary pathway of water loss is by way of the ________________ in the form of ________________.

   c) Smaller amounts of water are also lost in ________________ and ________________.

3. a) For healthy people, water intake should ________________ water output.

   b) If sweating increases, what must be done to compensate? ________________

   c) If fluid intake is excessive, however, ________________ will increase to return the body's water content to normal.

REGULATION OF WATER INTAKE AND OUTPUT

1. The part of the brain that regulates water content of the body is the ________________.
2. The term *osmolarity* means ________________.
3. The specialized cells in the hypothalamus that detect changes in the osmolarity of body fluids are called ________________.
4. a) When the body is dehydrated, the osmolarity of body fluids ________________ (increases or decreases), because there is more dissolved material in proportion to water.

   b) In a state of dehydration, a person experiences the sensation of ________________ and will ________________.

   c) As water is absorbed by the digestive tract, the osmolarity of the body fluids ________________ (increases or decreases) toward normal.

5. a) The hormone ________________ is produced by the hypothalamus and stored in the ________________ pituitary gland.

   b) The function of ADH is to ________________ by the kidneys.

   c) In what type of situation is ADH secreted? ________________

   d) In this situation, ADH effects will cause urinary output to ________________.

6. a) The hormone ________________ is produced by the adrenal cortex and increases the reabsorption of sodium ions by the kidney tubules.

   b) How does this affect the reabsorption of water by the kidneys? ________________ (increases or decreases)

   c) State two types of situations that stimulate secretion of aldosterone.

      1) ________________________________

      2) ________________________________

7. If there is too much water in the body, the secretion of ADH will ________________ (increase or decrease), and urinary output will ________________ (increase or decrease).
8. If blood volume or BP increases, the heart will produce the hormone ________________, which will ________________ (increase or decrease) the excretion of sodium ions and water.

ELECTROLYTES

1. When electrolytes are in water, they dissociate into their ________________.

2. Most electrolytes are ________________ (organic or inorganic) molecules that include ________________, ________________, and ________________.

3. a) Anions are ________________ ions.

   b) Give two examples. ________________ and ________________

   c) Cations are ________________ ions.

   d) Give two examples. ________________ and ________________

4. a) By their presence in body fluids, electrolytes help create the ________________ of these fluids and help regulate the ________________ of water between water compartments.

   b) Water will move by osmosis to a compartment with a ________________ (greater or lesser) concentration of electrolytes.

5. Some electrolytes are structural components of tissues, such as ________________, or are part of proteins, such as ________________.

6. Some electrolytes can be stored in the body, such as ________________ and ________________ in bones and ________________ and ________________ in the liver.

7. Match each body fluid with the statement that describes its electrolyte composition.

   Use each letter once.

   1) Intracellular fluid ________________
   2) Tissue fluid ________________
   3) Plasma ________________

   A. The most abundant cation is sodium, the most abundant anion is chloride, and there are few protein anions.
   B. The most abundant cation is sodium, the most abundant anion is chloride, and there are many protein anions.
   C. The most abundant cation is potassium, and the most abundant anions are phosphate and proteins.

REGULATION OF ELECTROLYTE INTAKE AND OUTPUT

1. We take in electrolytes as part of ________________ and ________________.

2. a) Electrolytes are lost from the body in ________________, ________________, and ________________.

   b) The most abundant electrolytes in sweat are ________________ and ________________.

   c) Electrolytes are present in urine when their blood levels are ________________ (lower or higher) than the body's need for them.

3. Match each hormone with its proper effect(s) on electrolytes.

   Use each letter once. One answer line will have two correct letters.

   1) Aldosterone ________________
   2) Calcitonin ________________
   3) Parathyroid hormone ________________
   4) ANP ________________

   A. Increases the reabsorption of calcium and phosphorus from the bones to the blood
   B. Increases the removal of calcium and phosphorus from the blood and into bones
   C. Increases the reabsorption of sodium by the kidneys
   D. Increases the excretion of potassium by the kidneys
   E. Increases the excretion of sodium by the kidneys

ACID-BASE BALANCE

1. The pH of body fluids is regulated by three mechanisms: the ______________ in body fluids, the ______________ system, and the ______________.
2. a) The normal pH range of blood is ______________ to ______________.
   b) The normal pH range of intracellular fluid is ______________ to ______________.
   c) The normal pH range of tissue fluid is similar to that of ______________ but is slightly more variable.
3. a) On the pH scale, a value of 7.0 is considered ______________.
   b) A pH below 7 is considered ______________, and a pH above 7 is considered ______________.
   c) Therefore, the pH of blood is slightly ______________, and the pH of intracellular fluid is slightly ______________ or neutral.

BUFFER SYSTEMS

1. a) A buffer system consists of two chemicals: a weak ______________ and a weak ______________.
   b) The chemicals of a buffer system react with ______________ acids or bases to change them to substances that ______________ (will or will not) have a great effect on pH.
2. The three buffer systems that help prevent drastic changes in the pH of the blood or other body fluids are the ______________, ______________, and ______________ systems.
3. a) The bicarbonate buffer system consists of the weak acid ______________ and the weak base ______________.
   b) Complete the following equation, in which the strong acid hydrochloric acid will be buffered:

   $HCl + NaHCO_3 \rightarrow$ ______________ + ______________

   c) Which product of the reaction has no effect on pH? ______________
   d) Which product has only a slight effect on pH? ______________
   e) Complete the following equation, in which the strong base sodium hydroxide will be buffered:

   $NaOH + H_2CO_3 \rightarrow$ ______________ + ______________

   f) Which product of this reaction has no effect on pH? ______________
   g) Which product has only a slight effect on pH? ______________
4. a) The phosphate buffer system consists of the weak acid ______________ and the weak base ______________.
   b) Complete the following equation, in which the strong acid hydrochloric acid will be buffered:

   $HCl + Na_2HPO_4 \rightarrow$ ______________ + ______________

   c) Which product of this reaction has no effect on pH? ______________
   d) Which product has only a slight effect on pH? ______________
   e) Complete the following equation, in which the strong base sodium hydroxide will be buffered:

   $NaOH + NaH_2PO_4 \rightarrow$ ______________ + ______________

   f) Which product of this reaction has no effect on pH? ______________
   g) Which product has only a slight effect on pH? ______________

5. a) In the protein buffer system, an amino acid may act as either an ________________ or a ________________.

   b) The ________________ group of an amino acid acts as an acid because it can donate a ________________ ion to the intracellular fluid.

   c) The ________________ group of an amino acid acts as a base because it can pick up an excess ________________ ion from the intracellular fluid.

6. How long does it take for a buffer system to react to prevent a great change in pH?

____

RESPIRATORY MECHANISMS TO REGULATE PH

1. The respiratory system has an effect on the pH of body fluids because it regulates the amount of ________________ present in these fluids.

2. a) Carbon dioxide affects the pH of body fluids because it reacts with water to form ________________.

   b) Complete the following equation to show the dissociation of carbonic acid:

   $CO_2 + H_2O \rightarrow H_2CO_3 \rightarrow$ ________________ + ________________

   c) Which of the ions formed will lower the pH of body fluids? ________________

3. Match each respiratory pH imbalance with the statements that apply to each.

   Use each letter once. Each answer line will have four correct letters.

   1) Respiratory acidosis ________________

   2) Respiratory alkalosis ________________

   A. Caused by an increase in the rate of respiration
   B. Caused by a decrease in the rate or efficiency of respiration
   C. More CO_2 is retained within the body
   D. More CO_2 is exhaled
   E. Fewer H^+ ions are produced
   F. More H^+ ions are produced
   G. The pH of body fluids decreases
   H. The pH of body fluids increases

4. To compensate for a metabolic acidosis, the respiratory rate will ________________ (increase or decrease) to exhale ________________ (more or less) CO_2 to ________________ (increase or decrease) the formation of H^- ions to ________________ (raise or lower) the pH toward normal.

5. To compensate for a metabolic alkalosis, the respiratory rate will ________________ (increase or decrease) to exhale ________________ (more or less) CO_2 to ________________ (increase or decrease) the formation of H^+ ions to ________________ (raise or lower) the pH toward normal.

6. How long does it take for the respiratory system to respond to changes in pH?

____

RENAL MECHANISMS TO REGULATE PH

1. Cells of the kidneys can obtain hydrogen (H^+) ions and bicarbonate (HCO_3^-) ions, to excrete or retain, from the reaction of ______________ and ______________ to form ______________.
2. To compensate for a state of acidosis, the kidneys will excrete ______________ ions and will conserve ______________ ions and ______________ ions.
3. To compensate for a state of alkalosis, the kidneys will excrete ______________ ions and ______________ ions and will conserve ______________ ions.
4. How long does it take the kidneys to respond to pH changes? ______________
5. Number the pH-regulating mechanisms in the order of their greatest capacity (NOT speed) to correct an ongoing pH imbalance.

   ______________ Buffer systems

   ______________ Respiration

   ______________ Kidneys

CAUSES AND EFFECTS OF PH CHANGES

1. State a specific cause of respiratory acidosis. ______________
2. State a specific cause of respiratory alkalosis. ______________
3. State two specific causes of metabolic acidosis. ______________

   and ______________
4. State two specific causes of metabolic alkalosis. ______________

   and ______________
5. a) A state of acidosis has its greatest detrimental effects on the ______________.

   b) State two possible consequences of acidosis. ______________

   and ______________
6. a) A state of alkalosis affects both the ______________ and ______________ nervous systems.

   b) State two possible consequences of alkalosis ______________

   and ______________

CROSSWORD PUZZLE

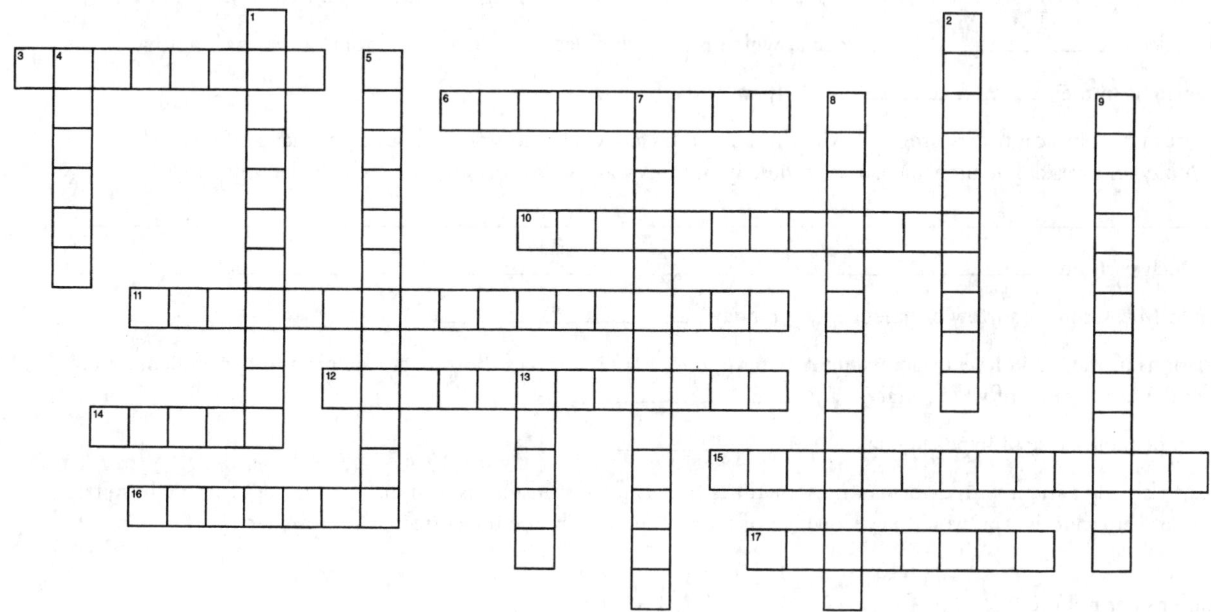

ACROSS

3. __________ group of an amino acid; can donate a hydrogen ion to the fluid
6. Metabolic __________; may be caused by overingestion of bicarbonate medications
10. Low blood calcium level
11. Fluid sites (two words)
12. Low blood sodium level
14. Abnormal increase in the amount of tissue fluid
15. High blood calcium level
16. Positive ions
17. Metabolic __________; may be caused by kidney disease

DOWN

1. Low blood potassium level
2. Concentration of dissolved materials present in a fluid
4. Negative ions
5. Chemicals that dissolve in water and dissociate into their (+) and (–) ions
7. Detect changes in the osmolarity of blood fluids
8. High blood sodium level
9. High blood potassium level
13. __________ group of an amino acid; can pick up an excess hydrogen ion from the fluid

CLINICAL APPLICATIONS

1. Mrs. A has congestive heart failure in which her right ventricle is not pumping efficiently. As a result, Mrs. A has what is called ______________ edema, which is an accumulation of tissue fluid that causes swelling that is most apparent in the ______________ (part of the body).

2. a) As part of his physical fitness program, Mr. K runs 5 miles a day. He did so on a hot August day and then felt very dizzy and weak. The probable cause of these symptoms was the loss of water and which electrolyte?

   ____

   b) By which pathway? ______________

   c) Should Mr. K run tomorrow or take it easy for a day? ______________

3. a) Variations in the blood level of potassium may have serious consequences. Both hypokalemia and hyperkalemia may affect the functioning of which vital organ? ______________

   b) State a possible cause of hypokalemia. ______________

4. Mr. S is an elderly patient with diabetes who sometimes forgets to administer his insulin. He is brought to the hospital in a state of ketoacidosis. You would expect which of these reports? (Choose the correct answer for each.)

Blood pH	a) 7.40	b) 7.35	c) 7.30
Respirations per minute	a) 40	b) 20	c) 10
Urine pH	a) 4.5	b) 6.5	c) 7.0

MULTIPLE CHOICE TEST #1

Choose the correct answer for each question.

1. Filtration forces water out of capillaries into tissue spaces, and the name for this water changes from:
   a) tissue fluid to lymph
   b) plasma to tissue fluid
   c) tissue fluid to plasma
   d) intracellular fluid to plasma

2. Most of the water output from the body is in the form of:
   a) sweat b) exhaled water vapor c) feces d) urine

3. Most of the water intake for the body is in the form of:
   a) fruits b) metabolic water c) beverages d) bread and cereal

4. All of these are cations except:
   a) bicarbonate ions b) sodium ions c) calcium ions d) potassium ions

5. The function of ADH is to:
   a) decrease the reabsorption of water by the kidneys
   b) increase the reabsorption of sodium ions by the kidneys
   c) increase the water secreted in sweat
   d) increase the reabsorption of water by the kidneys

6. An electrolyte is a substance that, in solution, dissociates into its:
   a) positive atoms
   b) positive and negative ions
   c) negative molecules
   d) positive and negative bonds

7. Water will move by osmosis to an area where there are more:
   a) cells b) electrolytes c) water molecules d) membranes

8. Regulation of the water balance of the body is a function of the:
   a) thalamus b) cerebrum c) medulla d) hypothalamus

9. To compensate for a state of dehydration, urinary output will:
   a) decrease b) increase c) remain the same d) increase then decrease

10. The direct effect on the kidneys of the hormone aldosterone is to:
 a) decrease the reabsorption of sodium and the excretion of potassium
 b) increase the reabsorption of sodium and potassium
 c) increase the reabsorption of sodium and the excretion of potassium
 d) increase the excretion of sodium and potassium

11. Proteins are significant anions in which of these fluids?
 a) tissue fluid and lymph
 b) tissue fluid and intracellular fluid
 c) plasma and intracellular fluid
 d) lymph and cerebrospinal fluid

12. Extracellular fluid includes all of these except:
 a) water in capillaries b) water within cells c) cerebrospinal fluid d) tissue fluid

13. The absorption of calcium ions by the small intestine and kidneys is increased by:
 a) parathyroid hormone b) aldosterone c) calcitonin d) ADH

14. The normal pH range of blood is:
 a) 7.25 to 7.35 b) 7.40 to 7.50 c) 7.30 to 7.40 d) 7.35 to 7.45

15. The mechanism with the greatest capacity to correct an ongoing pH imbalance is the:
 a) respiratory system b) buffer systems c) kidneys d) digestive system

16. The mechanism that works most rapidly to correct a pH imbalance is the:
 a) kidneys b) respiratory system c) buffer systems d) none of these works rapidly

17. The bicarbonate buffer system buffers HCl by reacting to form:
 a) NaCl and H_2CO_3 b) H_2O and H_2CO_3 c) H_2O and $NaHCO_3$ d) NaCl and H_2O

18. An amino acid is able to buffer a strong acid when:
 a) the carboxyl group picks up an excess hydrogen ion
 b) the amine group picks up an excess hydrogen ion
 c) the carboxyl group gives off a hydrogen ion
 d) the amine group gives off a hydrogen ion

19. A state of acidosis affects the:
 a) digestive system, preventing the digestion of proteins
 b) CNS, causing muscle spasms and convulsions
 c) CNS, causing confusion and coma
 d) respiratory system, causing the respiratory rate to decrease

20. The respiratory system will help compensate for a metabolic acidosis by:
 a) decreasing the respiratory rate to exhale less CO_2
 b) increasing the respiratory rate to exhale more CO_2
 c) decreasing the respiratory rate to retain more CO_2
 d) increasing the respiratory rate to retain more CO_2

21. To compensate for acidosis, the kidneys will excrete:
 a) more hydrogen ions b) more sodium ions c) fewer hydrogen ions d) more bicarbonate ions

22. A patient with untreated diabetes who is in a state of ketoacidosis will:
 a) be breathing slowly
 b) excrete an alkaline urine
 c) be breathing rapidly
 d) excrete a urine with a pH of 7

23. The hormone that increases renal excretion of sodium ions is:
 a) ADH b) ANP c) ABC d) AND

24. In tissue fluid and plasma, the most abundant cation is:
 a) sodium b) potassium c) calcium d) chloride

25. The intracellular cation that is essential for the repolarization of neurons and muscle cells is:
 a) sodium b) potassium c) calcium d) chloride

MULTIPLE CHOICE TEST #2

Read each question and the four answer choices carefully. When you have made a choice, follow the instructions to complete your answer.

1. Which statement is NOT true of hormone effects on fluid-electrolyte balance?
   a) PTH decreases the absorption of calcium ions by the small intestine.
   b) ADH increases the reabsorption of water by the kidneys.
   c) Aldosterone increases the reabsorption of sodium ions by the kidneys.
   d) ANP increases the excretion of sodium ions by the kidneys.

   Reword your choice to make it a correct statement.

   ____

2. Which statement is NOT true of water compartments?
   a) Water found in arteries and veins is called plasma.
   b) The tissue fluid of the CNS is called cerebrospinal fluid.
   c) The water found within cells is called intercellular fluid.
   d) Intracellular fluid is about two thirds of the total body water.

   Reword your choice to make it a correct statement.

   ____

3. Which statement is NOT true of water intake and output?
   a) The metabolic water produced in cell respiration contributes to daily water intake.
   b) Most water lost from the body is in the form of feces.
   c) Most water intake is in the form of beverages.
   d) The loss of water in sweat varies with the amount of daily exercise.

   Reword your choice to make it a correct statement.

   ____

4. Which statement is NOT true of the regulation of the water content of the body?
   a) The osmoreceptors that detect changes in the water content of body fluids are in the hypothalamus.
   b) A state of dehydration will bring about a decrease in urinary output.
   c) The hormones that have the greatest effect on the water content of the body are aldosterone and PTH.
   d) Strenuous exercise that increases sweating will result in a sensation of thirst.

   Reword your choice to make it a correct statement.

   ____

5. Which statement is NOT true of electrolytes?
   a) Anions are negative ions such as chloride ions.
   b) In water, electrolytes dissociate into their positive and negative ions.
   c) Some electrolytes are part of structural components of the body such as bones.
   d) The electrolyte concentration in body fluids helps regulate the process of filtration.

   Reword your choice to make it a correct statement.

   ____

6. Which statement is NOT true of buffer systems?
   a) The phosphate buffer system is one of the renal mechanisms for regulation of pH.
   b) A buffer system consists of a weak acid and a weak base.
   c) The buffer systems work very slowly to correct pH imbalances.
   d) The bicarbonate buffer system is important in extracellular fluid.

   Reword your choice to make it a correct statement.

   ____

7. Which statement is NOT true of the respiratory system and acid-base balance?
   a) If more CO_2 is retained in body fluids, more hydrogen ions will be formed.
   b) Respiratory compensation for metabolic acidosis involves decreasing respirations.
   c) Respiratory acidosis may be the result of a severe pulmonary disease.
   d) The respiratory system begins to compensate for a pH imbalance within a few minutes.

   Reword your choice to make it a correct statement.

   ___

8. Which statement is NOT true of the kidneys and acid-base balance?
   a) The kidneys begin to compensate for a pH imbalance within a few seconds to minutes.
   b) To compensate for acidosis, the kidneys excrete more hydrogen ions.
   c) To compensate for alkalosis, the kidneys retain more hydrogen ions within the body.
   d) The kidneys have the greatest capacity to compensate for an ongoing pH imbalance.

   Reword your choice to make it a correct statement.

   ___

9. Which statement is NOT true of the effects of pH imbalances?
   a) Alkalosis affects peripheral neurons and causes muscle spasms.
   b) Untreated alkalosis may progress to convulsions.
   c) Acidosis affects the CNS and causes confusion.
   d) Untreated acidosis may progress to hyperactivity.

   Reword your choice to make it a correct statement.

   ___

10. Which statement is NOT true of the functions of minerals?
    a) Iron is part of hemoglobin and myoglobin.
    b) Calcium is necessary for blood clotting.
    c) Phosphorus is part of DNA and RNA.
    d) Cobalt is part of the hormone thyroxine.

    Reword your choice to make it a correct statement.

    ___

MULTIPLE CHOICE TEST #3

Each question is a series of statements concerning a topic in this chapter. Read each statement carefully and select all of the correct statements.

1. Which of the following statements are true of fluid and electrolyte balance?
   a) A person's greatest daily water loss is most often by way of sweating.
   b) Interstitial fluid is the largest water compartment and intracellular fluid is the smallest.
   c) Chloride is the major anion in intracellular fluid, and sodium is the major cation.
   d) The retention of Na^+ ions is increased by aldosterone, which also increases the excretion of K^+ ions.
   e) Sulfur is needed for some proteins, and phosphorus is needed for DNA and RNA.
   f) ANP, secreted by the atria of the heart, increases the excretion of Na^+ ions in urine.
   g) A drop in blood pressure caused by severe hemorrhage will stimulate the secretion of both ADH and aldosterone.
   h) ADH, secreted by the posterior pituitary, increases the absorption of water by the small intestine.
   i) By the process of osmosis, and dependent on the electrolytes present, water enters and leaves cells.
   j) Metabolic water is constantly produced, but it is the smallest water intake volume.
   k) Electrolytes such as magnesium and calcium can be stored in bones.
   l) The liver can store electrolytes such as sodium and potassium.

2. Which of the following statements are true of acid-base balance?
   a) An amino acid acts as a base when its amino group picks up an excess H^+ ion.
   b) Anything that decreases the respiratory rate is a cause of respiratory alkalosis.
   c) The kidneys have the greatest capacity to buffer an ongoing pH change.
   d) If the pH of the blood is 7.2, the blood is slightly alkaline but is physiologically too acidic.
   e) The carbonic buffer system responds to increasing acidity by producing more carbonic acid, a weak acid.
   f) Confusion sometimes leading to coma is a symptom of alkalosis.
   g) The more CO_2 in the body fluids, the lower the pH of the fluid.
   h) Proteins are the most important buffer system in tissue fluid.
   i) The respiratory system responds to metabolic alkalosis by increasing the rate and depth of respiration to exhale more CO_2.
   j) The phosphate buffer system is one of the mechanisms used by the kidneys.
   k) Cells of the renal tubules are able to obtain bicarbonate ions from the reaction of CO_2 and water.
   l) The kidneys are the only organs that can remove H^+ ions from the body to counteract a decreasing pH of body fluids.

Chapter 20

The Reproductive Systems

This chapter describes the male and female reproductive systems, which produce the gametes that unite in fertilization to become a new individual. The female reproductive system has yet another function, that of providing a site for the developing embryo-fetus.

MEIOSIS

1. Meiosis is the cell division process that produces gametes, which are ________________ and ________________.

2. a) The process of meiosis begins with one cell that has the ________________ number of chromosomes, which is ________________ for people.
   b) In meiosis, one cell divides ________________ to form ________________ new cells, each with the ________________ number of chromosomes, which is ________________ for humans.

3. a) When meiosis takes place in the ovaries, the process is called ________________.
   b) When meiosis takes place in the testes, the process is called ________________.

4. Match each process of meiosis with the proper descriptive statements.

   Use each letter once. Each answer line will have four correct letters.

   1) Spermatogenesis ________________________
   2) Oogenesis ________________________

   A. Begins at puberty and ends at menopause
   B. Begins at puberty and continues throughout life
   C. Requires the hormones FSH and testosterone
   D. Is a continuous process
   E. Requires the hormones FSH and estrogen
   F. Is a cyclical process
   G. Four functional cells are produced
   H. Only one functional cell is produced

5. a) If fertilization occurs, the ________________ chromosomes of the sperm and the ________________ chromosomes of the egg will give the fertilized egg the diploid number of chromosomes, which is ________________.
   b) State the term used for a fertilized egg. ________________

6. The following diagrams depict spermatogenesis (left) and oogenesis (right).

   Label the cells indicated.

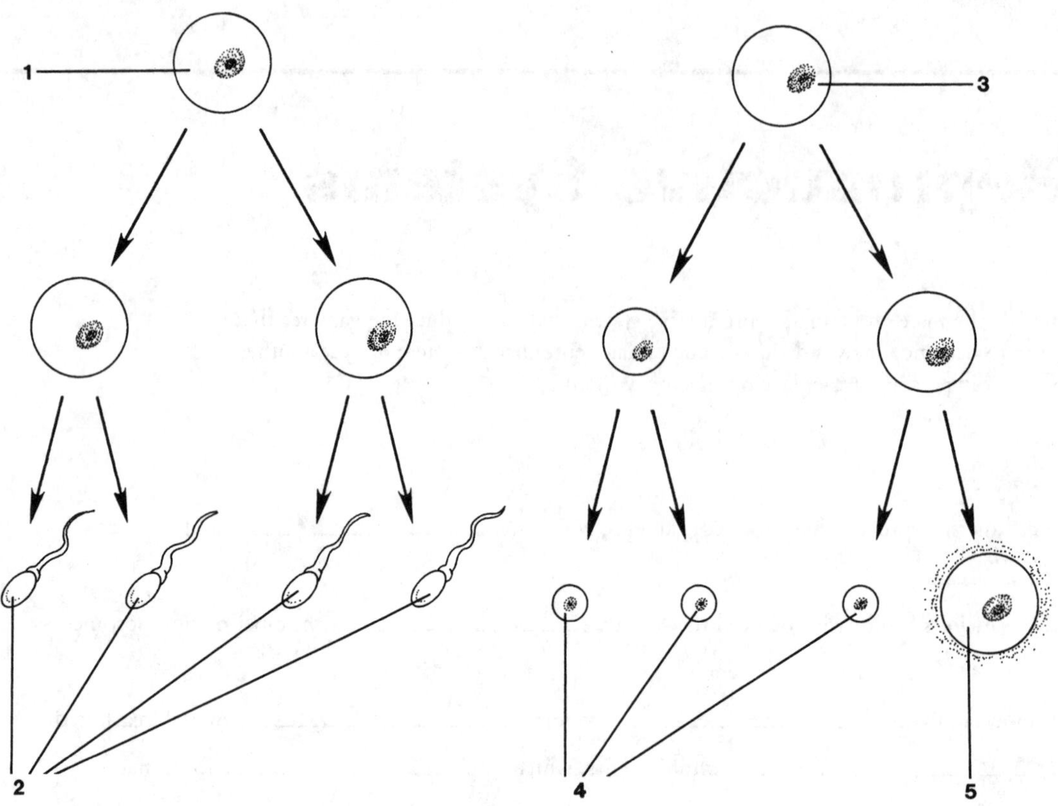

MALE REPRODUCTIVE SYSTEM

Testes and Sperm

1. a) The testes are suspended outside the abdominal cavity within the ____________________.

   b) Viable sperm are produced only when the temperature of the testes is slightly ____________________ than body temperature.

2. Within each testis, the interstitial cells produce the hormone ____________________, and the process of ____________________ takes place in the seminiferous tubules.

3. a) The stimulus for secretion of testosterone is the hormone ____________________ from the ____________________ gland.

   b) Testosterone is responsible for ____________________ of sperm.

   c) Testosterone also regulates the development of the male secondary sex characteristics. State two of these characteristics. ____________________ and ____________________

4. The following diagram depicts a sperm cell.

Label the parts indicated and name the part with each of these functions.

a) Provides motility ____________________

b) Contains the 23 chromosomes ____________________

c) Contains mitochondria that produce ATP ____________________

d) Contains enzymes to digest the membrane of the egg cell ____________________

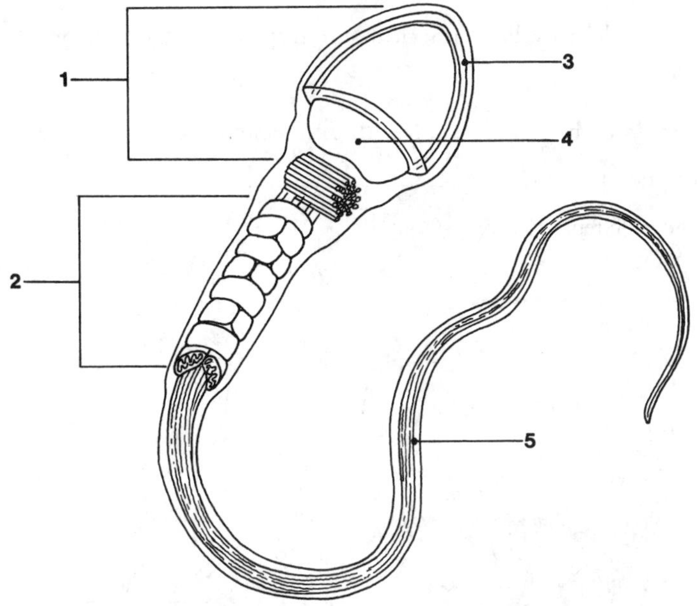

5. The following diagram depicts sections through a testis.

Label the parts indicated.

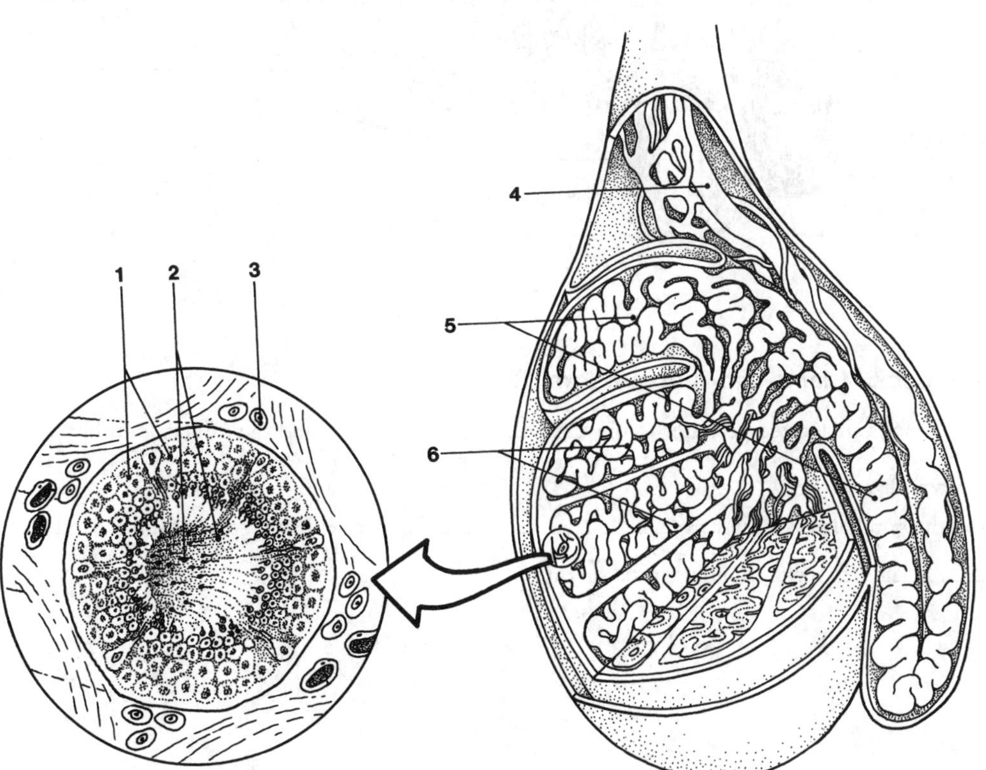

Epididymis, Ductus Deferens, and Ejaculatory Duct

1. a) The epididymis is a long tube that is coiled on the ____________ side of the testis.

   b) Within the epididymis, sperm complete their maturation, and their ____________ become functional.

2. a) The ductus deferens is also called the ____________ and carries sperm from the ____________ to the ____________.

   b) The ductus deferens enters the abdominal cavity through the ____________ canal and then curves over and down behind the ____________.

   c) The tissue in the wall of the ductus deferens that contracts to propel sperm to the ejaculatory duct is ____________.

3. The ejaculatory duct receives sperm from the ____________ and the secretion of the ____________ and empties these into the ____________.

4. The following diagram depicts the male reproductive system.

   Label the parts indicated.

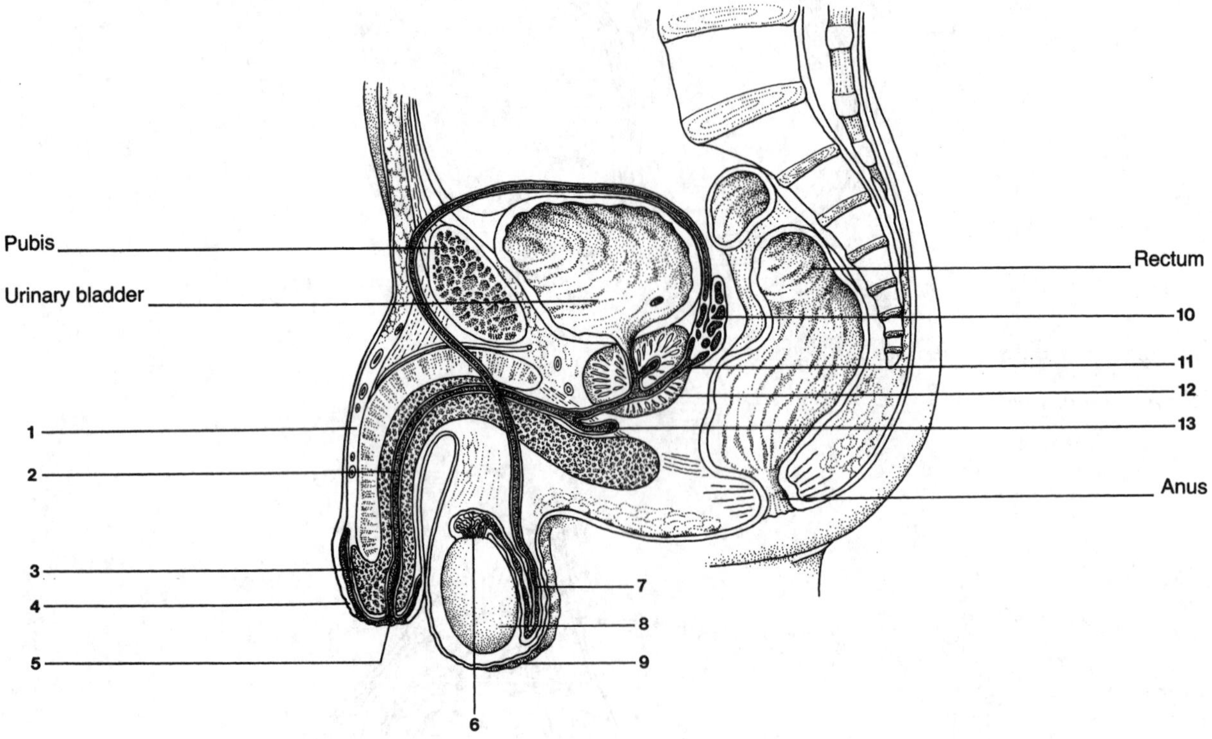

Seminal Vesicles, Prostate Gland, and Bulbourethral Glands

1. a) The two seminal vesicles are located posterior to the ____________, and their ducts join the ____________.

   b) The secretion of the seminal vesicles has an ____________ pH to enhance sperm motility and contains ____________ to nourish the sperm.

2. a) The prostate gland surrounds the first inch of the ____________________, just below the ____________________.

b) The slightly acidic secretion of the prostate gland contains citric acid, which is important for ____________________ (process) in the ____________________ (organelle) of sperm.

c) The ____________________ muscle of the prostate gland contracts as part of ____________________ to expel semen from the urethra.

3. a) The bulbourethral glands are also called ____________________.

b) The secretion of these glands has an ____________________ pH and is secreted into the ____________________ just before ejaculation.

4. The alkaline secretions of the male glands are important to neutralize the ____________________ pH of the female ____________________.

5. a) Semen consists of these secretions and ____________________.

b) The average pH of semen is ____________________.

Urethra and Penis

1. a) The urethra carries semen from the ____________________ to the exterior.

b) The first inch of the urethra is within the ____________________, and the longest portion of the urethra is within the ____________________.

2. The three masses of cavernous tissue within the penis consist of ____________________ muscle and connective tissue that contains large vascular channels called ____________________.

3. During sexual stimulation, ____________________ impulses cause dilation of the penile arteries, and the ____________________ fill with blood, resulting in an erection.

4. The erect penis is capable of penetrating the female vagina to deposit semen. The expulsion of semen from the urethra is called ____________________ and is brought about by contraction of the ____________________ gland and peristalsis of the ____________________.

Male Reproductive Hormones

1. a) Sperm production in the testes is initiated by the hormone ____________________ from the ____________________ gland.

b) The maturation of sperm requires the hormone ____________________ from the testes.

2. The secretion of testosterone is stimulated by the hormone ____________________ from the ____________________ gland.

3. The hormone inhibin is secreted by the ____________________, and its function is to decrease the secretion of ____________________.

4. Secretion of FSH and LH is stimulated by ____________________ from the ____________________.

FEMALE REPRODUCTIVE SYSTEM

Ovaries

1. The ovaries are located in the ____________________ cavity on either side of the ____________________.
2. The thousands of primary follicles in each ovary contain potential ____________________.
3. a) The growth of ovarian follicles is stimulated by the hormone ____________________ from the ____________________ gland.

   b) This hormone also stimulates the follicle cells to secrete the hormone ____________________.
4. A graafian follicle is a ____________________ follicle that ruptures and releases its ovum when stimulated by the hormone ____________________ from the ____________________ gland.
5. The hormone LH causes the ruptured follicle to become the ____________________, which then begins to secrete the hormone ____________________, as well as estrogen.
6. The hormone inhibin is also secreted by the ovary and decreases the secretion of ____________________.
7. a) The hormone relaxin is secreted by the ____________________.

   b) Relaxin inhibits contractions of the ____________________, which will help facilitate implantation.
8. Secretion of FSH and LH is stimulated by ____________________ from the ____________________.
9. The following diagram depicts the female reproductive system.

   Label the parts indicated.

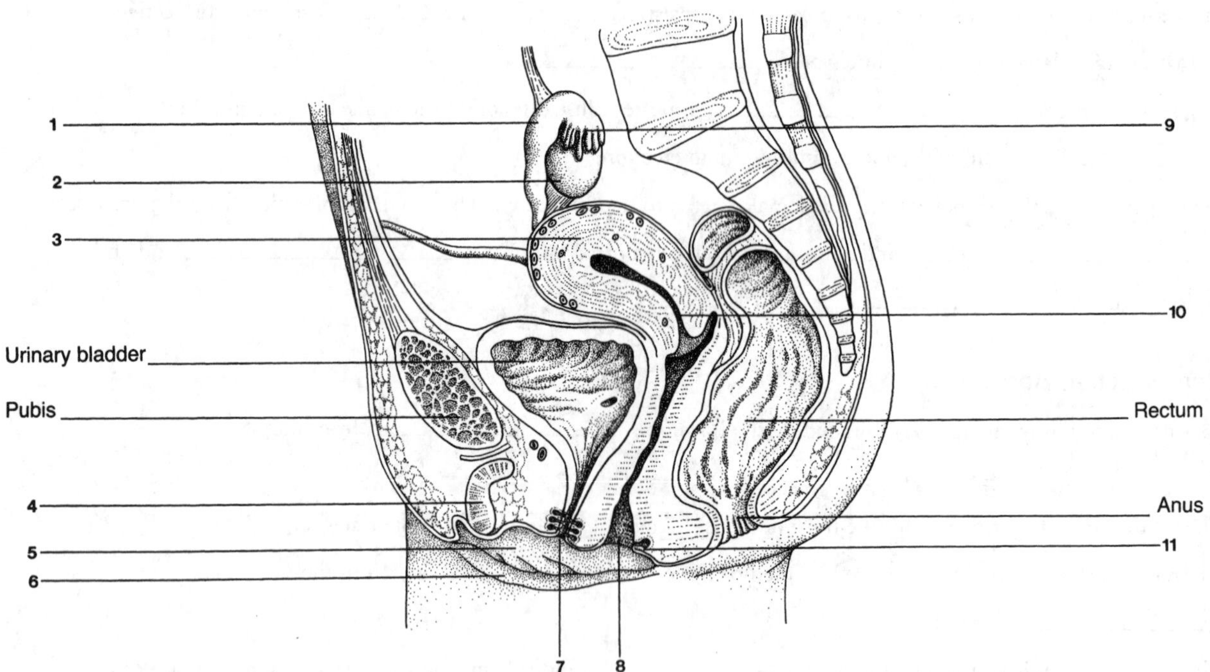

Fallopian Tubes

1. The fallopian tubes may also be called __________________ or __________________.
2. a) The lateral end of a fallopian tube encloses the __________________.
   b) The fringe like projections of this end of the tube are called __________________ and help pull the __________________ into the tube.
3. To propel the ovum toward the uterus, the __________________ layer of the fallopian tube contracts in peristaltic waves, and the __________________ epithelium sweeps the ovum.
4. The union of sperm and egg is called __________________ and usually takes place in the fallopian tube.
5. If a zygote does not reach the uterus but continues to develop within the fallopian tube, the pregnancy is called an __________________ pregnancy.

Uterus

1. The uterus is located in the __________________ cavity, medial to the __________________ and superior to the __________________.
2. The body of the uterus is the large central portion. Above the body is the __________________ of the uterus, and the lower portion that opens into the vagina is the __________________.
3. a) The epimetrium is a fold of the __________________ that covers the upper surface of the uterus.
   b) The smooth muscle layer of the uterus is called the __________________.
   c) The lining of the uterus is called the __________________.
4. a) The endometrium consists of two layers. The __________________ layer is the permanent, thin vascular layer next to the myometrium, and the __________________ layer is regenerated and lost in each menstrual cycle.
   b) Name the two hormones that promote the growth of blood vessels in the functional layer.
   __________________ and __________________
5. a) What is the function of the endometrium? ____
   b) What is the function of the myometrium? ____
   c) Name the two hormones that inhibit contraction of the myometrium during pregnancy. __________________ and __________________
   d) Name the hormone that stimulates strong contractions of the myometrium at the end of pregnancy.
   ____

6. The following diagram depicts the female reproductive organs.

   Label the parts indicated.

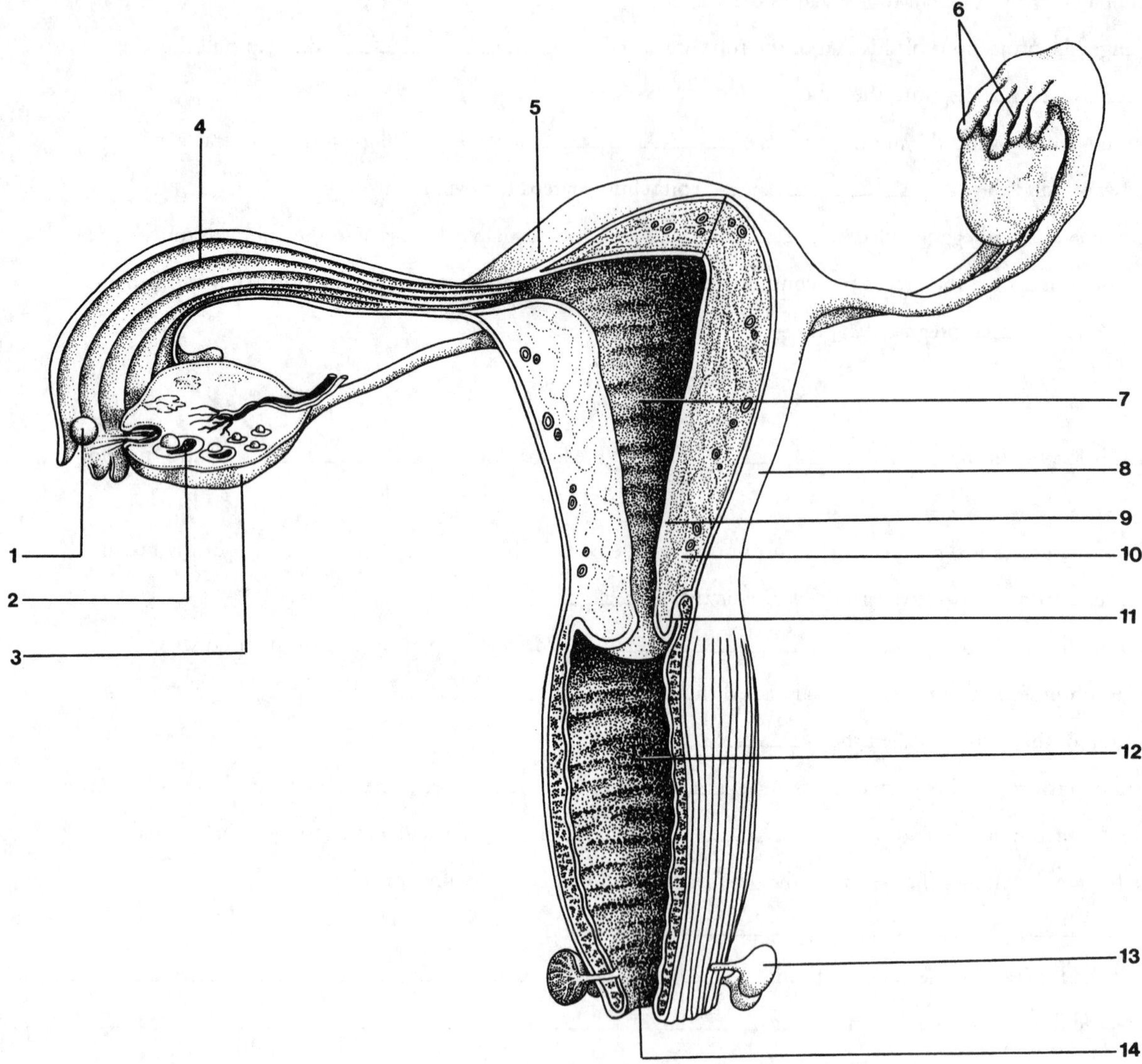

Vagina and Vulva

Match each structure with the proper descriptive statements.

Use each letter once. One answer line will have three correct letters.

1) Vagina ______________________
2) Clitoris ______________________
3) Labia majora and minora ______________________
4) Bartholin's glands ______________________
5) Vestibule ______________________

A. The secretion keeps the mucosa of the vestibule moist
B. Receives sperm from the penis during sexual intercourse
C. Paired folds that cover the vestibule to prevent drying of the mucous membranes
D. The birth canal and the exit for menstrual blood
E. A small mass of erectile tissue that responds to sexual stimulation
F. The area that contains the urethral and vaginal openings
G. Contains bacteria (normal flora) that maintain an acidic pH that inhibits the growth of pathogens

Mammary Glands

1. a) The alveolar glands of the mammary glands produce ____________ after pregnancy.

   b) Milk enters the ____________ ducts, which converge and empty at the ____________ of the breast.

2. During pregnancy, the alveolar glands are prepared for milk production by the hormones ____________ and ____________, which are secreted by the ____________.

3. a) After pregnancy, the hormone ____________ from the ____________ gland stimulates production of milk.

   b) The hormone responsible for the release of milk from the mammary glands is ____________ from the ____________ gland.

4. a) The pigmented area around the nipple of the breast is called the ____________.

   b) Within the breast, the alveolar glands are surrounded by ____________ tissue.

5. a) Milk contains fatty acids and the sugar ____________ for nourishment of the infant.

   b) The ____________ present in milk provide immunity for the infant to the diseases the mother is immune to.

   c) This type of immunity is ____________ (active or passive).

   d) The oligosaccharides in milk are for nourishment of the ____________ in the ____________ (location).

6. The following diagram depicts the mammary gland.

   Label the parts indicated.

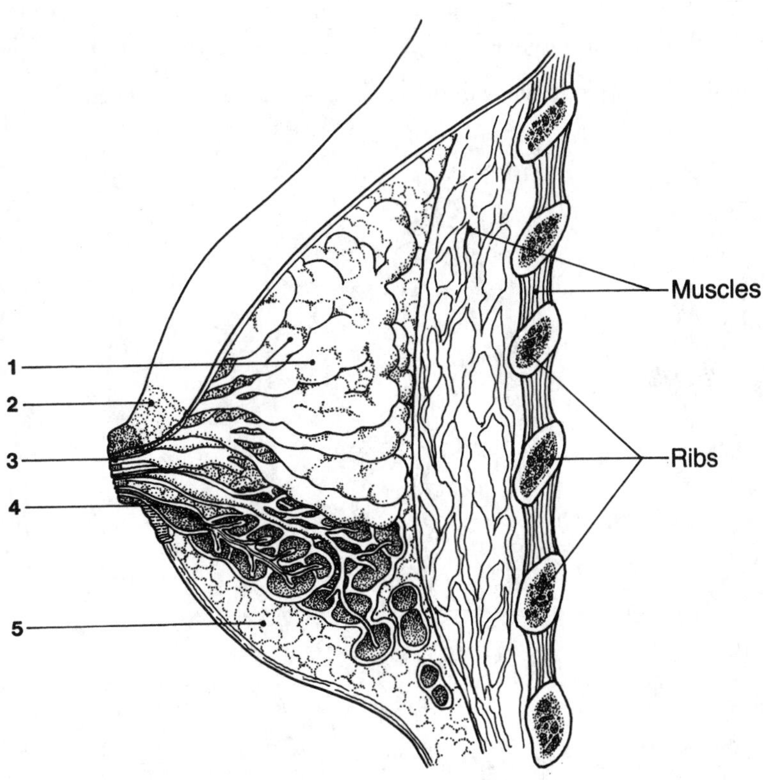

The Menstrual Cycle

1. The menstrual cycle requires hormones from the ____________________ and ____________________ and involves changes in the ____________________ and ____________________ (organs). The average menstrual cycle is approximately ____________________ days.

2. The hormones involved in the menstrual cycle are ____________________ and ____________________ from the anterior pituitary gland, ____________________ from the ovarian follicle, and ____________________ from the corpus luteum.

3. The three phases of the menstrual cycle are the ____________________ phase, the ____________________ phase, and the ____________________ phase.

4. a) The menstrual phase begins with the loss of the ____________________ layer of the endometrium; this is called ____________________.

   b) During this phase, the hormone ____________________ stimulates the growth of several ovarian follicles.

5. a) During the follicular phase, the hormones ____________________ and ____________________ stimulate the growth and maturation of the ovum.

   b) The growth of blood vessels in the endometrium is stimulated by the hormone ____________________.

   c) In ovulation, the hormone ____________________ causes a mature ovarian follicle to ____________________ and release its ovum.

6. a) During the luteal phase, the hormone ____________________ causes the ruptured follicle to become the ____________________ and secrete the hormone ____________________.

   b) The hormone ____________________ stimulates further growth of blood vessels in the endometrium and promotes the storage of ____________________.

7. If the ovum is not fertilized, a decrease in the hormone ____________________ causes the loss of the functional layer of the endometrium, and the cycle begins again with the ____________________ phase.

CROSSWORD PUZZLE

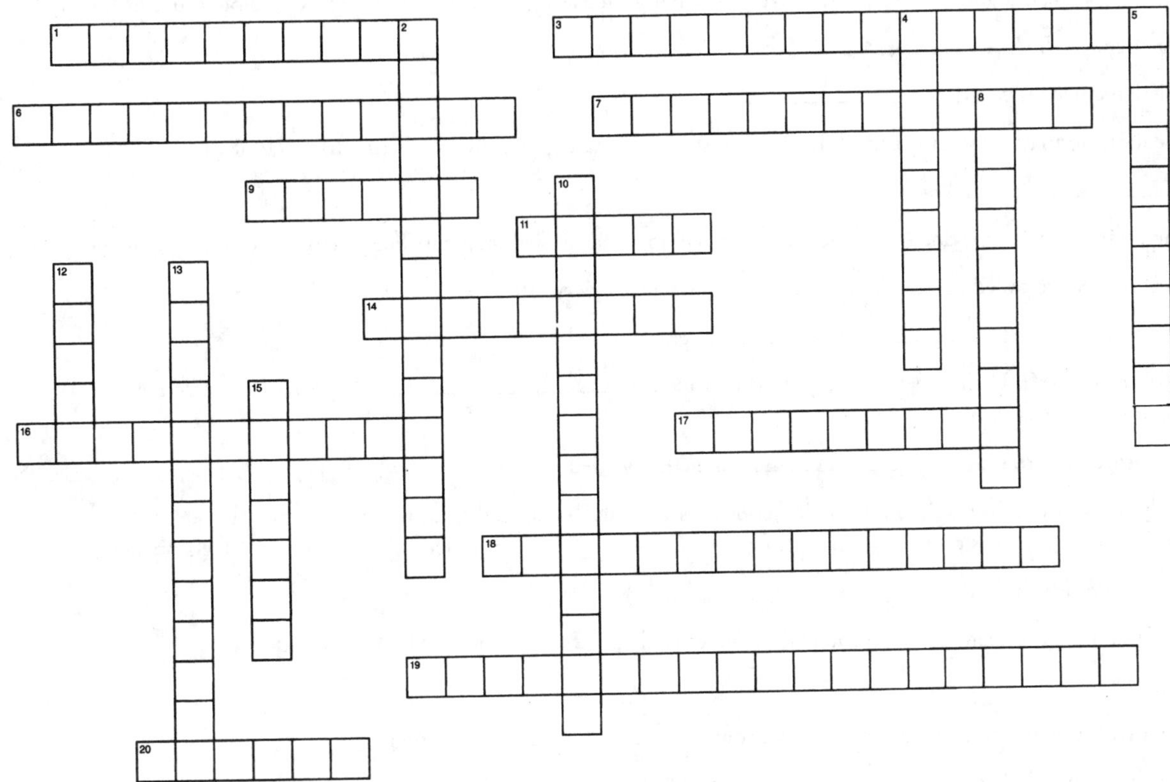

ACROSS

1. Smooth muscle layer of the uterus
3. Mature ovarian follicle (two words)
6. Opening in the abdominal wall for the spermatic cord (two words)
7. Muscular gland just below the urinary bladder in males (two words)
9. A fertilized egg
11. Suturing and severing of the fallopian tubes is called a __________ ligation
14. The vas deferens is sutured and cut
16. X-ray technique used to evaluate breast tissue for abnormalities
17. Cessation of the menstrual cycle (at ages 45 to 55)
18. Process of meiosis as it takes place in the testes
19. Contain sperm-generating cells (two words)
20. Narrow, lower end of the uterus

DOWN

2. Average is 28 days (two words)
4. Process of meiosis as it takes place in the ovaries
5. Lining of the uterus
8. Cessation of menses (during childbearing years)
10. Vas deferens (synonym) (two words)
12. Female external genital structures
13. Its structure ensures that the ovum will be kept moving toward the uterus (two words)
15. Down syndrome is the most common example

CLINICAL APPLICATIONS

1. a) Mr. C is 68 years old and tells his doctor that he is having increasing difficulty urinating. A possible cause of this is enlargement of the ______________ gland.
   b) This enlargement is called ______________.
   c) This condition may make urination difficult because the prostate gland surrounds the first inch of the ______________.
2. a) Mr. and Mrs. D ask their doctor for the most effective method of birth control. The doctor tells Mrs. D that the surgical procedure called ______________ would be very effective.
   b) In this procedure, the ______________ are cut and tied.
   c) The doctor tells Mr. D that the surgical procedure called a ______________ is also a very effective method of contraception.
   d) In this procedure, the ______________ are cut and tied off.
3. a) Sexually transmitted diseases (STDs) are infectious diseases that are spread from person to person by sexual activity. The STD that is caused by a bacterium and, if untreated, may have serious or even fatal effects on the heart or nervous system is ______________.
   b) One of the most prevalent STDs in the United States is chlamydial infection; chlamydia is a type of ______________.
   c) Women with this infection are at greater risk for an ______________ pregnancy.
   d) The infants born to these infected women are at risk for developing ______________ or ______________.
   e) The STD genital herpes is caused by a ______________.
   f) Newborns who acquire this herpes infection from their mothers during birth are at higher risk for damage to the ______________ system, or even death.
4. a) Breast cancer can often be successfully treated if detected early. A detection method that women can do themselves is ______________.
   b) A procedure that can detect even very small tumors, and that is recommended for older women, is ______________.

MULTIPLE CHOICE TEST #1

Choose the correct answer for each question.

1. The process of meiosis produces __________ cells, each with the __________ number of chromosomes.
   a) 4/haploid b) 2/haploid c) 4/diploid d) 2/diploid
2. The process of spermatogenesis produces:
   a) sperm cells, in cycles of 28 days
   b) sperm cells, from puberty to age 50
   c) only one functional sperm cell and three polar bodies
   d) sperm cells, from puberty throughout life
3. The process of oogenesis produces:
   a) an egg cell with the diploid number of chromosomes
   b) an egg cell approximately every 28 days
   c) egg cells and is stimulated by the hormones estrogen and progesterone
   d) egg cells from puberty throughout life

4. The hormones directly necessary for sperm production are:
   a) FSH and LH
   b) FSH and testosterone
   c) testosterone and inhibin
   d) LH and testosterone

5. The hormones directly necessary for egg cell production are:
   a) FSH and estrogen
   b) FSH and LH
   c) LH and progesterone
   d) estrogen and progesterone

6. The hormones directly necessary for the growth of blood vessels in the endometrium are:
   a) estrogen and progesterone
   b) FSH and estrogen
   c) FSH and relaxin
   d) LH and progesterone

7. The male reproductive duct that carries sperm from the epididymis into the abdominal cavity is the:
   a) ductus deferens b) urethra c) inguinal duct d) ejaculatory duct

8. The part of a sperm cell that contains the 23 chromosomes is the:
   a) flagellum b) acrosome c) middle piece d) head

9. The male reproductive gland that contributes to ejaculation is the:
   a) seminal vesicle b) Cowper's gland c) prostate gland d) bulbourethral gland

10. The male reproductive duct that carries semen through the penis to the exterior is the:
    a) ejaculatory duct b) urethra c) epididymis d) ductus deferens

11. The parts of the testes that produce sperm are the:
    a) interstitial cells b) rete testis c) seminiferous tubules d) epididymides

12. The alkaline fluid of semen is important to neutralize the:
    a) acidic pH of the uterus
    b) acidic pH of the vagina
    c) alkaline pH of the vagina
    d) alkaline pH of the fallopian tube

13. The expulsion of semen from the urethra is called:
    a) secretion b) ejaculation c) erection d) excretion

14. The layer of the uterus that will become the maternal portion of the placenta is the:
    a) epimetrium b) myometrium c) endometrium d) serosa

15. The tissues of the fallopian tube that propel the ovum toward the uterus are:
    a) striated muscle and ciliated epithelium
    b) smooth muscle and squamous epithelium
    c) striated muscle and cuboidal epithelium
    d) smooth muscle and ciliated epithelium

16. The ovum matures and the endometrium develops during this phase of the menstrual cycle:
    a) the menstrual phase b) the luteal phase c) the follicular phase d) the ovarian phase

17. The vascular layer of the endometrium that is not lost in menstruation is the:
    a) basilar layer b) myometrial layer c) functional layer d) nonfunctional layer

18. One function of the hormone LH is to:
    a) cause rupture of the corpus luteum
    b) cause rupture of a graafian follicle
    c) cause a graafian follicle to secrete progesterone
    d) cause the corpus luteum to become a follicle

19. After pregnancy, the mammary glands are stimulated to produce milk by the hormone:
    a) estrogen b) progesterone c) oxytocin d) prolactin

20. The parts of the vulva that cover the urethral and vaginal openings are the:
    a) clitoris and Bartholin's glands
    b) labia majora and minora
    c) labia minora and clitoris
    d) labia majora and Bartholin's glands

21. During the luteal phase of the menstrual cycle, the ruptured ovarian follicle becomes the:
    a) graafian follicle b) primary follicle c) corpus albicans d) corpus luteum

22. During the menstrual phase of the menstrual cycle, the functional layer of the endometrium is:
    a) regenerated by the basilar layer
    b) lost in menstruation
    c) stimulated to grow by estrogen
    d) stimulated to grow by progesterone

23. The site of fertilization is usually the:
    a) fallopian tube b) vagina c) uterus d) ovary

24. A zygote is a:
 a) fertilized egg with 46 chromosomes
 b) mature ovum with 23 chromosomes
 c) mature ovum with 46 chromosomes
 d) fertilized egg with 23 chromosomes

25. The hormone that stimulates release of milk from the mammary glands is:
 a) estrogen
 b) progesterone
 c) oxytocin
 d) prolactin

MULTIPLE CHOICE TEST #2

Read each question and the four answer choices carefully. When you have made a choice, follow the instructions to complete your answer.

1. Which statement is NOT true of the locations of the female reproductive organs?
 a) The uterus is medial to the ovaries and inferior to the urinary bladder.
 b) The ducts of Bartholin's glands open into the vaginal orifice.
 c) The fallopian tubes extend from the ovaries to the uterus.
 d) The vagina is anterior to the rectum and posterior to the urethra.

 Reword your choice to make it a correct statement.

 ____

2. Which statement is NOT true of the locations of the male reproductive organs?
 a) The ductus deferens enters the abdominal cavity through the inguinal canal.
 b) The testes are located in the scrotum medial to the upper thighs.
 c) The urethra extends through the seminal vesicles and penis.
 d) The epididymis is on the posterior side of a testis.

 Reword your choice to make it a correct statement.

 ____

3. Which statement is NOT true of spermatogenesis and oogenesis?
 a) Oogenesis is a cyclical process, and spermatogenesis is a continuous process.
 b) Both processes begin at puberty, but spermatogenesis ends at menopause.
 c) Spermatogenesis requires the hormones FSH and testosterone.
 d) Oogenesis requires the hormones FSH and estrogen.

 Reword your choice to make it a correct statement.

 ____

4. Which statement is NOT true of the male reproductive ducts?
 a) The urethra carries semen to the exterior.
 b) The epididymis carries sperm from the testis to the ductus deferens.
 c) The ejaculatory duct carries sperm to the urethra.
 d) The ductus deferens carries sperm from the urethra to the ejaculatory duct.

 Reword your choice to make it a correct statement.

 ____

5. Which statement is NOT true of the male reproductive glands?
 a) The secretion of the prostate gland enters the urethra and supplies an energy source for sperm.
 b) The secretion of the seminal vesicles is alkaline and contains glycogen.
 c) The bulbourethral glands secrete an alkaline fluid into the urethra.
 d) The prostate gland also contributes to ejaculation of semen.

 Reword your choice to make it a correct statement.

 ____

6. Which statement is NOT true of the menstrual cycle?
   a) The follicular phase ends with ovulation.
   b) The menstrual phase begins with the loss of the functional layer of the endometrium.
   c) During the luteal phase, the corpus luteum begins to secrete progesterone.
   d) The growth of blood vessels in the endometrium begins in the menstrual phase.

   Reword your choice to make it a correct statement.

   ___

7. Which statement is NOT true of the uterus?
   a) The basilar layer regenerates the functional layer during each menstrual cycle.
   b) The cervix is the narrow superior part that opens into the vagina.
   c) The myometrium contracts for labor and delivery of the baby.
   d) The fundus is the upper part between the openings of the fallopian tubes.

   Reword your choice to make it a correct statement.

   ___

8. Which statement is NOT true of the mammary glands?
   a) The lactiferous ducts carry milk to the nipple.
   b) Milk production is stimulated by the hormone prolactin.
   c) The alveolar glands produce milk after pregnancy.
   d) The release of milk is stimulated by the hormone estrogen.

   Reword your choice to make it a correct statement.

   ___

9. Which statement is NOT true of the ovaries and fallopian tubes?
   a) Fimbriae are on the end of the fallopian tube that encloses the ovary.
   b) Fertilization of the ovum usually takes place in the fallopian tube.
   c) An ectopic pregnancy occurs when the zygote becomes implanted in the uterus.
   d) The ovum is swept toward the uterus by ciliated epithelium that lines the fallopian tube.

   Reword your choice to make it a correct statement.

   ___

10. Which statement is NOT true of the testes?
    a) To produce viable sperm, the temperature of the scrotum must be slightly higher than body temperature.
    b) Spermatogenesis takes place in the seminiferous tubules.
    c) The interstitial cells produce testosterone when stimulated by LH.
    d) Two of the target organs of testosterone are the testes themselves.

    Reword your choice to make it a correct statement.

    ___

11. Which statement is NOT true of male reproductive hormones?
    a) Testosterone is responsible for the development of male secondary sex characteristics.
    b) Spermatogenesis is initiated by FSH.
    c) Secretion of testosterone is stimulated by LH.
    d) Inhibin decreases the secretion of testosterone.

    Reword your choice to make it a correct statement.

    ___

12. Which statement is NOT true of female reproductive hormones?
    a) Inhibin is produced by the corpus luteum.
    b) Estrogen is responsible for development of female secondary sex characteristics.
    c) Relaxin inhibits contractions of the epimetrium.
    d) Progesterone promotes growth of blood vessels in the endometrium.

    Reword your choice to make it a correct statement.

    ___

MULTIPLE CHOICE TEST #3

Each question is a series of statements concerning a topic in this chapter. Read each statement carefully and select all of the correct statements.

1. Which of the following statements are true of the male reproductive system?
   a) The longest part of the urethra passes through the prostate gland.
   b) The secretion of the seminal vesicles contains fatty acids for the nourishment of sperm.
   c) The interstitial cells of the testes produce testosterone.
   d) The epididymis carries sperm from a testis to the urethra.
   e) The smooth muscle of the prostate gland contributes to the ejaculation of semen.
   f) Dilation of the arteries of the penis is brought about by sympathetic impulses.
   g) The ductus deferens passes through the inguinal canal and empties into a seminal vesicle.
   h) FSH stimulates secretion of testosterone.
   i) Meiosis in the testes is initiated by the hormone LH.
   j) The chromosomes of a sperm cell are found with the mitochondria in the middle piece.

2. Which of the following statements are true of the female reproductive system?
   a) The fallopian tube is lined with flagella to sweep the egg to the uterus.
   b) The myometrium of the uterus is the smooth muscle layer.
   c) Bartholin's glands secrete mucus into the cervix of the uterus.
   d) The uterus is medial to the ovaries and lateral to the fallopian tubes.
   e) The corpus luteum is an endocrine gland that develops from a graafian follicle.
   f) The labia majora and minora cover the urethral and vaginal openings.
   g) At the end of pregnancy, the vagina becomes the birth canal.
   h) After ovulation, the corpus luteum secretes FSH first, then LH.
   i) The mammary glands are stimulated to produce milk by the hormone prolactin from the hypothalamus.
   j) The endometrium is the vascular uterine lining that changes in thickness during each menstrual cycle.

Chapter 21

Human Development and Genetics

This chapter describes the fundamentals of development and genetics. Development is the study of the growth of a zygote from embryo to fetus to functioning human being. This takes place during the 40 weeks of gestation. Genetics is the study of the inheritance of particular characteristics.

HUMAN DEVELOPMENT

Fertilization

1. The ________________ of the sperm cell contains enzymes to digest the membrane of the egg cell to allow the entry of the sperm.
2. a) Fertilization is the union of the nuclei of the ________________ and ________________.
   b) Each of these nuclei contains ________________ chromosomes, and the zygote, therefore, has ________________ chromosomes, the ________________ number for humans.
   c) Fertilization usually takes place within the female ________________.
3. The 23 pairs of chromosomes in the zygote consist of 22 pairs called ________________ and one pair called the ________________.
4. Men have the sex chromosomes called ________________, and women have the sex chromosomes called ________________.

Implantation

1. a) The single-celled zygote divides by the process of ________________ to form two cells.
   b) Further mitotic divisions are called ________________ and produce four cells, and then eight cells, and so on.
2. a) A solid sphere of embryonic cells is called a ________________.
   b) As mitosis continues, the solid sphere becomes hollow and is called the ________________.
3. a) The inner cell mass of the blastocyst contains the potential ________________, and the cells are called ________________.
   b) The outer layer of cells of the blastocyst is called the ________________, and it secretes enzymes to digest the surface of the ________________ to permit ________________ of the blastocyst in the uterus.
4. The trophoblast will eventually become the embryonic membrane called the ________________, which in turn will develop into the fetal portion of the ________________.

5. The following diagram depicts the processes of fertilization through implantation.

Label the parts indicated.

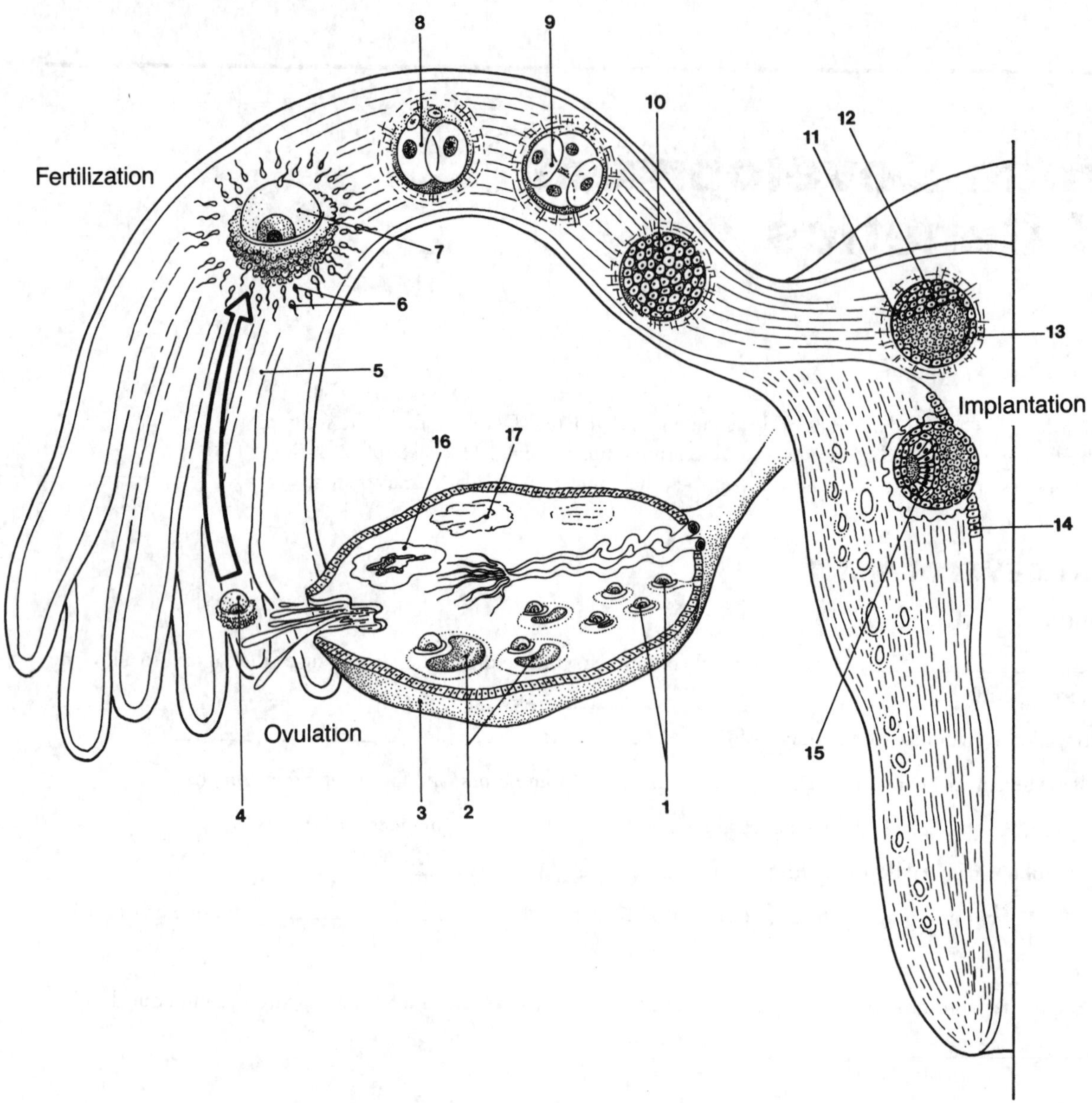

Embryo and Embryonic Membranes

1. About 2 weeks after fertilization, the embryonic disc develops three primary layers of cells. These layers are called the ____________, ____________, and ____________.

2. Name one tissue or part of the body that develops from each primary layer.
   1) Endoderm ____________
   2) Mesoderm ____________
   3) Ectoderm ____________

3. Match each embryonic membrane with the proper descriptive statements.

   Use each letter once. Each answer line will have two correct letters.

   1) Amnion ____________
   2) Chorion ____________
   3) Yolk sac ____________

   A. Contains fluid that absorbs shock around the embryo
   B. Forms the first blood cells for the embryo
   C. Develops small projections called chorionic villi
   D. Contains the fetal blood vessels that become part of the placenta
   E. Contains cells that will become oogonia or spermatogonia
   F. Contains fluid into which fetal urine is excreted

4. a) The period of embryonic growth ends at the ____________ week of gestation.
   b) The period of fetal growth extends from week ____________ to week ____________ of gestation.

5. The following diagrams depict embryos at 2 weeks and 5 weeks of gestation.

Label the parts indicated.

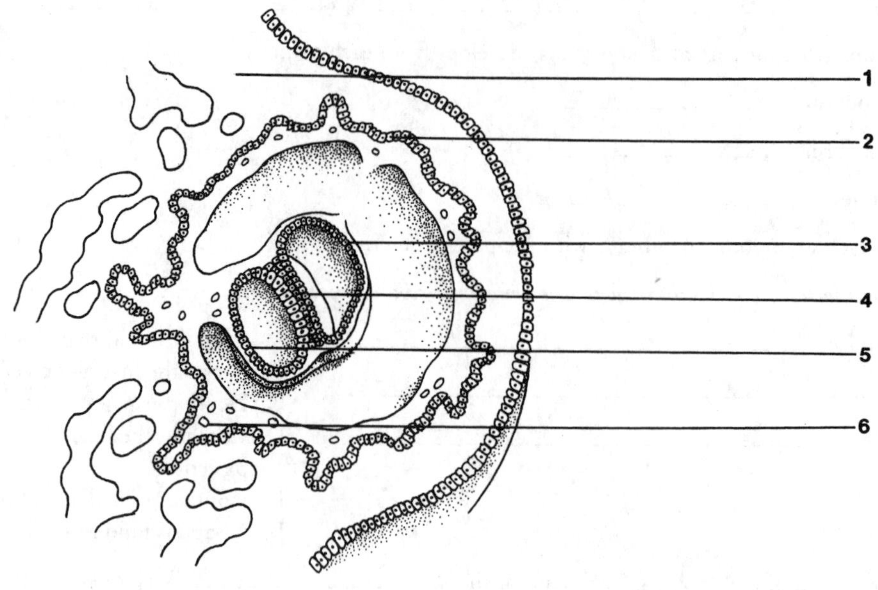

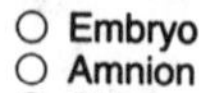

- O Embryo
- O Amnion
- O Chorion
- O Yolk sac

Two weeks

Five weeks

Placenta and Umbilical Cord

1. a) The maternal part of the placenta is formed by the ________________ of the uterus.
   b) The fetal part of the placenta is formed by the ________________.
   c) The fetus is connected to the placenta by the ________________.
   d) It is in the placenta that ________________ of materials take place between fetal blood and maternal blood.

2. In the placenta, fetal blood vessels are within maternal ______________, and there is no mixing of fetal and maternal blood.

3. The umbilical cord contains two umbilical ______________ and one umbilical ______________.

4. a) The umbilical arteries carry blood from the ______________ to the ______________.

   b) This blood contains ______________ and ______________ that diffuse into the maternal blood sinuses for elimination by the mother.

5. a) The umbilical vein carries blood from the ______________ to the ______________.

   b) This blood contains ______________ and ______________ from the maternal blood to be brought to the fetus.

6. After the birth of a baby, the umbilical cord is cut and the placenta is delivered as the ______________.

7. a) The placenta is the source of the hormones necessary to maintain pregnancy. The first hormone secreted is called hCG and is produced by the ______________ of the embryo.

   b) Under the influence of hCG, the ______________ in the maternal ovary continues to secrete the hormones ______________ and ______________.

8. The two most abundant hormones secreted by the placenta itself are ______________ and ______________, and their secretion continues throughout gestation.

9. a) During pregnancy, estrogen and progesterone prevent the development of ovarian follicles by inhibiting the secretion of ______________ and ______________ from the anterior pituitary gland.

   b) Estrogen and progesterone also prepare the ______________ for lactation after the birth of a baby.

   c) Progesterone is very important to prevent contractions of the ______________, which might otherwise result in a miscarriage.

   d) The other placental hormone that inhibits these contractions is ______________.

Parturition and Labor

1. a) The term *parturition* means ______________.

   b) The term *labor* refers to the ______________ that occurs during birth.

2. a) Toward the end of gestation, the myometrium begins to contract weakly because the secretion of ______________ by the placenta begins to decrease.

   b) At this time, how is the fetus usually positioned in the uterus? ______________

3. a) During the first stage of labor, contractions of the uterus force the ______________ into the cervix.

   b) How does this change the diameter of the cervical opening? ______________

   c) What happens to the amniotic sac? ______________

   d) What happens to the amniotic fluid? ______________

4. a) During the second stage of labor, the hormone ______________ causes stronger contractions of the uterus.

   b) These strong contractions result in the delivery of the ______________.

5. a) During the third stage of labor, the uterus continues to contract and delivers the ______________.

   b) Further contractions of the uterus are very important to ______________ the uterine blood vessels to prevent hemorrhage in the mother.

The Infant at Birth

1. a) The baby becomes independent of the mother when the ________________ is cut.

   b) The baby begins to breathe as the blood level of ________________ rises and stimulates the respiratory center in the ________________.

2. a) In the baby's heart, the ________________ is closed as more blood returns to the left atrium.

   b) Just outside the heart, the ________________ constricts.

   c) Both of these changes now permit more blood to flow to the baby's ________________.

3. A newborn baby may have mild jaundice because the ________________ is not yet mature enough to excrete ________________ efficiently.

GENETICS

Chromosomes and Genes

1. a) The human diploid number of chromosomes is ________________.

   b) The chromosomes are actually 23 pairs called ________________ chromosomes.

   c) One member of each pair has come from the father and is called ________________.

   d) The other member of the pair has come from the mother and is called ________________.

   e) The chromosome pairs designated 1 to 22 are called ________________.

   f) The remaining pair is called the ________________, which are XX for ________________ and XY for ________________.

2. a) The hereditary material of chromosomes is ________________.

   b) The genetic code is the sequence of ________________ in the DNA.

   c) A gene is the DNA code for ________________.

3. Match each genetic term with its proper definition.

   Use each letter once.

   1) Alleles ________________
   2) Genotype ________________
   3) Phenotype ________________
   4) Homozygous ________________
   5) Heterozygous ________________
   6) Dominant allele ________________
   7) Recessive allele ________________

   A. The expression of the alleles that are present; the appearance of the individual
   B. Will appear in the phenotype only if two are present in the genotype
   C. Will appear in the phenotype even if only one is present in the genotype
   D. Having two similar alleles (genes) for one trait
   E. The alleles that are present; the genetic makeup
   F. The possibilities for the expression of a gene
   G. Having two different alleles (genes) for one trait

Genetics Problems

1. **Inheritance of dominant-recessive traits**

   1) a) Mom has blond hair (recessive) and Dad has brown hair (dominant). If Dad is homozygous for this trait, what are the possibilities for their children? Use this Punnett square:

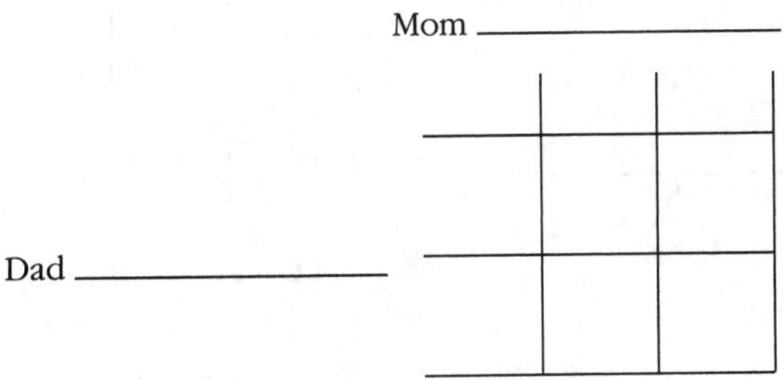

   b) Each child has a ________________ % chance of having brown hair and a ________________ % chance of having blond hair.

   2) a) Mom has blond hair and Dad is heterozygous for brown hair. What are the possibilities for their children? Use this Punnett square:

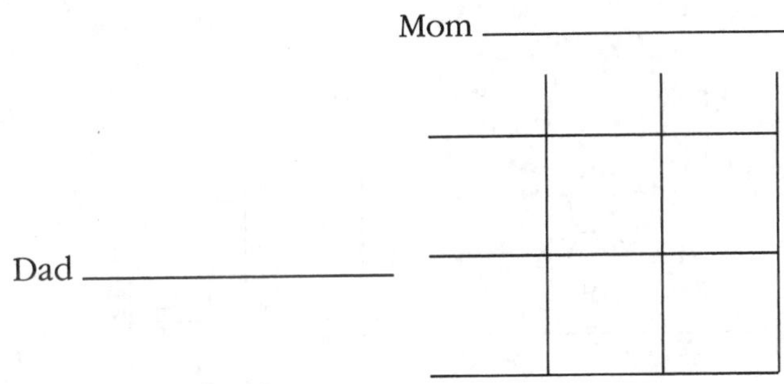

   b) Each child has a ________________ % chance of having brown hair and a ________________ % chance of having blond hair.

2. **Inheritance of blood type**—A and B alleles are co-dominant, and the O allele is recessive.

   1) a) Mom has type O blood and Dad is heterozygous type A. What are the possible blood types for their children? Use this Punnett square:

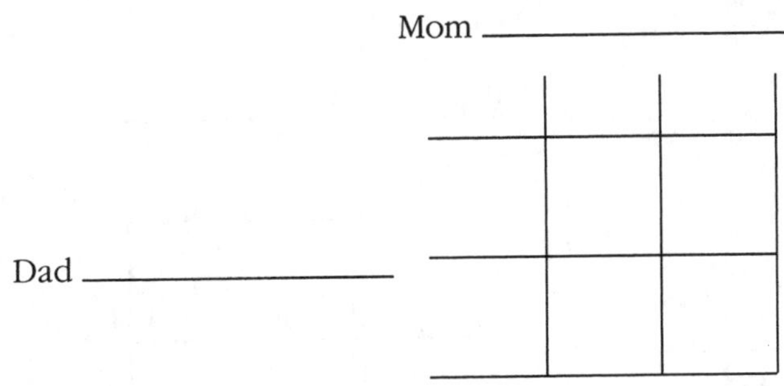

   b) Each child has a ________________ % chance of having type O blood and a ________________ % chance of having type A blood.

2) a) Both Mom and Dad have type AB blood. What are the possible blood types for their children? Use this Punnett square:

Mom ______________

Dad ______________

b) Each child has a ______________ % chance of having type A blood, a ______________ % chance of having type B, a ______________ % chance of having type AB, and a ______________ % chance of having type O blood.

3. **Inheritance of sex-linked traits**—The recessive gene is found only on the X chromosome, and there is no corresponding gene on the Y chromosome.

1) a) Mom is a carrier of red-green color blindness and Dad is red-green color blind. What are the possibilities for their children? Use this Punnett square:

Mom ______________

Dad ______________

b) Each daughter has a ______________ % chance of being color blind and a ______________ % chance of being a carrier of this trait. Each son has a ______________ % chance of being color blind and a ______________ % chance of having normal color vision.

2) a) Mom is red-green color blind and Dad has normal color vision. What are the possibilities for their children? Use this Punnett square:

Mom ______________

Dad ______________

b) Each daughter has a ______________ % chance of being color blind and a ______________ % chance of being a carrier of this trait. Each son has a ______________ % chance of being color blind and a ______________ % chance of having normal color vision.

CROSSWORD PUZZLE

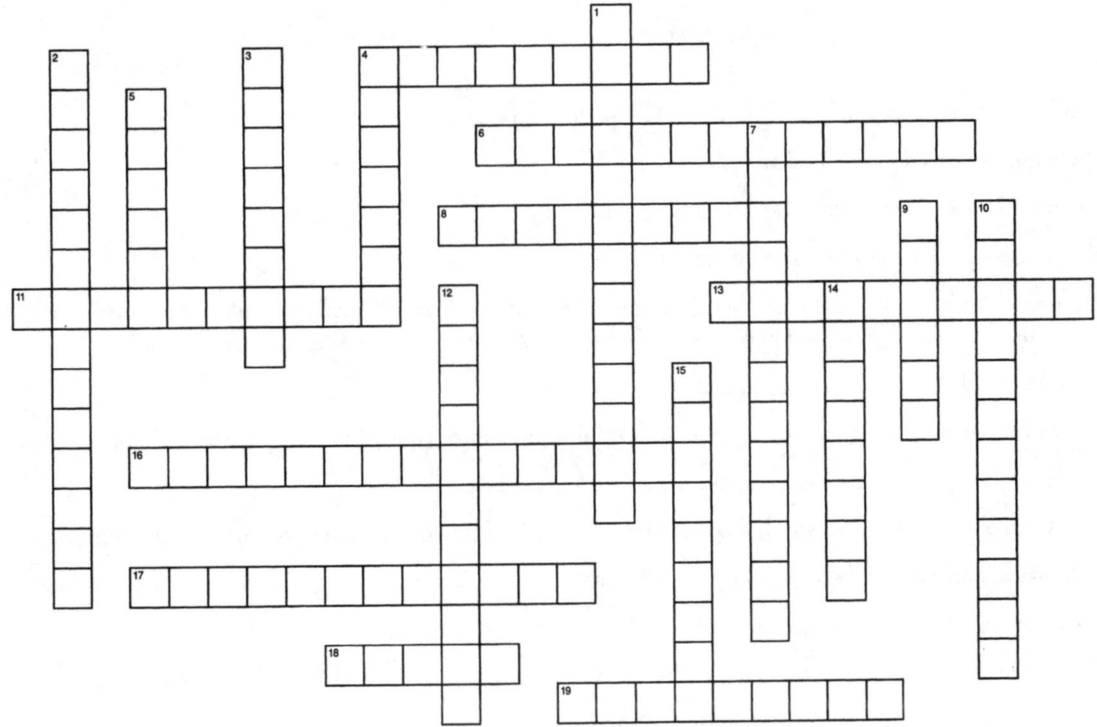

ACROSS

4. There are 22 pairs of these
6. Connects the fetus to the placenta (two words)
8. Anything that may cause developmental abnormalities in an embryo-fetus
11. Two alleles are the same
13. Present at birth
16. Consist of maternal and paternal chromosomes of the same type (two words)
17. Happens 7 to 8 days after fertilization
18. Sequence of events that occur during birth
19. How the alleles are expressed

DOWN

1. Provides a cushion for the fetus (two words)
2. Either XX or XY (two words)
3. Mitotic divisions following fertilization of the egg
4. Possibilities for a gene
5. The __________ stage; lasts until the eighth week after fertilization
7. Chromosome abnormalities may be detected with this procedure
9. Membrane that surrounds the embryo
10. Two alleles are different
12. Formal term for birth
14. Actual genetic makeup
15. Usually lasts 40 weeks

CLINICAL APPLICATIONS

1. a) Mrs. J delivers her first baby at 36 weeks of gestation. The most serious problem for premature infants usually involves the ____________________ (organs), which may not yet have produced sufficient quantities of surfactant.

   b) Mrs. J's baby has an Apgar score of 9, which means (choose one answer):

   1) the baby requires immediate medical attention

   2) the baby must be closely monitored because it is showing distress

   3) the baby is responding well to its independent existence

2. a) Mrs. C is pregnant with her first child, and her doctor asks her which diseases she had as a child. The viruses that cause certain mild diseases of childhood may be harmful to a fetus if a pregnant woman acquires the disease. Such viruses are called ____________________.

   b) Name two of these viral diseases that are usually not harmful to the mother but which may cause developmental defects in a fetus. ____________________ and ____________________

3. a) Mrs. L is 41 years old and pregnant with her third child. She has read that older women are more likely to have a child with Down syndrome. In this syndrome, the child has ____________________ of chromosome 21 and as a result has some degree of ____________________ disability.

   b) Mrs. L's doctor tells her that the chromosome number of the fetus can be determined by a procedure called ____________________ that is performed at 16 to 18 weeks of gestation.

4. a) A cesarean section is the delivery of an infant by way of an incision through the ____________________ and ____________________.

   b) State two reasons why a cesarean section may be preferable to a vaginal delivery.

   1) ____________________

   2) ____________________

MULTIPLE CHOICE TEST #1

Choose the correct answer for each question.

1. The embryonic stage that becomes implanted in the uterus is the:
   a) 4-cell stage b) morula c) blastocyst d) 16-cell stage

2. The most superficial parts of the body develop from this primary layer:
   a) ectoderm b) mesoderm c) endoderm d) outerderm

3. The chromosome number of a zygote is:
   a) 23, the haploid number c) 23, the diploid number
   b) 46, the haploid number d) 46, the diploid number

4. The average gestation period is approximately __________ weeks long, and the first __________ weeks are considered the period of embryonic growth.
   a) 46 / 12 b) 38 / 16 c) 42 / 4 d) 40 / 8

5. Both fetal and maternal tissues contribute to the formation of the:
   a) amnion b) chorion c) umbilical cord d) placenta

6. The fetal vessel that carries oxygen and nutrients from the placenta to the fetus is the:
   a) aorta b) umbilical vein c) umbilical artery d) chorionic vein

7. The fluid that surrounds the embryo-fetus and serves as a cushion is:
   a) plasma b) amniotic fluid c) intracellular fluid d) chorionic fluid

8. The first blood cells for the fetus are formed by the:
   a) yolk sac b) chorion c) red bone marrow d) spleen

9. The embryonic membrane that contributes to the formation of the placenta is the:
   a) amnion b) umbilical cord c) yolk sac d) chorion

10. In the placenta, how are materials exchanged between mother and fetus?
    a) maternal blood enters fetal circulation
    b) osmosis and filtration
    c) diffusion and active transport
    d) fetal blood enters maternal circulation

11. During the third stage of labor:
    a) the infant is delivered
    b) the cervix dilates
    c) the placenta is delivered
    d) amniotic fluid exits through the birth canal

12. The second stage of labor includes all of these except:
    a) rupture of the amniotic sac
    b) delivery of the infant
    c) strong contractions of the myometrium
    d) secretion of oxytocin by the posterior pituitary gland

13. The first stage of labor includes all of these except:
    a) dilation of the cervix
    b) rupture of the amniotic sac
    c) weak contractions of the myometrium
    d) delivery of the infant

14. Pregnancy is maintained by these hormones produced by the chorion and placenta:
    a) estrogen, FSH, and relaxin
    b) hCG, estrogen, and progesterone
    c) oxytocin, hCG, and progesterone
    d) FSH, estrogen, and progesterone

15. A gene is the genetic code for one:
    a) cell b) protein c) DNA d) RNA

16. With respect to a particular trait, the term *genotype* means:
    a) the sex chromosomes that are present
    b) the genetic makeup of the individual
    c) the appearance of the individual
    d) the haploid number of chromosomes

17. A pair of homologous chromosomes includes:
    a) two maternal chromosomes of the same number
    b) two paternal chromosomes of the same number
    c) a maternal and paternal chromosome of different numbers
    d) a maternal and paternal chromosome of the same number

18. A person who is heterozygous for a particular trait may have:
    a) two dominant genes
    b) two recessive genes
    c) a dominant gene and a recessive gene
    d) two sex-linked genes

19. With respect to a particular trait, the term *phenotype* refers to:
    a) the appearance of the individual
    b) the genetic makeup of the individual
    c) the diploid number of chromosomes
    d) the number of alleles present

20. The hereditary material of cells is contained in the:
    a) protein in the chromosomes
    b) DNA in the cell membrane
    c) protein in the nucleus
    d) DNA in the chromosomes

MULTIPLE CHOICE TEST #2

Read each question and the four answer choices carefully. When you have made a choice, follow the instructions to complete your answer.

1. Which statement is NOT true of the stages of labor?
   a) The amniotic sac ruptures during the second stage.
   b) The placenta is delivered in the third stage.
   c) The second stage ends with delivery of the infant.
   d) The cervix dilates during the first stage.

   Reword your choice to make it a correct statement.

   ---

2. Which statement is NOT true of embryonic membranes?
   a) The human yolk sac does not contain nutrients for the embryo.
   b) The chorion will become part of the placenta.
   c) The amnion contains amniotic fluid to cushion the fetus.
   d) The first blood cells for the fetus are formed by the chorion.

   Reword your choice to make it a correct statement.

   ---

3. Which statement is NOT true of the growth of the embryo-fetus?
   a) All the organ systems are established by the end of 8 weeks of gestation.
   b) The heart begins to beat in the embryo stage.
   c) The period of embryonic growth lasts from week 1 to week 18.
   d) The last fetal organs to become functional are the lungs.

   Reword your choice to make it a correct statement.

   ---

4. Which statement is NOT true of the placenta and umbilical cord?
   a) The umbilical vein carries oxygenated blood from the placenta to the fetus.
   b) There is no mixing of fetal and maternal blood in the placenta.
   c) The umbilical arteries carry blood with waste products from the fetus to the placenta.
   d) Within the placenta, fetal blood vessels are within the maternal arteries.

   Reword your choice to make it a correct statement.

   ---

5. Which statement is NOT true of fertilization and implantation?
   a) The zygote undergoes a series of mitotic cell divisions called cleavage.
   b) The embryonic stage that undergoes implantation is the morula.
   c) The trophoblast of the blastocyst produces enzymes to digest the endometrium.
   d) The zygote contains 23 chromosomes from the egg and 23 chromosomes from the sperm.

   Reword your choice to make it a correct statement.

   ---

6. Which statement is NOT true of the hormones of pregnancy?
   a) The chorion produces hCG that stimulates the corpus luteum during early gestation.
   b) No egg cells are produced during pregnancy because estrogen and progesterone inhibit the secretions of the anterior pituitary gland.
   c) Premature contractions of the myometrium are prevented by estrogen.
   d) The mammary glands are prepared for lactation by estrogen and progesterone.

   Reword your choice to make it a correct statement.

   ---

7. Which statement is NOT true of chromosomes and genes?
   a) A gene is the DNA code for one protein.
   b) A homologous pair of chromosomes consists of one maternal and one paternal chromosome of the same number.
   c) Human cells have 22 pairs of autosomes and one pair of sex chromosomes.
   d) The sex chromosomes are XX in men and XY in women.

   Reword your choice to make it a correct statement.

   ____

8. Which statement is NOT true of the terminology of genetics?
   a) The expression of the alleles in the appearance of the individual is the genotype.
   b) To be homozygous for a trait means to have two similar alleles.
   c) To be heterozygous for a trait means to have two different alleles.
   d) Alleles are the possibilities for the ways a gene may be expressed in an individual.

   Reword your choice to make it a correct statement.

   ____

9. Which statement is NOT true of inheritance?
   a) If a gene has more than two possible alleles, as in blood types, then there will be more than two possible phenotypes.
   b) A dominant gene is one that appears in the phenotype of a heterozygous individual.
   c) Recessive genes may not appear in one generation but may reappear in individuals in the next generation.
   d) A recessive gene is one that appears in the phenotype of a heterozygous individual.

   Reword your choice to make it a correct statement.

   ____

10. Which statement is NOT true of inheritance of sex-linked traits?
   a) Men cannot be carriers of sex-linked traits.
   b) Women who are carriers of sex-linked traits may pass the gene to a son or a daughter.
   c) A son who inherits a sex-linked trait has inherited the gene from his father.
   d) The genes for sex-linked traits are found only on the X chromosome.

   Reword your choice to make it a correct statement.

   ____

MULTIPLE CHOICE TEST #3

Each question is a series of statements concerning a topic in this chapter. Read each statement carefully and select all of the correct statements.

1. Which of the following statements are true of human development?
   a) Implantation of a blastocyst usually occurs the day after fertilization.
   b) The period of fetal growth is from week 30 to week 40 of gestation.
   c) Amniotic fluid is a cushion for the fetus, and the yolk sac becomes part of the fetal liver.
   d) The foramen ovale and ductus venosus permit blood to bypass the fetal lungs.
   e) Strong contractions of the endometrium are stimulated by oxytocin.
   f) The fetal part of the placenta is formed from the chorion.
   g) The umbilical vein carries nutrients from the placenta to the fetus.
   h) Mesoderm forms “middle” structures such as bones and muscles.
   i) Progesterone from the placenta inhibits contractions of the myometrium during pregnancy.
   j) During the third stage of labor, the infant is delivered.

2. Which of the following statements are true of human genetics?
   a) A heterozygous phenotype for a trait may look the same as a homozygous phenotype.
   b) The ABO blood group has two possible alleles and three possible blood types.
   c) Human cells have 46 chromosomes in 22 pairs of autosomes and 1 pair of sex chromosomes.
   d) A segment of DNA that is the code for one protein is considered a gene.
   e) A dominant trait will appear in the phenotype if one or two genes for it are present in the genotype.
   f) A child of brown-eyed parents can never have blue eyes.
   g) Chromosomes are made of DNA and protein; the DNA is the hereditary material.
   h) Dominant traits are always more frequent in a population than their recessive counterparts.
   i) Two different phenotypes may have the same genotype.
   j) A son acquires a sex-linked trait from his father, but a daughter acquires a sex-linked trait from her mother.

Chapter 22

An Introduction to Microbiology and Human Disease

This chapter explains some of the basic principles of microbiology, describes our microbiota (or normal flora), and is an introduction to human diseases caused by microorganisms.

CLASSIFICATION OF MICROORGANISMS

1. Name the group of microorganisms described by each statement.
   a) Organisms that are not cells, and all are parasites ____________________
   b) Multicellular animals that may be vectors of disease ____________________
   c) Very simple single-celled organisms; most are decomposers ____________________
   d) Single-celled animals, some of which are parasites ____________________
   e) Multicellular animals, such as those that cause trichinosis ____________________
   f) Yeasts and molds, some of which are pathogens ____________________
2. a) The use of genus and species names for living things is called ____________________
   b) Which of these names is the more inclusive (i.e., larger) category? ____________________
   c) Which of these names is always capitalized? ____________________
   d) Which of these names is more like your own first name? ____________________

NORMAL FLORA—MICROBIOTA

1. a) The microorganisms that are on or in the body for short periods of time are called ____________________.
   b) The microorganisms usually on or in most of us in specific sites are called ____________________.
   c) A microorganism that is usually harmless but may become a pathogen in certain circumstances is called an ____________________.

2. Match each body site (microbiome) with the description of its normal flora.

Use each letter once.

1) Blood ______
2) Skin ______
3) Oral cavity ______
4) Tissue fluid ______
5) Esophagus ______
6) Stomach ______
7) Nasal cavities ______
8) Lungs ______
9) Vagina ______
10) Urinary bladder ______
11) Small intestine ______
12) Large intestine ______

A. Microbes in inhaled air are swept by ciliated epithelium to the pharynx.
B. Should be free of microorganisms
C. Bacteria create an acidic pH that inhibits the growth of pathogens.
D. The part with the largest flora is the ileum.
E. Hydrochloric acid kills most bacteria that enter.
F. Pathogens that reach the alveoli are destroyed by macrophages.
G. Contains microbes swallowed with saliva or food
H. A huge flora produces vitamins and inhibits the growth of pathogens.
I. Microbial growth is limited by the flaking of the stratum corneum.
J. Microbial growth is kept in check by lysozyme.
K. Is virtually free of bacteria, as is the upper urethra
L. Pathogens that get through breaks in the skin are destroyed by wandering macrophages.

INFECTIOUS DISEASE

1. a) The total of the body's defenses against pathogens is called ______.
   b) The ability of a pathogen to cause disease is called ______.
2. State two of the terms used for an infection in which the person shows no symptoms. ______ and ______
3. In the course of an infectious disease, the time between the entry of the pathogen and the appearance of symptoms is called the ______.
4. Match each type of infection with its proper definition.

Use each letter once.

1) Self-limiting ______
2) Endogenous ______
3) Clinical ______
4) Nosocomial ______
5) Localized ______
6) Systemic ______
7) Septicemia ______
8) Secondary ______
9) Acute ______
10) Chronic ______

A. Made possible by a primary infection that lowers host resistance
B. Progresses slowly, or may last a long time
C. Lasts a certain length of time and is usually followed by recovery
D. Bacteria in the blood; always serious
E. Confined to one area of the body
F. Caused by the person's own normal flora in an abnormal site
G. Acquired in a hospital or other institution
H. Begins abruptly or is severe
I. Has spread throughout the body from an initial site
J. Symptoms are present

EPIDEMIOLOGY

1. a) A disease that is present in a population, with an expected number of cases, is called ____________________.
   b) When more than the usual number of cases occurs, the disease is said to be ____________________.
   c) When a disease is epidemic in several countries, it is said to be a ____________________.
2. a) The way a pathogen enters a host is called the ____________________.
   b) Name five of these. ____________________, ____________________, ____________________, ____________________, and ____________________
   c) The way a pathogen leaves a host is called the ____________________.
   d) Name five of these. ____________________, ____________________, ____________________, ____________________, and ____________________
3. With respect to epidemiology, a pathogen travels from one person's ____________________ to the next host's ____________________.
4. a) Any person or animal that is a source of a pathogen may be called a ____________________.
   b) A disease that is usually an animal disease is called a ____________________.
   c) A person who has recovered from a disease but continues to shed the pathogen is called a ____________________.
5. a) A disease that may be directly or indirectly transmitted from host to host is called ____________________.
   b) The communicable diseases that are easily spread from host to host by casual contact are called ____________________.
   c) Diseases that cannot be transmitted from host to host are called ____________________.

METHODS OF CONTROL OF MICROBES

1. a) A chemical that destroys microbes on inanimate objects is called a ____________________.
   b) A chemical that destroys microbes on a living being is called an ____________________.
2. a) A process that destroys all living organisms is called ____________________.
   b) Pathogens in foods such as milk may be killed by the process called ____________________.
   c) The process that destroys microorganisms in city water supplies is called ____________________.

THE PATHOGENS

Bacteria

1. State the shape of each.
   a) Bacillus ____________________
   b) Coccus ____________________
   c) Spirillum ____________________

2. a) Chemicals that are used to treat bacterial infections are called ________________.

   b) An antibiotic that affects only a few kinds of bacteria is called ________________.

   c) An antibiotic that affects many kinds of bacteria is called ________________.

   d) Bacteria that are no longer affected by a particular antibiotic are said to be ________________ to it.

   e) The laboratory procedure to determine the proper antibiotic to use to treat an infection is called ________________.

3. Match each bacterial structure or characteristic with the correct statement.

   Use each letter once.

   1) Binary fission ________________
   2) Endotoxin ________________
   3) Toxins ________________
   4) Spores ________________
   5) Cell wall ________________
   6) Capsule ________________
   7) Flagella ________________
   8) Anaerobic ________________
   9) Aerobic ________________
   10) Facultatively anaerobic ________________

   A. Gives the cell its shape and is the basis for the Gram stain
   B. Provide motility
   C. The cell division process by which most bacteria reproduce
   D. Chemical products that are poisonous to host cells
   E. The toxic cell walls of gram-negative bacteria
   F. Bacteria that reproduce only in the absence of oxygen
   G. Inhibits phagocytosis by host white blood cells
   H. A dormant stage resistant to heat and drying
   I. Bacteria that reproduce in the presence or absence of oxygen
   J. Bacteria that reproduce only in the presence of oxygen

4. a) Name three bacterial diseases that are zoonoses. ________________, ________________, and ________________

   b) Name two bacterial diseases that are considered sexually transmitted diseases. ________________ and ________________

   c) Name five other bacterial diseases, at least three of which are largely prevented by vaccinating children.

   ________________, ________________, ________________, ________________, and ________________

Viruses

1. a) A virus is made of either ________________ or ________________, surrounded by a ________________.

   b) Viruses are obligate intercellular parasites, which means that they must be inside ________________ to ________________.

2. a) Name two viral diseases that are sexually transmitted. ________________ and ________________

   b) Name two viral diseases that are zoonoses. ________________ and ________________

   c) Name three viral diseases that are largely prevented by vaccinating children. ________________, ________________, and ________________

Fungi

1. a) Most fungi are saprophytes, which means that their food is ____________.

   b) The mold-like fungi reproduce by forming ____________.

2. a) Diseases caused by fungi are called ____________.

   b) If on the skin or a mucous membrane, the disease is called a ____________ mycosis.

   c) If the fungi affect deeper organs, the disease is called a ____________ mycosis.

3. a) The fungi that in small numbers may be part of the resident flora are the ____________.

   b) Name the body sites where yeasts may reside. ____________

   c) A common trigger for mucosal yeast infections is the use of ____________ that suppress the normal bacterial flora.

Protozoa

1. a) Protozoa are classified as ____________.

   b) Most protozoa are inhabitants of ____________.

2. a) Some protozoan parasites form ____________, which are dormant forms that can survive outside a host.

   b) Some human protozoan parasites inhabit the intestines and are spread by the ____________ route in contaminated ____________ or ____________.

Worms and Arthropods

1. a) Name the worm infestation that may be prevented by thorough cooking of pork and wild game.

   ____

   b) The most common worm infestation in much of North America is probably ____________.

2. Name four diseases spread by arthropod vectors, and name the vectors.

   1) ____________
   2) ____________
   3) ____________
   4) ____________

CROSSWORD PUZZLE

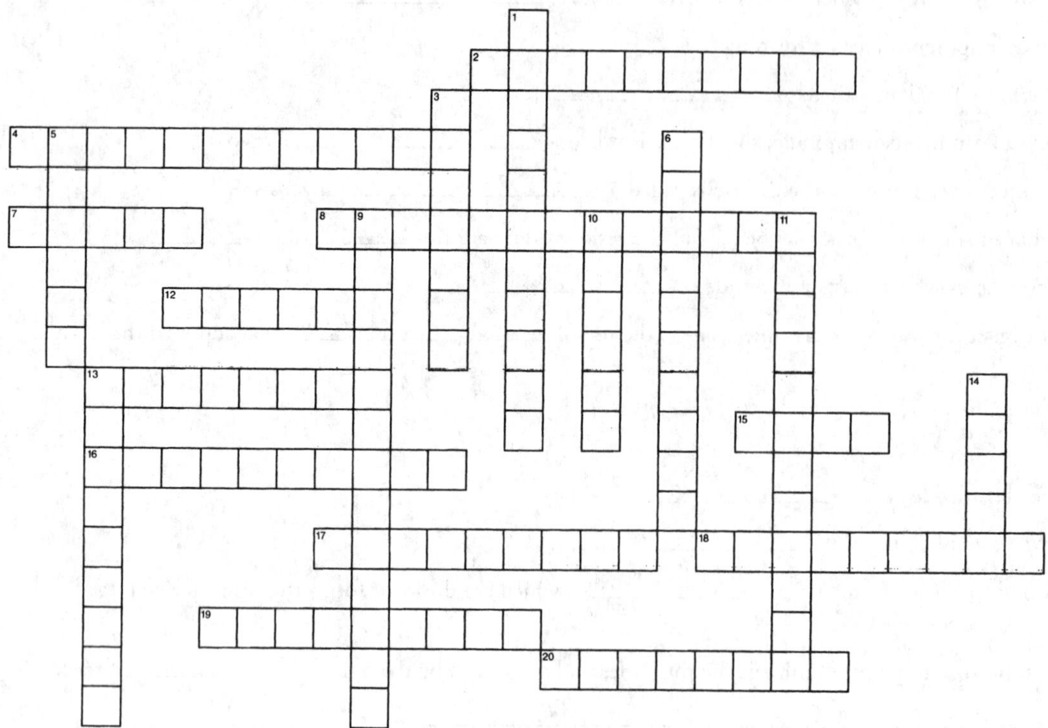

ACROSS

2. The time before symptoms appear; the ________ period
4. The study of the spread of disease
7. An antibiotic effective against many kinds of bacteria; ________ spectrum
8. An infection caused by a usually harmless microorganism
12. A mosquito is this for malaria
13. Spiral-shaped bacteria
15. The cell walls of gram-negative bacteria: ________ toxin
16. A disease that is easily spread from host to host
17. A living source of infection
18. A staining procedure helpful in the identification of bacteria
19. Antibodies that neutralize a toxin
20. Diseases of animals that may be transmitted to people

DOWN

1. The harboring of worm parasites
3. An infection caused by a fungus
5. The way a pathogen leaves a host; the ________ of exit
6. An infection acquired in a hospital
9. The way a pathogen enters a host
10. An antibiotic effective against a few kinds of bacteria; ________ spectrum
11. A disease that may be directly or indirectly transmitted from host to host
13. An infection made possible by a primary infection
14. Spherical bacteria

MULTIPLE CHOICE TEST #1

Choose the correct answer for each question.

1. A contagious disease is always:
a) communicable b) self-limiting c) noncommunicable d) secondary

2. An infection in which the person shows no symptoms is called:
a) asymptomatic b) subclinical c) inapparent d) all of these

3. A virus can reproduce only in:
a) the blood b) tissue fluid c) a cell d) dead tissue

4. Some bacilli may survive unfavorable environments by forming:
a) spores b) capsules c) cell walls d) toxins

5. Endotoxin is which part of a gram-negative bacterial cell?
a) cell membrane b) cell wall c) flagella d) capsule

6. Most bacteria do not survive passage through the stomach because of the presence of:
a) water b) mucus c) hydrochloric acid d) antibodies

7. Besides inhibiting the growth of pathogens, the resident flora of the colon produce:
a) hormones b) antibodies c) carbohydrates d) vitamins

8. The time between the entry of the pathogen and the appearance of the symptoms is called the:
a) latent period b) incubation period c) dormant stage d) self-limiting stage

9. A mycosis is an infection caused by:
a) protozoa b) fungi c) bacteria d) worms

10. Pathogens in inhaled air usually do not reach the lungs because of the presence of ________________ in the trachea.
a) ciliated epithelium b) muscle tissue c) air d) cartilage

11. In women, the normal flora of the vagina creates a(n) ________________ pH that prevents the growth of pathogens.
a) neutral b) alkaline c) acidic d) very alkaline

12. When a disease is usually present in a population, it is said to be:
a) self-limiting b) epidemic c) endemic d) pandemic

13. An infection that begins suddenly and spreads quickly throughout the body may be called:
a) acute and localized b) chronic and systemic c) chronic and localized d) acute and systemic

14. An infection that is caused by the person's own normal flora in an abnormal site is called:
a) epidemic b) symptomatic c) nosocomial d) endogenous

15. Pathogens in foods such as cheese may be destroyed by a process called:
a) sterilization b) pasteurization c) disinfection d) chlorination

16. Rod-shaped bacteria that are capable of movement are:
a) bacilli with flagella b) cocci with capsules c) cocci with flagella d) bacilli with capsules

17. Viruses cause disease by doing what in cells?
a) reproducing in them b) activating the cells c) becoming dormant d) deactivating the cells

18. Bacteria that can reproduce either in the presence or absence of oxygen are called:
a) anaerobic b) aerobic c) facultatively anaerobic d) very talented

19. Which of these is NOT a portal of entry for pathogens?
a) nose b) unbroken skin c) reproductive tract d) mouth

20. An infection acquired following a previous infection that lowers a person's resistance is called:
a) secondary b) self-limiting c) symptomatic d) endemic

MULTIPLE CHOICE TEST #2

Read each question and the four choices carefully. When you have made a choice, follow the directions to complete your answer.

1. Which statement is NOT true of epidemiology?
   a) Any body opening is a potential portal of entry for pathogens.
   b) Reservoirs of some infections may be people or animals.
   c) Some pathogens may cross the placenta and infect a fetus.
   d) All communicable diseases are contagious.

   Reword your choice to make it a correct statement.

   ___

2. Which statement is NOT true of types of infections?
   a) Septicemia refers to the presence of bacteria in the blood.
   b) A nosocomial infection is one acquired at home or at work.
   c) A secondary infection follows a first infection in which the person's resistance was lowered.
   d) A person with a subclinical infection has no symptoms.

   Reword your choice to make it a correct statement.

   ___

3. Which statement is NOT true of normal flora?
   a) The skin has a large bacterial population.
   b) Bacteria in the colon produce enough vitamin C to meet our needs.
   c) Ciliated epithelium sweeps pathogens and mucus out of the trachea.
   d) The vaginal flora maintains an acidic pH that inhibits the growth of pathogens.

   Reword your choice to make it a correct statement.

   ___

4. Which statement is NOT true of bacterial structure?
   a) Flagella provide movement for some bacteria.
   b) The Gram stain is based on the chemistry of the cell wall.
   c) A spore is a form that is susceptible to heat and drying.
   d) Bacilli are rod shaped.

   Reword your choice to make it a correct statement.

   ___

5. Which statement is NOT true of viruses?
   a) Viruses must be in the host's blood to reproduce.
   b) A virus is made of DNA or RNA and a protein shell.
   c) Some viruses are spread by vectors.
   d) Some viruses may become dormant in the host and cause infection years later.

   Reword your choice to make it a correct statement.

   ___

6. Which statement is NOT true of fungi?
   a) In small numbers, yeasts may be part of the normal flora of the skin.
   b) Yeasts are unicellular fungi.
   c) The name ringworm refers to a fungal infection of the brain.
   d) Systemic mycoses may occur in someone with a chronic pulmonary disease.

   Reword your choice to make it a correct statement.

   ___

7. Which statement is NOT true of protozoa?
   a) Intestinal protozoa are spread by the fecal–oral route from host to host.
   b) Some protozoa form cysts, which quickly die outside a host.
   c) Malaria is an important disease caused by a protozoan.
   d) Most protozoa live in water and are not pathogens.

   Reword your choice to make it a correct statement.

   ____

8. Which statement is NOT true of worms and arthropods?
   a) Tapeworms live in the small intestine and absorb digested nutrients.
   b) A mosquito is the vector of malaria and the Zika virus.
   c) Pinworm and trichinosis are worm infestations found in North America.
   d) Lice and ticks spread disease when they bite to obtain skin for food.

   Reword your choice to make it a correct statement.

   ____

9. Which statement is NOT true of bacteria?
   a) Some bacteria cause disease by producing toxins.
   b) An antibiotic that is effective against many kinds of bacteria is called broad-spectrum.
   c) Anaerobic bacteria reproduce only in the presence of oxygen.
   d) A Gram stain is helpful in the identification of bacteria.

   Reword your choice to make it a correct statement.

   ____

10. Which statement is NOT true of the treatment of disease?
    a) Antiviral medications are effective against all viruses.
    b) Antibiotics destroy bacteria by interrupting chemical reactions.
    c) Side effects of an antibiotic usually indicate that it is interrupting human chemical reactions.
    d) Bacterial resistance to antibiotics is a genetic capability.

    Reword your choice to make it a correct statement.

    ____

MULTIPLE CHOICE TEST #3

Each question is a series of statements concerning a topic in this chapter. Read each statement carefully and select all of the correct statements.

1. Which of the following statements are true of diseases and infections?
   a) Aerobic bacteria require oxygen to reproduce.
   b) Endogenous infections are caused by the person's own flora.
   c) A reservoir of infection is always an insect such as a flea.
   d) A nosocomial infection is one that is acquired by way of the respiratory tract.
   e) A disease that has a vector does not have reservoirs.
   f) An endemic disease is expected to be present in a population.
   g) Septicemia is an infection of the liver.
   h) A fomite is a nonliving object that may be a source of infection.
   i) An acute infection is characterized by rapid recovery.
   j) A secondary infection is one that comes from food.
   k) A self-limiting disease is never communicable.
   l) The mouth is a possible portal of entry, but not a portal of exit.
   m) Inapparent is a synonym for inactive or dormant.
   n) A contagious disease is always communicable, but a communicable disease is not always contagious.

2. Which of the following statements are true of pathogens and the diseases they cause?
   a) Endotoxin shock is a serious disease caused by gram-negative bacteria.
   b) For bacteria, “bacillus” is to “rod” as “coccus” is to “cube.”
   c) Viruses cause disease by reproducing within cells.
   d) Molds are fungi, and some may cause serious lung disease.
   e) Protozoan parasites are often spread in water.
   f) Anaerobic bacteria live on dead tissue and therefore cannot cause human disease.
   g) A zoonosis is a disease of zoo animals.
   h) “Spore” is to bacteria as “cyst” is to protozoa.
   i) Viruses secrete enzymes to bore holes in cell membranes in order to enter cells.
   j) Worm parasites may have life cycles that require two hosts.
   k) A superficial mycosis is a bacterial skin disease.
   l) Insect vectors often spread pathogens when they bite to obtain blood.

Answer Key

Chapter 1

LEVELS OF ORGANIZATION

1. a) chemical
   b) any three of these: carbohydrates, lipids, proteins, nucleic acids
   c) any three of these: water, oxygen, carbon dioxide, calcium, iron (or any of the minerals)
2. a) organ system
   b) any three of the 11 organ systems: respiratory, skeletal, circulatory, nervous, digestive, muscular, reproductive, etc.
3. a) cellular b) tissue c) organ
4. 1) D, 1 2) A, 3 3) B, 2 4) C, 4
5. a) anatomy, physiology
   b) bones support (or protect) the body—they are very strong (or hard)
6. a) microbiota or normal flora (resident flora)
   b) microbiomes; any three of nasal cavities, oral cavity, skin surface, intestines, vagina

METABOLISM AND HOMEOSTASIS

1. a) chemical, physical b) metabolic rate
2. health
3. stable (the same)
4. externally, internally
5. 1) food is digested into simple chemicals the body cells can use
   2) oxygen enters the blood to become available to cells
   3) take aspirin (take a nap)
   4) blood will clot to prevent extensive blood loss
   5) shivering will occur to produce heat (put on a coat)
6. 1, 5, 2, 4, 3
7. A positive feedback mechanism requires an external event to stop it (does not contain its own brake).
8. 1) thyroid gland
   2) cells decrease energy production
   3) hypothalamus and pituitary gland
   4) increases secretion of thyroxine
   5) metabolic rate increases
   6) decrease secretion of hormones
   7) heat gain mechanisms activated
   8) fever
   9) white blood cells

TERMS OF LOCATION

1. below, or lower than; superior
2. closer to the origin; distal
3. toward the side; medial
4. extending from the main part; central
5. toward the surface; deep
6. in front of, or toward the front; posterior
7. behind, or toward the back; ventral
8. within, or inside of; external

BODY PARTS AND AREAS

1. 1) cranial
   2) orbital
   3) nasal
   4) buccal
   5) frontal
   6) temporal
   7) cervical
   8) axillary
   9) umbilical
   10) volar
   11) deltoid
   12) pectoral
   13) mammary
   14) brachial
   15) iliac
   16) inguinal
   17) femoral
   18) patellar
   19) plantar
   20) popliteal
   21) parietal
   22) occipital
   23) lumbar
   24) sacral
   25) gluteal
   26) perineal

2. 1) a) lateral b) shoulder c) neck
   2) a) proximal b) thigh c) knee
   3) a) posterior
      b) back of the head
      c) small of the back
   4) a) ventral b) chest c) navel
   5) a) superior
      b) palm of the hand
      c) sole of the foot
   6) a) superficial b) skin
   7) a) external b) lungs
   8) peripheral, central

BODY CAVITIES

1. 1) C, G 2) A, D, H 3) B, C
   4) F 5) E, I, J
2. 1) thoracic cavity
   2) abdominal cavity
   3) pelvic cavity
   4) cranial cavity
   5) spinal cavity

BODY SECTIONS

1. 1) B 2) D 3) E 4) C 5) A
2. 1) midsagittal plane
   2) sagittal plane
   3) transverse plane
   4) frontal (coronal) plane
   5) longitudinal section
   6) cross-section

ORGAN SYSTEMS

1. 1) muscular
   2) circulatory
   3) reproductive
   4) skeletal
   5) urinary
   6) respiratory
   7) nervous
   8) endocrine
   9) lymphatic
   10) digestive
   11) integumentary
2. 1) circulatory
   2) urinary
   3) nervous
   4) integumentary
   5) respiratory
   6) muscular
   7) skeletal
   8) reproductive or endocrine
   9) digestive or endocrine
   10) lymphatic or circulatory
   11) endocrine
   12) reproductive or endocrine
   13) respiratory
   14) digestive
   15) muscular
   16) nervous
   17) endocrine
   18) circulatory
   19) integumentary
   20) digestive

CROSSWORD PUZZLE

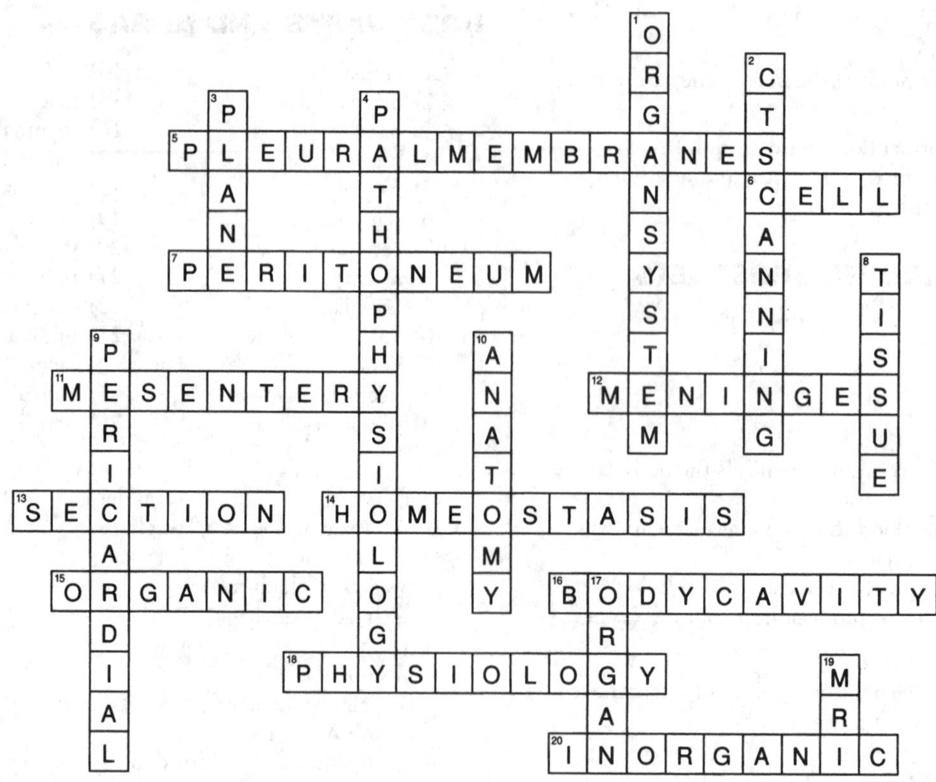

CLINICAL APPLICATIONS

1. 1) the skin (and subcutaneous tissue)
   2) muscles of the abdominal wall
   3) peritoneum
2. meninges; brain and spinal cord
3. 1) the *kidneys* have stopped functioning
   2) the *heart* has stopped beating
   3) a blood vessel of the *lung*
   4) an unconscious state due to disease of the *liver*
   5) an erosion of the *stomach* lining
4. 1) stomach—upper left
   2) liver—upper right and left
   3) small intestine—all four
   4) large intestine—all four

MULTIPLE CHOICE TEST #1

1. a	6. b	11. d	16. d	21. d
2. d	7. c	12. b	17. b	22. c
3. d	8. d	13. b	18. c	23. b
4. a	9. a	14. b	19. b	24. d
5. b	10. c	15. a	20. d	25. a

MULTIPLE CHOICE TEST #2

The letter of the correct choice is followed by the statement necessary to complete the answer. When an incorrect statement is to be corrected to make it true, the new (true) words are underlined.

1. d—The muscles are part of the organ level.
2. c—The blood is a tissue.
3. a
4. c—Nerve tissue is specialized to generate and transmit impulses.
5. c—Internal changes do have effects on homeostasis.
6. a—lines the abdominal cavity
7. b and e—The eyes and ears are lateral to the midline of the head.
8. d—midtransverse section
9. d—A group of cells with similar structure and function is called a tissue.
10. a—The diaphragm separates the thoracic and abdominal cavities.
11. b—External changes do affect it.
12. b—The orbital area is superior to the oral area.

MULTIPLE CHOICE TEST #3

1. a, d, e, and f are correct
2. d, e, h, and i are correct
3. a, c, and e are correct

Chapter 2

ELEMENTS, ATOMS, AND BONDS

1. element
2. Fe—iron C—carbon
   Na—sodium Mg—magnesium
   Cl—chlorine Co—cobalt
   Ca—calcium H—hydrogen
   I—iodine N—nitrogen
   Cu—copper Mn—manganese
   O—oxygen K—potassium
   P—phosphorus S—sulfur
   Zn—zinc F—fluorine
3. protons, neutrons, and electrons
4. a) electron, proton
   b) neutron
   c) protons and neutrons
   d) protons and electrons
   e) electrons
5. ion
6. molecule
7. ionic
8. covalent
9. a) any one of these: oxygen, carbon dioxide, nitrogen
   b) water
   c) any one of these: sodium chloride, calcium chloride
10. ionic
11. covalent
12. 1) electron orbitals 2) protons
    3) protons 4) electron orbitals
    5) electron transfer
13. disulfide
14. hydrogen
15. a) synthesis b) decomposition

INORGANIC COMPOUNDS

1. water
2. a) dissolve
   b) any one of these: transport of nutrients in the blood, excretion of wastes in urine, senses of taste and smell
3. a) friction
   b) one of these: mucus in the digestive tract, synovial fluid in joints
4. excess body heat may be lost in the process of sweating
5. a) plasma b) intracellular fluid
   c) lymph d) tissue fluid
   e) 1. tissue fluid 2. intracellular fluid
   3. lymph 4. plasma
6. a) as a gas b) O_2 c) cell respiration, ATP (energy)
7. a) cell respiration b) CO_2 c) acidic

CELL RESPIRATION

1. $CO_2 + H_2O + ATP + heat$
2. CO_2 is exhaled; H_2O becomes part of intracellular water
3. ATP provides energy for cellular activities; heat contributes to a constant body temperature

TRACE ELEMENTS

1. 1) C 2) A 3) B, E, H 4) D
   5) B, F 6) E, I 7) G

ACIDS, BASES, AND PH

1. a) 0–14 b) 7 c) H^+, OH^-
   d) acidic e) alkaline (basic)
2. a) pH 7.35-7.45 b) alkaline
3. buffers
4. a) carbonic acid, base
   b) sodium bicarbonate, acid
5. a) sodium chloride and carbonic acid
   b) great c) NaCl, no effect d) H_2CO_3, slight

ORGANIC COMPOUNDS

1. 1) C 2) H 3) K 4) A 5) F 6) J 7) B 8) I 9) D 10) G 11) E
2. 1) D 2) G 3) I 4) B 5) J 6) K, N 7) C 8) F 9) A, L 10) E 11) H, M

CHEMICAL STRUCTURE—REVIEW

1. carbon, hydrogen, oxygen, nitrogen
2. a) amino acid b) covalent
3. a) proteins
   b) any three of these:
      1) enzymes to catalyze reactions
      2) antibodies for defense against pathogens
      3) hemoglobin to carry oxygen
      4) muscle contraction
      5) some hormones
      6) structure of cells and tissues
4. carbon, hydrogen, oxygen
5. $C_6H_{12}O_6$
6. a) glucose b) source of energy
   c) 1) glycogen—storage form of excess glucose in the liver
      2) starch—an energy source in plant food
      3) cellulose—fiber to provide exercise for the colon; food for the beneficial bacteria; promotes peristalsis

ENZYMES

1. catalysts
2. 1) enzyme–substrate complex
   2) product
   3) substrate
   4) enzyme unchanged
   5) denatured enzyme
   6) active site
   7) toxic ion blocks active site
3. shape
4. H^+, active site

DNA, RNA, AND ATP

1. pentose sugar, phosphate group, a nitrogenous base
2. 1) B, D 2) A, F 3) C, E
3. T is always paired with A; C is always paired with G
4. 1) C, E 2) A 3) B, D
5. 1) adenine
   2) ribose
   3) phosphate groups
   4) ATP
   5) ADP
   6) energy
   7) muscle contraction
   8) mitochondria

CROSSWORD PUZZLE

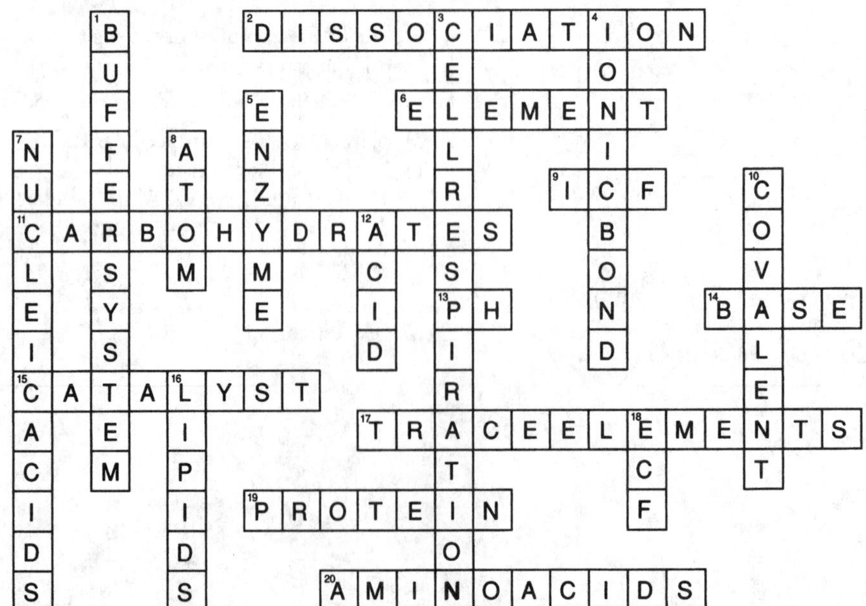

CLINICAL APPLICATIONS

1. a) decrease, acidosis b) a
2. d
3. b
4. a

MULTIPLE CHOICE TEST #1

1. b	6. b	11. a	16. c	21. a
2. b	7. c	12. b	17. d	22. c
3. a	8. c	13. e	18. d	23. b
4. d	9. d	14. d	19. b	24. d
5. a	10. b	15. d	20. a	25. d

MULTIPLE CHOICE TEST #2

The letter of the correct choice is followed by the statement necessary to complete the answer. When an incorrect statement is to be corrected to make it true, the new (true) words are underlined.

1. b—Water is a molecule made of two elements, hydrogen and oxygen.
2. b—Covalent bonds are <u>not</u> weakened when in a water solution.
3. d—Intracellular fluid is water found <u>within</u> cells.
4. c—Glucose is a <u>carbohydrate</u> (or sugar) molecule obtained from food.
5. a—The heat energy produced <u>provides a constant body temperature</u>.
6. b—Iron is part of hemoglobin; calcium is part of bones and teeth; iodine is part of thyroxine.
7. b—The normal pH range of blood is <u>7.35–7.45</u>.
8. d—The excretion of waste products in urine depends on the <u>solvent</u> action of water.
9. c—Glycogen is a form of energy storage in the liver; true fats are a form of energy; storage in adipose tissue; pentose sugars are part of DNA and RNA.
10. a—Hemoglobin transports <u>oxygen</u> in RBCs.
11. a—All enzymes are <u>proteins</u>.
12. c—<u>DNA</u> is the genetic code for our hereditary characteristics.
13. d—It contains <u>three phosphate groups</u>.
14. a—Synthesis reactions involve the <u>formation</u> of bonds.
15. c—Water molecules are cohesive because of the presence of <u>hydrogen</u> bonds.
16. a—Saturated fatty acids have the maximum number of <u>hydrogen</u> atoms.

MULTIPLE CHOICE TEST #3

1. a, d, g, and h are correct
2. b, c, d, e, f, g, and l are correct
3. a, b, e, and f are correct

Chapter 3

CELL STRUCTURE

1. 1) B, E, G, I 2) D, H 3) A, C, F, J
2. a) phospholipids, proteins, cholesterol
   b) 1) cholesterol 2) proteins 3) proteins 4) phospholipids 5) proteins
3. 1) cilia
   2) microvilli
   3) cell membrane
   4) Golgi apparatus
   5) smooth endoplasmic reticulum
   6) centrioles
   7) proteasome
   8) nucleus
   9) rough endoplasmic reticulum
   10) ribosomes
   11) mitochondrion
   12) lysosome
4. 1) E, 7 2) A, 1 3) J, 9 4) B, 3 5) G, 4 6) C, 8 7) F, 6 8) H, 11 9) D, 5 10) I, 10 11) K, 2

CELLULAR TRANSPORT MECHANISMS

1. 1) C, 2 2) F, 1 3) A, 5 4) E, 7 5) G, 3 6) D, 4 7) B, 6
2. 1) low O_2 2) high O_2 3) low CO_2 4) high CO_2
3. 1) active transport 2) osmosis 3) active transport
4. 1) the same 2) a higher 3) a lower
5. 1) isotonic 2) hypotonic 3) hypertonic

DNA AND THE GENETIC CODE

1. a) chromosomes b) 46
2. two, double helix
3. thymine, guanine
4. protein
5. amino acids
6. a) three b) triplet c) codon

RNA AND PROTEIN SYNTHESIS

1. one
2. messenger, mRNA
3. uracil
4. nucleus, ribosomes
5. transfer, tRNA
6. codon

7. peptide, ribosomes
8. a) 1) mRNA 2) structural proteins 3) enzymes
   b) DNA → mRNA is transcription; mRNA → protein is translation
9. 1) nucleus
   2) DNA
   3) mRNA
   4) tRNA
   5) amino acids
   6) peptide bond
   7) ribosome

MITOSIS AND MEIOSIS

1. 1) A, B, G, H 2) C, D, E, F
2. each DNA molecule makes a copy of itself
3. 1) B, D, G 3) C, E
   2) A, F 4) H, I
4. a) outer layer of skin—to replace cells that are worn off the skin surface
   b) red bone marrow—to replace RBCs that live only 120 days (or) stomach lining—to replace cells destroyed by the hydrochloric acid in gastric juice
5. a) nerve cells or muscle cells
   b) loss of these cells will have a permanent effect on the functioning of the organ of which they are a part
6. a) interphase b) cytokinesis c) anaphase
   d) metaphase e) prophase f) telophase
   Sequence: a, e, d, c, f, b

CROSSWORD PUZZLE

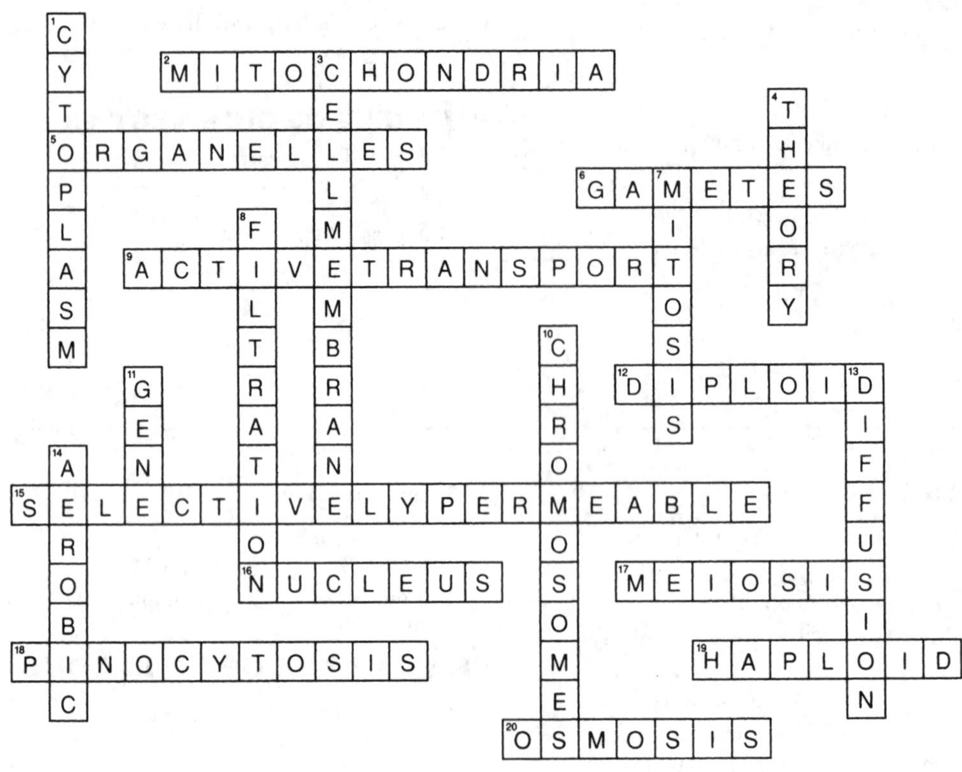

CLINICAL APPLICATIONS

1. red bone marrow
2. a) the same b) gain, osmosis, swell
3. a) mitosis b) protein synthesis

MULTIPLE CHOICE TEST #1

1. b	6. b	11. d	16. d	21. b
2. b	7. a	12. c	17. b	22. d
3. a	8. b	13. a	18. a	23. c
4. d	9. c	14. c	19. b	24. b
5. d	10. c	15. d	20. b	25. a

MULTIPLE CHOICE TEST #2

The letter of the correct choice is followed by the statement necessary to complete the answer. When an incorrect statement is to be corrected to make it true, the new (true) words are underlined.

1. c—active transport—the movement of molecules from an area of lesser concentration to an area of greater concentration
2. d—ribosomes—the site of protein synthesis
3. b—DNA exists as a double strand of nucleotides called a double helix.
4. c—Amino acids are bonded to one another by peptide bonds.

5. d—anaphase—The spindle fibers pull each set of chromosomes toward opposite poles of the cell.
6. c—It is selectively permeable, meaning that only certain substances may pass through.
7. a—Most nerve cells usually do not reproduce themselves.
8. c—osmosis—the absorption of water by the small intestine
9. a—Human cells in a hypertonic solution would shrivel. (or) Human cells in an isotonic solution would remain undamaged.
10. d—Proteasomes have enzymes to destroy misfolded proteins.

MULTIPLE CHOICE TEST #3

1. a, c, d, e, h, i, and j are correct
2. a, e, f, g, and h are correct
3. None are correct

Chapter 4

EPITHELIAL TISSUE

1. 1) columnar epithelium
   2) cuboidal epithelium
   3) simple squamous epithelium
   4) ciliated epithelium
   5) stratified squamous epithelium
2. 1) E, 1, 5, 8 2) A, 3, 9 3) F, 7
   4) D, 2, 6 5) C, 4, 11, 13 6) B, 10, 12
3. 1) A, 2 2) C, 3 3) B, D, 1

CONNECTIVE TISSUE

1. 1) B, 4, 8 4) G, 2, 9 7) F, 6, 10
   2) E, 1 5) A, 3, 11
   3) D, 7 6) C, 5, 12
2. A) blood B) adipose tissue
   C) areolar tissue D) fibrous connective tissue
   E) bone F) cartilage

MUSCLE TISSUE

1. 1) SM 5) SK 9) SM 13) SM
   2) SK 6) SK 10) SK 14) C
   3) SM 7) C 11) C
   4) C 8) SM 12) SK
2. A) skeletal muscle B) smooth muscle
   C) cardiac muscle

NERVE TISSUE

1. neurons, impulses
2. 1) cell body 2) nucleus
   3) dendrites 4) axon
3. a) away from b) toward
4. a) Schwann cells b) oligodendrocytes
5. a) synapse b) neurotransmitters
6. brain and spinal cord (or peripheral nerves)
7. any two of these: sensation, movement, learning, memory

MEMBRANES

1. 1) B, C, D, F, H, J, L 2) A, E, G, I, K
2. a) 1. parietal pleura 2. visceral pleura
   3. peritoneum 4. mesentery
   5. visceral pericardium
   6. parietal pericardium
   b) parietal, visceral
   c) peritoneum, mesentery
   d) parietal, visceral
3. 1) F 3) A 5) C 7) D
   2) G 4) B 6) E

CROSSWORD PUZZLE

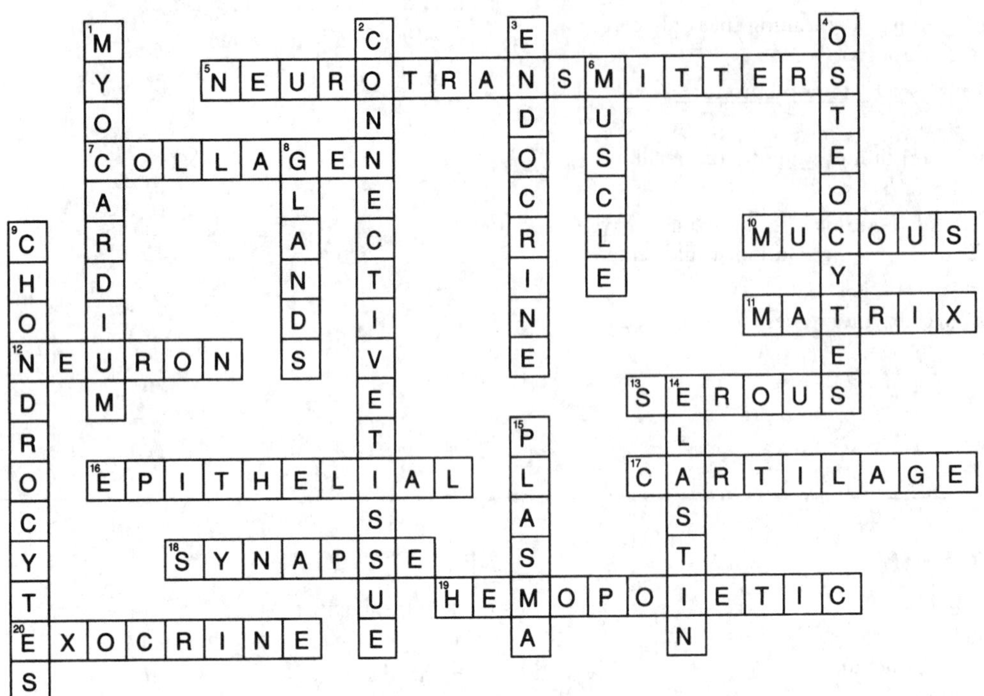

CLINICAL APPLICATIONS

1. blood supply
2. capillaries
3. a) skeletal b) heart
4. peritoneum, abdominal
5. synovial, cartilage

MULTIPLE CHOICE TEST #1

1. b	6. b	11. a	16. c	21. d
2. a	7. d	12. b	17. d	22. d
3. d	8. c	13. c	18. d	23. b
4. d	9. c	14. d	19. a	24. c
5. c	10. a	15. a	20. b	25. a

MULTIPLE CHOICE TEST #2

The letter of the correct choice is followed by the statement necessary to complete the answer. When an incorrect statement is to be corrected to make it true, the new (true) words are underlined.

1. d—Columnar epithelium that absorbs nutrients is found in the small intestine.
2. b—Cardiac muscle that pumps blood is found in the heart.
3. a—peritoneum—lines the abdominal cavity
4. d—Bones are moved by skeletal muscle.
5. d—An example of an endocrine gland is the thyroid gland. (or) An example of an exocrine gland is a salivary gland (sweat gland).
6. a—Smooth muscle in the iris of the eye changes the size of the pupil.
7. d—The axon of a neuron carries impulses away from the cell body.
8. b—Nutrients and waste products are transported by blood plasma.
9. a—Adipose tissue stores fat as a potential energy source.
10. c—Stratified squamous epithelium of the outer layer of the skin has dead cells on the surface.

MULTIPLE CHOICE TEST #3

1. b, c, g, and h are correct
2. a, c, d, e, h, and i are correct
3. a, b, and e are correct
4. a, d, and f are correct
5. All are correct

Chapter 5

GENERAL STRUCTURE AND FUNCTIONS

1. 1) B, 3 2) C, 1 3) A, 2

EPIDERMIS

1. a) stratum germinativum b) stratum corneum
2. a) keratin
   b) water, bacteria, and chemicals (or water)
3. mitosis, stratum germinativum
4. tissue fluid, blister
5. a) bone marrow b) lymphocytes, lymph nodes
   c) antibodies d) defensins
6. a) melanin b) ultraviolet rays
   c) protects living skin cells from further exposure to UV rays
7. stratum germinativum, touch
8. 1) stratum corneum
   2) dermis
   3) mitosis
   4) stratum germinativum
   5) capillary
   6) Langerhans cell
   7) melanocyte
   8) sensory neuron
   9) Merkel cell

DERMIS

1. 1) E 4) A 6) I 8) H 10) K
   2) C 5) F 7) D 9) G 11) B
   3) J
2. a) pain, heat, cold, and itch
   b) touch and pressure
   c) the skin of the palm has more receptors for touch
3. a) nails b) keratin
4. a) cerumen b) sebum
   c) sweat d) sebum
5. a) eyes or nose b) head c) keratin
6. arterioles
7. a) D, cholesterol, UV rays
   b) calcium and phosphorus
8. 1) epidermis
   2) dermis
   3) subcutaneous tissue
   4) sweat gland
   5) hair shaft
   6) stratum corneum
   7) stratum germinativum
   8) papillary layer
   9) sebaceous gland
   10) hair root
   11) hair follicle
   12) sensory receptor
   13) nerve
   14) vein
   15) artery

SUBCUTANEOUS TISSUE

1. superficial fascia
2. a) dermis and the muscles
   b) areolar connective tissue and adipose tissue
3. a) pathogens, breaks in the skin
   b) histamine, inflammation
4. a) fat, energy
   b) cushions bony prominences and provides insulation from cold

MAINTENANCE OF BODY TEMPERATURE

1. eccrine sweat, arterioles
2. a) warm b) evaporation
3. a) constrict b) decrease, retained
4. a) dilate b) increase, lost
5. smooth muscle

BURNS

1. 1) B 2) C 3) A
2. a) stratum corneum
   b) infection and dehydration

CROSSWORD PUZZLE

CLINICAL APPLICATIONS

1. a) cancer, sunlight (UV rays)
   b) use a sunscreen on exposed skin
2. a) third b) infection c) dehydration
3. eccrine sweat, rise (increase)
4. eczema, allergic

MULTIPLE CHOICE TEST #1

1. d	6. c	11. d	16. c	21. b
2. b	7. d	12. c	17. b	22. c
3. a	8. b	13. b	18. c	23. a
4. a	9. a	14. d	19. c	24. d
5. b	10. c	15. a	20. d	25. d

MULTIPLE CHOICE TEST #2

The letter of the correct choice is followed by the statement necessary to complete the answer. When an incorrect statement is to be corrected to make it true, the new (true) words are underlined.

1. b—Adipose tissue stores fat as a form of potential energy.
2. a—The epidermis is made of stratified squamous epithelium.
3. d—Secretion of sweat by eccrine sweat glands is important to lose excess body heat.
4. c—Melanocytes produce melanin that protects living skin layers from ultraviolet rays.
5. d—In sweating, excess body heat is lost in the process of evaporation of sweat.
6. c—The receptors in the dermis are for the cutaneous senses of pain, heat, cold, pressure, itch, and touch.
7. d—It is located between the stratum germinativum and the subcutaneous tissue.
8. b—Cholesterol in the skin is converted to vitamin D.
9. b—The epidermis is the outer layer and the dermis is the inner layer.
10. c—melanin

MULTIPLE CHOICE TEST #3

1. a, c, and d are correct
2. b, d, e, and f are correct
3. c, d, e, and f are correct
4. b, c, and e are correct

Chapter 6

FUNCTIONS OF THE SKELETON

1. red bone marrow
2. calcium
3. a) supports b) muscles
4. a) mechanical injury
   b) the skull protects the brain (or) the ribs protect the heart and lungs

BONE TISSUE

1. 1) C 2) D, F 3) A, E 4) B

CLASSIFICATION OF BONES

1. 1) D, 2, 4 2) C, 1, 3 3) A, 1, 3 4) B, 1, 3
2. cartilage
3. periosteum, fibrous
4. blood vessels
5. a) tendons, ligaments
   b) fibrous connective tissue
6. 1) proximal epiphysis
   2) diaphysis
   3) distal epiphysis
   4) yellow bone marrow
   5) marrow cavity
   6) compact bone
   7) spongy bone
   8) haversian systems
   9) haversian canal
   10) osteocytes
   11) periosteum

EMBRYONIC GROWTH OF BONE

1. a) ossification b) osteoblasts
2. a) fibrous connective tissue b) third, fibroblasts
3. a) fontanels
   b) 1. frontal bone
      2. anterior fontanel
      3. parietal bone
      4. posterior fontanel
      5. mandible
      6. sphenoid fontanel
      7. temporal bone
      8. occipital bone
4. fontanels permit compression of the infant's skull during birth
5. cartilage
6. diaphysis, epiphysis
7. epiphyseal discs
8. 1) cartilage production
   2) bone replaces cartilage
   3) bone replaces cartilage
   4) cartilage production
9. cartilage, bone
10. a) osteoclasts
    b) yellow, adipose

FACTORS THAT AFFECT BONE GROWTH AND MAINTENANCE

1. 1) D 3) A 5) C
   2) A 4) C 6) B
2. 1) D, anterior pituitary gland
   2) F, pancreas
   3) A, thyroid gland
   4) B, thyroid gland
   5) E, parathyroid glands
   6) C, ovaries or testes
   7) calcitonin
   8) parathyroid hormone
   9) estrogen or testosterone
3. a) the mother, the father
   b) enzymes, cartilage and bone
4. a) bearing weight b) calcium
   c) become brittle and fracture easily

THE SKELETON

1. a) axial, skull, rib cage
   b) appendicular, arms and legs
2. a) rib cage b) skull
   c) hip bones d) scapula and clavicle
3. 1) clavicle
   2) scapula
   3) humerus
   4) radius
   5) ulna
   6) carpals
   7) metacarpals
   8) phalanges
   9) tarsals
   10) metatarsals
   11) phalanges
   12) frontal bone
   13) zygomatic bone
   14) maxilla
   15) mandible
   16) sternum
   17) rib
   18) vertebra
   19) hip bone
   20) sacrum
   21) coccyx
   22) femur
   23) patella
   24) tibia
   25) fibula

THE SKULL

1. 1) sphenoid bone
   2) lacrimal bone
   3) frontal bone
   4) coronal suture
   5) nasal bone
   6) zygomatic bone
   7) ethmoid bone
   8) conchae
   9) vomer
   10) maxilla
   11) mandible
   12) nasal bone
   13) lacrimal bone
   14) zygomatic bone
   15) maxilla
   16) frontal bone
   17) parietal bone
   18) coronal suture
   19) sphenoid bone
   20) squamosal suture
   21) temporal bone
   22) lambdoidal suture
   23) occipital bone
   24) external auditory meatus
   25) condyloid process
   26) mastoid process
   27) mandible

2. 1) mandible and maxillae
   2) temporal bone
   3) zygomatic bone
   4) maxillae and palatine bones
   5) ethmoid bone
   6) vomer and ethmoid bone
   7) lacrimal bone
   8) occipital bone, spinal cord
   9) sphenoid bone
   10) nasal bones
   11) a) maxillae, frontal, sphenoid, and ethmoid bones
       b) lighten the skull and provide resonance for the voice
3. parietal and temporal
4. parietal and occipital
5. parietal and frontal
6. parietal
7. a) malleus, incus, and stapes
   b) hearing

THE VERTEBRAL COLUMN

1. support, spinal cord
2. a) 1. atlas
      2. axis
      3. first thoracic
      4. intervertebral disc
      5. first lumbar
      6. sacrum
      7. coccyx
      8. cervical vertebrae
      9. thoracic vertebrae
      10. lumbar vertebrae
      11. sacrum
      12. coccyx
   b) 1) 7 2) 12 3) 5
      4) 5, sacrum 5) 4 or 5, coccyx
3. a) atlas, axis b) pivot, side to side
4. posterior, thoracic
5. sacrum
6. a) body b) cartilage
   c) absorb shock and permit movement
   d) symphysis

THE RIB CAGE

1. 1) first thoracic vertebra
   2) costal cartilage
   3) body of sternum
   4) xiphoid process
   5) floating ribs
   6) first rib
   7) manubrium
   8) true ribs
   9) seventh rib
   10) eighth rib
   11) false ribs
   12) 12th rib
   13) 12th thoracic vertebra
2. seven, three, two
3. heart and lungs
4. liver and spleen
5. up and out, expand (enlarge), lungs

THE SHOULDER AND ARM

1. scapula and clavicle
2. scapula
3. a) manubrium, scapula
   b) braces the scapula (keeps it back)
4. head, glenoid fossa, ball and socket
5. a) radius and ulna b) ulna
6. a) humerus and ulna b) hinge
7. a) pivot b) palm up to palm down
8. a) carpals b) eight c) gliding
9. five, carpals, phalanges
10. a) saddle b) rotation or gripping
11. a) two b) three c) 14 d) hinge
12. 1) acromial end
    2) acromion process
    3) coracoid process
    4) head of humerus
    5) humerus
    6) capitulum
    7) head of radius
    8) radius
    9) scaphoid
    10) trapezium
    11) trapezoid
    12) capitate
    13) lunate
    14) triquetrum
    15) pisiform
    16) hamate
    17) metacarpals
    18) phalanges
    19) clavicle
    20) sternal end
    21) glenoid fossa
    22) scapula
    23) trochlea
    24) semilunar notch
    25) ulna

THE HIP AND LEG

1. hip bone
2. acetabulum
3. a) ilium, ischium, and pubis b) ilium
   c) sacroiliac d) pubis
   e) pubic symphysis f) ischium
4. a) femur b) ball and socket
5. a) tibia and fibula b) tibia
   c) muscle attachment
6. a) femur and tibia b) hinge
7. a) tarsals b) seven c) calcaneus
8. five, tarsals, phalanges
9. a) 14 b) two c) three
10. 1) ilium
    2) sacroiliac joint
    3) acetabulum
    4) ischium
    5) pubis
    6) sacrum
    7) coccyx
    8) pubic symphysis
11. 1) acetabulum
    2) head of femur
    3) greater trochanter
    4) neck of femur
    5) lesser trochanter
    6) femur
    7) medial condyles
    8) patella
    9) lateral condyles
    10) head of fibula
    11) tibial tuberosity
    12) tibia
    13) fibula
    14) medial malleolus
    15) lateral malleolus
    16) talus
    17) navicular
    18) cuneiforms
    19) cuboid
    20) metatarsals
    21) phalanges
    22) cuboid
    23) navicular
    24) talus
    25) calcaneus
    26) lateral malleolus

JOINTS (ARTICULATIONS)

1.

Name of Joint	Movement Possible	Examples (name two bones)
1) Suture	None—immovable	a. frontal and parietal bones b. parietal and temporal bones
2) Symphysis	Slight movement	a. two vertebrae b. two pubic bones
3) Hinge	Movement in one plane	a. humerus and ulna b. femur and tibia
4) Gliding	Sliding movement	a. two carpals b. sacrum and ilium
5) Saddle	Movement in all planes	a. carpometacarpal of the thumb
6) Ball and socket	Movement in all planes	a. scapula and humerus b. hip bone and femur
7) Pivot	Rotation	a. atlas and axis b. radius and ulna
8) Condyloid	Movement in one plane with some lateral movement	a. mandible and temporal bone

2. 1) acetabulum
   2) head of femur
   3) ball-and-socket joint
   4) trochlea of humerus
   5) semilunar notch of ulna
   6) hinge joint
   7) atlas
   8) odontoid process of axis
   9) pivot joint
   10) carpal
   11) carpal
   12) gliding joint
   13) body of vertebra
   14) intervertebral disc
   15) body of vertebra
   16) symphysis joint
   17) metacarpal of thumb
   18) carpal
   19) saddle joint

SYNOVIAL JOINT STRUCTURE

1. 1) E 2) C 3) A 4) B 5) D
2. 1) bursa 2) articular cartilage
   3) synovial membrane 4) joint capsule
   5) synovial fluid (joint cavity)

CROSSWORD PUZZLE

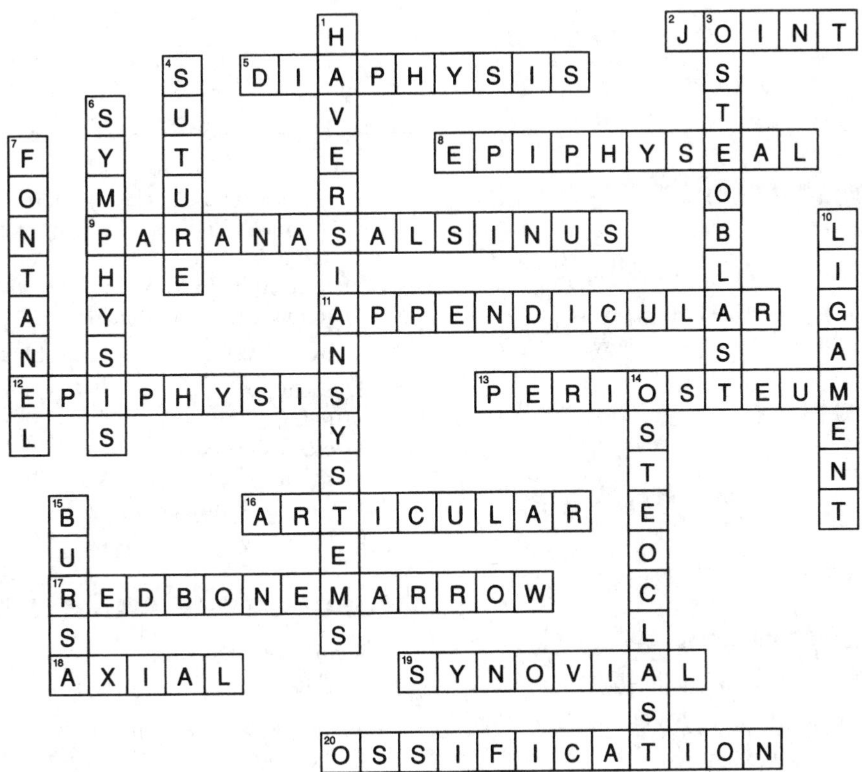

CLINICAL APPLICATIONS

1. a) skin b) epiphyseal disc c) osteoblasts
2. a) spontaneous (pathologic)
   b) osteoporosis c) neck of the femur
3. a) lateral b) lung
4. a) intervertebral
   b) the ruptured disc is compressing spinal nerves that supply the leg
5. 1) cheek bone 2) back of the head
   3) lower leg near the ankle 4) knee cap
   5) collarbone and top of the breastbone 6) lower jaw
   7) finger bones

MULTIPLE CHOICE TEST #1

1. d	6. c	11. c	16. c	21. c
2. b	7. b	12. b	17. b	22. d
3. b	8. a	13. c	18. b	23. b
4. d	9. a	14. d	19. b	24. a
5. d	10. b	15. a	20. b	25. a

MULTIPLE CHOICE TEST #2

The letter of the correct choice is followed by the statement necessary to complete the answer. When an incorrect statement is to be corrected to make it true, the new (true) words are underlined.

1. c—Blood cells are produced by red bone marrow found in spongy bone.
2. d—Calcitonin decreases the reabsorption of calcium from bones.
3. b—New bone matrix is produced by cells called osteoblasts.
4. d—Vitamin D is necessary for the absorption of calcium by the small intestine.
5. b—The rib cage protects the heart and lungs (or liver and spleen).
6. b—Synovial fluid decreases friction within the joint cavity.
7. a—The hinge joint between the femur and tibia permits movement in one plane at the knee.
8. b—The only movable joint is the condyloid joint between the temporal bone and the mandible.
9. d—The hip bones articulate with the sacrum.
10. b—The only weight-bearing bone in the lower leg is the tibia.
11. b—During inhalation the ribs are pulled up and out to expand the lungs.
12. d—There are eight carpals in each wrist and 14 phalanges in each hand.

MULTIPLE CHOICE TEST #3

1. d, e, g, and i are correct
2. c, e, g, and i are correct
3. c, e, f, and k are correct

Chapter 7

MUSCLES AND MOVEMENT

1. bones, tendons
2. a) move the skeleton b) produce body heat
3. a) respiratory system b) skeletal system
   c) nervous system d) circulatory system

MUSCLE STRUCTURE

1. muscle cell
2. a) they become shorter b) more
3. fascia, periosteum
4. a) joint b) origin c) insertion d) insertion

MUSCLE ARRANGEMENTS

1. a) opposite b) the same (or stabilizing)
2. a) pull b) push c) pull
3. 1) to decrease the angle of a joint; extension
   2) to move a body part away from the midline; adduction
   3) to turn the palm down; supination
   4) to lower the foot; dorsiflexion
4. a) frontal, cerebrum b) cerebellum
5. 1) extension 7) flexion
   2) radius 8) radius
   3) biceps—relaxed 9) biceps—contracted
   4) scapula 10) scapula
   5) triceps—contracted 11) triceps—relaxed
   6) humerus 12) humerus

MUSCLE TONE AND EXERCISE

1. slight contraction
2. heat, body temperature
3. a) movement b) no movement c) both
   d) isotonic e) heart, respiratory
4. a) improves coordination (or) helps maintain posture
   b) permits rapid muscle responses (or) produces body heat

MUSCLE SENSE

1. knowing where our muscles are without having to look at them
2. stretch receptors or proprioceptors
3. brain
4. a) parietal b) cerebellum, coordination
5. permits us to accomplish everyday activities without having to constantly watch our movements

ENERGY SOURCES FOR MUSCLE CONTRACTION

1. a) ATP, creatine phosphate, glycogen
   b) ATP c) glycogen
2. creatinine, kidneys
3. glucose
4. a) Glucose + $O_2 \rightarrow CO_2 + H_2O$ + ATP + Heat
   b) ATP
5. a) myoglobin b) hemoglobin c) iron
6. a) oxygen debt b) lactic acid, fatigue
7. liver, pyruvic acid
8. a) water: becomes part of intracellular water
   b) heat: contributes to a constant body temperature
   c) carbon dioxide: is exhaled

MUSCLE FIBER—MICROSCOPIC STRUCTURE

1. neuron and a muscle cell
2. 1) B, C 2) D, E 3) A, F
3. 1) D 2) C 3) B 4) A
4. 1) motor neuron
   2) axon terminal
   3) mitochondrion
   4) vesicle of acetylcholine
   5) synapse (synaptic cleft)
   6) sarcolemma
   7) sarcomere
   8) acetylcholine
   9) sodium ion outside sarcolemma
   10) ACh receptor
   11) cholinesterase
   12) sodium ion inside sarcolemma

SARCOLEMMA—ELECTRICAL EVENTS

1. a) positive, negative b) outside, inside
   c) sodium and potassium pumps
2. a) sodium b) negative, positive
   c) repolarization, potassium d) action potential
3. 1) sodium ions
   2) potassium ions
   3) sodium ions enter the cell
   4) wave of depolarization from the point of stimulus
   5) potassium ions leave the cell
   6) wave of repolarization

CONTRACTION—SLIDING FILAMENT MECHANISM

1. 1) F 2) I 3) D 4) J 5) A 6) G 7) C 8) H 9) B 10) E
2. 1) T tubule
   2) sarcoplasmic reticulum
   3) myosin filament
   4) actin filament
   5) troponin
   6) tropomyosin
   7) actin
   8) calcium ions
   9) calcium ions bonded to troponin
   10) myosin cross-bridges pulling actin
3. tetanus
4. rapid nerve impulses sustain (or prolong) the state of contraction
5. Movement—sustained contractions are essential for useful movement

   Disease—abnormal sustained contractions result in muscle spasms (or) muscles are unable to relax voluntarily

EXERCISE—HOMEOSTASIS

1. 1) C, E 2) B 3) D, F 4) A 5) G

MAJOR MUSCLES OF THE BODY

1. a) 1. masseter
   2. sternocleidomastoid
   3. deltoid
   4. pectoralis major
   5. biceps brachii
   6. brachioradialis
   7. rectus femoris
   8. sartorius
   9. vastus lateralis
   10. vastus medialis
   11. tibialis anterior
   12. external oblique
   13. rectus abdominis
   14. adductor longus
   15. gastrocnemius
   16. soleus

   b) 1. triceps brachii
   2. latissimus dorsi
   3. gluteus medius
   4. gluteus maximus
   5. biceps femoris
   6. soleus
   7. Achilles tendon
   8. trapezius
   9. deltoid
   10. teres major
   11. semimembranosus
   12. gastrocnemius
2. any three of these pairs: biceps brachii and triceps brachii; quadriceps femoris and hamstring group; gastrocnemius and tibialis anterior; deltoid and pectoralis major; pectoralis major and latissimus dorsi; rectus abdominis and sacrospinalis group; adductor group and gluteus medius. There are others that are correct, if they have opposite functions.
3. deltoid, vastus lateralis (part of the quadriceps femoris), and gluteus medius
4. 1) flexion of head
   2) extension of head
   3) flexion of arm
   4) abduction of arm
   5) extension of arm
   6) adduction of arm
   7) flexion of forearm
   8) extension of forearm
   9) flexion of leg
   10) extension of leg
   11) flexion of thigh
   12) extension of thigh
   13) abduction of thigh
   14) adduction of thigh

CROSSWORD PUZZLE

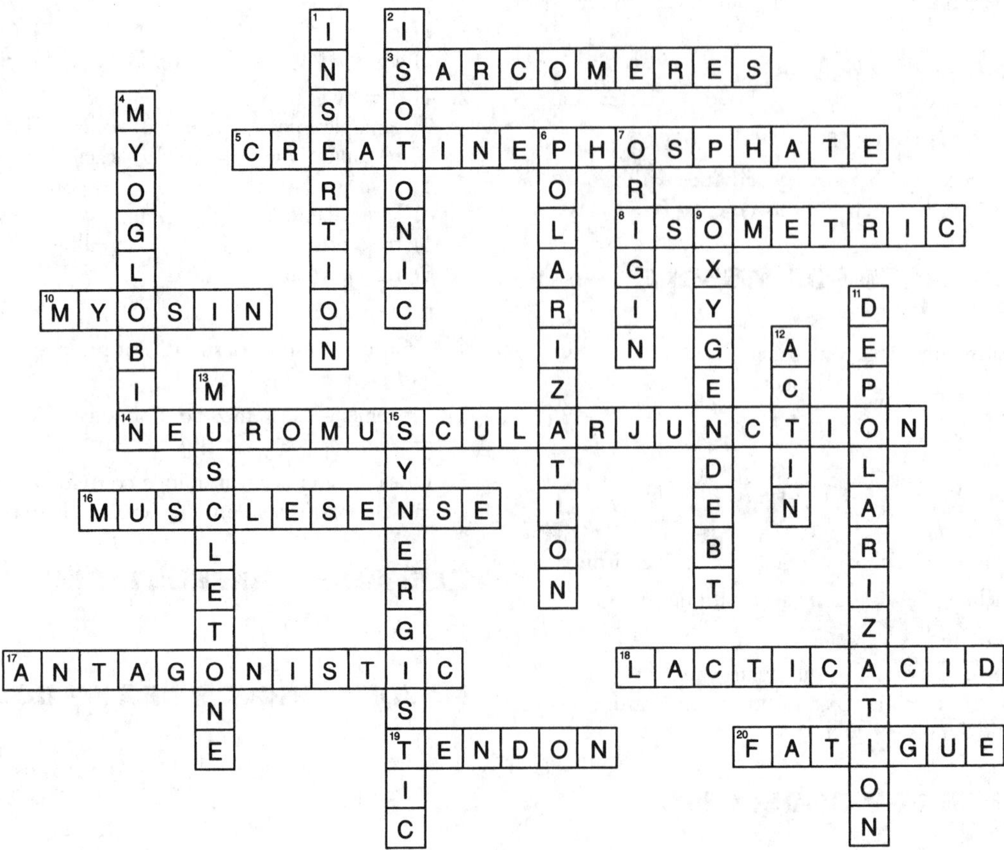

CLINICAL APPLICATIONS

1. a) nerves b) paralyzed
   c) become smaller d) atrophy
2. a) bacteria b) paralysis, acetylcholine
   c) spasms, relax
3. a) his mother (the gene is on the X chromosome)
   b) fibrous connective tissue or by fat

MULTIPLE CHOICE TEST #1

1. d	6. c	11. a	16. d	21. c
2. d	7. b	12. c	17. c	22. b
3. d	8. a	13. d	18. b	23. a
4. c	9. b	14. a	19. d	24. d
5. c	10. d	15. a	20. d	25. a

MULTIPLE CHOICE TEST #2

The letter of the correct choice is followed by the statement necessary to complete the answer. When an incorrect statement is to be corrected to make it true, the new (true) words are underlined.

1. d—Unconscious muscle sense is integrated by the cerebellum.
2. c—Muscles are attached to bones by tendons, which are made of fibrous connective tissue.
3. a—The deltoid is the shoulder muscle that abducts the arm.
4. c—Muscles that are paralyzed will atrophy, which means to become smaller from disuse.
5. d—Muscle tone does depend on nerve impulses to muscle fibers.
6. b—The direct energy source for contraction is ATP.
7. b—Acetylcholine makes the sarcolemma very permeable to sodium ions.
8. b—Isotonic exercise is aerobic because it involves contraction with movement.
9. b—Some oxygen is stored in muscle by the protein myoglobin.
10. d—During depolarization, the inside of the sarcolemma becomes positive.

MULTIPLE CHOICE TEST #3

1. a, c, and e are correct
2. all are correct
3. a, b, and f are correct
4. b, d, e, and f are correct
5. d, e, f, h, and k are correct

Chapter 8

NERVOUS SYSTEM DIVISIONS

1. brain and spinal cord
2. cranial nerves and spinal nerves
3. peripheral nervous system
4. alimentary tube (stomach and intestines)

NERVE TISSUE

1. neuron
2. a) cell body b) axon c) dendrites
3. a) 1. synaptic knobs
   2. axon
   3. myelin sheath
   4. Schwann cell
   5. nucleus
   6. cell body
   7. dendrite
   8. synapse
   9. cell body
   10. nucleus
   11. dendrites
   12. neurolemma
   13. axon
   14. myelin sheath
   15. axon terminals

   b) axon c) cell body d) dendrites
   e) synapse f) sensory neuron begins in receptors; motor neuron ends in muscle
4. a) Schwann cells b) oligodendrocytes
   c) provides electrical insulation for neurons
5. a) nuclei and cytoplasm b) regeneration
6. a) neurotransmitter, axon c) cholinesterase
   b) inactivator chemical
7. 1) axon of presynaptic neuron
   2) vesicles of neurotransmitter
   3) neurotransmitter
   4) dendrite of postsynaptic neuron
   5) inactivator
   6) receptor site
   7) sodium ion enters cell
   8) neurotransmitter inactivated

NEURONS, NERVES, AND NERVE TRACTS

1. 1) B, D 2) C, F 3) A 4) G 5) E

THE NERVE IMPULSE

1. a) positive, negative b) sodium
   c) potassium and negative d) polarization
2. a) sodium, into b) negative, positive
3. a) potassium, out of, repolarization
   b) positive, negative
   c) sodium and potassium pumps
   d) action potential

THE SPINAL CORD AND SPINAL NERVES

1. 1) transmits impulses to and from the brain
   2) is the center for the spinal cord reflexes
2. a) vertebrae
   b) foramen magnum, first and second lumbar vertebrae
3. a) 1. dorsal root ganglion
   2. dorsal root
   3. white matter
   4. central canal
   5. gray matter
   6. ventral root

   b) gray matter
   c) dorsal root ganglion
   d) sensory, sensory
   e) motor, motor
   f) white matter
   g) sensory, toward
   h) motor, away from
   i) central canal
4. 1) eight 2) 12 3) five 4) five 5) one
5. 1) thoracic
   2) lumbar and sacral
   3) cervical (and first thoracic)
   4) cervical

SPINAL CORD REFLEXES

1. an involuntary response to a stimulus
2. 1, 5, 4, 3, 2
3. 1) stretch receptor
   2) quadriceps femoris
   3) sensory neuron
   4) femoral nerve
   5) dorsal root ganglion
   6) dorsal root
   7) synapse
   8) gray matter
   9) white matter
   10) ventral root
   11) motor neuron
4. stretching of a muscle, contraction of the same muscle
5. a) keep the body upright (maintain posture)
   b) protect the body from potential harm
   c) no conscious decision is required; that would take too much time

THE BRAIN

1. 1) H, N
   2) B, L, O
   3) K
   4) A, P
   5) C, J
   6) E, I, Q, S, T
   7) M, R
   8) G
   9) F
   10) D
2. 1) C, F 2) A, B 3) E, G 4) D
3. 1) corpus callosum
   2) frontal lobe
   3) thalamus
   4) third ventricle
   5) hypothalamus
   6) pituitary gland
   7) brainstem
   8) midbrain
   9) pons
   10) medulla
   11) parietal lobe
   12) occipital lobe
   13) cerebellum
   14) fourth ventricle

4. 1) frontal lobe
   2) temporal lobe
   3) pons
   4) medulla
   5) parietal lobe
   6) occipital lobe
   7) cerebellum
5. cerebral cortex, cell bodies
6. a) lateral
   b) fourth
   c) third
   d) cerebral aqueduct
7. 1) cerebral cortex
   2) corpus callosum
   3) lateral ventricle
   4) basal ganglia
   5) temporal lobe
   6) optic tracts
   7) hypothalamus
   8) third ventricle
   9) thalamus
   10) longitudinal fissure

MENINGES AND CEREBROSPINAL FLUID

1. three
2. a) dura mater
   b) skull and vertebral canal
   c) arachnoid
   d) pia mater, brain and spinal cord
3. arachnoid membrane, pia mater, cerebrospinal fluid
4. choroid plexuses, ventricles
5. 1) ventricles
   2) subarachnoid space
   3) central canal
   4) subarachnoid space
6. 1) exchanges nutrients and waste products between the CNS and blood
   2) absorbs shock around (cushions) the CNS
7. a) arachnoid villi, cranial venous sinuses
   b) the same as
8. 1) dura mater
   2) arachnoid membrane
   3) pia mater
   4) central canal
   5) gray matter of spinal cord
   6) white matter of spinal cord
   7) roots of a spinal nerve
   8) spinal nerve
   9) subarachnoid space
9. 1) dura mater
   2) arachnoid membrane
   3) subarachnoid space
   4) pia mater
   5) lateral ventricle
   6) choroid plexus
   7) choroid plexus in third ventricle
   8) arachnoid villus
   9) cranial venous sinus
   10) cerebral aqueduct
   11) choroid plexus in fourth ventricle
   12) central canal

CRANIAL NERVES

1. 1) optic
   2) acoustic
   3) facial and glossopharyngeal
   4) olfactory
   5) acoustic
   6) facial and glossopharyngeal
   7) oculomotor, abducens, and trochlear
   8) vagus
   9) vagus
   10) vagus and accessory
   11) hypoglossal
   12) accessory
   13) oculomotor
   14) vagus and glossopharyngeal
   15) trigeminal
   16) facial
   17) trigeminal
2. 1) olfactory
   2) optic
   3) oculomotor
   4) trochlear
   5) trigeminal
   6) abducens
   7) facial
   8) acoustic
   9) glossopharyngeal
   10) vagus
   11) accessory
   12) hypoglossal

THE AUTONOMIC NERVOUS SYSTEM (ANS)

1. smooth, cardiac, glands
2. hypothalamus
3. a) sympathetic, stressful
   b) parasympathetic, non-stressful
4. 1) P
   2) S
   3) S
   4) P
   5) P
   6) S
   7) P
   8) S
5. 1) thoracic spinal cord
   2) sympathetic preganglionic neuron
   3) sympathetic postganglionic neuron
   4) sympathetic ganglia
   5) lumbar spinal cord
   6) brain
   7) sacral spinal cord
   8) parasympathetic preganglionic neurons
   9) parasympathetic ganglion
   10) parasympathetic postganglionic neurons

6.

Visceral Effector	Parasympathetic Response	Sympathetic Response
1) stomach and intestines (glands)	increase secretions	decrease secretions
2) stomach and intestines (smooth muscle)	increase peristalsis	slow peristalsis
3) heart	decreases rate (to normal)	increases rate
4) iris	constricts pupil	dilates pupil
5) urinary bladder	contracts	relaxes
6) bronchioles	constrict (to normal)	dilate
7) salivary glands	increase secretion	decrease secretion
8) blood vessels in skeletal muscle	none	vasodilation
9) blood vessels in skin and viscera	none	vasoconstriction
10) sweat glands	none	increase secretion
11) liver	none	changes glycogen to glucose

CROSSWORD PUZZLE

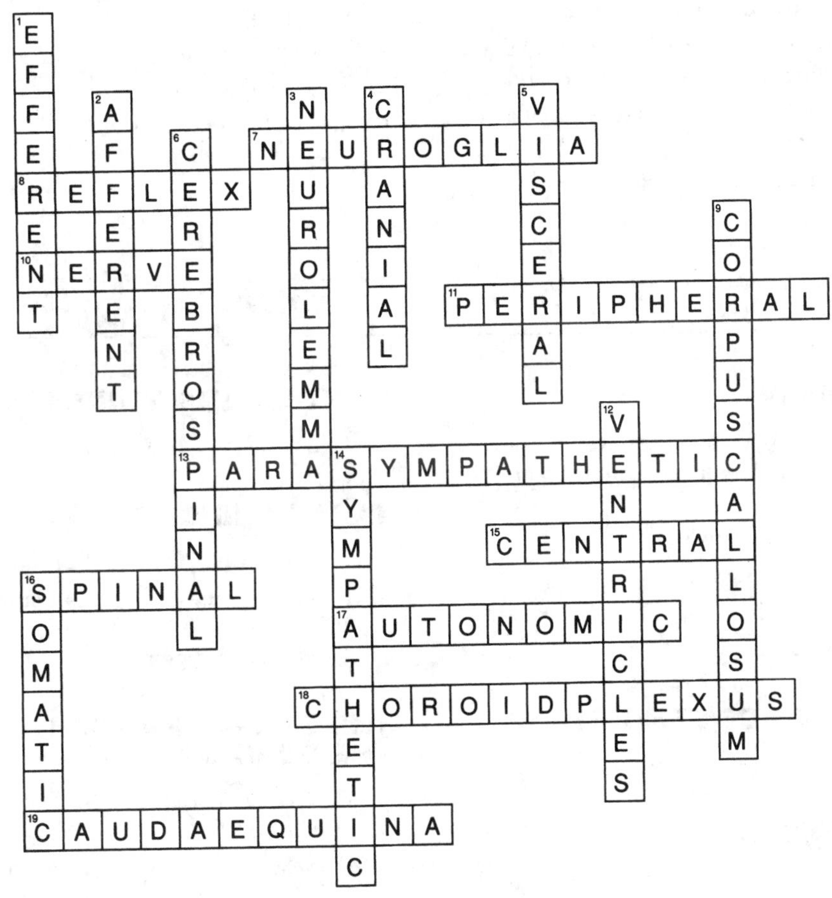

CLINICAL APPLICATIONS

1. a) meningitis b) cerebrospinal fluid
   c) cloudy, bacteria, white blood cells
2. a) frontal, left b) motor speech area
3. a) spinal shock b) sensation
   c) paralysis below the injury
   d) motor impulses from the brain cannot get to the muscles
4. a) medulla b) occipital c) cerebellum

MULTIPLE CHOICE TEST #1

1. c	7. b	13. c	19. b	25. b
2. a	8. d	14. b	20. c	26. d
3. a	9. d	15. d	21. c	27. a
4. d	10. d	16. a	22. a	28. c
5. b	11. c	17. b	23. c	29. d
6. a	12. c	18. b	24. a	30. b

MULTIPLE CHOICE TEST #2

The letter of the correct choice is followed by the statement necessary to complete the answer. When an incorrect statement is to be corrected to make it true, the new (true) words are underlined.

1. c—CSF is reabsorbed into the blood in the cranial venous sinuses.
2. d—Regulates coordination (or) equilibrium (or) muscle tone.
3. d—Sensory neurons may also be called afferent neurons. (or) Motor neurons may also be called efferent neurons.
4. c—visual area
5. b—The stimulus is detected by receptors.
6. a—visual and auditory reflexes (or) equilibrium
7. c—The spinal cord extends from the foramen magnum to the disc between the first and second lumbar vertebrae.
8. b—Each spinal nerve has two roots; the dorsal root and the ventral, or motor, root.
9. c—The nucleus is located within the cell body.
10. b—The pia mater is the innermost layer and is on the surface of the brain and spinal cord.
11. d—sense of smell
12. b—The parasympathetic division causes constriction of the bronchioles and increased peristalsis.
13. c—The parasympathetic division contains the vagus nerves.
14. b—During depolarization, sodium ions rush into the cell.

MULTIPLE CHOICE TEST #3

1. b, c, and d are correct
2. a, b, c, and g are correct
3. Only j is correct
4. b, d, and f are correct
5. a, c, e, and g are correct
6. d, e, and h are correct

Chapter 9

SENSORY PATHWAY

1. 1) receptors 2) sensory neurons
   3) sensory tracts 4) sensory area in the brain
2. a) changes, impulses
   b) stimulus, electrical nerve impulses
3. receptors, central nervous system
4. central nervous system, brain
5. cerebrum (or) cerebral cortex

CHARACTERISTICS OF SENSATIONS

1. 1) D 2) A 3) B 4) C 5) E

CUTANEOUS SENSES

1. a) dermis of the skin
   b) changes in the external environment
2. a) pain, heat, cold, itch
   b) touch, pressure
3. parietal
4. internal organ, cutaneous
5. heart—left shoulder (or) gallbladder—right shoulder

MUSCLE SENSE

1. proprioceptors (stretch receptors), stretching
2. a) parietal b) cerebellum
3. enables us to accomplish everyday activities without having to look at our muscles to see where a body part is

HUNGER AND THIRST

1. 1) B, C, E 2) A, D, F

TASTE AND SMELL

1. 1) B, D, E, G, H, J 2) A, C, F, I
2. 1) olfactory nerve 4) taste buds
   2) olfactory bulb 5) tongue
   3) olfactory receptors

THE EYE—STRUCTURES OUTSIDE THE EYEBALL

1. 1) D 2) G 3) F 4) C
   5) A 6) E 7) B
2. 1) eyelid 4) conjunctiva
   2) lacrimal gland 5) lacrimal sac
   3) lacrimal ducts 6) nasolacrimal duct
3. a) bones of the orbit b) six
   c) move the eyeball
4. oculomotor, abducens, trochlear

THE EYE—EYEBALL

1. a) sclera, fibrous connective b) choroid layer
   c) retina
2. 1) F 2) R 3) C 4) L 5) A 6) Q 7) H 8) P 9) E 10) B 11) D 12) K 13) G 14) O 15) M 16) N 17) I 18) J
3. 1) superior rectus muscle
   2) ciliary body
   3) suspensory ligaments
   4) conjunctiva
   5) cornea
   6) iris
   7) lens
   8) pupil
   9) aqueous humor
   10) anterior chamber
   11) posterior chamber
   12) vitreous humor
   13) fovea
   14) optic nerve
   15) retinal artery and vein
   16) optic disc
   17) retina
   18) choroid layer
   19) sclera
   20) canal of Schlemm
   21) inferior rectus muscle
4. 1) optic 2) oculomotor 3) sympathetic
5. a) occipital b) binocular c) right side up

SUMMARY OF VISION

1. 1, 4, 6, 2, 7, 3, 5

THE EAR

1. hearing and equilibrium
2. outer ear, middle ear, inner ear

OUTER EAR

1. a) cartilage
   b) no, because it is not movable
2. a) temporal b) ceruminous c) cerumen

MIDDLE EAR

1. temporal, air
2. 1) B 2) E 3) A 4) C 5) D
3. 1) pinna (auricle)
   2) ear canal
   3) eardrum
   4) malleus
   5) incus
   6) semicircular canals
   7) stapes
   8) vestibule
   9) vestibular nerve (8th)
   10) cochlear nerve (8th)
   11) cochlea
   12) eustachian tube

INNER EAR

1. temporal, membrane, fluid
2. 1) semicircular canals 2) cochlea
   3) utricle and saccule (or) vestibule
3. a) impulses b) acoustic (or) 8th cranial nerve
4. a) organ of Corti b) fluid (endolymph)
   c) temporal
5. a) vestibule b) otoliths c) gravity, head
6. a) three b) movement
7. a) cerebellum, midbrain
   b) cerebrum (temporal lobe)
8. 1) semicircular canals
   2) ampulla
   3) utricle
   4) bony labyrinth
   5) perilymph
   6) membranous labyrinth
   7) crista
   8) saccule
   9) vestibular nerve (8th)
   10) cochlear nerve (8th)
   11) tympanic canal
   12) cochlear duct
   13) vestibular canal
   14) cochlea
   15) round window
   16) oval window
9. 1) otoliths
   2) hair cells
   3) hair cells bent by gravity
   4) 8th cranial nerve
   5) hair cells
   6) 8th cranial nerve
   7) hair cells bend opposite to movement
   8) hair cells bend in response to stopping

SUMMARY OF HEARING

1. 1, 5, 3, 8, 2, 4, 6, 7, 9
2. round window

ARTERIAL RECEPTORS

1. blood pressure, carotid sinus, aortic sinus
2. pH, oxygen, and carbon dioxide levels; carotid body, aortic body
3. a) heart rate or respiration b) medulla

CROSSWORD PUZZLE

```
            O  O
            LACRIMALGLANDS
   R        F  G     E
  SEMICIRCULARCANALS N
   T  O E   C  N     S
   I  R F VITREOUS
   N  N E   O  F
   A  E R   R  CONJUNCTIVA
      A R   Y  O         Q
        E   R  R    T    U
      RODS  EUSTACHIANTUBE
 IRIS       C  I O  S    O
    C       E    C  T    U
    L  RHODOPSIN H  E    S
 CONES      T    L  B
    R       O    E  U
TYMPANICMEMBRANE A  D
            S       S
```

CLINICAL APPLICATIONS

1. a) glaucoma, aqueous humor
   b) retina or optic nerve
2. a) conduction
   b) eardrum or auditory bones
3. conjunctiva, conjunctivitis
4. a) distant objects, near vision (reading)
   b) lens
5. 1) a—no chance, because a son receives the Y chromosome from his father, not the X chromosome
   2) a—no chance, because a daughter would need two genes for color blindness
   3) carrier

MULTIPLE CHOICE TEST #1

1. d	7. a	13. c	19. c	25. b
2. d	8. a	14. d	20. d	26. d
3. c	9. d	15. c	21. a	27. c
4. c	10. b	16. d	22. a	28. d
5. b	11. a	17. b	23. b	29. a
6. b	12. b	18. b	24. d	30. d

MULTIPLE CHOICE TEST #2

The letter of the correct choice is followed by the statement necessary to complete the answer. When an incorrect statement is to be corrected, the new (true) words are underlined.

1. c—The eardrum is at the end of the ear canal.
2. b—The auditory areas are in the temporal lobes.
3. c—If too much light strikes the eye, the iris will constrict the pupil.
4. c—The eardrum transmits vibrations to the malleus, incus, and stapes.
5. d—The cranial nerves for the sense of smell are the olfactory nerves.
6. a—The receptors for pain are free nerve endings in the dermis.
7. c—The effect of a previous sensation on a current sensation is called contrast.
8. d—The eyeball is moved from side to side by the extrinsic muscles.
9. b—The cerebellum and the midbrain regulate the reflexes that keep us upright.
10. d—Chemoreceptors in the aortic and carotid bodies detect changes in the blood levels of oxygen and carbon dioxide.

MULTIPLE CHOICE TEST #3

1. a, b, c, d, and e are correct
2. d, f, g, and k are correct
3. c, d, e, f, h, and j are correct
4. a, b, d, e, and f are correct
5. b, c, d, and e are correct

Chapter 10

ENDOCRINE GLANDS

1. a) hormones, blood b) target c) receptors
2. amines, proteins, and steroids
3. a) sphenoid b) larynx
   c) thyroid gland d) duodenum, spleen
   e) kidneys f) uterus
   g) scrotum
4. 1) anterior pituitary
   2) thyroid gland
   3) adrenal medulla
   4) testes
   5) hypothalamus
   6) posterior pituitary
   7) parathyroid glands
   8) thymus
   9) adrenal cortex
   10) pancreas
   11) ovaries

THE PITUITARY GLAND

1. anterior pituitary, posterior pituitary
2. a) hypothalamus b) releasing, hypothalamus

Posterior Pituitary Gland

1. vasopressin, kidneys
2. a) increase, water
   b) decreases, increases
3. decreased
4. a) uterus and mammary glands
   b) smooth muscle (myometrium), baby, placenta
   c) milk
   d) hypothalamus

Anterior Pituitary Gland

1. a) 1) amino acids, proteins 2) mitosis
   3) fats
   b) GHRH c) GHIH
   d) 1) mitosis
   2) bone and muscle
   3) protein synthesis
   4) liver and other viscera
   5) increase use of fats for energy
   6) ATP
2. a) thyroid gland b) thyroxine and T_3
   c) TRH
3. a) adrenal cortex b) cortisol c) CRH
4. a) mammary glands b) milk c) hypothalamus
5. a) ovaries, testes b) egg cells, estrogen
   c) sperm d) GnRH
6. a) ovaries, testes
   b) 1) ovulation
   2) corpus luteum, progesterone
   c) testosterone d) GnRH e) gonadotropic

THYROID GLAND

1. a) 1) proteins
   2) carbohydrates, fats, and excess amino acids
   3) physical, mental
   b) iodine
   c) TSH, anterior pituitary
   d) 1) glucose
   2) fats
   3) excess amino acids
   4) ATP
   5) protein synthesis
   6) bone and muscle
   7) liver and viscera
   8) brain
   9) reproductive organs
2. a) bones b) calcium and phosphate
   c) decreased d) hypercalcemia

PARATHYROID GLANDS

1. a) bones, small intestine, kidneys
   b) 1) calcium and phosphate
   2) small intestine
   3) calcium
   4) D
   c) increased, decreased
   d) hypocalcemia, hypercalcemia
2. 1) calcium is retained in bone matrix
   2) hypocalcemia
   3) kidneys reabsorb calcium, activate vitamin D
   4) small intestine absorbs calcium
   5) calcium reabsorbed from bones
   6) hypercalcemia
   7) hypercalcemia inhibits PTH secretion

PANCREAS

1. a) islets of Langerhans
   b) glucagon, insulin
2. a) liver
   b) 1) glycogen 2) fats and amino acids
   c) increased, food d) hypoglycemia
3. a) 1) glucose
   2) skeletal
   3) glucose, energy
   4) amino acids and fatty acids; proteins and fats
   b) decreased c) hyperglycemia
4. 1) liver changes glycogen to glucose and converts amino acids to carbohydrates
   2) hyperglycemia
   3) liver and skeletal muscles change glucose to glycogen
   4) cells use glucose for energy production

ADRENAL GLANDS

1. adrenal cortex, adrenal medulla
2. epinephrine and norepinephrine, medulla
3. cortex, aldosterone
4. cortex, cortisol

Adrenal Medulla

1. a) blood vessels b) vasoconstriction
2. a) 1) increases rate
      2) vasodilation
      3) vasoconstriction
      4) decreases peristalsis
      5) dilation
      6) conversion of glycogen to glucose
      7) increases use
   b) sympathetic
   c) sympathetic, stressful
3. 1) vasoconstriction in skin
   2) vasoconstriction in viscera
   3) vasoconstriction in skeletal muscle
   4) decrease peristalsis
   5) vasodilation in skeletal muscle
   6) increase rate and force of contraction
   7) dilate bronchioles
   8) glycogen converted to glucose
   9) increase use of fats for energy
   10) increase cell respiration

Adrenal Cortex

1. a) kidneys
   b) sodium, potassium
   c) 1) hydrogen (or K^+)
      2) sodium and bicarbonate
      3) water
   d) blood pressure (or pH)
   e) sodium, potassium, blood pressure
   f) 1) sodium ions reabsorbed
      2) potassium ions excreted
      3) bicarbonate ions reabsorbed
      4) water reabsorbed
      5) hydrogen ions excreted
      6) blood volume, BP, and pH
2. a) 1) amino acids, fats
      2) glucose
      3) anti-inflammatory
   b) ACTH, anterior pituitary
   c) 1) increase use of fats
      2) ATP
      3) increase use of excess amino acids
      4) ATP
      5) conserves glucose
      6) glucose stored as glycogen
      7) glucose used by the brain
      8) limits inflammation to what is useful

OVARIES

1. a) uterus, mammary glands (or) bones, adipose tissue
   b) 1) ovum
      2) endometrium
      3) growth of duct system of mammary glands (or) fat deposition
      4) closure of the epiphyseal discs
   c) FSH, anterior pituitary
2. a) uterus, mammary glands
   b) 1) blood vessels, glycogen
      2) secretory cells
   c) LH, anterior pituitary
3. a) anterior pituitary, hypothalamus
   b) FSH, GnRH

TESTES

1. a) testes, bones, muscles (or) reproductive organs
   b) 1) sperm
      2) growth of reproductive organs (or) larynx, facial hair, muscles
      3) closure of the epiphyseal discs
   c) LH, anterior pituitary
2. a) anterior pituitary b) FSH, spermatogenesis
   c) increased testosterone

OTHER HORMONES

1. a) pineal gland b) sleep
2. a) phospholipids b) locally
3. three of these: blood clotting, pain mechanisms, digestive secretions, reproduction, inflammation, vasodilation or vasoconstriction

MECHANISMS OF HORMONE ACTION

1. a) receptors b) target
2. 1) hormone
   2) cyclic AMP
   3) enzymes
   4) secretion of a product; protein synthesis (or) a change in membrane permeability
3. 1) lipids 3) genes, protein synthesis
   2) receptor, nucleus 4) enzymes

ENDOCRINE DISORDERS

Hypersecretion Disorders

1. 1) A, 3 2) D, 1 3) B, 2 4) C, 4

Hyposecretion Disorders

1. 1) D, 4 2) C, 5 3) A, 1 4) E, 3 5) B, 2

STIMULUS FOR SECRETION—SUMMARY

1. 1) E 5) F 9) B 13) K
   2) A 6) G 10) H 14) L
   3) J 7) M 11) M 15) N
   4) D 8) I 12) C 16) L

CROSSWORD PUZZLE

1 ISLETSOFLANGERHANS
2 GLUCONEOGENESIS
3 GLYCOGENOLYSIS
4 GLYCOGENESIS
5 SYMPATHOMIMETIC
6 RELEASINGFACTORS
7 BETA
8 HYPERGLYCEMIA
9 CORPUSLUTEUM
10 HYPERCALCEMIA
11 PROSTAGLANDINS
12 RENIN-ANGIOTENSIN
13 HORMONE
14 HYPOGLYCEMIA
15 TARGETORGAN
16 CATECHOLAMINES
17 ALPHA
18 NEGATIVEFEEDBACK
19 HYPOCALCEMIA
20 HYPOPHYSIS

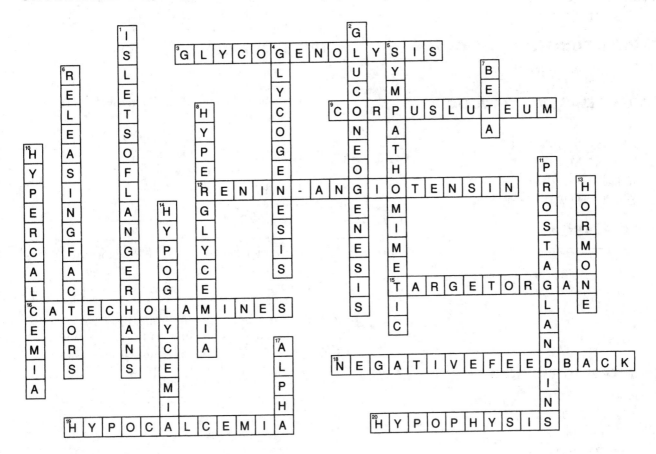

CLINICAL APPLICATIONS

1. a) "Do any of your close relatives have diabetes?"
   b) diabetes mellitus
   c) type 2 (or) non-insulin dependent
   d) lose weight and try to stay at normal weight
2. thyroxine
3. parathyroid, PTH
4. a) adrenal cortex
   b) aldosterone, cortisol

MULTIPLE CHOICE TEST #1

1. c	7. b	13. a	19. a	25. b
2. d	8. d	14. b	20. c	26. c
3. b	9. c	15. c	21. c	27. a
4. a	10. c	16. d	22. d	28. d
5. a	11. d	17. c	23. b	29. c
6. b	12. d	18. b	24. a	30. a

MULTIPLE CHOICE TEST #2

The letter of the correct choice is followed by the statement necessary to complete the answer. When a statement is to be corrected to make it true, the new (true) words are underlined.

1. a—Insulin causes the liver to change glucose to glycogen.
2. a—GH increases the rate of mitosis in growing bones.
3. d—FSH and LH are called gonadotropic hormones because their target organs are the ovaries and testes.
4. b—insulin—hyperglycemia
5. d—Epinephrine causes the liver to change glycogen to glucose.
6. c—Aldosterone increases the reabsorption of sodium ions.
7. d—These hormones are actually produced by the hypothalamus.
8. b—The thyroid gland is inferior to the larynx on the front of the trachea.
9. d—Secondary sex characteristics in men and women are regulated by testosterone and estrogen, respectively.
10. b—Steroid hormones exert their effects by increasing the process of protein synthesis.

MULTIPLE CHOICE TEST #3

1. a, b, d, and f are correct
2. None are correct
3. b, c, and e are correct
4. a, b, c, and d are correct

Chapter 11

GENERAL FUNCTIONS OF BLOOD

1. 1) D, G 2) A, E, F 3) B, C

CHARACTERISTICS OF BLOOD

1. 4 to 6
2. a) 38%–48% b) 52%–62%
3. a) 7.35–7.45 b) alkaline
4. a) thickness (or resistance to flow)
   b) cells, plasma proteins (albumin)

BLOOD PLASMA

1. 91
2. a) dissolve b) nutrients, waste products
3. a) bicarbonate b) HCO_3^-
4. 1) A, D, F
   2) A, C, G
   3) B, E, H

HEMOPOIETIC TISSUES

1. blood cells
2. red bone marrow, flat, irregular
3. a) stem cell b) mitosis
   c) RBCs, WBCs, platelets
4. a) spleen, lymph nodes, thymus
   b) lymphocytes
5. 1) lymph nodes 4) spleen
   2) lymph nodes 5) red bone marrow
   3) red bone marrow

RED BLOOD CELLS

1. a) erythrocytes, red bone marrow
   b) a nucleus
   c) biconcave discs (or) thinner in the middle than at the edge
2. a) yolk sac b) liver, spleen
3. a) hemoglobin b) iron
4. a) pulmonary (lungs), oxyhemoglobin
   b) systemic, reduced hemoglobin
5. 1) red blood cells 3) agranular WBC
   2) granular WBCs 4) platelets
6. a) oxygen b) low blood oxygen level
   c) low oxygen in tissues d) erythropoietin, RBC production
7. a) normoblast b) reticulocyte
   c) mature RBCs, oxygen
8. a) protein, iron b) vitamin B_{12}, DNA
   c) stomach d) vitamin B_{12}
9. a) 120 b) liver, spleen, red bone marrow
   c) liver, hemoglobin
   d) amino acids, protein synthesis
10. a) bilirubin b) liver, bile
    c) feces
11. a) whites of the eyes (or) light skin
    b) jaundice
12. 1) RBCs circulate 120 days
    2) iron
    3) returned to RBM to make new RBCs
    4) bilirubin
    5) colon bacteria
    6) urine
    7) globin
    8) protein synthesis
13. 1) 4.5–6.0 million
    2) 38–48
    3) 12–18

RED BLOOD CELL TYPES

1. ABO, Rh
2. A, B, AB, O
3.

	Type A	Type B	Type AB	Type O
Antigens present on RBCs	A	B	A and B	neither A nor B
Antibodies present in plasma	anti-B	anti-A	neither anti-A nor anti-B	both anti-A and anti-B

4. 1) A antigen 4) B antigen
   2) B antibody 5) A antibody
   3) A antigen 6) B antigen
5. a) positive; negative b) no
   c) if an Rh negative person receives (or is exposed to) Rh positive blood
6. a) hemolysis b) kidneys, free hemoglobin

WHITE BLOOD CELLS

1. a) leukocytes b) granular, agranular
2. a) neutrophils, eosinophils, basophils
   b) an immature neutrophil
   c) lymphocytes, monocytes
3. nuclei
4. pathogens, immunity

5. 1) A 2) B 3) E 4) F 5) C, G
   6) D, H
6. 1) 5,000–10,000 2) leukocytosis, infection
   3) leukopenia, radiation (or) exposure to certain chemicals or antibiotics
   4) a) 55% to 70%
      b) 20% to 35%
      c) 3% to 8%
      d) 1% to 3%
      e) 0.5% to 1%
7. a) all the cells of an individual
   b) foreign c) rejection

PLATELETS

1. thrombocytes, red bone marrow
2. megakaryocytes, thrombopoietin
3. a) prevention of blood loss
   b) vascular spasm, platelet plugs, chemical clotting
4. a) smooth muscle b) serotonin, damage
   c) smaller, blood clot
5. a) capillaries b) stick to the edges
6. 150,000–300,000
7. thrombocytopenia

CHEMICAL BLOOD CLOTTING

1. rough, rough
2. liver, blood plasma
3. a) K b) bacteria, colon (large intestine)
4. calcium, bones
5. a) platelets, damaged tissues
   b) prothrombin activator
   c) prothrombin activator, thrombin
   d) thrombin, fibrin
   e) fibrin
6. a) retraction, repair b) fibrinolysis
7. a) simple squamous b) heparin
   c) liver, thrombin d) vicious, positive feedback
8. a) thrombus b) embolism
9. 1) platelet factors 6) thrombin
   2) calcium 7) fibrinogen
   3) prothrombin activator 8) calcium
   4) prothrombin 9) fibrin
   5) calcium

CROSSWORD PUZZLE

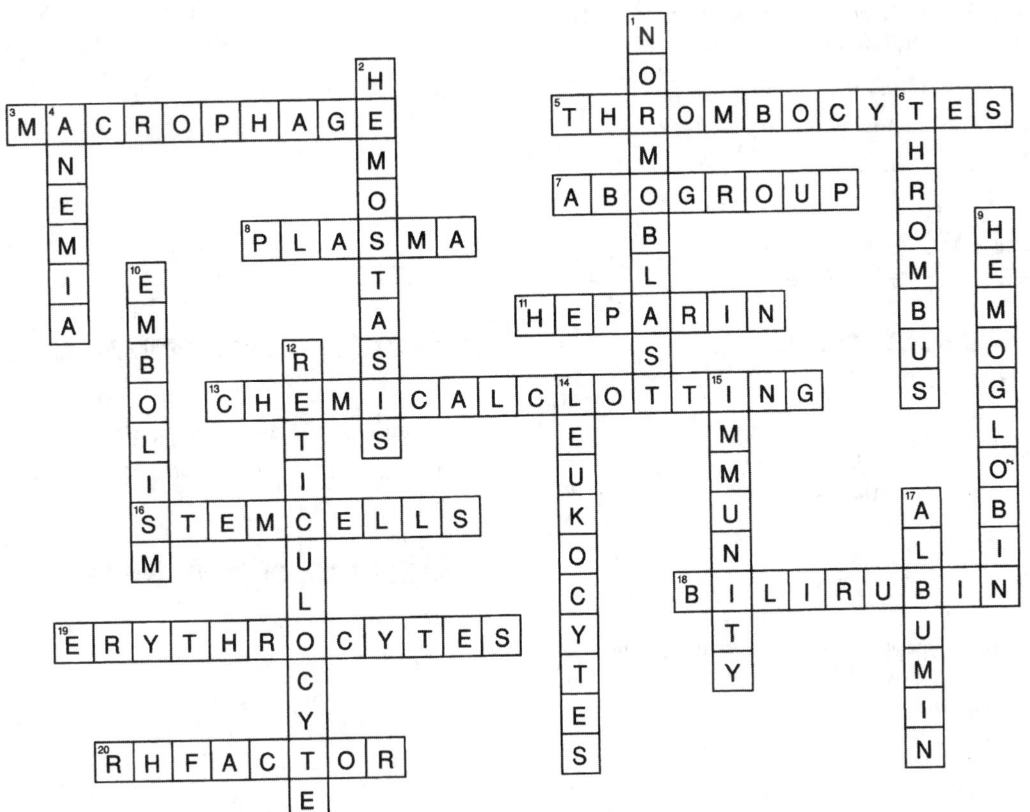

CLINICAL APPLICATIONS

1. b—the WBC count is high
2. a) Factor 8 (or VIII)
   b) Factor 8 (obtained from donated blood)
   c) boy—neither parent has hemophilia, so the mother must be a carrier of the gene
3. a) erythroblastosis fetalis
   b) negative, positive, positive, positive
   c) RhoGAM will destroy any Rh-positive RBCs in maternal circulation to prevent maternal production of Rh antibodies
4. a) HLA types
   b) there will be less chance of rejection of the donated organ
   c) genetic (or hereditary)

MULTIPLE CHOICE TEST #1

1. b	6. d	11. b	16. a	21. b
2. a	7. b	12. c	17. c	22. c
3. b	8. a	13. b	18. d	23. d
4. c	9. b	14. b	19. a	24. d
5. b	10. d	15. a	20. d	25. b

MULTIPLE CHOICE TEST #2

The letter of the correct choice is followed by the statement necessary to complete the answer. When an incorrect statement is to be corrected to make it true, the new (true) words are underlined.

1. d—Oxygen is transported within RBCs.
2. b—Eosinophils help detoxify foreign proteins.
3. c—The intrinsic factor is produced by the stomach lining to prevent digestion of vitamin B_{12}.
4. b—Prothrombin, fibrinogen, and other clotting factors are synthesized by the liver.
5. c—Type B has B antigens on the RBCs. (or) Type A has A antigens on the RBCs.
6. b—Hb—12–18 grams/100 mL
7. a—The normal life span of RBCs is 120 days.
8. b—Gamma globulins are antibodies produced by lymphocytes.
9. a—A normoblast is the last stage with a nucleus.
10. d—Heparin is produced by basophils.
11. b—Platelet plugs are useful only in capillaries.
12. d—anemia—decreased RBC count

MULTIPLE CHOICE TEST #3

1. a, e, and h are correct
2. b, c, and d are correct
3. d, e, h, and l are correct

Chapter 12

CARDIAC MUSCLE TISSUE

1. a) muscle fibers (myocytes)
   b) ATP
2. a) sarcomeres b) myosin, actin
3. a) depolarization, repolarization
   b) sodium ions, potassium ions
   c) nerve impulses
   d) intercalated discs
   e) atria, ventricles
4. a) atrial natriuretic peptide (ANP); stretching of the heart muscle (increased BP)
   b) sodium ions, water
   c) vasodilation
   d) lower (decrease) BP
   e) heat

FUNCTION AND LOCATION

1. pump blood
2. a) lungs, thoracic b) diaphragm

PERICARDIAL MEMBRANES

1. three
2. fibrous, fibrous connective tissue
3. a) parietal b) visceral, epicardium
4. prevent friction

CHAMBERS OF THE HEART

1. four, atria, ventricles
2. cardiac muscle, myocardium
3. a) endocardium, valves b) smooth, clotting
4. a) ventricles b) interatrial septum
   c) interventricular septum

CHAMBERS—VESSELS AND VALVES

1. 1) D, 1 2) A, 4 3) B, 3 4) C, 2
2. 1) superior vena cava
   2) right pulmonary artery
   3) right pulmonary veins
   4) coronary arteries
   5) right atrium
   6) inferior vena cava
   7) right ventricle
   8) aorta
   9) left pulmonary artery
   10) left pulmonary veins
   11) left atrium
   12) left ventricle
3. 1) C, 3 2) B, 2 3) A, 4 4) D, 1
4. 1) superior vena cava
   2) right pulmonary
   3) pulmonary semilunar valve
   4) endocardium
   5) right atrium
   6) tricuspid valve
   7) inferior vena cava
   8) aorta
   9) pulmonary artery
   10) left pulmonary artery
   11) pulmonary veins
   12) left atrium
   13) bicuspid (mitral) artery valve
   14) aortic semilunar valve
   15) left ventricle
   16) papillary muscles
   17) interventricular septum
   18) epicardium
   19) right ventricle
   20) chordae tendineae
5. a) myocardium, ventricle
   b) fibrous connective, flaps of the AV valves
   c) contract

CIRCULATION THROUGH THE HEART

1. a) lungs, body b) body, lungs
2. 1, 2, 6, 7, 4, 9, 8, 10, 3, 5

CORONARY VESSELS

1. a) myocardium (heart) b) oxygen
2. a) aorta b) right atrium
3. a) heart muscle cells begin to die
   b) a blood clot (or atherosclerosis) in a coronary vessel

CARDIAC CYCLE AND HEART SOUNDS

1. heartbeat
2. a) contraction b) relaxation
3. diastole, diastole
4. 1, 3, 6, 2, 4, 7, 5, 8
5. flows passively
6. is pumped
7. a) two b) AV c) semilunar
   d) heart murmur

CARDIAC CONDUCTION PATHWAY

1. a) impulses b) no c) rate
2. 1) SA node 2) AV node
   3) bundle of His 4) bundle branches
   5) Purkinje fibers
3. 1) SA node 2) AV node
   3) bundle of His 4) bundle branches
   5) Purkinje fibers
4. a) rapid b) sodium, depolarize
5. SA node, AV node
6. bundle of His, bundle branches, Purkinje fibers
7. a) ECG b) arrhythmia c) fibrillation

HEART RATE

1. a) 60–80 beats per minute b) pulse
2. higher, smaller
3. lower

CARDIAC OUTPUT

1. one minute
2. blood pressure, oxygen
3. a) one beat b) 60–80
4. stroke volume × pulse
5. 4.5 liters
   cardiac output = 75 × 60
6. 60 bpm
   6000 = 100 × pulse
   6000/100 = pulse
7. 12 liters
   cardiac output = 100 × 120
8. 10 liters
   cardiac reserve = 16 – 6

REGULATION OF HEART RATE

1. a) medulla
   b) accelerator and inhibiting centers
2. a) increase b) vagus, decrease
3. a) pressoreceptors, carotid, aortic
   b) chemoreceptors, carotid, aortic
4. a) glossopharyngeal
   b) vagus
5. 1) pressoreceptors, sinus
   2) glossopharyngeal
   3) accelerator, sympathetic
   4) increase, raise
6. epinephrine, adrenal medulla
7. 1) glossopharyngeal nerves
   2) vagus nerve (sensory)
   3) vagus nerve (parasympathetic)
   4) sympathetic nerves
   5) carotid sinus and body
   6) aortic arch
   7) aortic sinus and body

CROSSWORD PUZZLE

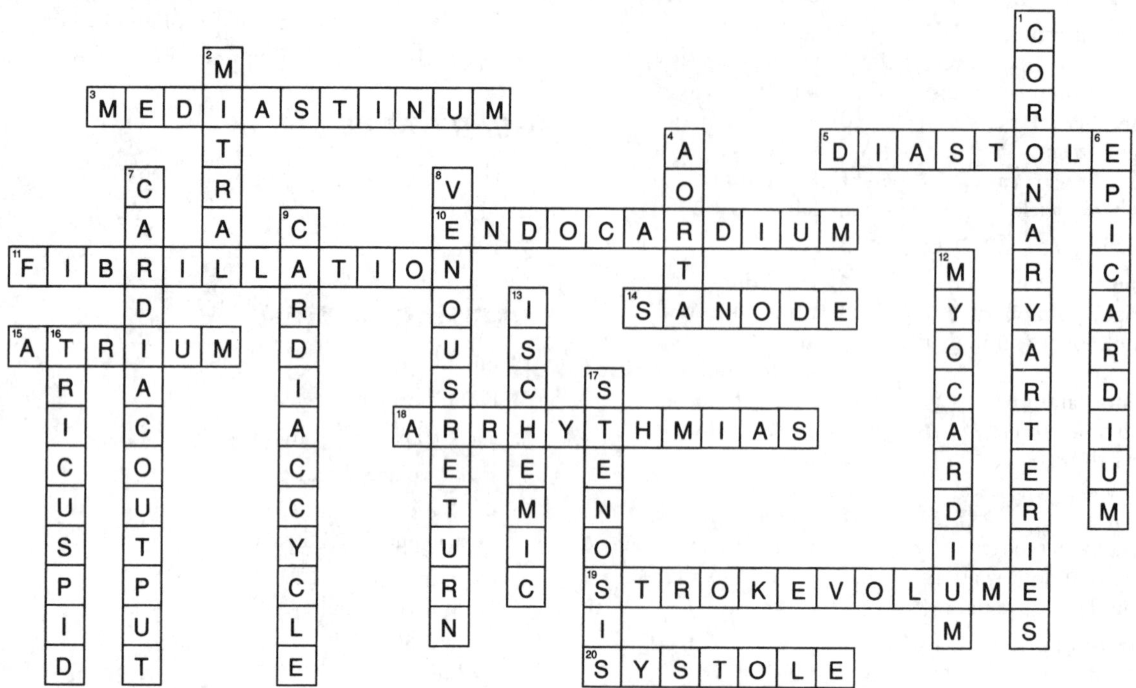

CLINICAL APPLICATIONS

1. a) myocardial infarction (heart attack)
   b) 1) heredity (family history)
      2) high-fat diet
      3) smoking (or) little exercise
2. a) rapid, uncoordinated contractions
   b) pumping, decrease
3. a) SA node b) slower
   c) decrease
4. 1) heart murmur
   2) myocardial infarction
   3) coronary atherosclerosis

MULTIPLE CHOICE TEST #1

1. a	6. d	11. b	16. d	21. b
2. d	7. b	12. c	17. c	22. b
3. b	8. c	13. b	18. d	23. a
4. c	9. a	14. b	19. d	24. a
5. a	10. d	15. a	20. a	25. c

MULTIPLE CHOICE TEST #2

The letter of the correct choice is followed by the statement necessary to complete the answer. When an incorrect statement is to be corrected to make it true, the new (true) words are underlined.

1. d—The visceral pericardium may also be called the epicardium.
2. b—The left ventricle pumps blood into the aorta.
3. a—The mitral valve prevents backflow of blood from the left ventricle to the left atrium.
4. d—When the atria are in systole, the ventricles are in diastole.
5. c—The SA node initiates each heartbeat.
6. a—Cardiac output is the amount of blood pumped by a ventricle in 1 minute.
7. b—Sympathetic impulses to the heart increase the heart rate. (or) Parasympathetic impulses to the heart decrease the heart rate.
8. b—The heart rate will increase in response to sympathetic impulses.
9. b—The coronary sinus empties blood from the myocardium into the right atrium.
10. d—An athlete's heart rate is low because the heart's stroke volume is higher.

MULTIPLE CHOICE TEST #3

1. a, d, e, g, and h are correct
2. b, c, f, g, and l are correct

Chapter 13

BLOOD VESSELS—STRUCTURE AND FUNCTIONS

1. a) capillaries b) arteries c) veins
2. a) lining, simple squamous epithelial
   b) prevents abnormal clotting
3. a) smooth muscle, elastic connective tissue
   b) help maintain diastolic blood pressure
4. a) outer, fibrous connective tissue
   b) prevents rupture of blood vessels
5. 1) A, E, F 2) B, C, D
6. a) anastomoses b) alternate pathways
7. a) simple squamous epithelium
   b) exchanges
8. a) precapillary sphincters
   b) dilate, increase, oxygen
9. a) sinusoids
   b) liver, spleen, red bone marrow
10. 1) tunica externa
    2) tunica media
    3) endothelial lining
    4) valve
    5) arteriole
    6) endothelial cells
    7) capillary network
    8) precapillary sphincter
    9) venule

EXCHANGES IN CAPILLARIES

1. 1) A, E, G 2) C, D 3) B, F
2. a) oxygen b) plasma, nutrients c) CO_2
   d) tissue fluid, waste products

PATHWAYS OF CIRCULATION

Pathway—Pulmonary Circulation

1. a) right, lungs b) left
2. oxygen, carbon dioxide, capillaries, alveoli

Pathway—Systemic Circulation

1. a) left, body b) right
2. capillaries, cells (or tissue fluid)
3. 1) abdominal
   2) ascending
   3) thoracic
   4) aortic arch
4. 1) heart
   2) thigh
   3) arm
   4) bronchioles
   5) brain
   6) kidneys
   7) liver
   8) chest wall
   9) lower leg
   10) shoulder
   11) forearm
   12) small intestine
   13) esophagus
   14) brain
   15) abdominal organs
   16) knee
   17) hip
   18) foot
5. 1) occipital
   2) internal carotid
   3) vertebral
   4) brachiocephalic
   5) aortic arch
   6) celiac
   7) hepatic
   8) splenic
   9) left gastric
   10) superior mesenteric
   11) abdominal aorta
   12) common iliac
   13) internal iliac
   14) external iliac
   15) femoral
   16) popliteal
   17) anterior tibial
   18) posterior tibial
   19) maxillary
   20) facial
   21) external carotid
   22) common carotid
   23) subclavian
   24) pulmonary
   25) axillary
   26) intercostal
   27) brachial
   28) renal
   29) gonadal
   30) inferior mesenteric
   31) radial
   32) ulnar
   33) deep volar arch
   34) superficial volar arch
6. 1) neck
   2) armpit
   3) leg and thigh
   4) lower body
   5) hip
   6) forearm
   7) brain
   8) upper body
   9) kidney
   10) shoulder
   11) arm
   12) thigh
7. 1) anterior facial
   2) superior vena cava
   3) axillary
   4) cephalic
   5) hemiazygos
   6) intercostal
   7) inferior vena cava
   8) brachial
   9) basilic
   10) gonadal
   11) superior mesenteric
   12) volar arch
   13) volar digital
   14) superior sagittal sinus
   15) inferior sagittal sinus
   16) straight sinus
   17) transverse sinus
   18) vertebral
   19) external jugular
   20) internal jugular
   21) subclavian
   22) brachiocephalic
   23) pulmonary
   24) hepatic
   25) hepatic portal
   26) left gastric
   27) renal
   28) splenic
   29) inferior mesenteric
   30) internal iliac
   31) external iliac
   32) femoral
   33) great saphenous
   34) popliteal
   35) small saphenous
   36) anterior tibial
   37) dorsal arch

HEPATIC PORTAL CIRCULATION

1. digestive organs and spleen
2. superior mesenteric, splenic
3. sinusoids, inferior vena cava
4. a) capillaries b) digestive organs and spleen, liver
5.

	Example 1	**Example 2**	**Example 3**
1)	glucose	alcohol	bilirubin
2)	small intestine	stomach	spleen
3)	excess stored as glycogen	detoxifies	excretes into bile

FETAL CIRCULATION

1. placenta
2. a) oxygen—CO_2, nutrients, waste products
   b) diffusion, active transport
3. umbilical arteries, umbilical vein
4. a) fetus, placenta
   b) CO_2, waste products
5. a) placenta, fetus b) oxygen, nutrients
6. ductus venosus, inferior vena cava
7. a) foramen ovale, right, left b) lungs
8. a) pulmonary artery, aorta
   b) permits most fetal blood to bypass the lungs
9. a) constricts b) closed, constricts
   c) pulmonary
10. 1) aorta
    2) ductus arteriosus
    3) pulmonary artery
    4) foramen ovale
    5) inferior vena cava
    6) ductus venosus
    7) umbilical arteries
    8) umbilical vein
    9) placenta
    10) navel of fetus

VELOCITY OF BLOOD FLOW

1. decreases
2. decreases, increases
3. a) capillaries b) slowest
   c) slow flow allows enough time for exchanges of nutrients, wastes, and gases

BLOOD PRESSURE

1. blood vessels
2. mm Hg (millimeters of mercury)
3. a) left ventricle b) contracting
   c) relaxing
4. aorta (or arteries), veins
5. a) arterial, venous b) filtration, rupture
   c) nutrients
6. a) right ventricle b) prevent

Maintenance of Systemic Blood Pressure

1. a) heart b) decrease, decrease
   c) 1) smooth muscle
      2) legs, compress (or squeeze)
      3) thoracic cavity, breathing, heart
2. increase
3. a) constriction b) increase
   c) decrease
4. a) systole, diastole b) decreases, increases
5. a) blood cells, albumin (or plasma proteins)
   b) decrease c) anemia, liver, albumin
6. a) decrease b) increased heart rate (or) vasoconstriction
7. a) 1) vasoconstriction
      2) increased heart rate, force of contraction
      3) kidneys, volume
      4) kidneys, water
      5) sodium ions and water, decreases
   b) 1) vasoconstriction
      2) increases heart rate and force
      3) increases reabsorption of Na^+ ions and water follows
      4) increases water reabsorption
      5) increases excretion of Na^+ ions and water follows
      6) raises BP
      7) lowers BP

Regulation of Blood Pressure

1. 1) forcefully, increase, Starling's law
   2) filtration
   3) 1) renin
      2) angiotensinogen
      3) angiotensin I
      4) angiotensin II
      5) vasoconstriction
      6) aldosterone
      7) increased Na^+ and H_2O reabsorption
      8) increased BP
2. 1) medulla
      a) vasoconstrictor, vasodilator
      b) pressoreceptors
   2) sympathetic
      a) more, raise
      b) fewer, lower
3. 1) pressoreceptors in carotid and aortic sinuses stimulated
   2) inhibit vasomotor center
   3) vasodilation decreases peripheral resistance
   4) stimulate cardioinhibitory center
   5) heart rate slows, decreases cardiac output
   6) pressoreceptors in carotid and aortic sinuses inhibited
   7) stimulate vasomotor center
   8) vasoconstriction increases peripheral resistance
   9) stimulate cardioaccelerator center
   10) heart rate and cardiac output increase

Hypertension

1. 130–140, 80–90
2. a) arteriosclerosis b) brain, kidneys
3. hypertrophy
4. a) angina b) oxygen

CROSSWORD PUZZLE

CLINICAL APPLICATIONS

1. a) 130–140, 80–90
   b) hypertrophy c) angina, oxygen
2. 1) varicose veins
   2) aneurysm
   3) arteriosclerosis
   4) atherosclerosis
3. 1) b (the very rapid pulse and fast breathing rate)
   2) b
   3) The faster respiratory rate makes the respiratory pump more effective in increasing venous return to maintain cardiac output.

MULTIPLE CHOICE TEST #1

1. a	6. d	11. b	16. a	21. d
2. a	7. d	12. d	17. d	22. c
3. c	8. b	13. d	18. a	23. a
4. c	9. b	14. b	19. d	24. a
5. b	10. c	15. c	20. a	25. d

MULTIPLE CHOICE TEST #2

The letter of the correct choice is followed by the statement necessary to complete the answer. When an incorrect statement is to be corrected to make it true, the new (true) words are underlined.

1. c—The vein that takes blood into the liver is the portal vein.
2. a—The foramen ovale permits blood to flow from the right atrium to the left atrium to bypass the fetal lungs.
3. b—The lining of both arteries and veins is very smooth to prevent abnormal clotting.
4. d—Decreased venous return will result in decreased cardiac output.
5. b—renal artery—kidney
6. c—Carbon dioxide diffuses from tissues to the blood.
7. d—The carotid pulse is felt at the side of the neck.
8. d—The tissue in arteries and veins that constricts or dilates is smooth muscle.
9. b—Norepinephrine causes vasoconstriction, which raises BP.
10. d—BP in the veins is lower than BP in the capillaries.

MULTIPLE CHOICE TEST #3

1. b, d, g, and h are correct
2. b, c, d, e, g, and h are correct
3. e, f, and h are correct

Chapter 14

LYMPH AND LYMPH VESSELS

1. lymph
2. a) filtration, tissue fluid
   b) lymph capillaries
3. 1) D 2) F 3) A 4) E 5) B 6) C

LYMPH NODES AND NODULES

1. lymphatic, lymphocytes, stem cells
2. a) plasma cells b) macrophages
3. 1) B, C, 1, 2 2) A, D, 3, 4
4. 1) right lymphatic duct
   2) right subclavian vein
   3) cervical nodes
   4) left subclavian vein
   5) thoracic duct
   6) axillary nodes
   7) spleen
   8) cisterna chyli
   9) inguinal nodes

SPLEEN

1. a) abdominal, diaphragm, stomach
   b) ribs
2. platelets
3. phagocytize, antibodies
4. red blood cells, bilirubin
5. a) lymph nodes and nodules
   b) liver and red bone marrow
6. red blood cells

THYMUS

1. fetus—child
2. below (inferior to)
3. lymphocytes
4. a) antigens, immunity b) competence
5. a) self-recognition b) self-tolerance

IMMUNITY

1. pathogens, prevent
2. a) self, foreign
   b) bacteria, viruses, fungi, protozoa, malignant cells
3. 1) E, F, G, H, I, J 2) A, B, C, D

INNATE IMMUNITY

1. barriers, cells, chemicals
2. a) stratum corneum, subcutaneous, mucous
   b) ciliated c) HCl
   d) lysozyme e) sebum
   f) white blood (WBCs)
3. 1) E
   2) D, G
   3) B, F
   4) A
   5) C
4. 1) A, D
   2) E
   3) C
   4) B
5. a) damage
   b) histamine, leukotrienes
   c) redness, swelling, heat, pain
   d) spread, repair
6. 1) stratum corneum
   2) Langerhans cell
   3) keratinocytes produce defensins
   4) subcutaneous tissue and WBCs
   5) ciliated epithelium (respiratory)
   6) HCl in gastric juice
   7) lysozyme in tears
   8) lysozyme in saliva
   9) foreign antigen
   10) activates lymphocytes
   11) macrophage
   12) neutrophil
   13) activates lymphocytes
   14) perforins
   15) histamine and leukotrienes
   16) blocks viral reproduction
   17) lyses cellular antigen
   18) vasodilation
   19) increased capillary permeability
   20) tissue fluid and WBCs

ADAPTIVE IMMUNITY

1. a) T cells, B cells
   b) red bone marrow
   c) thymus, red bone marrow
   d) spleen, lymph nodes and nodules
2. a) protein b) specific
   c) gamma globulins, immune globulins
3. a) cell-mediated, antibody-mediated (humoral)
   b) foreign
   c) macrophages, helper T cells

CELL-MEDIATED IMMUNITY

1. 1) E 2) A 3) D 4) C 5) B

ANTIBODY-MEDIATED (HUMORAL) IMMUNITY

1. 1) C 2) A 3) E 4) F 5) B
   6) D
2. 1) macrophage
   2) helper T cell
   3) memory T cell
   4) cytotoxic T cell
   5) produces cytokines to attract macrophages
   6) macrophage
   7) helper T cell
   8) activated B cell
   9) memory B cell
   10) plasma cell
   11) antibodies
   12) antigen–antibody complex
   13) macrophage
   14) complement fixation—lysis of cellular antigen

3. a) antibodies b) macrophages
4. a) macrophages b) lysis (rupture)
5. a) yes b) slow c) small
   d) may develop the disease
6. a) memory b) fast c) large
   d) will not get the disease e) vaccines

TYPES OF IMMUNITY

1. 1) C, 1 2) E, 2 3) A, 2 4) D, 3
   5) B, 3
2. a) population, immune to the disease
   b) hosts

CROSSWORD PUZZLE

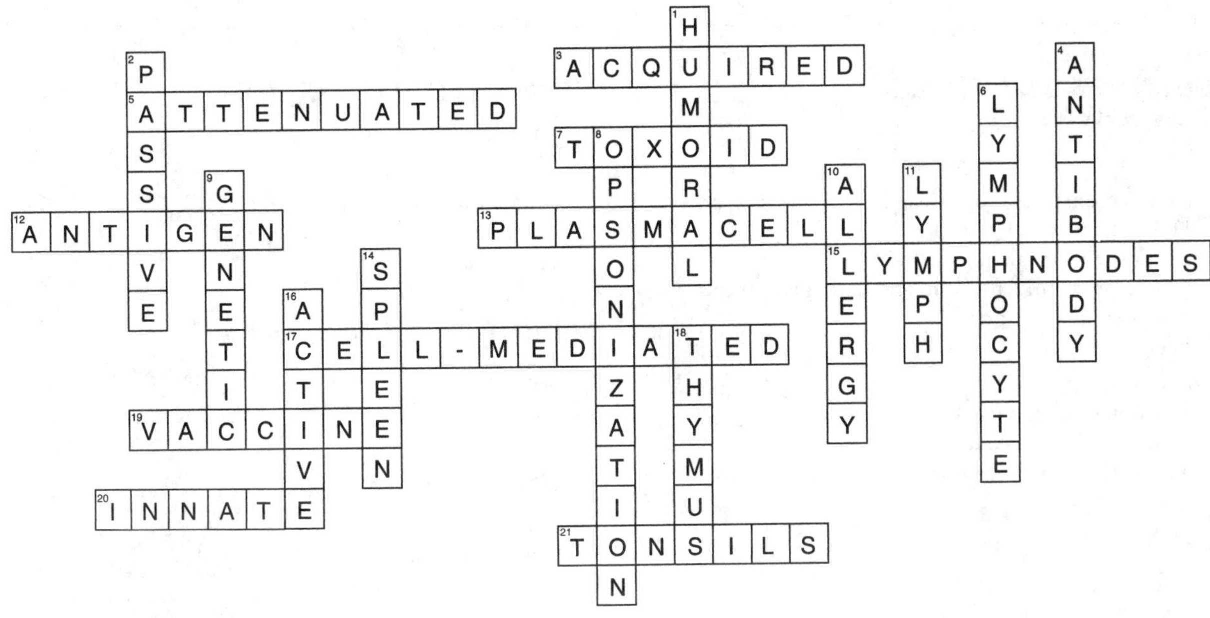

CLINICAL APPLICATIONS

1. a) virus, HIV b) helper
   c) opportunistic
   d) 1) sexual contact
      2) contact with blood
      3) placental transmission
2. a) antibodies, memory
   b) The vaccine contains a harmless form of the measles virus, which allows the body to prepare its defenses against the real measles virus.
3. a) cervical lymph nodes b) macrophages
4. a) histamine b) allergens
   c) no d) no
   e) asthma (or) anaphylactic shock

MULTIPLE CHOICE TEST #1

1. d
2. b
3. c
4. d
5. a
6. d
7. c
8. b
9. d
10. c
11. b
12. d
13. a
14. c
15. d
16. b
17. d
18. c
19. b
20. a
21. c
22. d
23. b
24. a
25. a
26. d
27. b
28. d
29. b
30. a

MULTIPLE CHOICE TEST #2

The letter of the correct choice is followed by the statement necessary to complete the answer. When an incorrect statement is to be corrected to make it true, the new (true) words are underlined.

1. b—Antibodies are produced in the spleen by cells called fixed plasma cells.
2. d—The three major paired groups are the cervical, axillary, and inguinal nodes.
3. a—They are found below the epithelium of mucous membranes.
4. c—Lymph contains the antibodies produced by the plasma cells.
5. a—The right lymphatic duct returns lymph to the right subclavian vein.
6. d—A vaccine provides artificially acquired active immunity.
7. c—T cells are the lymphocytes produced by the thymus.
8. a—When an antibody is produced, it will bond to a specific foreign antigen.
9. c—Mast cells and basophils produce leukotrienes and histamine.
10. b—Repeated responses become more efficient.

11. b—The foreign antigen is recognized by macrophages and helper T cells.
12. b—Opsonization is the labeling of foreign antigens by antibodies.
13. c—The memory cells of adaptive immunity are lymphocytes.
14. d—Histamine makes capillaries more permeable.

MULTIPLE CHOICE TEST #3

1. b, c, e, f, and g are correct
2. All are correct
3. e, f, g, and h are correct
4. a, d, and h are correct

Chapter 15

NOSE, NASAL CAVITIES, AND PHARYNX

1. hairs
2. a) nasal septum b) ciliated
   c) 1) warms the air
   2) moistens the air
   3) cilia sweep mucus, dust, and pathogens to the pharynx
   d) smell
3. a) nasal cavities
   b) 1) frontal, sphenoid
   2) ethmoid, or maxillae
   c) lighten the skull (or) provide resonance for the voice
4. 1) B, C, F, G, I
   2) D, E
   3) A, H
5. larynx

LARYNX

1. a) pharynx, trachea b) speaking
2. 1) E 2) A, F 3) B 4) D 5) C
3. vagus, accessory
4. together, exhaled
5. 1) epiglottis
   2) thyroid cartilage
   3) vocal cords
   4) cricoid cartilage
   5) tracheal cartilages

TRACHEA AND BRONCHIAL TREE

1. larynx, primary bronchi
2. a) cartilage b) keep the trachea open
   c) ciliated epithelium
3. a) trachea b) two, three
4. a) bronchioles, cartilage b) alveoli
5. 1) frontal sinus
   2) nasal cavity
   3) nasopharynx
   4) adenoid
   5) palatine tonsil
   6) oropharynx
   7) epiglottis
   8) larynx
   9) laryngopharynx
   10) trachea
   11) primary bronchi
   12) bronchioles
   13) mediastinum
   14) visceral pleura
   15) pleural space
   16) parietal pleura
   17) diaphragm
   18) pulmonary capillaries
   19) alveoli

THE PATHWAY OF AIR

1. 1, 6, 8, 5, 2, 4, 10, 3, 7, 9, 11
2. 1) primary bronchi
   2) secondary bronchi
   3) bronchioles
   4) alveoli

PLEURAL MEMBRANES

1. a) visceral b) parietal
2. a) prevents friction b) together

LUNGS

1. a) rib cage
   b) mediastinum, heart c) diaphragm
2. a) hilus
   b) primary bronchus, pulmonary artery and veins
3. a) simple squamous epithelium
   b) simple squamous epithelium
   c) thin, diffusion (exchange) of gases
   d) elastic connective tissue
4. a) diffusion of gases
   b) pulmonary surfactant, inflation
   c) alveolar type II cells
5. 1) surfactant and tissue fluid
   2) cell of alveolus
   3) capillary endothelium (cell)
   4) type I alveolar cell
   5) type II alveolar cell
   6) elastic fibers
   7) alveolar macrophage
   8) red blood cells

MECHANISM OF BREATHING

1. a) ventilation b) inhalation, exhalation
2. medulla and pons
3. 1) intercostal, intercostal
   2) diaphragm, phrenic
4. air pressure
5. 1) B, D 2) C, F 3) A, E

INHALATION (INSPIRATION)

1. 1, 4, 3, 6, 2, 5, 7
2. contraction, lungs

EXHALATION (EXPIRATION)

1. 1, 3, 5, 4, 2, 6
2. contraction
3. internal intercostal, down and in, abdominal, upward

PULMONARY VOLUMES

1. 1) F 2) C 3) E 4) B
   5) A 6) D
2. gas exchange
3. a) alveoli, gas exchange b) anatomic
   c) physiologic
4. a) compliance
   b) pneumonia, asthma, TB, etc.
   c) fractured ribs, pleurisy, ascites
   d) alveolar ventilation, physiologic

EXCHANGE OF GASES

1. a) alveoli, blood b) systemic capillaries
2. oxygen, carbon dioxide
3. a) 21%, 0.04% b) 16%, 4.5%
4. partial pressure, P
5. a) 1) high PO_2
   2) low PO_2; arrow from air to blood
   3) low PCO_2
   4) high PCO_2; arrow from blood to air
   b) external
   c) heart, left, body
6. a) 1) low PO_2
   2) high PO_2; arrow from blood to cells
   3) high PCO_2
   4) low PCO_2; arrow from cells to blood
   b) internal
   c) heart, right, lungs

TRANSPORT OF GASES IN THE BLOOD

1. a) red blood cells
   b) iron, hemoglobin
   c) lungs
2. 1) low PO_2 2) high PCO_2
   3) high temperature
3. bicarbonate (HCO_3^-), plasma
4. hydrogen (H^+)

NERVOUS REGULATION OF RESPIRATION

1. medulla, pons
2. a) inspiration b) forced exhalation
3. apneustic, pneumotaxic, exhalation
4. 12–20
5. baroreceptors, inspiration
6. a) emotional b) voluntary
   c) holding one's breath (or) singing, breathing faster
7. a) medulla b) pharynx, larynx, or trachea
   c) explosive, mouth
8. a) medulla b) nasal cavities
   c) explosive, nose
9. 1) cortex
   2) hypothalamus
   3) pneumotaxic center
   4) pneumotaxic center interrupts apneustic center
   5) apneustic center
   6) apneustic center prolongs inhalation
   7) expiration center
   8) impulses for forced exhalation
   9) inspiration center
   10) impulses to external intercostal muscles
   11) impulses to diaphragm
   12) impulses from baroreceptors depress the inspiration center

CHEMICAL REGULATION OF RESPIRATION

1. a) O_2 and CO_2 b) pH
2. 1) A, B 2) C, D
3. a) CO_2 b) pH
   c) residual, does not
4. 1) carotid and aortic bodies
   2) medulla
   3) inspiration center
   4) increase rate and depth of respiration
   5) more O_2 available to enter blood
   6) more CO_2 exhaled, increased pH

RESPIRATION AND ACID–BASE BALANCE

1. CO_2
2. 1) B, D, E 2) A, C, F
3. any severe pulmonary disease: emphysema, pneumonia
4. prolonged hyperventilation
5. a) untreated diabetes mellitus, severe diarrhea or vomiting
   b) increase, exhale
6. a) overingestion of antacid medication
   b) decrease, retain

CROSSWORD PUZZLE

		[1]V	E	N	T	[2]I	L	A	T	I	O	N		[3]G									[4]E	
						N								L					[5]V		[6]S		P	
						T		[7]P						O			[8]A	C	I	D	O	S	I	S
				[9]L	A	R	Y	N	X					T					T		F		G	
						A		E				[10]P		T					A		T		L	
						P		U				A		I					L		P		O	
					[11]P	U	L	M	O	N	A	R	Y	S	U	R	F	A	C	T	A	N	T	
						L		O				T							A		L		T	
						M		N				I							P		A		I	
[12]I		[13]A	L	V	E	O	L	I				A							A		T		S	
N						N		[14]A	L	K	A	L	O	S	I	S			C		E			
T						I						P							I					
R		[15]B	R	O	N	C	H	I	A	L	T	R	E	E					T		[16]R			
A												E			[17]E	M	P	H	Y	S	E	M	A	
[18]P	H	R	E	N	I	C	N	E	R	V	E	S									S			
L												S									I			
E											[19]P	U	L	M	O	N	A	R	Y	E	D	E	M	A
U												R									U			
R		[20]T	I	D	A	L	V	O	L	U	M	E									A			
A																					L			
L																					A			
																					I			
																					R			

CLINICAL APPLICATIONS

1. a) elastic connective, passive b) exhale
2. a) surfactant b) inflated
3. a) air, intrapleural b) collapse
   c) pneumothorax
4. a) pulmonary b) pulmonary
   c) pulmonary edema

MULTIPLE CHOICE TEST #1

1. c	7. a	13. d	19. b	25. c
2. b	8. b	14. c	20. a	26. c
3. d	9. d	15. b	21. d	27. c
4. b	10. c	16. a	22. c	28. c
5. a	11. d	17. d	23. b	29. d
6. a	12. c	18. c	24. a	30. a

MULTIPLE CHOICE TEST #2

The letter of the correct choice is followed by the statement necessary to complete the answer. When an incorrect statement is to be corrected to make it true, the new (true) words are underlined.

1. b—Air is brought to the alveoli by the bronchioles.
2. d—The epiglottis covers the top of the larynx during swallowing.
3. c—The blood that enters pulmonary capillaries has a high PCO_2.
4. b—The motor nerves to the diaphragm are called the phrenic nerves.
5. a—The apneustic center in the pons helps bring about inhalation. (or) The pneumotaxic center in the pons helps bring about exhalation.
6. c—Hypoxia is detected by the carotid and aortic chemoreceptors.
7. b—Hemoglobin in RBCs is able to transport oxygen because it contains iron.
8. c—Vital capacity is the deepest inhalation followed by the most forceful exhalation.
9. a—The normal range of respirations per minute is 12–20.
10. b—The respiratory system helps compensate for metabolic acidosis by increasing the respiratory rate. (or) The respiratory system helps compensate for metabolic alkalosis by decreasing the respiratory rate.

MULTIPLE CHOICE TEST #3

1. c, d, e, and f are correct
2. a, e, f, and h are correct
3. a, b, and d are correct
4. b and d are correct
5. a, c, and e are correct

Chapter 16

DIVISIONS OF THE DIGESTIVE SYSTEM

1. 1) A, C, F, G 2) B, D, E
2. oral cavity, stomach, small intestine
3. small intestine
4. 1) teeth
   2) oral cavity
   3) tongue
   4) sublingual gland
   5) submaxillary gland
   6) parotid gland
   7) liver
   8) gallbladder
   9) common bile duct
   10) duodenum
   11) pyloric sphincter
   12) ascending colon
   13) jejunum
   14) ileum
   15) cecum
   16) appendix
   17) esophagus
   18) lower esophageal sphincter
   19) stomach
   20) transverse colon
   21) pancreas
   22) descending colon
   23) sigmoid colon
   24) rectum
   25) anal canal

TYPES OF DIGESTION AND END PRODUCTS OF DIGESTION

1. 1) B, C 2) A, D
2. 1) fatty acids and glycerol
   2) amino acids
   3) monosaccharides
   3. vitamins, minerals, water

ORAL CAVITY AND PHARYNX

1. a) mouth b) hard palate
2. mechanical, chewing
3. a) deciduous, 20 b) permanent, 32
4. 1) C, E 2) A 3) F 4) B, D
5. a) taste b) pharynx
6. 1) sublingual 2) parotid 3) submandibular
7. ducts, oral cavity
8. a) amylase, maltose b) tasted, swallowed
   c) lysozyme
9. swallowing, medulla

ESOPHAGUS

1. pharynx, stomach
2. a) lower esophageal sphincter
   b) stomach contents, esophagus

TYPICAL STRUCTURE OF THE ALIMENTARY TUBE

1. 1) A, D, F 2) C, G 3) E, H, J 4) B, I
2. 1) mucosa
   2) submucosa
   3) external muscle layer
   4) serosa
   5) epithelium
   6) lacteal
   7) capillary network
   8) lymph nodule
   9) smooth muscle
   10) Meissner's plexus
   11) circular smooth muscle
   12) Auerbach's plexus
   13) longitudinal smooth muscle

STOMACH

1. esophagus, small intestine
2. reservoir
3. a) rugae
   b) gastric pits, gastric juice
4. 1) C 2) A 3) B 4) D
5. a) hydrochloric acid
   b) pepsin
6. a) parasympathetic
   b) G or enteroendocrine, gastrin
7. a) external muscle b) stomach, duodenum
   c) duodenum, stomach
8. 1) esophagus
   2) cardiac orifice
   3) lesser curvature
   4) pyloric sphincter
   5) duodenum
   6) fundus
   7) longitudinal muscle layer
   8) circular muscle layer
   9) oblique muscle layer
   10) body
   11) greater curvature
   12) rugae
   13) pylorus

LIVER AND GALLBLADDER

1. a) diaphragm, abdominal
   b) lobule, sinusoids
   c) hepatic, portal, bile
2. a) bile, fats b) mechanical
3. hepatic, cystic, common bile, duodenum
4. bilirubin, excess cholesterol, feces
5. a) liver b) stores bile, concentrates bile
6. a) enteroendocrine, duodenum
   b) secretin c) cholecystokinin
7. 1) gallbladder
   2) cystic duct
   3) common bile duct
   4) duodenum
   5) opening of common bile duct
   6) liver
   7) hepatic duct
   8) stomach
   9) pyloric sphincter
   10) pancreas
   11) main pancreatic duct

PANCREAS

1. duodenum, spleen
2. common bile
3. a) trypsin b) lipase, fatty acids and glycerol
   c) amylase, maltose
4. stomach, duodenum

SMALL INTESTINE

1. a) large intestine b) stomach, large intestine
2. a) duodenum, jejunum, ileum
   b) duodenum
3. a) external muscle layer
   b) enteric nervous system
   c) lymph nodules
4. bile, enzyme pancreatic juice
5. a) glands b) food (chyme)
6. a) disaccharides, monosaccharides
   b) amino acids
7. 1) A 2) B 3) C, D
8. capillary network, lacteal
9. 1) A, 1 2) B, 4 3) A, 1 4) B, 4
   5) A, 1 6) A, 1 7) A, 2 8) A, 3
10. a) B_{12}, intrinsic b) D, parathyroid
11. a) liver b) left subclavian
12. 1) plica circulares
    2) microvilli
    3) absorptive cell
    4) goblet cell
    5) lacteal
    6) capillary network
    7) enteroendocrine cell
    8) villus
    9) intestinal gland

REVIEW OF DIGESTION

1.

		Food Type	
Organ	**Carbohydrates**	**Fats**	**Proteins**
Salivary glands	Amylase—digests starch to maltose	None	None
Stomach	None	None	Pepsin—digests protein to polypeptides
Liver	None	Bile salts—emulsify fats	None
Pancreas	Amylase—digests starch to maltose	Lipase—digests emulsified fats to fatty acids and glycerol	Trypsin—digests polypeptides to peptides
Small intestine	Sucrase, maltase, and lactase—digest disaccharides to monosaccharides	None	Peptidases—digest peptides to amino acids

2. 1) amylase
   2) amylase
   3) sucrase, maltase, lactase
   4) bile
   5) lipase
   6) pepsin
   7) trypsin
   8) peptidases

LARGE INTESTINE

1. colon, ileum, anus
2. cecum
3. 1, 6, 2, 5, 4, 7, 3
4. 1) water, minerals, vitamins
   2) undigested food (feces)
5. a) bacteria b) pathogens
   c) vitamins, vitamin K
6. a) spinal cord b) rectum
7. 1, 4, 2, 3
8. external anal sphincter
9. 1) transverse colon
   2) haustra
   3) taeniae coli
   4) hepatic flexure
   5) ascending colon
   6) ileum
   7) ileocecal valve
   8) appendix
   9) cecum
   10) splenic flexure
   11) descending colon
   12) sigmoid colon
   13) rectum
   14) anal canal
   15) anus

LIVER—OTHER FUNCTIONS

1. 1) J
   2) I
   3) H
   4) G
   5) M
   6) B
   7) C
   8) L
   9) E
   10) A
   11) D
   12) F
   13) N
   14) K

CROSSWORD PUZZLE

CLINICAL APPLICATIONS

1. a) gallbladder b) cholesterol
   c) remove the gallbladder surgically
2. a) bacteria b) antibiotics
3. a) yellow b) bilirubin, liver
   c) virus, contaminated food or water
4. 1) pyloric stenosis
   2) lactose intolerance
   3) peritonitis

MULTIPLE CHOICE TEST #1

1. a	7. d	13. b	19. a	25. a
2. d	8. c	14. d	20. b	26. d
3. c	9. b	15. c	21. c	27. b
4. b	10. a	16. d	22. d	28. c
5. d	11. a	17. a	23. d	29. c
6. d	12. c	18. b	24. c	30. c

MULTIPLE CHOICE TEST #2

The letter of the correct choice is followed by the statement necessary to complete the answer. When an incorrect statement is to be corrected to make it true, the new (true) words are underlined.

1. a—The parotid glands are in front of the ears. (or) The sublingual glands are below the floor of the mouth.
2. d—The process of beta-oxidation permits the use of fatty acids for energy production.
3. b—Lipase from the pancreas digests fats to fatty acids and glycerol.
4. b—Increased secretion of saliva is a parasympathetic response. (or) Decreased secretion of saliva is a sympathetic response.
5. c—Each digestive enzyme digests only one type of food.
6. b—Blood vessels and nerves are found in the pulp cavity of a tooth.
7. d—The lower esophageal sphincter prevents the backup of stomach contents.
8. a—Hydrochloric acid destroys most microorganisms that enter the stomach because it has a pH of 1–2.
9. c—Villi are folds of the mucosa of the small intestine.
10. a—The common bile duct is formed by the hepatic duct and the cystic duct.
11. c—Enzymes are produced to complete the digestion of proteins and carbohydrates (or) disaccharides.
12. b—The ileum of the small intestine opens into the cecum.

MULTIPLE CHOICE TEST #3

1. b, f, and j are correct
2. a, e, g, and j are correct
3. c, f, g, and h are correct

Chapter 17

BODY TEMPERATURE

1. a) 96.5° to 99.5°F, 36° to 38°C b) 98.6°F or 37°C
2. infants, elderly

HEAT PRODUCTION

1. a) cell respiration b) ATP
2. a) thyroxine, thyroid b) cell respiration
3. epinephrine, adrenal medulla
4. a) ATP b) skeletal muscles, muscle tone c) liver d) blood
5. peristalsis (or) production of enzymes
6. increases

HEAT LOSS

1. a) skin b) respiratory c) urinary, digestive
2. a) radiation, conduction, convection b) convection c) radiation, conduction d) cooler
3. a) blood b) vasodilation, vasoconstriction
4. a) evaporates b) eccrine c) high, low d) dehydration

REGULATION OF BODY TEMPERATURE

1. hypothalamus
2. a) blood b) temperature receptors
3. 1) D, E, F 2) A, B, C

FEVER

1. high, pyrogens
2. a) bacteria, viruses b) inflammation
3. a) higher, cold b) shivering
4. a) lower, warm b) sweating
5. pathogens (bacteria), white blood cells
6. a) lose their shape (become inactive) b) brain, neurons

METABOLISM

1. metabolism
2. a) catabolism b) anabolism c) anabolism d) catabolism e) enzymes
3. $C_6H_{12}O_6 + O_2 \rightarrow CO_2 + H_2O + ATP + Heat$
4. glycolysis, the Krebs (citric acid) cycle, the cytochrome (electron) transport system
5. 1) A 2) B, C 3) A 4) B, C 5) A 6) B 7) C 8) A, B 9) C 10) C 11) A, B 12) B 13) A 14) A
6. 1) niacin 2) riboflavin 3) thiamine 4) iron, copper

PROTEINS AND FATS AS ENERGY SOURCES

1. a) amine (NH_2) b) Krebs cycle c) acetyl, pyruvic acid
2. a) pyruvic acid b) beta-oxidation, acetyl
3. a) ketones b) amino acids
4. 1) glucose 2) ATP 3) ATP 4) pyruvic acid 5) carbon dioxide 6) acetyl CoA 7) carbon dioxide 8) ATP 9) hydrogen ions 10) metabolic water 11) ATP 12) amino acids 13) urea 14) glycerol 15) fatty acid 16) acetyl groups

ENERGY AVAILABLE FROM FOOD

1. Calories, kilocalories
2. 4, 9, 4

SYNTHESIS USES OF FOODS

1. a) glycogen, skeletal muscles b) pentose sugars, DNA, RNA
2. nonessential, proteins
3. There are many possible answers:
   1) hemoglobin—RBCs
   2) antibodies—WBCs
   3) keratin—epidermal cells
   4) collagen—fibroblasts (tendons, ligaments)
   5) melanin—melanocytes
   6) albumin—liver
   7) myosin—muscle cells
   8) and others
4. a) true fats (triglycerides) b) phospholipids c) cholesterol
5. 1) estrogen or testosterone—ovaries or testes
   2) cortisol or aldosterone—adrenal cortex

METABOLIC RATE

1. a) heat b) kilocalories
2. basal metabolic rate (BMR)
3. children, the elderly
4. testosterone, estrogen
5. sympathetic, epinephrine, norepinephrine
6. skeletal muscles

CROSSWORD PUZZLE

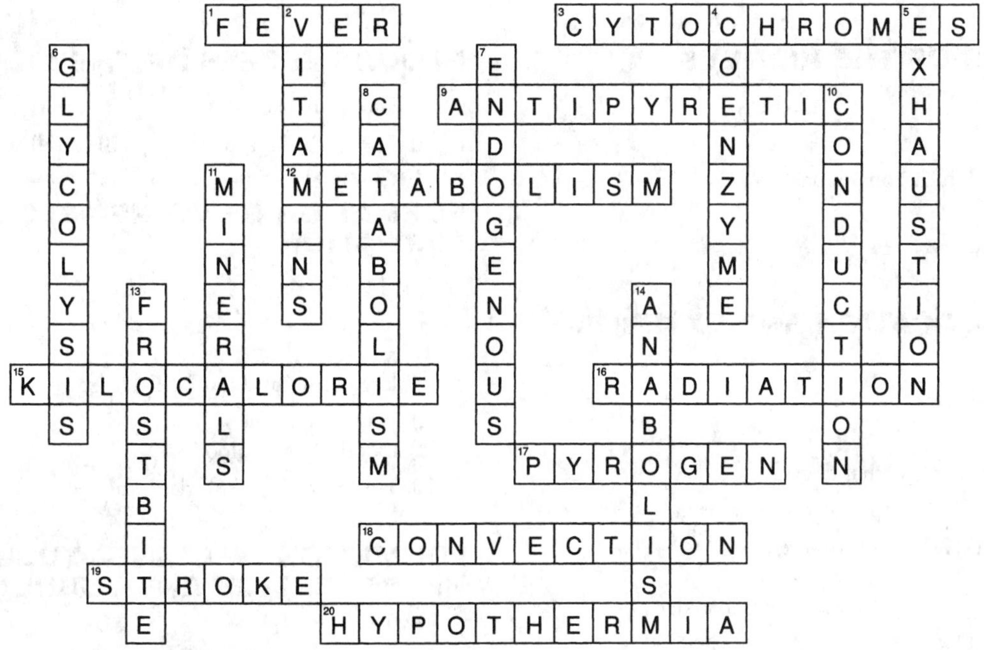

CLINICAL APPLICATIONS

1. a) frostbite b) ruptures, oxygen
2. a) heat stroke b) volume (BP), rises sharply
3. a) 35,000 b) 1000
   c) 35,000/1000 = 35 days
   d) exercise to expend more calories

MULTIPLE CHOICE TEST #1

1. b	6. d	11. a	16. b	21. d
2. b	7. a	12. b	17. c	22. d
3. a	8. b	13. c	18. d	23. d
4. c	9. c	14. d	19. b	24. d
5. c	10. d	15. a	20. b	25. d

MULTIPLE CHOICE TEST #2

The letter of the correct choice is followed by the statement necessary to complete the answer. When an incorrect statement is to be corrected to make it true, the new (true) words are underlined.

1. d—Almost half the body's total heat at rest is produced by the liver and skeletal muscles.
2. a—Radiation is an effective heat loss mechanism only when the environment is cooler than body temperature.
3. b—More heat will be lost as a result of vasodilation in the dermis.
4. d—Shivering may occur as muscle tone increases.
5. b—High fevers may cause brain damage because enzymes become denatured.
6. d—Men and young children usually have higher metabolic rates than do women and elderly people.
7. b—The complete breakdown of glucose in cell respiration does require oxygen.
8. c—Fatty acids and glycerol are used to synthesize true fats to be stored in adipose tissue.
9. a—A gram of protein yields 4 kilocalories of energy.
10. d—Most of the ATP produced in cell respiration is produced in the cytochrome (electron) transport system.

MULTIPLE CHOICE TEST #3

1. a, b, h, i, and j are correct
2. a, c, e, h, and l are correct

Chapter 18

FUNCTIONS OF THE KIDNEYS

1. waste products
2. minerals
3. hydrogen ions, bicarbonate ions
4. water
5. tissue fluid

KIDNEYS—LOCATION AND EXTERNAL ANATOMY

1. spinal, behind
2. ribs
3. a) adipose b) renal fascia
4. a) hilus b) renal artery, renal vein
5. 1) vertebra: T-12
   2) vertebra: L-5
   3) sacrum
   4) urethra in penis
   5) kidney
   6) ureter
   7) ilium
   8) urinary bladder
   9) urethra

KIDNEYS—INTERNAL STRUCTURE

1. 1) B, H 2) C, F, G 3) A, D, E
2. 1) renal pelvis
   2) renal artery
   3) renal vein
   4) ureter
   5) renal pyramid
   6) papilla
   7) renal cortex
   8) calyx
   9) renal cortex
   10) renal medulla
   11) papilla
   12) renal corpuscle
   13) renal tubule
   14) papillary duct

THE NEPHRON

1. a) functional
   b) renal corpuscle, renal tubule
2. glomerulus, Bowman's capsule
3. a) afferent, efferent b) efferent arteriole
4. a) pores b) permeable c) renal filtrate
5. 3, 2, 4, 1
6. renal pelvis
7. peritubular capillaries, efferent
8. 1) proximal convoluted tubule
   2) glomerulus
   3) Bowman's capsule (podocyte, inner)
   4) Bowman's capsule (outer)
   5) distal convoluted tubule
   6) efferent arteriole
   7) afferent arteriole
   8) loop of Henle
   9) collecting tubule
   10) peritubular capillaries
   11) juxtaglomerular cells

BLOOD VESSELS OF THE KIDNEY

1. a) abdominal aorta b) inferior vena cava
2. a) 1, 7, 3, 5, 2, 4, 6, 8 b) capillaries

FORMATION OF URINE—GLOMERULAR FILTRATION

1. renal corpuscle
2. a) glomerulus, Bowman's capsule
   b) renal filtrate
   c) large proteins, blood cells, large
3. a) yes b) yes c) 2
   d) blood plasma, blood
4. a) 1 minute b) decrease c) increase

FORMATION OF URINE—TUBULAR REABSORPTION AND TUBULAR SECRETION

1. renal tubule, peritubular capillaries
2. a) 99% b) urine
3. 1) osmosis
   2) active transport
   3) active transport
   4) passive transport
   5) pinocytosis
   6) active transport
4. a) limit b) high, high c) all
5. a) aldosterone
   b) parathyroid hormone (PTH)
6. a) filtrate
   b) creatinine, ammonia, medications
   c) hydrogen
7. a) sodium, water b) volume, pressure
8. a) water b) concentrated c) less, dilute
9. a) sodium, water
   b) decrease volume, decrease BP
10. 1, 4, 9, 7, 8, 3, 5, 2, 10, 6, 11
11. Active transport—1) glucose
    2) amino acids 3) positive ions
    Osmosis—1) water
    Pinocytosis—1) small proteins
    Passive transport—1) negative ions
    Secretion—1) creatinine 2) medications
    3) H^+ 4) ammonia
12. 1) ADH
    2) increases reabsorption of water
    3) PTH
    4) increases reabsorption of Ca^{+2}
    5) aldosterone
    6) increases reabsorption of Na^+
    7) increases excretion of K^+
    8) ANP
    9) decreases reabsorption of Na^+

THE KIDNEYS AND ACID–BASE BALANCE

1. CO_2 and H_2O, H_2CO_3 (carbonic acid)
2. a) hydrogen (H^+), bicarbonate (HCO_3^-)
   b) raise
3. a) bicarbonate (HCO_3^-), hydrogen (H^+)
   b) lower

OTHER FUNCTIONS OF THE KIDNEYS

1. a) decreases
   b) juxtaglomerular, afferent
   c) angiotensin II
   d) increases vasoconstriction, increases secretion of aldosterone
   e) raise
2. a) erythropoietin, low oxygen in tissues
   b) red bone marrow, RBCs
3. D, calcium, phosphate

ELIMINATION OF URINE

1. a) behind
   b) hilus (or pelvis), lower posterior, urinary bladder
   c) smooth muscle
2. a) pubic b) prostate c) uterus
3. reservoir, contract
4. a) rugae b) transitional c) expansion
   d) trigone e) ureters, urethra
5. a) detrusor b) internal urethral, involuntary
6. 1) ureter
   2) parietal peritoneum
   3) detrusor muscle
   4) rugae
   5) opening of ureter
   6) trigone
   7) internal urethral sphincter
   8) external urethral sphincter
   9) urethra
   10) parietal peritoneum
   11) rugae
   12) ureter
   13) detrusor muscle
   14) opening of ureter
   15) trigone
   16) internal urethral sphincter
   17) prostate gland
   18) prostatic urethra
   19) membranous urethra
   20) external urethral sphincter
   21) cavernous urethra
   22) cavernous tissue of penis
7. a) urinary bladder, exterior
   b) anterior c) prostate, penis, semen
   d) external urethral, voluntary
8. a) urination b) spinal cord
9. a) 1, 5, 4, 3, 2, 6 b) external urethral sphincter

CHARACTERISTICS OF URINE

1. a) 1 to 2 liters b) excessive sweating or diarrhea
   c) excessive consumption of fluids
2. a) straw, amber b) clear
3. a) 1.010 to 1.025
   b) dissolved materials, concentrating
   c) concentrated
   d) excessive sweating or diarrhea
4. 6, 4.6 to 8.0
5. water, solvent
6. 1) B 2) C 3) A
7. b

ABNORMAL CONSTITUENTS OF URINE

1. 1) E 2) D 3) C 4) B
   5) A

CROSSWORD PUZZLE

CLINICAL APPLICATIONS

1. a) bacteria b) colon c) kidney
   d) urinary bladder
2. a) kidney stones b) increase c) dilute
3. a) no
   b) no—although blood levels of nitrogenous wastes are elevated, the kidneys are still producing urine at approximately half the low normal rate
4. a) 1) postrenal
      2) intrinsic renal
      3) postrenal
      4) prerenal
      5) intrinsic renal
   b) hemodialysis
   c) waste products, excess minerals

MULTIPLE CHOICE TEST #1

1. c	6. b	11. c	16. b	21. d
2. b	7. c	12. d	17. a	22. d
3. b	8. a	13. d	18. b	23. b
4. a	9. a	14. b	19. c	24. d
5. d	10. d	15. a	20. d	25. d

MULTIPLE CHOICE TEST #2

The letter of the correct choice is followed by the statement necessary to complete the answer. When an incorrect statement is to be corrected to make it true, the new (true) words are underlined.

1. b—Reabsorbed materials enter the peritubular capillaries.
2. c—Aldosterone increases the reabsorption of sodium ions.
3. a—The stimulus is stretching of the detrusor muscle of the urinary bladder by accumulating urine.
4. b—The kidneys are lateral to the spinal column.
5. c—The loops of Henle and collecting tubules are found in the renal medulla.
6. c—Blood cells and large proteins remain in the blood in the glomerulus.
7. d—Glucose is present in urine only if the blood glucose level is too high.
8. d—The voluntary external urethral sphincter can close the urethra.
9. a—The kidneys change inactive forms of vitamin D to the active form.
10. d—The kidneys respond to a decreasing pH of the blood by excreting more hydrogen ions.

MULTIPLE CHOICE TEST #3

1. a, d, e, g, h, and l are correct
2. d, f, and h are correct
3. a, b, c, e, f, and h are correct

Chapter 19

WATER COMPARTMENTS

1. intracellular, two-thirds
2. a) extracellular
   b) 1) tissue fluid
   2) plasma
   3) lymph
3. 1) C 2) A 3) D 4) B
4. osmosis, filtration

WATER INTAKE AND OUTPUT

1. a) drinking liquids b) food
   c) cell respiration
2. a) urinary system, urine b) skin, sweat
   c) exhaled water vapor, feces
3. a) equal b) drink more liquids
   c) urinary output

REGULATION OF WATER INTAKE AND OUTPUT

1. hypothalamus
2. the concentration of dissolved materials in a fluid
3. osmoreceptors
4. a) increases b) thirst, drink liquids
   c) decreases
5. a) ADH, posterior
   b) increase the reabsorption of water
   c) dehydration
   d) decrease
6. a) aldosterone b) increases
   c) 1) low blood pressure
   2) low blood sodium level
7. decrease, increase
8. ANP, increase

ELECTROLYTES

1. positive and negative ions
2. inorganic, salts, acids, bases
3. a) negative b) Cl^-, HCO_3^-, HPO_4^{-2}
   c) positive d) Na^+, K^+, Ca^{+2}
4. a) osmolarity, osmosis b) greater
5. bones, enzymes
6. calcium and phosphorus (or magnesium)
   iron and copper
7. 1) C 2) A 3) B

REGULATION OF ELECTROLYTE INTAKE AND OUTPUT

1. food, beverages
2. a) urine, sweat, feces
   b) sodium, chloride c) higher
3. 1) C, D 2) B 3) A 4) E

ACID–BASE BALANCE

1. buffer systems, respiratory, kidneys
2. a) 7.35–7.45 b) 6.8–7.0
   c) blood
3. a) neutral b) acidic, alkaline
   c) alkaline, acidic

BUFFER SYSTEMS

1. a) acid, base b) strong, will not
2. bicarbonate, phosphate, protein
3. a) H_2CO_3, $NaHCO_3$
   b) $NaCl + H_2CO_3$
   c) NaCl
   d) H_2CO_3
   e) $H_2O + NaHCO_3$
   f) H_2O
   g) $NaHCO_3$
4. a) NaH_2PO_4, Na_2HPO_4
   b) $NaCl + NaH_2PO_4$
   c) NaCl
   d) NaH_2PO_4
   e) $H_2O + Na_2HPO_4$
   f) H_2O
   g) Na_2HPO_4
5. a) acid, base b) carboxyl (COOH), hydrogen
   c) amine (NH_2), hydrogen
6. less than a second

RESPIRATORY MECHANISMS TO REGULATE PH

1. CO_2
2. a) H_2CO_3 b) $H^+ + HCO_3^-$ c) H^+
3. 1) B, C, F, G 2) A, D, E, H
4. increase, more, decrease, raise
5. decrease, less, increase, lower
6. a few minutes

RENAL MECHANISMS TO REGULATE PH

1. CO_2 and H_2O, H_2CO_3 (carbonic acid)
2. hydrogen, sodium, bicarbonate
3. sodium, bicarbonate, hydrogen
4. a few hours to days
5. 3, 2, 1

CAUSES AND EFFECTS OF PH CHANGES

1. emphysema, pneumonia, asthma
2. prolonged hyperventilation
3. kidney disease, untreated diabetes mellitus, vomiting or diarrhea
4. overingestion of antacids, vomiting of stomach contents
5. a) central nervous system
   b) confusion and disorientation, coma and death
6. a) central, peripheral
   b) muscle twitches, convulsions

CROSSWORD PUZZLE

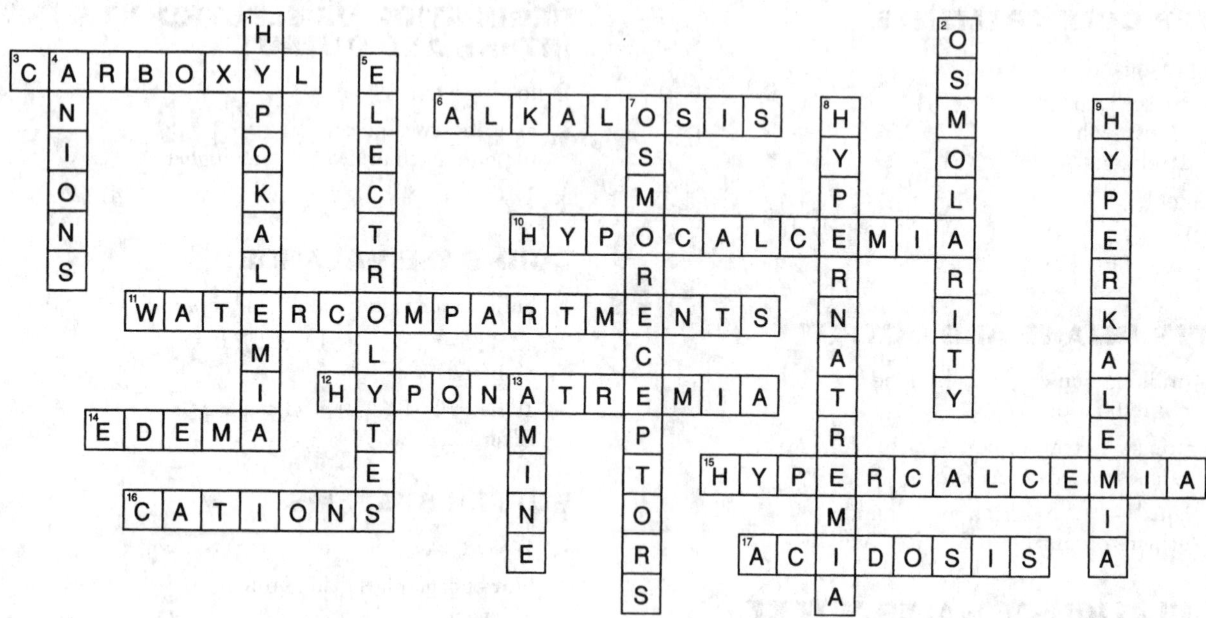

CLINICAL APPLICATIONS

1. systemic, lower legs and ankles
2. a) sodium
   b) sweating
   c) take it easy
3. a) the heart
   b) kidney disease (or) excessive vomiting or diarrhea
4. 1) c
   2) a (this is respiratory compensation for acidosis)
   3) a (the kidneys are excreting more hydrogen ions)

MULTIPLE CHOICE TEST #1

1. b	6. b	11. c	16. c	21. a
2. d	7. b	12. b	17. a	22. c
3. c	8. d	13. a	18. b	23. b
4. a	9. a	14. d	19. c	24. a
5. d	10. c	15. c	20. b	25. b

MULTIPLE CHOICE TEST #2

The letter of the correct choice is followed by the statement necessary to complete the answer. When an incorrect statement is to be corrected to make it true, the new (true) words are underlined.

1. a—PTH increases the absorption of calcium ions by the small intestine.
2. c—The water found within cells is called intracellular fluid.
3. b—Most water lost from the body is in the form of urine.
4. c—The hormones that have the greatest effect on the water content of the body are aldosterone and ADH.
5. d—The electrolyte concentration in body fluids helps regulate the process of osmosis.
6. c—The buffer systems work very rapidly to correct pH imbalances.
7. b—Respiratory compensation for metabolic acidosis involves increasing respirations.
8. a—The kidneys begin to compensate for a pH imbalance within a few hours to days.
9. d—Untreated acidosis may progress to coma.
10. d—Iodine is part of the hormone thyroxine.

MULTIPLE CHOICE TEST #3

1. d, e, f, g, i, j, and k are correct
2. a, c, d, e, g, j, k, and l are correct

Chapter 20

MEIOSIS

1. egg, sperm
2. a) diploid, 46 b) twice, four, haploid, 23
3. a) oogenesis b) spermatogenesis
4. 1) B, C, D, G 2) A, E, F, H
5. a) 23, 23, 46 b) zygote
6. 1) primary spermatocyte
   2) sperm cells
   3) primary oocyte
   4) polar bodies
   5) ovum

MALE REPRODUCTIVE SYSTEM

Testes and Sperm

1. a) scrotum b) lower
2. testosterone, spermatogenesis (meiosis)
3. a) LH, anterior pituitary b) maturation
   c) two of these: growth of reproductive organs, growth of the larynx, growth of muscles, growth of facial and body hair
4. a) flagellum
   b) head
   c) middle piece
   d) acrosome
   1) head
   2) middle piece
   3) acrosome
   4) nucleus
   5) flagellum
5. 1) spermatogonia
   2) sperm cells
   3) interstitial cells
   4) vas deferens
   5) epididymis
   6) seminiferous tubules

Epididymis, Ductus Deferens, and Ejaculatory Duct

1. a) posterior b) flagella
2. a) vas deferens, epididymis, ejaculatory duct
   b) inguinal, urinary bladder
   c) smooth muscle
3. ductus deferens, seminal vesicles, urethra
4. 1) penis
   2) urethra
   3) glans penis
   4) prepuce
   5) urethral orifice
   6) epididymis
   7) vas deferens
   8) testis
   9) scrotum
   10) seminal vesicle
   11) ejaculatory duct
   12) prostate gland
   13) bulbourethral gland

Seminal Vesicles, Prostate Gland, and Bulbourethral Glands

1. a) urinary bladder, ejaculatory ducts
   b) alkaline, fructose
2. a) urethra, urinary bladder
   b) energy production, mitochondria
   c) smooth, ejaculation
3. a) Cowper's glands b) alkaline, urethra
4. acidic, vagina
5. a) sperm cells b) 7.4

Urethra and Penis

1. a) ejaculatory ducts b) prostate gland, penis
2. smooth, blood sinuses
3. parasympathetic, blood sinuses
4. ejaculation, prostate, reproductive ducts

Male Reproductive Hormones

1. a) FSH, anterior pituitary b) testosterone
2. LH, anterior pituitary
3. testes, FSH
4. GnRH, hypothalamus

FEMALE REPRODUCTIVE SYSTEM

Ovaries

1. pelvic, uterus
2. egg cells (ova)
3. a) FSH, anterior pituitary b) estrogen
4. mature, LH, anterior pituitary
5. corpus luteum, progesterone
6. FSH
7. a) corpus luteum b) myometrium
8. GnRH, hypothalamus
9. 1) fallopian tube
   2) ovary
   3) uterus
   4) clitoris
   5) labium minora
   6) labium majora
   7) urethra
   8) vagina
   9) fimbriae
   10) cervix
   11) Bartholin's gland

FALLOPIAN TUBES

1. uterine tubes, oviducts
2. a) ovary b) fimbriae, ovum
3. smooth muscle, ciliated
4. fertilization
5. ectopic

UTERUS

1. pelvic, ovaries, urinary bladder
2. fundus, cervix
3. a) peritoneum b) myometrium
   c) endometrium
4. a) basilar, functional
   b) estrogen and progesterone

5. a) becomes the maternal placenta
   b) contracts for delivery of the baby and placenta
   c) progesterone, relaxin   d) oxytocin
6. 1) ovum
   2) mature follicle
   3) ovary
   4) fallopian tube
   5) fundus
   6) fimbriae
   7) uterine cavity
   8) body of uterus
   9) endometrium
   10) myometrium
   11) cervix
   12) vagina
   13) Bartholin's gland
   14) vaginal orifice

Vagina and Vulva

1. 1) B, D, G   2) E   3) C   4) A   5) F

Mammary Glands

1. a) milk   b) lactiferous, nipple
2. estrogen, progesterone, placenta
3. a) prolactin, anterior pituitary
   b) oxytocin, posterior pituitary
4. a) areola   b) adipose
5. a) lactose   b) antibodies (IgA)   c) passive
   d) bacteria (normal flora, microbiota), intestines of the infant
6. 1) glandular tissue
   2) areola
   3) nipple
   4) lactiferous duct
   5) adipose tissue

The Menstrual Cycle

1. anterior pituitary gland, ovaries, uterus, ovaries, 28
2. FSH, LH, estrogen, progesterone
3. menstrual, follicular, luteal
4. a) functional, menstruation   b) FSH
5. a) FSH, estrogen   b) estrogen
   c) LH, rupture
6. a) LH, corpus luteum, progesterone
   b) progesterone, glycogen
7. progesterone, menstrual

CROSSWORD PUZZLE

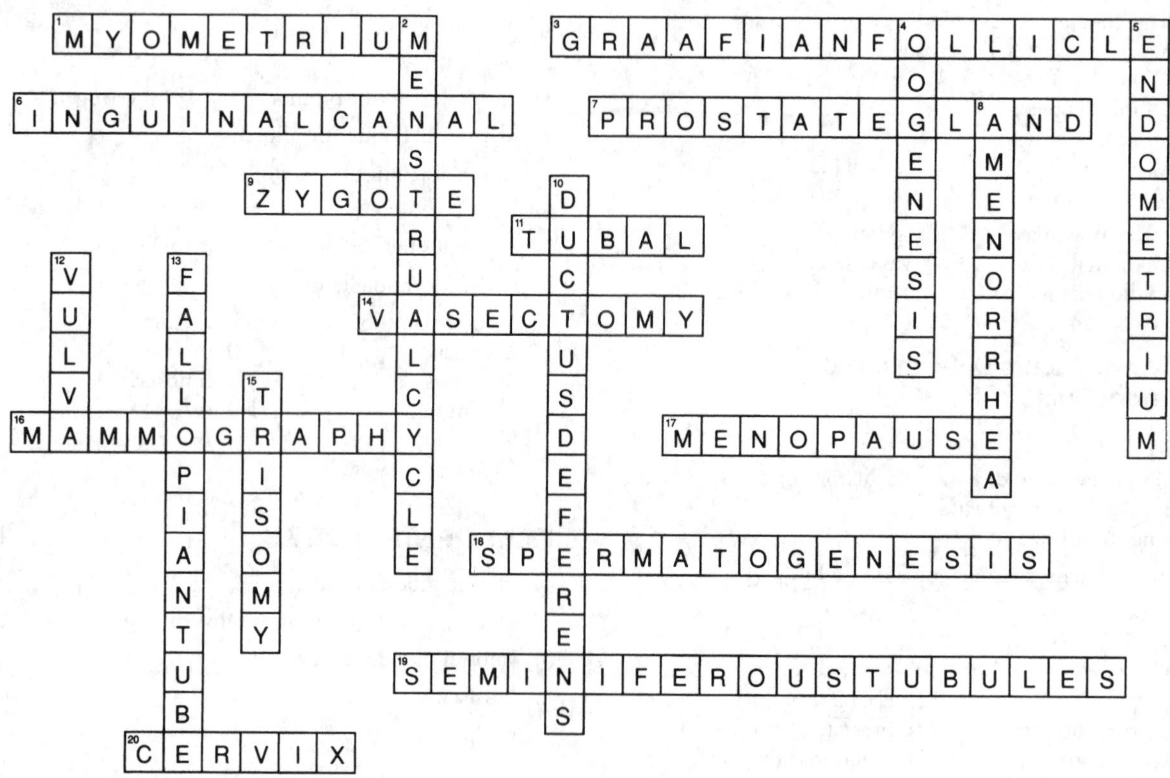

CLINICAL APPLICATIONS

1. a) prostate   b) prostatic hypertrophy
   c) urethra
2. a) tubal ligation   b) fallopian tubes
   c) vasectomy   d) ductus deferens
3. a) syphilis   b) bacteria
   c) ectopic   d) conjunctivitis, pneumonia
   e) virus   f) nervous
4. a) breast self-examination
   b) mammography

MULTIPLE CHOICE TEST #1

1. a	6. a	11. c	16. c	21. d
2. d	7. a	12. b	17. a	22. b
3. b	8. d	13. b	18. b	23. a
4. b	9. c	14. c	19. d	24. a
5. a	10. b	15. d	20. b	25. c

MULTIPLE CHOICE TEST #2

The letter of the correct choice is followed by the statement necessary to complete the answer. When an incorrect statement is to be corrected to make it true, the new (true) words are underlined.

1. a—The uterus is medial to the ovaries and superior to the urinary bladder.
2. c—The urethra extends through the prostate gland and penis.
3. b—Both processes begin at puberty, but oogenesis ends at menopause.
4. d—The ductus deferens carries sperm from the epididymis to the ejaculatory duct.
5. b—The secretion of the seminal vesicles is alkaline and contains fructose.
6. d—The growth of blood vessels in the endometrium begins in the follicular phase.
7. b—The cervix is the narrow inferior part that opens into the vagina.
8. d—The release of milk is stimulated by the hormone oxytocin.
9. c—An ectopic pregnancy occurs when the zygote becomes implanted in the fallopian tube.
10. a—To produce viable sperm, the temperature of the scrotum must be slightly lower than body temperature.
11. d—Inhibin decreases the secretion of FSH.
12. c—Relaxin inhibits contractions of the myometrium.

MULTIPLE CHOICE TEST #3

1. c and e are correct
2. b, e, f, g, and j are correct

Chapter 21

HUMAN DEVELOPMENT

Fertilization

1. acrosome
2. a) egg, sperm b) 23, 46, diploid
   c) fallopian tube
3. autosomes, sex chromosomes
4. XY, XX

Implantation

1. a) mitosis b) cleavage
2. a) morula b) blastocyst
3. a) embryo, embryonic stem cells
   b) trophoblast, endometrium, implantation
4. chorion, placenta
5. 1) primary follicles
   2) maturing follicles
   3) ovary
   4) ovum
   5) fallopian tube
   6) sperm
   7) ovum (zygote)
   8) two-cell stage
   9) four-cell stage
   10) morula
   11) blastocyst
   12) inner cell mass
   13) trophoblast
   14) endometrium
   15) embryonic disc
   16) corpus luteum
   17) corpus albicans

Embryo and Embryonic Membranes

1. ectoderm, mesoderm, endoderm
2. 1) lining of digestive organs, lungs
   2) skeletal muscles, bones
   3) epidermis, nervous system
3. 1) A, F 2) C, D 3) B, E
4. a) eighth b) 9, 40
5. 1) endometrium
   2) chorion
   3) yolk sac
   4) embryonic disc
   5) amnion
   6) chorionic villus
   7) yolk sac
   8) head of embryo
   9) amniotic sac
   10) umbilical cord
   11) arm bud
   12) leg bud
   13) placenta

Placenta and Umbilical Cord

1. a) endometrium b) chorion
   c) umbilical cord d) exchanges
2. blood sinuses
3. arteries, vein
4. a) fetus, placenta b) CO_2, waste products
5. a) placenta, fetus b) O_2, nutrients
6. afterbirth
7. a) chorion
   b) corpus luteum, estrogen, progesterone
8. estrogen, progesterone
9. a) FSH, LH b) mammary glands
   c) myometrium d) relaxin

Parturition and Labor

1. a) birth b) sequence of events
2. a) progesterone b) head downward
3. a) amniotic sac b) dilates (widens) it
   c) it ruptures d) it flows out through the birth canal
4. a) oxytocin b) baby
5. a) afterbirth (placenta) b) compress

The Infant at Birth

1. a) umbilical cord b) CO_2, medulla
2. a) foramen ovale b) ductus arteriosus
   c) lungs
3. liver, bilirubin

GENETICS

Chromosomes and Genes

1. a) 46 b) homologous pairs of
   c) paternal d) maternal
   e) autosomes f) sex chromosomes, women, men
2. a) DNA b) bases c) one protein
3. 1) F 2) E 3) A 4) D 5) G
   6) C 7) B

Genetics Problems

1. 1) a)

Mom—bb

Dad—BB	b	b
B	Bb	Bb
B	Bb	Bb

b) 100% chance—brown hair, 0% chance—blond hair

2) a)

Mom—bb

Dad—Bb	b	b
B	Bb	Bb
b	bb	bb

b) 50% chance—brown hair, 50% chance—blond hair

2. 1) a)

Mom—OO

Dad—AO	O	O
A	AO	AO
O	OO	OO

b) 50% chance—type O, 50% chance—type A

2) a)

Mom—AB

Dad—AB	A	B
A	AA	AB
B	AB	BB

b) 25% chance—type A, 25% chance—type B, 50% chance—type AB, 0% chance—type O

3. 1) a)

Mom—XcX

Dad—XcY	Xc	X
Xc	XcXc	XcX
Y	XcY	XY

b) daughter: 50% chance—color blind, 50% chance—carrier
son: 50% chance—color blind, 50% chance—normal color vision

2) a)

Mom—XcXc

Dad—XY	Xc	Xc
X	XcX	XcX
Y	XcY	XcY

b) daughter: 0% chance—color blind, 100% chance—carrier
son: 100% chance—color blind, 0% chance—normal color vision

CROSSWORD PUZZLE

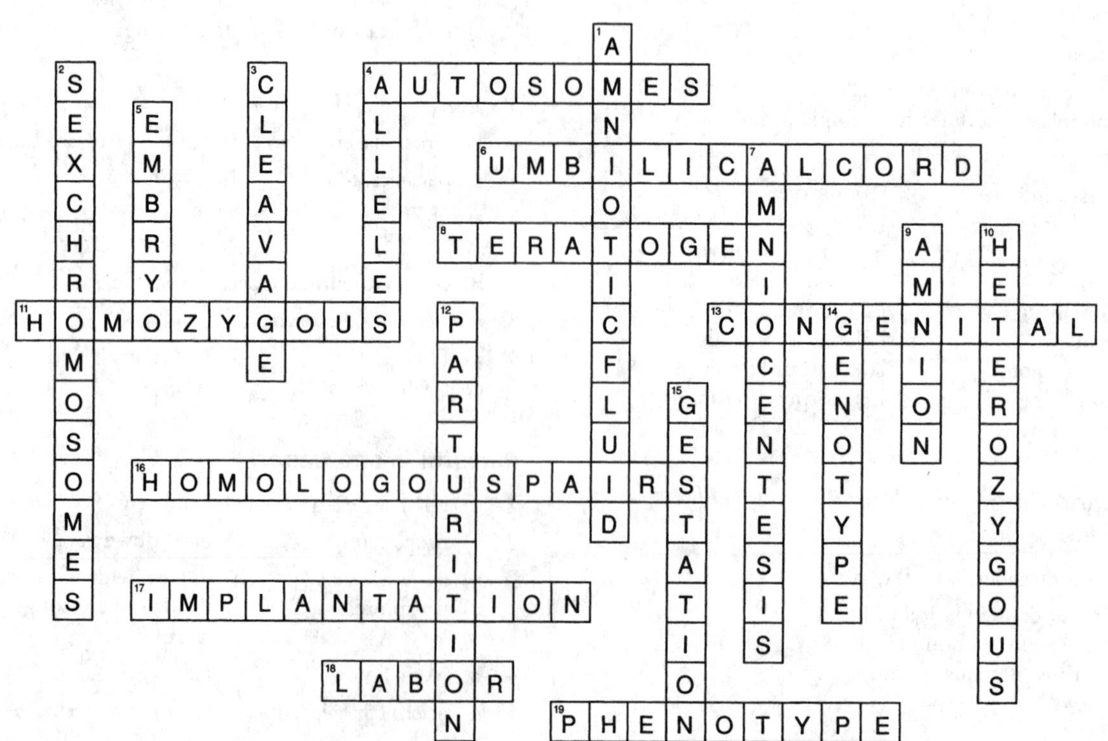

CLINICAL APPLICATIONS

1. a) lungs b) 3
2. a) teratogens
   b) German measles, chickenpox, Zika virus
3. a) three, mental (cognitive b) amniocentesis
4. a) abdominal wall, uterus
   b) 1) fetal distress (or) poor fetal position (breech birth)
      2) maternal pelvic outlet is too small to permit a vaginal delivery

MULTIPLE CHOICE TEST #1

1. c	5. d	9. d	13. d	17. d
2. a	6. b	10. c	14. b	18. c
3. d	7. b	11. c	15. b	19. a
4. d	8. a	12. a	16. b	20. d

MULTIPLE CHOICE TEST #2

The letter of the correct choice is followed by the statement necessary to complete the answer. When an incorrect statement is to be corrected to make it true, the new (true) words are underlined.

1. a—The amniotic sac ruptures during the first stage.
2. d—The first blood cells for the fetus are formed by the yolk sac.
3. c—The period of embryonic growth lasts from week 1 to week 8.
4. d—Within the placenta, fetal blood vessels are within the maternal blood sinuses.
5. b—The embryonic stage that undergoes implantation is the blastocyst.
6. c—Premature contractions of the myometrium are prevented by progesterone (relaxin).
7. d—The sex chromosomes are XX in women and XY in men.
8. a—The expression of the alleles in the appearance of the individual is the phenotype.
9. d—A recessive gene is one that appears in the phenotype only if the individual is homozygous.
10. c—A son who inherits a sex-linked trait has inherited the gene from his mother.

MULTIPLE CHOICE TEST #3

1. f, g, h, and i are correct
2. a, c, d, e, and g are correct

Chapter 22

CLASSIFICATION OF MICROORGANISMS

1. a) viruses b) arthropods c) bacteria
   d) protozoa e) worms f) fungi
2. a) binomial nomenclature b) genus
   c) genus d) species

NORMAL FLORA—MICROBIOTA

1. a) transient flora b) resident flora
   c) opportunist
2. 1) B 4) L 7) A 10) K
   2) I 5) G 8) F 11) D
   3) J 6) E 9) C 12) H

INFECTIOUS DISEASE

1. a) resistance b) virulence
2. asymptomatic, subclinical, inapparent
3. incubation period
4. 1) C 4) G 7) D 10) B
   2) F 5) E 8) A
   3) J 6) I 9) H

EPIDEMIOLOGY

1. a) endemic b) epidemic c) pandemic
2. a) portal of entry
   b) breaks in the skin, nose, mouth; eye; reproductive tract; urinary tract; bites of vectors
   c) portal of exit
   d) respiratory droplets, cutaneous contact, blood from wounds or bites of vectors, urine/feces, sexual contact
3. portal of exit, portal of entry
4. a) reservoir b) zoonosis c) carrier
5. a) communicable b) contagious
   c) noncommunicable

METHODS OF CONTROL OF MICROBES

1. a) disinfectant b) antiseptic
2. a) sterilization b) pasteurization
   c) chlorination

THE PATHOGENS

Bacteria

1. a) rod b) spherical c) spiral
2. a) antibiotics b) narrow-spectrum
   c) broad-spectrum d) resistant
   e) culture and sensitivity testing
3. 1) C 4) H 7) B 10) I
   2) E 5) A 8) F
   3) D 6) G 9) J
4. a) plague, Lyme disease, tularemia, anthrax, leptospirosis, salmonella food poisoning, Rocky Mountain spotted fever, listeriosis
   b) gonorrhea, syphilis
   c) tetanus (v), diphtheria (v), whooping cough (v), *Haemophilus* meningitis (v), tuberculosis, cholera (v), typhoid

Viruses

1. a) DNA or RNA, protein shell
   b) living cells, reproduce
2. a) AIDS, herpes simplex, hepatitis B, human papilloma virus
   b) rabies, encephalitis, bird flu
   c) measles, mumps, rubella, polio, hepatitis B

Fungi

1. a) dead organic matter b) spores
2. a) mycoses b) superficial c) systemic
3. a) yeasts b) skin, mouth, colon, vagina
   c) antibiotics

Protozoa

1. a) unicellular animals b) water
2. a) cysts b) fecal-oral, water or food

Worms and Arthropods

1. a) trichinosis
   b) pinworm
2. plague—flea; Lyme disease—tick; malaria—mosquito; yellow fever—mosquito; Rocky Mountain spotted fever—tick; encephalitis—mosquito; Zika virus—mosquito

CROSSWORD PUZZLE

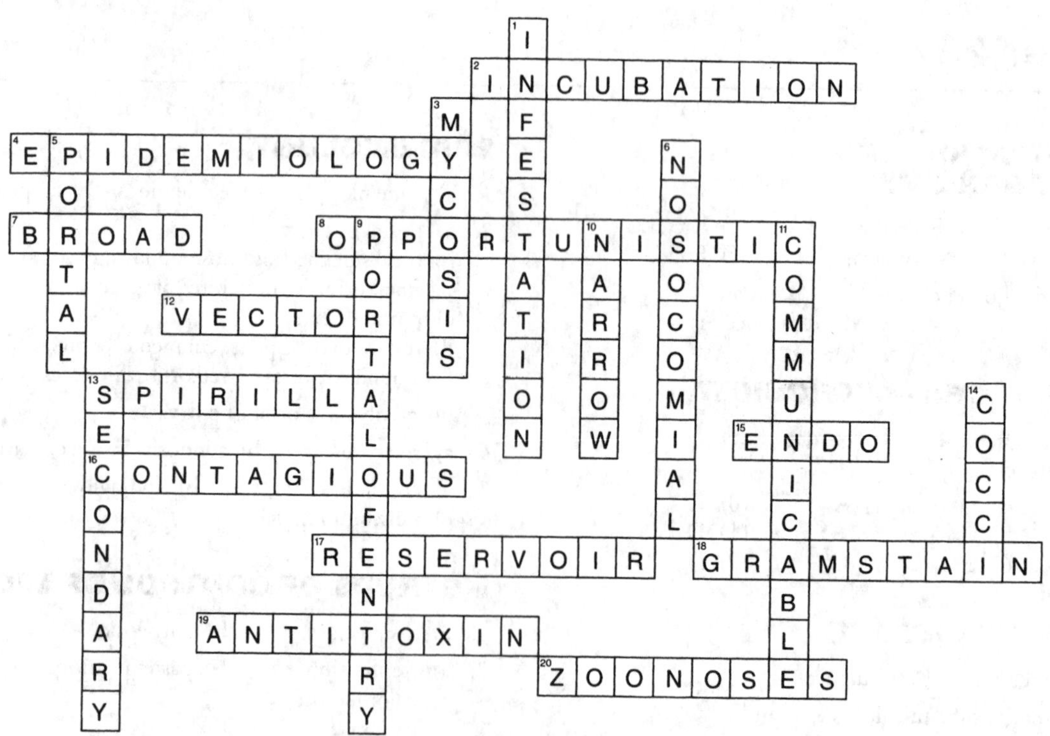

MULTIPLE CHOICE TEST #1

1. a	5. b	9. b	13. d	17. a
2. d	6. c	10. a	14. d	18. c
3. c	7. d	11. c	15. b	19. b
4. a	8. b	12. c	16. a	20. a

MULTIPLE CHOICE TEST #2

The letter of the correct choice is followed by the statement necessary to complete the answer. When an incorrect statement is to be corrected to make it true, the new (true) words are underlined.

1. d—All contagious diseases are communicable.
2. b—A nosocomial infection is one acquired in a hospital.
3. b—Bacteria in the colon produce enough vitamin K to meet our needs.
4. c—A spore is a form that is resistant to heat and drying.
5. a—Viruses must be in the host's cells to reproduce.
6. c—The name *ringworm* refers to a fungus infection of the skin.
7. b—Some protozoa form cysts, which survive for long periods outside a host.
8. d—Lice and ticks spread disease when they bite to obtain blood for food.
9. c—Anaerobic bacteria reproduce only in the absence of oxygen.
10. a—Antiviral medications are effective against some viruses.

MULTIPLE CHOICE TEST #3

1. a, b, f, h, and n are correct
2. a, c, d, e, h, j, and l are correct